ELEMENTS OF MATERIALS SCIENCE AND ENGINEERING

THIRD EDITION

LAWRENCE H. VAN VLACK
The University of Michigan
Ann Arbor, Michigan

ADDISON-WESLEY PUBLISHING COMPANY
Reading, Massachusetts
Menlo Park, California
London · Amsterdam
Don Mills, Ontario · Sydney

This book is in the
ADDISON-WESLEY SERIES IN METALLURGY AND MATERIALS

Consulting Editor
Morris Cohen

Printed in the United States of America. Published simultaneously in Canada. Library of Congress Catalog Card No. 74-30702.

ISBN 0-201-08073-7
DEFGHIJ-MA-79

PREFACE

The teaching of materials science and engineering evolved rapidly between the date of the first edition of this book (1959) and the second edition (1964). For the most part, the changes that occurred helped to shape the concept that the properties and behavior of a material are closely related to the internal structure of that material:

1. In order to modify properties, appropriate changes must be made in the internal structure.
2. If processing or service conditions alter the structure, the characteristics of the material are affected.

With this concept well established, there was no technical urgency to revise the second edition. However, a decade does bring about changes in prerequisite science courses, changes in curricular patterns, changes in instructor personnel, and accumulated experience of teaching procedures for materials science and engineering. As a result, it is now desirable to adapt to these with a third edition.

This edition was written for the same academic level as the previous edition. That is, it follows the introductory college science courses. The previous general format has also been maintained, whereby initial attention is given to the engineering properties of interest, and pertinent chemical bonding principles are reviewed. Structural characteristics are considered on the atomic and electronic levels in Chapters 3 through 5 to provide the basis for the properties of single-phase (1) metals in Chapter 6, (2) polymers in Chapter 7, and (3) ceramics in Chapter 8. Multiphase materials are presented in Chapters 9 through 11, with sequential emphasis on phase diagrams, microstructure, and methods of developing the desired properties through microstructural control. The major importance of corrosion warrants a separate Chapter 12. As previously, the last chapter addresses itself to composite materials as (1) a basis of integrating the principles from earlier chapters, and (2) a means of introducing the more complex materials and materials systems that can be available to the engineer and scientist.

This edition differs from the earlier edition in several ways. Most important to the student are the teaching aids which were developed as a result of classroom experience. These include chapter previews, and a review and study section at the end of each chapter, which contains a summary section, terms for review, discussion topics, and study problems. Expanded use is made of example problems, both in number and through the accompanying comments to reinforce textual concepts. Suggestions were solicited from several hundred instructors who are users of the second edition. Those received ranged from data corrections to suggestions for rearranging topics within chapters. Essentially every suggestion was usable. As the author, I am deeply indebted to those colleagues across this country and in others.

This edition incorporates SI units throughout. However, since their use is not yet universal, it does not forego English units entirely. Specifically, temperatures are generally given in °C, with Fahrenheit equivalents shown parenthetically. Force, stress, and energy data are also presented in duplicate, because many engineers must be "bilingual" for a number of years as they communicate with their technical colleagues on the one hand and with nontechnical workers on the other. In general, the metric units have been used solely for electrical data.

As in the second edition, those sections and subsections are identified that may be deleted by the instructor if necessary. Those topics marked with a bullet (•) contain material which is not a primary prerequisite for later (unmarked) sections.

Unfortunately it is impossible to give public thanks to each of my faculty colleagues, to the office staff, and to innumerable students at The University of Michigan who have contributed in their way to this new edition. So that they may know that their contributions are greatly appreciated, I dedicate this volume to today's students, who will be tomorrow's engineers.

Special thanks are due Professor Morris Cohen of the Massachusetts Institute of Technology for his generous counsel on many points throughout the text and for the Foreword, which he wrote from his vantage point as Chairman of the National Academy of Science Committee that prepared the study "Materials and Man's Needs." As previously, I am grateful for the most pleasant and always helpful contacts with the Addison-Wesley personnel. Finally, and most important, Fran's patience and encouragement throughout the months of revision were indispensable.

August 1974 L. H. VV.

CONTENTS

FOREWORD

MATERIALS IN THE SCHEME OF THINGS

Materials are all about us; they are engrained in our culture and thinking as well as in our very existence. In fact, materials have been so intimately related to the emergence and ascent of man that they have given names to the Stone, Bronze, and Iron Ages of civilization. Naturally-occurring and manmade materials have become such an integral part of our lives that we often take them for granted, and yet materials rank with energy and information as basic resources of mankind. Although materials may appear to "do nothing," they are indeed the working substance of our society; they play a crucial role not only in our way of life but also in the well-being and security of nations.

But what are materials? How do we understand, manipulate, and use them? Questions like these were the subject of a comprehensive study by the National Academy of Sciences in 1971–73, and resulted in what has become known as the COSMAT Report* on "Materials and Man's Needs." Materials are, of course, a part of the matter in the universe, but more specifically *they are substances whose properties make them useful in structures, machines, devices, or products.* For example, these categories include metals, ceramics, semiconductors, polymers (plastics), glasses, dielectrics, fibers, wood, sand, stone, and many composites. The production and processing of these materials into finished goods account for about one-fifth of the jobs and gross national product in the United States.

Since the human body might be regarded as a structure or machine or device, we could also embrace foods, drugs, biomatter, fertilizers, etc., among the classes of materials, but it is presently customary to leave these materials to the life and agricultural sciences. For similar reasons, even though fossil fuels, water, and air likewise fall within the broad definition of materials, they are usually dealt with in other fields.

The materials of mankind can be visualized to flow in a vast *materials cycle*—a global cradle-to-grave system. Raw materials are taken from the earth by mining,

---

* COSMAT is an abbreviation for the Academy's Committee on the Survey of Materials Science and Engineering. The general thrust of this Foreword is taken from the COSMAT Report.

drilling, excavating, or harvesting, then converted into bulk materials like metal ingots, crushed stone, petrochemicals, and lumber, and subsequently fabricated into engineering materials, like electric wire, structural steel, concrete, plastics, and plywood, for meeting end-product requirements in society. Eventually, after due performance in the service of man, these materials find their way back to earth as scrap, or preferably re-enter the cycle for reprocessing and further use before their ultimate disposal. An important aspect of the materials-cycle concept is that it reveals many strong interactions among materials, energy, and the environment, and that all three must be taken into account in national planning and technological assessment. These considerations are becoming especially critical because of mounting shortages in energy and materials just at a time when the inhabitants of this planet are manifesting deeper concern for the quality of their living space. As a case in point, if scrap aluminum can be effectively recycled, it will require only about one-twentieth the energy needed for an equivalent tonnage of primary aluminum from the ore, and the earth would be that much less scarred by the associated removal operations.

Clearly, then, in the development of human knowledge, one is not surprised to find a *science and engineering of materials* taking its place among all the other bodies of inquiry and endeavor that extend the reach of man. Simply stated, *materials science and engineering is concerned with the generation and application of knowledge relating the composition, structure, and processing of materials to their properties and uses.* We can see here a linkage that interrelates the structure, properties, processing, function, and performance of materials. However, if we wish to highlight the *materials science* part of this spectrum, we focus on understanding the nature of materials, leading to theories or descriptions that explain how structure relates to composition, properties, and behavior. On the other hand, the *materials engineering* part of the spectrum deals with the synthesis and use of both fundamental and empirical knowledge in order to develop, prepare, modify, and apply materials to meet specified needs.

It is evident that the distinction between materials science and materials engineering is primarily one of viewpoint or center of emphasis; there is no line of demarcation between the two domains, and we find increasing logic in adopting the combined name of *materials science and engineering.* Actually, this textbook joins the two, both in its title and coverage, by using the term *materials science* generically to include many aspects of *materials engineering.*

Figures I and II illustrate the selection and function of materials in two materials systems, each requiring numerous professional judgments concerning performance, reliability, maintainability, economic trade-offs, and environmental considerations. In the nuclear reactor shown, the fuel consists of uranium oxide pellets encased in a zirconium alloy that is heat resistant and does not waste neutrons. The neutron flux, and hence the operating temperature of the fuel, is controlled by a boron carbide device that can absorb neutrons as desired, but has to be corrosion-protected by stainless steel. The generated steam is contained in an outer high-pressure vessel

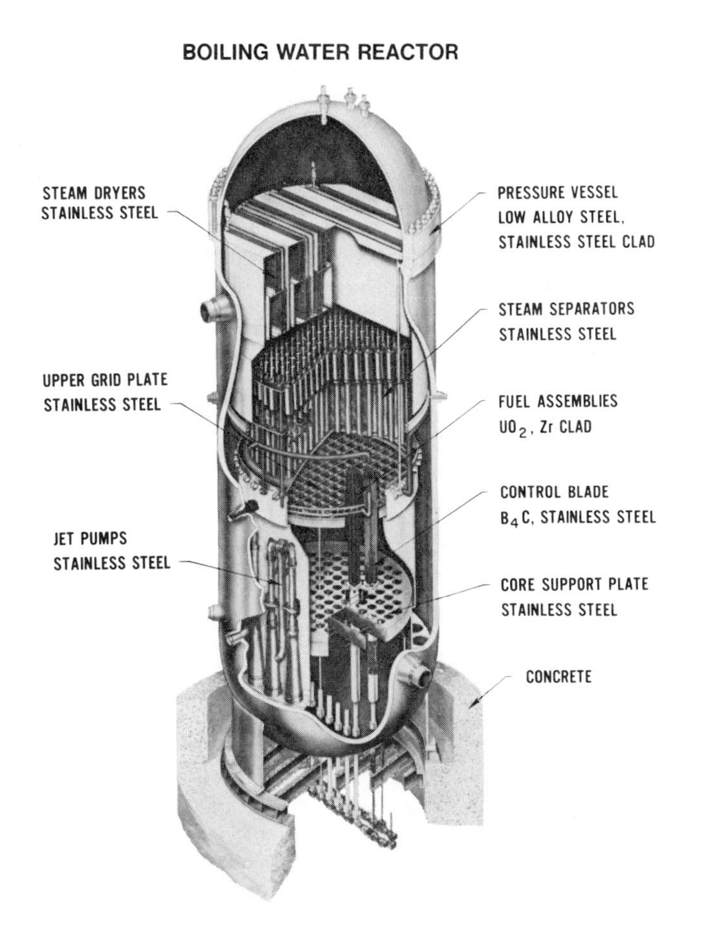

Fig. I. Materials system for energy conversion. Any integrated design such as a boiling-water nuclear reactor requires specific contributions from a wide variety of materials. The role, properties, and limitations of each material must be understood by the scientist or engineer if he is to design a workable, safe, cost-effective, and environmentally reliable system. (Courtesy General Electric Company.)

built of low-alloy steel for providing strength at reasonable cost, but clad with more expensive stainless steel for corrosion protection. This nuclear-energy conversion involves a complex materials system in which the selected materials must perform *inter*dependently, over and above their specific functions. This is a striking application of materials for producing energy. Conversely, about ten percent of the energy consumed in the United States goes into the production and forming of materials.

Another materials system, showing how materials enter into man's shelter, is indicated in Fig. II. Although housing may seem like a less sophisticated case of materials science and engineering than is the nuclear reactor, the example given here is, nevertheless, one of challenge and opportunity; for this house has been built almost entirely of *recycled* materials, thereby conserving the earth's resources and energy while reducing damage to the environment. In order to manufacture the recycled metals, glass, concrete, roofing shingles, sheathing paper, insulation, and wood products, it is necessary to use a high level of materials development and technology, to compete with conventional materials in cost as well as in quality.

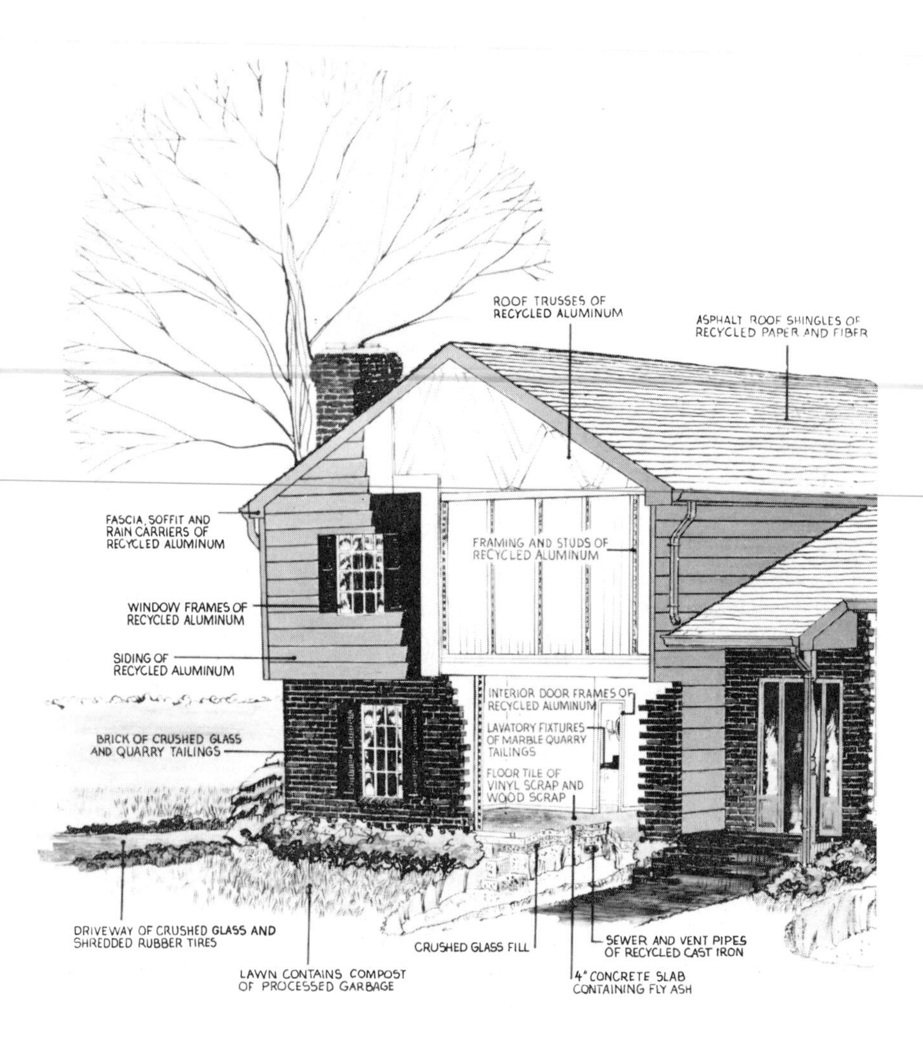

So we see that materials science and engineering is a purposeful enterprise, reaching down into the microworld of atoms and electrons and tying the condensed state of matter to the macroworld of material function and service to meet societal problems. The circular chart in Fig. III portrays a large section of man's knowledge, extending from basic sciences at the core, through applied sciences in the middle ring, to various engineering fields in the outer rim. In the center, we show physics and chemistry flanked by mathematics and mechanics; and on moving out radially, we pass through various applications-oriented disciplines. The part of this map to be visualized as materials science and engineering is the shaded sector at the right, which

Fig. II. Reusable materials. Recycling is very desirable and is receiving increased national emphasis. However, it cannot be an effective answer to our resource and environmental problems unless the recycled materials are comparable to virgin materials on the basis of appearance, properties, and service qualities—and without the penalty of added cost. These requirements present many technological challenges for the engineer. (Courtesy Reynolds Metals Company.)

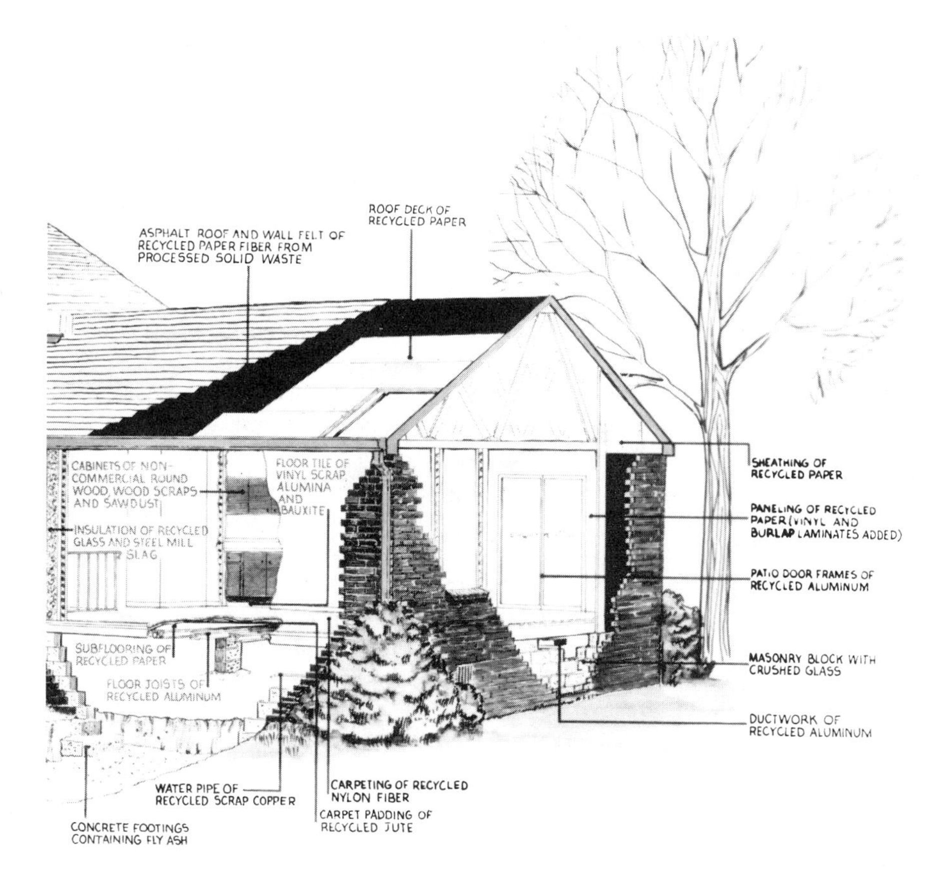

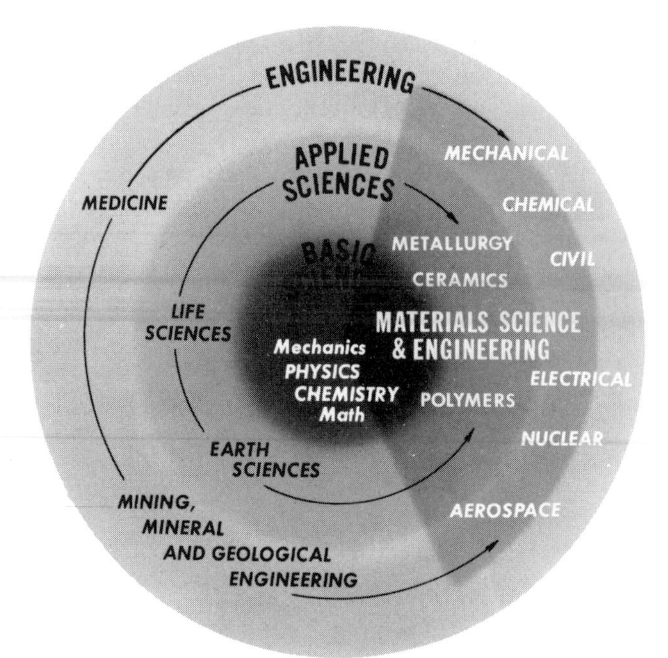

Fig. III. Materials science and engineering. This text draws from introductory science principles of physics and chemistry to relate *composition and structure* of materials to the *properties and service behavior* which are important to engineers. More than one-fourth of our national technical effort involves the development and technology of materials. This effort draws from the physical sciences and all branches of engineering. (Courtesy National Academy of Science.)

may be compared to other sectors designating the life sciences and the earth sciences. In its broad sense, materials science and engineering is a multidiscipline which embraces (but does not replace!) some disciplines (e.g., metallurgy and ceramics) and some subdisciplines (e.g., solid-state physics and polymer chemistry), and also overlaps several engineering disciplines.

There are, of course, many scientists and engineers who are materials specialists—metallurgists, ceramists, and polymer chemists—and who are wholly involved in materials science and engineering. Equally important, government data, when analyzed, reveal that one of every six hours of professional work done by *all other engineers* directly involves materials and their utilization. The time fractions are even higher for chemists and physicists. As a result, the equivalent of one-half million of the nearly two million scientists and engineers in this country contribute to this major segment of our national product and our national welfare.

In a real sense, materials science and engineering constitutes a framework in which professionals in many disciplines work creatively to probe nature's processes and, at the same time, advance knowledge in response to the pull of human needs.

Cambridge, Massachusetts Morris Cohen
June 1974

CHAPTER ONE

INTRODUCTION TO MATERIALS: SELECTED CHARACTERISTICS

PREVIEW

The theme of this text is that *the properties and behavior of materials depend on their internal structure.* Selected mechanical, thermal, and electrical parameters and consequences are defined in this chapter to serve as a basis for structure–property discussions in subsequent chapters. A brief consideration is also given to data presentation, their variation, and accuracy in calculations.

1–1 STRUCTURE ⟶ PROPERTIES

Every applied scientist and engineer—mechanical, civil, electrical, or other—is vitally concerned with the materials available to him. Whether his product is a bridge, a computer, a space vehicle, a heart pacemaker, or an automobile exhaust system, he must have an intimate knowledge of the properties and behavioral characteristics of the materials he proposes to use. Consider, for a moment, the variety of materials used in the manufacture of an automobile: iron, steel, glass, plastics, rubber—to name a few. And for steel alone there are as many as 2000 varieties or modifications. On what basis is the selection of the material for a particular part to be made?

In making his choice, the designer must take into account such properties as strength, electrical and/or thermal conductivity, density, and others. Further, he must consider the behavior of the material during processing and use (where formability, machinability, electrical stability, chemical durability, and radiation behavior are important), as well as cost and availability. For example (Fig. 1–1.1), the steel for transmission gears must machine easily in production, but must then be toughened

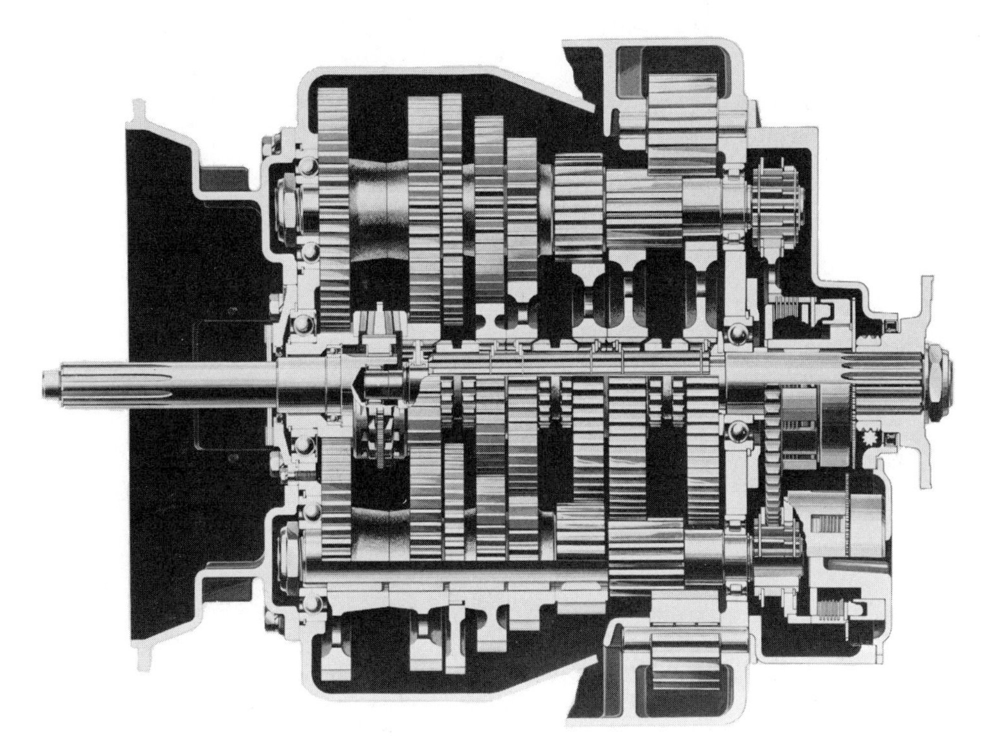

Fig. 1–1.1. Materials of engineering (electromechanical synchronizing transmission system). The steel in this engineering system must be machinable during production, and then have its properties changed before use, to toughen it. (Courtesy G. L. Myers, Spicer Transmission Division of Dana Corp.)

enough to withstand hard usage. Fenders must be made of a metal that is easily shaped, but that will resist deformation by impact. Electrical wiring must be able to withstand extremes of temperature, and semiconductors must have constant amperage/voltage characteristics over long periods of time.

Many improved designs depend on the development of completely new materials. For example, the transistor could not have been built with the materials available only a few years ago; the development of the lasers required new kinds of crystals and glasses; and although engineering designs for gas-turbine engines are far advanced, there still is a need for an inexpensive material that will resist high temperatures, for the turbine blades.

Since it is obviously impossible for the engineer or scientist to have detailed knowledge of the many thousands of materials already available, as well as to keep

Fig. 1-1.2. Electronic calculator. The performance of a calculator depends on the arrangement of the components of its internal circuit. Likewise, the behavior of a material depends upon the structure of its internal components. We shall see that these arrangements involve electron structures around atoms, the coordination of atoms, the structure of crystals, and the microstructure of adjacent crystals. The engineer and scientist can select and modify these internal structures to meet his design needs, just as the circuit designer modifies electrical components. To do this intelligently, we must know the relationships between structure and properties of materials. (Courtesy of the Arnold Engineering Company.)

abreast of new developments, he must have a firm grasp of the underlying principles that govern the properties of *all* materials. The principle which is of most value to engineers and scientists is that *properties of a material originate from the internal structures of that material.* This is analogous to saying that the operation of a TV set or other electronic product (Fig. 1–1.2) depends upon the components, devices, and circuits within that product. Anyone can twirl knobs, but the electronic technician must understand the internal circuits if he is going to repair a TV set efficiently; and the electrical engineer and the physicist must know the characteristics of each circuit element if they are to design or improve the performance of the final product.

The *internal structures* of materials involve atoms and the way atoms are associated with their neighbors in crystals, molecules, and microstructures. In the following chapters we shall devote much attention to these structures, because a technical person must understand them if he is going to produce and use materials, just as the mechanical engineer must understand the operation of an internal-combustion engine if he is going to design or improve a car for the demands of the next decade.

We shall start by becoming acquainted in this chapter with some property terms and measurements. Then we will review those science fundamentals (Chapter 2) that bear directly upon the above structure–property relationship. From there we will consider, in turn, the various structural features (from atoms to composites) on which properties depend. Whenever possible, we will solve problems (1) to illustrate structure and/or property principles, and (2) to extend the problem-solving capabilities that are so important to engineers and applied scientists.*

1–2 MECHANICAL BEHAVIOR

Deformation occurs when forces are applied to a material. We label the amount of deformation per unit length as *strain*, and the force per unit area as *stress.* Energy is absorbed by a material during deformation because a force has acted along the deformation distance. *Strength* is a measure of the level of the *stress* required to make

---

* This text will include metric units throughout. Since the celsius (centigrade) scale is already widely used in the United States, it will be the basis for all calculations. However, parenthetical additions of °F will be made where they may be useful for comparative purposes. Likewise many graphs will include an additional Fahrenheit ordinate. Metric units will also be the basis for calculations of electrical behavior.

In problems relating to mechanical properties, a dual set of units will be presented; this should aid the student in becoming "bilingual" in this respect. Once adopted by this country, the general transition to metric units should be expected to be quite fast; however, the engineer may find it necessary to communicate for some time with individuals who are not immediately fluent with metric units, but who possess an accumulation of valuable technical experience.

Conversions between English and *Système International* (SI), units are found in Appendix A. Data for calories are also included, since that metric unit is widely encountered in current technical literature. Since it is assumed that the reader can make adjustments such as g/cm^3 to kg/m^3 when necessary, the smaller g and cm will be used *whenever their use reduces the bulk which accompanies exponents.*

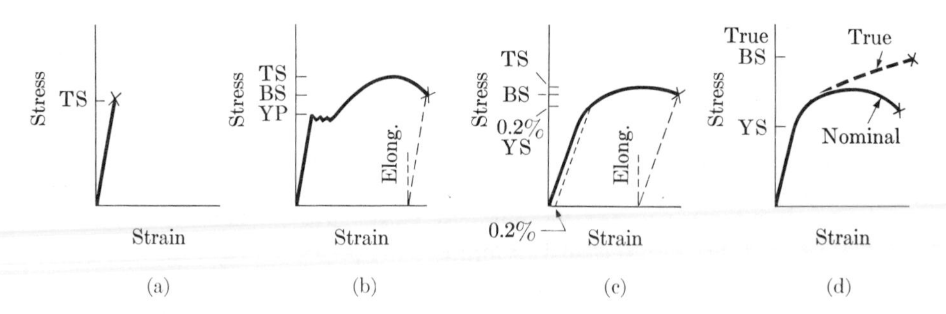

Fig. 1–2.1. Stress-strain diagrams. (a) Nonductile material with no plastic deformation (example: cast iron). (b) Ductile material with yield point (example: low-carbon steel). (c) Ductile material without marked yield point (example: aluminum). (d) True stress-strain curve versus nominal stress-strain curve. BS = breaking strength, TS = tensile strength, YS = yield strength, Elong. = elongation, X = fracture, YP = yield point.

a material fail. *Ductility* identifies the amount of *strain* at failure, while *toughness* refers to the amount of *energy* absorbed by a material during failure (Table 1–2.1).

The design engineer makes various specifications for these mechanical properties. Commonly he prescribes high strength, e.g., in a pipeline steel. He may also want high ductility to avoid *brittle** failure. Since the two (strength and ductility) tend to be incompatible, the engineer often must make trade-offs between the two to optimize his specifications. Also, there are various ways to define strength and ductility. For example, has a steel rod failed when it bends, or must it actually *fracture* to have failed? The answer, of course, depends on the requirements of the engineering design; but the contrast shows us the advantage of identifying at least two strengths—one for *initial yielding*, and one for *maximum load* a material may support. To do this, we will consider the stress–strain (σ–ϵ) diagrams of Fig. 1–2.1, giving attention in turn to deformation, to strength and hardness, and to toughness.

Deformation. Initially, strain is essentially proportional to the stress; furthermore, it is reversible. After the stress is removed, the strain disappears. The *modulus of elasticity* (*Young's modulus*) is the ratio between the stress σ and this reversible strain ϵ:

$$E = \sigma/\epsilon. \tag{1–2.1}$$

The units of Young's modulus E are in psi (English), or pascals (metric) as indicated in Table 1–2.1. The values of Young's moduli for selected materials are included in Appendix C. We shall see, in Chapter 2, that the *elastic modulus* (Young's modulus) is a measure of the interatomic bonding forces. The engineer must be fully aware of this property, since it relates directly to the rigidity of his engineering designs.

---

* Opposite of *tough*.

At higher stresses, permanent displacement can occur among the atoms within a material, so that much of the strain is not reversible when the applied stresses are removed. We call this *plastic strain* (as contrasted to *elastic strain* in the previous paragraph). This kind of strain is necessary during the processing of materials (e.g., during the rolling of aluminum plate, first to a thinner sheet, and then to a very thin foil). In product applications we commonly design to *avoid* plastic deformation, and therefore must design at stresses within the elastic (proportional) range of Fig. 1–2.1 (b) and (c).

Ductility, the total plastic strain prior to fracture, may be expressed as *elongation.* Like all strain, it is dimensionless, $(L_f - L_o)/L_o$ or $\Delta L/L_o$. From Fig. 1–2.2, however, note that, since elongation is commonly localized in the necked-down area, the amount of elongation depends on the *gage length.* Whenever reporting elongation, one must be specific about the gage length.

Table 1–2.1
MECHANICAL PROPERTIES OF MATERIALS

Property, or characteristic	Symbol	Definition (or comments)	Common units	
			English	SI
Stress	σ	Force/unit area (F/A)	psi* $lb_f/in.^2$	pascal* $(N\dagger/m^2)$
Strain	ϵ	Fractional deformation $(\Delta L/L)$	—	—
Elastic modulus	E	Stress/elastic strain	psi	pascal
Strength		Stress at failure	psi	pascal
Yield	YS	Resistance to plastic deformation		
Tensile	TS	Maximum strength (based on original dimensions)		
Ductility		Plastic strain at failure	—	—
Elongation	El	$(L_f - L_o)/L_o$		
Reduction of area	R of A	$(A_o - A_f)/A_o$		
Toughness		Energy for failure by fracture	ft-lb	joules
Hardness‡		Resistance to plastic indentation	Empirical units	

* 1 pascal (Pa) = 1 newton/m² = 0.145×10^{-3} psi; 1000 psi = 6.894 MPa.
† A load of 1 kg mass produces a force F of 9.8 newtons (N) by gravity.
‡ Three different procedures are commonly used to determine hardness values:

Brinell (BHN): A large indenter is used. The hardness is related to the diameter (1 to 4 mm) of the indentation.

Rockwell (R): A small indenter is used. The hardness is related to the penetration depth. Several different scales are available, based on the indenter size and the applied load.

Vickers (DPH): A small diamond pyramid and a very light load are used. The indentation size is measured under a microscope.

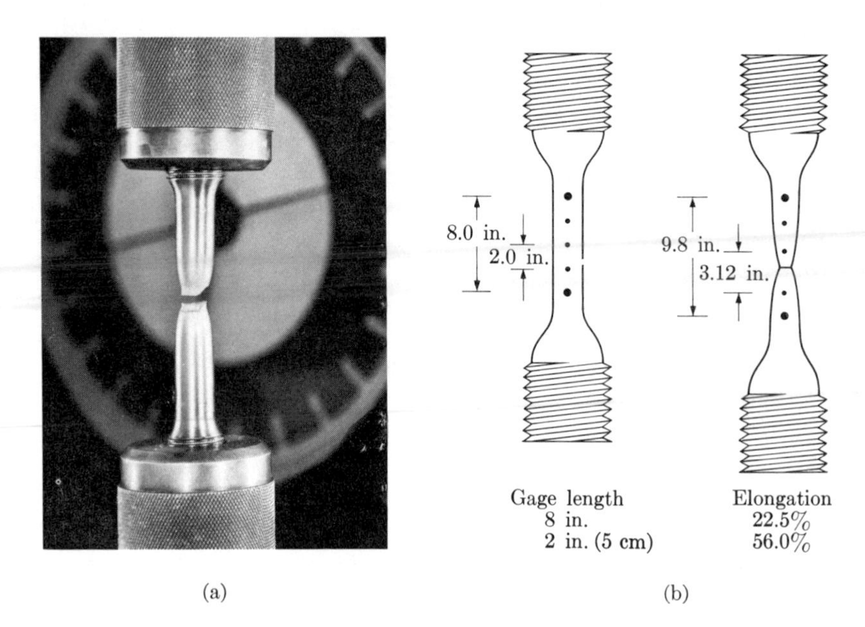

Fig. 1–2.2. Tensile tests. (a) Completed test on a round specimen. (U.S. Steel Corp.) (b) Elongation versus gage length. Since final deformation is localized, an elongation value is meaningless unless the gage length is indicated. For routine testing, a 2-in. (5-cm) gage length is common. Metric measurements (20.32 cm → 24.89 cm) and (5.08 cm → 7.92 cm) give the same values for elongation.

A second measure of ductility is *reduction in area*, $(A_o - A_f)/A_o$, at the point of fracture. Highly ductile materials are greatly reduced in cross section before breaking. The reduction in area is preferred by some engineers as a measure of ductility because it does not require a gage length, and because it can be used to determine the *true strain* at the point of fracture. (See Example 1–2.4(b).) One cannot establish an exact correlation between elongation and reduction in area, since the plastic deformation may be highly localized. Elongation is a measure of plastic "stretching," whereas reduction in area is a measure of plastic "contraction." Of course, a very ductile material will have high values of each, and a nonductile material has near-zero values of each.

Strength (and hardness). The ability of a material to resist plastic deformation is called the *yield strength* (YS), and is computed by dividing the force initiating the yield by the cross-sectional area. In materials such as some of the softer steels, the yield strength is marked by a definite *yield point* (Fig. 1–2.1(b)). In other materials, where the proportional limit is less obvious, it is common to define the yield strength as that stress required to give 0.2% (or some other tolerable value) plastic offset (Fig. 1–2.1(c)).

The *tensile strength* (TS) of a material is calculated by dividing the maximum load by the *original* cross-sectional area. This strength, like all other strengths, is expressed in the same units as stress. Note particularly that tensile strength is based on the original cross-sectional area. This is important, inasmuch as a ductile material will have its cross-sectional area somewhat reduced under maximum load. Although the *true*, or physical, *stress* in the material is based on actual area (Fig. 1–2.1(d)), the *nominal stress* is more important to the engineer, who must, of course, make his designs on the basis of initial dimensions.

Because the cross-sectional area of a ductile material may be reduced before it breaks, the *breaking strength* may be less than the *tensile strength*. Expressions for both are, by definition, based on original area (Fig. 1–2.1(c)).

Hardness is defined as the resistance of a material to penetration of its surface. As might be expected, the hardness and the strength of a material are closely related, as indicated in Fig. 1–2.3. *The Brinell hardness number* (BHN) is a hardness index calculated from the area of penetration by a large indenter. The indentation is made by a very hard steel or tungsten-carbide ball under a standardized load. The *Rockwell hardness* (R), another of several common indexes of hardness used by engineers, is related to BHN but is measured by the depth of penetration by a small standardized indenter. Several different Rockwell scales for materials of different hardness ranges have been established by selecting various indenter shapes and loads.

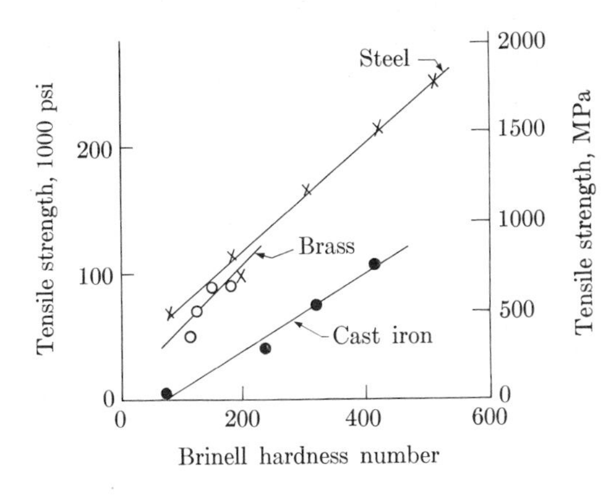

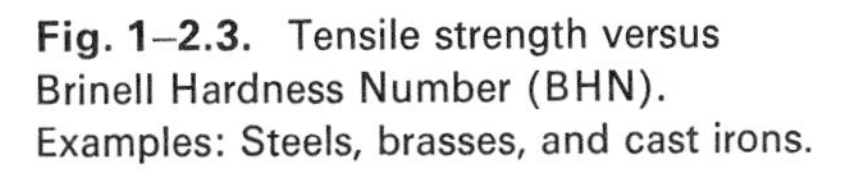
Fig. 1–2.3. Tensile strength versus Brinell Hardness Number (BHN). Examples: Steels, brasses, and cast irons.

Toughness. This is a measure of the *energy* required to break a material (Fig. 1–2.4). This is in contrast to *strength*, which is a measure of the *stress* required to deform or break a material. Energy, the product of force times distance, is expressed in joules, or foot-pounds; it is closely related to the area under the stress–strain curve. A ductile material with the same strength as a nonductile material will require more energy for breaking and be tougher. Standardized *Charpy* or *Izod* tests are two of several procedures used to measure toughness. They differ in the shape of the test

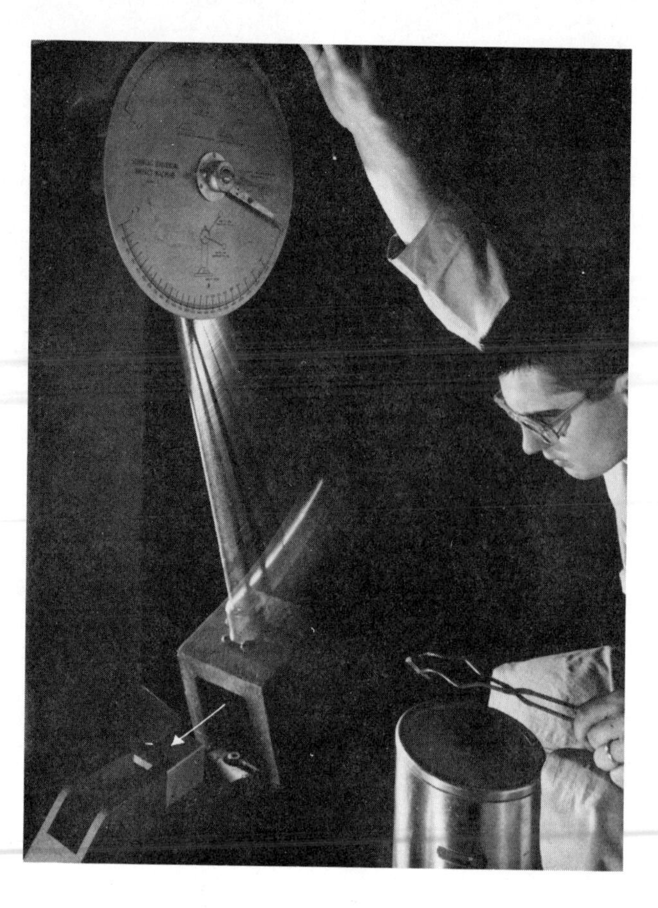

Fig. 1–2.4. Toughness test. The test specimen (arrow) is broken by the impact of the swinging pendulum. The amount of absorbed energy can be calculated from the arc of the follow-through swing. (U.S. Steel Corp.)

piece and in the method of applying the energy. Since toughness also depends on the geometry of stress concentrations, one must be careful to identify the test procedure which was used.

Example 1–2.1 Which part has the greater stress: (a) an aluminum bar of 0.97 in. × 1.21 in. (2.46 cm × 3.07 cm) cross section, under a load of 16,800 lb (7640 kg, and therefore a force of 75,000 N), or (b) a steel bar whose cross-sectional diameter is 0.505 in. (1.28 cm), under an 11,000-lb (5000-kg) load?

Solution

Units: $\dfrac{\text{Pounds}}{(\text{in.})(\text{in.})} = \text{psi}.$ $\qquad \dfrac{\text{Newtons}}{(\text{m})(\text{m})} = \text{Pascals}.$

Calculation:

a) $\dfrac{16{,}800}{(0.97)(1.21)} = 14{,}300 \text{ psi};$ $\qquad$ a) $\dfrac{(7640)(9.8)}{(0.0246)(0.0307)} = 100 \text{ MPa};$

b) $\dfrac{11{,}000}{(\pi/4)(0.505)^2} = 55{,}000 \text{ psi}.$ $\qquad$ b) $\dfrac{(5000)(9.8)}{(\pi/4)(0.0128)^2} = 380 \text{ MPa}.$ ◀

Example 1–2.2 A 2-in. (5.08 cm) gage length is marked on a copper rod. The rod is strained so that the gage marks are 2.27 in. (5.77 cm) apart. Calculate the strain.

Solution

Units:
$$\frac{(\text{in.} - \text{in.})}{\text{in.}} = \frac{\text{in.}}{\text{in.}} = \frac{\text{percent}}{100}.$$

Calculation:
$$\frac{2.27 - 2.00}{2.00} = \frac{0.135 \text{ in.}}{\text{in.}} = 13.5\%.$$

Comment. The metric solution gives the identical result. ◀

Example 1–2.3 If the average modulus of elasticity of the steel used is 205,000 MPa (30,000,000 psi), how much will a wire 2.5 mm (0.1 in.) in diameter and 3 meters (10 ft) long be deformed when it supports a load of 500 kg (= 1100 lb_f and 4900 N)?

Solution: Modulus of elasticity = stress/strain, or,

$$\text{Strain} = \frac{\sigma}{E}.$$

Units:
$$\text{m/m} = \frac{\text{N/m}^2}{\text{pascals}}. \qquad \text{in./in.} = \frac{\text{lb/in.}^2}{\text{psi}}.$$

Calculation:

$$\text{Strain} = \frac{4900/(\pi/4)(0.0025)^2}{205{,}000 \times 10^6} = 0.005 \text{ m/m}.$$

$$\text{Deformation} = (0.005 \text{ m/m})(300 \text{ cm}) = 1.5 \text{ cm}.$$

$$\text{Strain} = \frac{1100/(\pi/4)(0.1)^2}{30{,}000{,}000} = 0.005 \text{ in./in.}$$

$$\text{Deformation} = (0.005 \text{ in./in.})(120 \text{ in.}) = 0.6 \text{ in.}$$ ◀

•**Example 1–2.4** A copper wire has a nominal breaking strength of 43,000 psi (300 MPa) and a reduction of area of 77%. Calculate (a) the true tensile strength (breaking load/true area), and (b) the true strain ϵ_{tr} at the point of fracture. (The instantaneous strain $d\epsilon$ is equal to dl/l.)

Solution

a)
$$\frac{F}{A_0} = 43{,}000 \text{ psi}, \qquad F = 43{,}000 A_0,$$

$$\frac{F}{A_{tr}} = \frac{F}{(1 - 0.77)A_0} = \frac{43{,}000 A_0}{0.23 A_0}$$

$$= 187{,}000 \text{ psi} \qquad (= 1300 \text{ MPa}).$$

• Examples preceded by a bullet may be assigned at the discretion of the instructor. (*See* Preface.)

b) Since

$$d\epsilon \equiv \frac{dl}{l},$$

$$\epsilon_{tr} = \int_{l_0}^{l_f} \frac{dl}{l} = \ln \frac{l_f}{l_0}. \qquad \bullet(1\text{–}2.2)$$

Also,

$$A_0 l_0 = A_f l_f,$$

$$\epsilon_{tr} = \ln \frac{A_0}{A_f} = \ln \frac{A_0}{0.23 A_0}$$

$$= 1.47 \quad \text{or} \quad 147\%. \blacktriangleleft$$

1–3 THERMAL CHARACTERISTICS

Heat capacity. The distinction between temperature and heat content of a material is an important one for the engineer and scientist. *Temperature* T is a level of thermal activity, whereas the *heat content* is thermal energy. The two are related through heat capacity.

In the absence of any volume change, the *heat capacity* c is the change in heat content per °C. Not uncommonly in technical tables, specific heat is used in place of heat capacity. The *specific heat* of a material is defined as the ratio of the heat capacity of the material to that of water. Thus, with the heat capacity of water equalling 1 cal/g · °C (= 4.184 joules/g · °C = 1 Btu/lb · °F), one can make thermal calculations in the units of his choice.

Various heats of transformation are of importance in materials. The better known of these are the *heat of fusion* and the *heat of vaporization*, which are the heats required to produce melting and gasification, respectively. Each involves a change within the material from one atomic or molecular structure to another. We shall learn later that there are various structural changes possible *within solids*, and that these changes also require a change in heat or thermal-energy content of the material.

Thermal expansion. The expansion which normally occurs during the heating of a material arises from the more intense thermal vibrations of the atoms. To a first approximation, the increase in length, $\Delta L/L$, is proportional to the change in temperature, T:

$$\Delta L/L = \alpha_l \, \Delta T. \qquad (1\text{–}3.1)$$

Closer examination shows that the *linear expansion coefficient*, α_l, generally increases slightly with temperature (Fig. 1–3.1). The thermal expansion data listed in Appendix C are for 20°C (68°F).

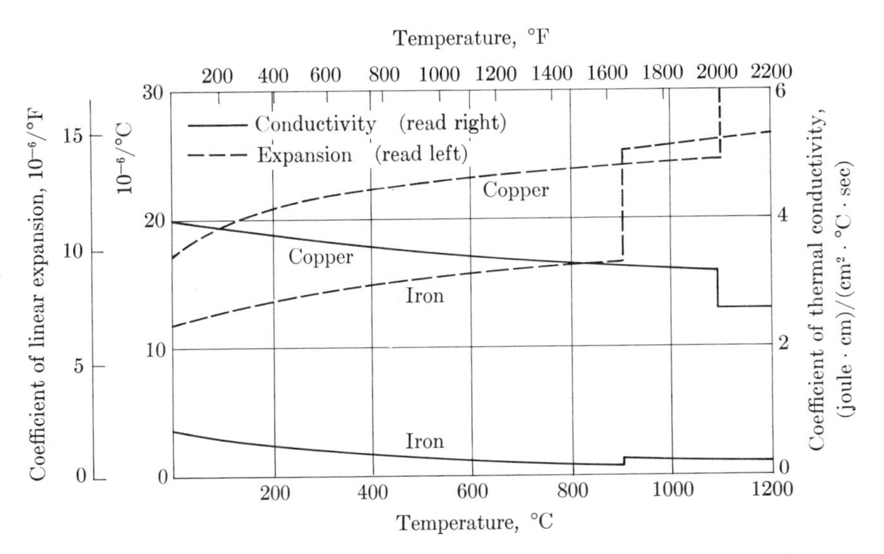

Fig. 1–3.1. Thermal properties versus temperature. The discontinuity for copper at 1084.5°C (1984°F) is a result of melting. Iron has a discontinuity because there is a rearrangement of the atoms at 912°C (1673°F). See Chapter 3.

The *volume expansion coefficient*, α_v bears the same relationship between the volume change, $\Delta V/V$, and temperature change, ΔT, as shown previously in Eq. (1–3.1). For all intents and purposes, α_v is three times the value of α_l. (See Study Problem 1–3.3(b).)*

The discontinuities in the expansion coefficients of Fig. 1–3.1 occur with changes in the atomic arrangements within the material. Specifically, copper melts at 1084.5°C (1984°F), and the iron atoms rearrange themselves from having eight neighbors below 912°C (1673°F) to having twelve neighbors above that temperature. Since structural changes such as these affect many engineering properties, we will pay specific attention to the atomic arrangements in later chapters.

---

* Materials such as graphite and mica are anisotropic; i.e., their properties vary with crystal direction. Therefore, a more general statement would be that

$$\alpha_v = \alpha_x + \alpha_y + \alpha_z, \tag{1–3.2(a)}$$

where the x, y, and z subscripts refer to the linear expansion coefficient in the three coordinate directions. When these are equal, as is the case in most engineering materials,

$$\alpha_v = 3\alpha_l. \tag{1–3.2(b)}$$

Thermal conductivity. Heat transfer through solids most commonly occurs by conduction. Thermal conductivity k is the proportionality constant relating the *heat flux* $\vec{Q}$ and the thermal gradient, $\Delta T/\Delta x$:

$$\vec{Q} = k\left(\frac{T_2 - T_1}{x_2 - x_1}\right). \qquad (1\text{–}3.3)^*$$

The preceding coefficient of thermal conductivity is also temperature-sensitive. However, unlike the coefficient of thermal expansion, this coefficient *decreases* as the temperature is raised above room temperature. (The reasons for this behavior will be discussed later.) The changes of atomic packing, which accompany melting and other atomic rearrangements arising from temperature variations, produce discontinuities in the thermal-conductivity values (Fig. 1–3.1).

The engineer is commonly interested in the unsteady state of thermal transfer, as well as the steady state. In the unsteady state, thermal transfer produces a temperature change, and thus decreases the thermal gradient. Under such conditions, *thermal diffusivity* h is important:

$$h = \frac{k}{c\rho} \qquad (1\text{–}3.4)$$

where k is the thermal conductivity, c is the heat capacity, and ρ is the density. A material with high heat requirements per unit volume, $c\rho$, has a low thermal diffusivity, simply because more joules (or Btu's) must be added to or removed from the material in order to change the temperature. The units applicable to thermal diffusivity, indicated below, will provide a basis for later considerations of atomic diffusivity.

$$\text{Thermal diffusivity} = \frac{\text{conductivity}}{\text{(heat capacity)(density)}}$$

$$= \frac{(\text{energy/cm}^2 \cdot \text{sec})/(^\circ\text{C/cm})}{(\text{energy/g} \cdot {}^\circ\text{C})(\text{g/cm}^3)} = \text{cm}^2/\text{sec}. \qquad (1\text{–}3.5)$$

Example 1–3.1 An aluminum wire is stressed 5000 psi (34.5 MPa) in tension. What temperature increase is required to change its length by the same amount?

* The units for k must have the dimensions to give

$$\left(\frac{\text{energy}}{\text{area} \cdot \text{time}}\right) \Big/ \left(\frac{\text{temperature}}{\text{thickness}}\right).$$

Unfortunately there is not always consistency when dimensions for conductivity data are presented in tables. The dimensions listed in the footnotes of Appendix C for conversion between various English and metric values are consistent with the above dimensions.

Solution: Obtain the elastic and expansion data from Appendix C.

$$\Delta L/L = \sigma/E = (34.5 \times 10^6 \text{ Pa})/(70{,}000 \times 10^6 \text{ Pa}) = \alpha\,\Delta T = (22.5 \times 10^{-6}/^\circ\text{C})\,(\Delta T),$$

$$\Delta T = 22^\circ\text{C}\ (= 40^\circ\text{F}).$$

$$\Delta L/L = \epsilon = (5000 \text{ psi})/(10^7 \text{ psi}) = (12.5 \times 10^{-6}/^\circ\text{F})\,\Delta T,$$

$$\Delta T = 40^\circ\text{F}\ (= 22^\circ\text{C}).$$ ◀

Example 1–3.2 A stainless steel plate 0.4 cm thick has circulating hot water on one side and a rapid flow of air on the other side, so that the two metal surfaces are 90°C and 20°C, respectively. How many joules are conducted through the plate per minute?

Solution: Obtain data from Appendix C.

$$\vec{Q} = (0.15 \text{ J} \cdot \text{cm/cm}^2 \cdot \text{s} \cdot {}^\circ\text{C})(70^\circ\text{C}/0.4 \text{ cm})(60 \text{ s/min})$$

$$= 1575 \text{ joules/cm}^2 \cdot \text{min}.$$

Comment. We shall learn later that stainless steel has lower thermal (and electrical) conductivity than other metals because it contains added alloying elements (Chapter 5). ◀

1–4 RESPONSES TO ELECTRIC FIELDS

Conductivity (and resistivity). Metals and semiconductors will conduct electrical charges when they are placed in an electrical field. The conductivity σ depends on the number of carriers n, the charge q carried by each, and the mobility μ of the charge carrier. The conductivity is the reciprocal of the *resistivity* ρ:

$$\frac{1}{\rho} = \sigma = nq\mu. \tag{1–4.1(a)}$$

The units are:

$$\text{ohm}^{-1} \cdot \text{cm}^{-1} = \left(\frac{\text{carriers}}{\text{cm}^3}\right)\left(\frac{\text{coul}}{\text{carrier}}\right)\left(\frac{\text{cm/sec}}{\text{volt/cm}}\right). \tag{1–4.1(b)}$$

In metals and in those semiconductors in which electrons are the charge carriers, the charge per carrier is 0.16×10^{-18} coul, or 0.16×10^{-18} amp · sec. The *mobility* may be considered as a net, or *drift velocity* $\bar{v}$ of the carrier, which arises from the *electric field* $\mathscr{E}$, where the units are (cm/sec) and (volts/cm), respectively:

$$\mu = \frac{\bar{v}}{\mathscr{E}}. \tag{1–4.2}$$

Mobility is commonly expressed as $\text{cm}^2/\text{volt} \cdot \text{sec}$. The relationships of Eqs. (1–4.1) and (1–4.2) will be particularly useful in Chapter 5 when we consider semiconductors.

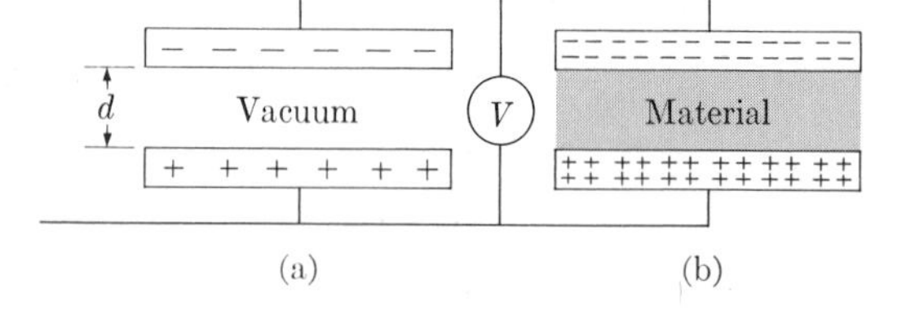

Fig. 1–4.1. Charge density $\mathscr{D}$ versus relative dielectric constant, κ. The presence of a material increases the charge density, $\mathscr{D}_m$, held by the capacitor plates in proportion to the relative dielectric constant, κ:

$$\mathscr{D}_m = \kappa \mathscr{D}_0.$$

Dielectric behavior. Electrical insulators do not, of course, transport electric charges. However, they are not inert to an electric field. We can show this by separating two electrode plates by a distance d and applying a voltage V between them (Fig. 1–4.1(a)). The *electric field* $\mathscr{E}$ is the voltage gradient:

$$\mathscr{E} = \frac{V}{d}. \tag{1–4.3}$$

Under these conditions, when there is nothing between the plates, the *charge density* $\mathscr{D}_0$ on each plate is proportional to the field $\mathscr{E}$. For each volt/m of field, there are 8.85×10^{-12} coulombs per square meter of electrode. (This charge density requires 55×10^6 electrons/m^2 for each volt/m, since the *electron charge* is 0.16×10^{-18} coulombs.)

$$\mathscr{D}_0 = (8.85 \times 10^{-12}\ \text{C/V} \cdot \text{m})\mathscr{E}. \tag{1–4.4}$$

If a material m is placed between the electrodes in Fig. 1–4.1, the charge density can be increased to $\mathscr{D}_m$ from the $\mathscr{D}_0$ just described. The ratio, $\mathscr{D}_m/\mathscr{D}_0$, is called the relative *dielectric constant* κ of the material which is used as the dielectric spacer between the electrodes:

$$\kappa = \frac{\mathscr{D}_m}{\mathscr{D}_0}. \tag{1–4.5}$$

The relative dielectric constant will be important to us in Chapters 7 and 8, in which we look at the *dielectric* properties of the plastics and ceramics which are used in capacitors.* The relative dielectric constant is always greater than 1.0 because both electrons and the positive and negative ions are displaced within the material when it is inside an electric field.

Example 1–4.1 A semiconductor has 10^{15} electrons that serve as charge carriers per cm^3, and a resistivity of 10 ohm · cm. (a) What is the mobility μ of the electrons? (b) What is the drift velocity $\bar{v}$ when the voltage gradient is 0.016 volts/mm?

---

* For a parallel-plate capacitor, the capacitance C is determined by

$$C = \kappa(8.85 \times 10^{-12}\ \text{C/V} \cdot \text{m})(A/d), \tag{1–4.6}$$

where A is the plate area and d is the spacing between plates.

Solution

a) $$\mu = 1/\rho nq = 1/(10\ \text{ohm}\cdot\text{cm})(10^{15}/\text{cm}^3)(0.16 \times 10^{-18}\ \text{A}\cdot\text{s})$$
$$= 625\ \text{cm}^2/\text{V}\cdot\text{s} = 625(\text{cm/s})/(\text{V/cm}).$$

b) $$\bar{v} = (625(\text{cm/s})/(\text{V/cm}))(0.16\ \text{V/cm}) = 100\ \text{cm/s}. \blacktriangleleft$$

Example 1–4.2 Two capacitor plates (3 cm × 2 cm each) are parallel and 0.22 cm apart, with nothing between them. What voltage is required to produce a charge of 0.24×10^{-10} coul on the electrodes?

Solution: Rearranging Eqs. (1–4.3) and (1–4.4) and using cm, we get

$$V = \frac{(0.24 \times 10^{-10}\ \text{C})/(3\ \text{cm})(2\ \text{cm})}{(8.85 \times 10^{-14}\ \text{C/V}\cdot\text{cm})/(0.22\ \text{cm})} = 10\ \text{volts}. \blacktriangleleft$$

1–5 MEASUREMENT OF PROPERTIES

Qualitative information. Schematic diagrams which show the effect of one variable on a dependent property are indispensable aids in translating complicated empirical relationships into qualitative terms. Figure 1–5.1, for example, illustrates the change in the strength of concrete in relation to the amount of water added. Concrete, of course, is strongest when a minimum amount of water is used, although there must be sufficient water to make the concrete workable.

Other variables may be shown schematically by the use of additional parameters. Figure 1–5.2 adds the parameter of time, t, to the relationship previously given in Fig. 1–5.1. Figure 1–5.2 tells us that (1) for any given addition of water to cement, the strength increases as the period of time increases, (2) for any given period of time, the strength is less if excess water was used, and (3) a given strength may be attained in less time if less water is used.

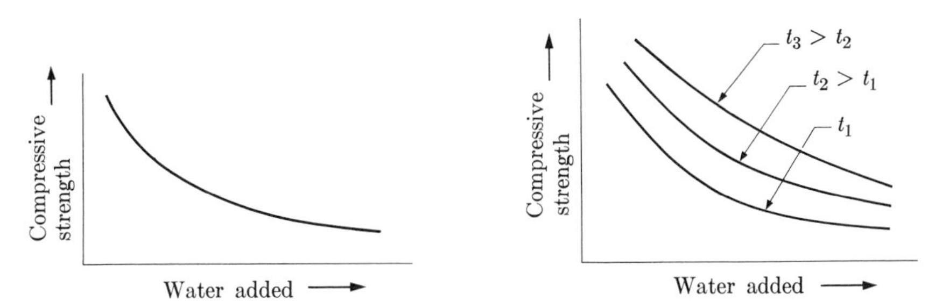

Fig. 1–5.1. Schematic representation of two variables. Strength of concrete versus water content. The water content is the independent variable.

Fig. 1–5.2. Schematic representation of three variables. Strength of concrete as related to time t and water content.

Schematic representations help the engineer to determine in advance what variables can be controlled to obtain the desired result. With such information he can anticipate possible modifications of his materials in production or in service.

Quantitative data. It is commonly more important to secure quantitative data concerning the properties of materials. Thus, from Fig. 1–5.3, the design engineer observes that concrete may have a compressive strength of 4000 psi (27.6 MPa) if the water-to-portland cement ratio is 0.6. However, to make the information complete, the parameter of time as well as data on particle size and temperature should be included, since each of these influences the quantitative relationships.

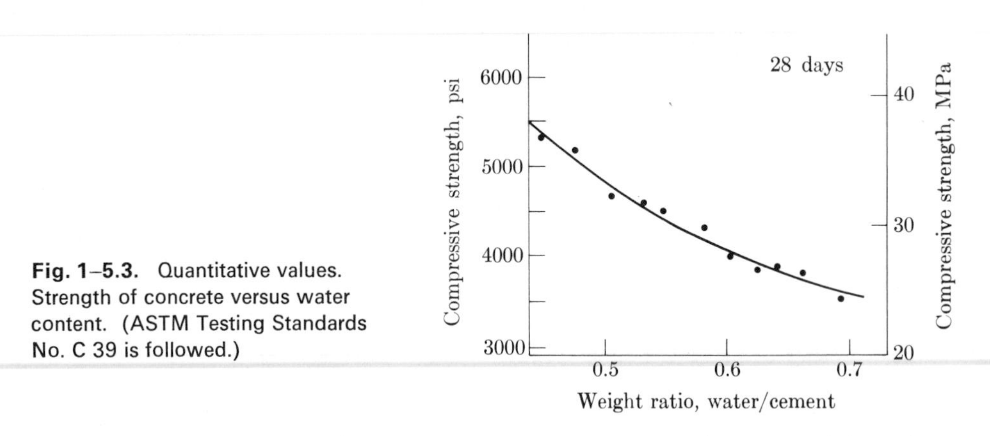

Fig. 1–5.3. Quantitative values. Strength of concrete versus water content. (ASTM Testing Standards No. C 39 is followed.)

Variance of data. All laboratory and industrial data contain a spread in values. Figure 1–5.4 shows the scatter obtained in the toughness testing of fifty samples of steel at 21°C (70°F). This occurred even though the samples and testing procedures were identical, so far as could be determined. The variation in toughness can arise from several sources: (1) undetected differences within the steel which was sampled; (2) differences in the preparation of the samples; (3) differences in the testing procedure.

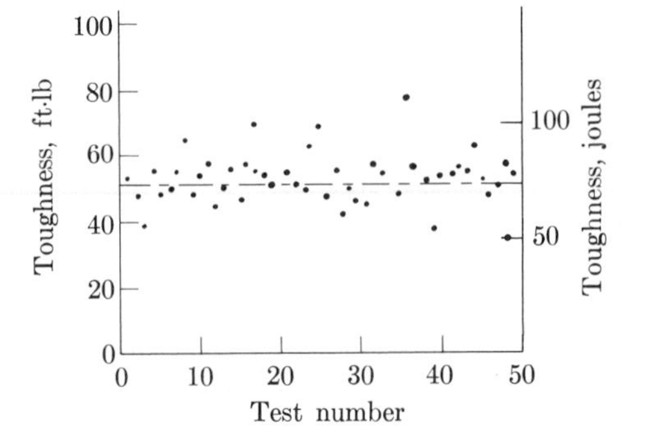

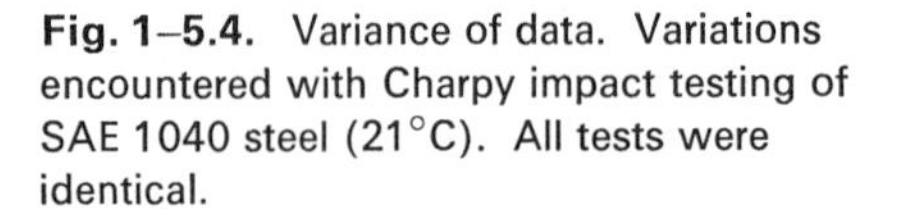

Fig. 1–5.4. Variance of data. Variations encountered with Charpy impact testing of SAE 1040 steel (21°C). All tests were identical.

Because of this range of data, results are commonly reported as the *mean* (average) value, or *median* (middle) test value. In Example 1–5.1, which cites 21 strength values for a plastic, the mean and median strengths are 20.7 MPa (3005 psi) and 20.9 MPa (3030 psi), respectively. Although the mean $\overline{X}$ and median $\overline{M}$ values are close to each other, they are not identical. The difference arises in this case because individual results are more dispersed at the lower strengths.

Reporting either the mean or median does not fully satisfy the design engineer or the applied scientist, because he does not know how strong or how weak individual samples may be; he therefore asks for the *standard deviation* (SD).* The standard deviation is a statistical measure of *scatter*. It is calculated as follows:

$$\text{SD} = \sqrt{\Sigma(X_i - \overline{X})^2/n}, \qquad \bullet(1\text{–}5.1)\dagger$$

where $\overline{X}$ is the mean and X_i are the individual values (up to n). In practice, the standard deviation has significance in that approximately two-thirds of the results lie within ± 1 SD, and 5% of statistical data lie outside ± 2 SD, when there is sufficient data to follow normal distribution laws.

Accuracy of calculations. It is commonly desirable to calculate properties as closely as the available data warrant. In this text, the available data include two or three, and occasionally four, *significant figures.* This is typical of data which are generally provided for commercial materials. Thus, it is possible to use a slide rule for most of our calculations because a standard slide rule can be read to one part in 500. With the advent of hand calculators, answer displays commonly contain nonsignificant figures. The practice of reporting nonsignificant figures should be avoided. The recommended practice is to round off the *final* answer (not intermediate values) to the number of significant figures possessed by the least accurate *noninteger* datum (+ one).§

Example 1–5.1 The test data for the breaking stresses of 21 samples of a plastic have been arranged in descending order in Fig. 1–5.5.

a) Determine the two measures of central tendency: median $\overline{M}$ and mean $\overline{X}$.

b) •Calculate the scatter or standard deviation, SD.

---

* The symbol σ is usually used for standard deviation; however, σ also applies to stress. So, to avoid possible confusion in problems such as Example 1–5.1, we will use SD as an alternative.

† See Preface for the bullet, •, notation.

§ The added figure has occasional value in revealing a possible bias within the previous number. The exception for integers is apparent when one determines the mean weight for a family of *three* individuals by using a scale with an accuracy of one part in 100. We would not expect the integer of *three* to dictate only one digit for the mean weight.

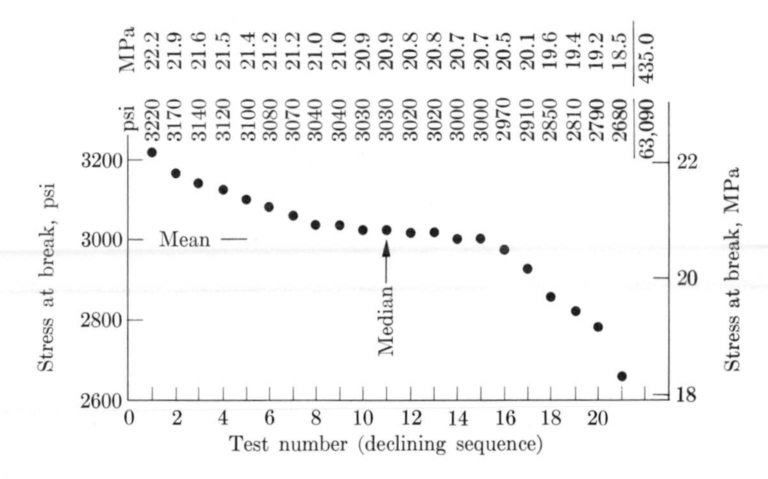

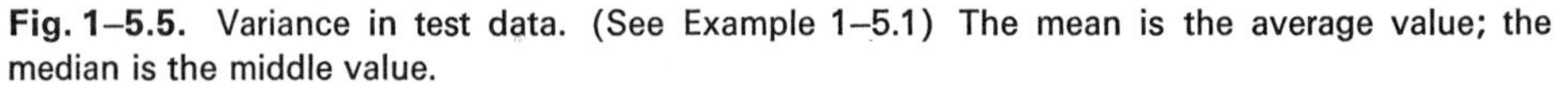

Fig. 1–5.5. Variance in test data. (See Example 1–5.1) The mean is the average value; the median is the middle value.

Solution: We may use either psi or MPa.

a) The 11th (median) value is 3030 psi (20.9 MPa). The mean value is:

$$\bar{X} = \frac{\Sigma X}{21} = \frac{63{,}090 \text{ psi}}{21} = 3005 \text{ psi}$$

or

$$\frac{435 \text{ MPa}}{21} = 20.7 \text{ MPa.}$$

•b)

$$\text{SD} = \sqrt{\Sigma(X - \bar{X})^2/21} = \sqrt{16.8/21} = 0.9 \text{ MPa}$$

or

$$\sqrt{350{,}000/21} = 130 \text{ psi.}$$

Comments. The data of Fig. 1–5.5 have a typical statistical variation, since 14 of the 21 values lie within 3005 ± 130 psi (20.7 ± 0.9 MPa). The median is higher than the mean because there is more dispersion among the low test data. ◀

REVIEW AND STUDY

SUMMARY

If technology is to meet the needs and desires of society, there must be an intelligent use of materials. This involves selecting the materials with the optimum characteristics, cost, and availability, and incorporating them into a design which is safe, reliable, and compatible with everyone's well-being. In order to provide the technology, we must select materials to meet specific requirements, such as strength, thermal and electrical conductivity, fabricability, etc. Therefore, we are interested in the various properties which give us a basis for our designs and products.

The properties and behavior of materials arise from their internal structures. If a specific set of properties is required, the materials must be chosen appropriately so that they have suitable structures of atoms, crystals, and other internal arrangements. Should the internal structure of a material be changed during processing or service, there will be corresponding changes in properties.

In this chapter, we have cited some—but not all—of the properties which require attention. The ones we have encountered, however, will enable us to discuss in later chapters the features of internal structures which (a) provide resistance to mechanical alteration by applied stresses; (b) account for the more common characteristics of a material when it is thermally agitated; and (c) relate to the behavior of a material within an electric field. These features are important because knowledge of them enables the designer to specify, and the technologist to produce, materials with optimum structures and characteristics for present and future products.

Materials recover from elastic strain after stress is removed because the atoms have maintained their original neighbors. With higher stresses, strain proceeds beyond the elastic range to give plastic deformation. Highly ductile materials are tough, in that they require considerable energy before failure.

The input of thermal energy to raise the temperature increases the thermal agitation within the material. This extra energy expands the structure. Thermal energy is transferred to cooler parts of the material by atom vibrations and (in metals) by electron conduction.

The diffusion of electrons accounts for the electrical conductivity of most conducting materials. The charges within an insulator do not migrate in electric fields, but are displaced, so that, in addition to simply being "isolators," dielectrics have useful functional electrical properties.

Finally, attention was given to the significance and accuracy of property data. Measures of central tendency (mean and median) and of scatter (standard deviation) should become familiar to the reader. Likewise, *unwarranted* accuracy which is incorporated during the process of calculation should be avoided.

TERMS FOR REVIEW

Brinell hardness number, BHN
Brittle
Charge density, $\mathscr{D}$
Conductivity
 electrical, σ
 thermal, k
Dielectric constant, κ
Drift velocity, $\bar{v}$
Ductility
Electric field, $\mathscr{E}$
Electron charge, q
Elongation
Gage length
Hardness
Heat capacity, c
Internal structure
Mean, $\overline{X}$
Median, $\overline{M}$
Mobility, μ
Modulus, elastic, E
Modulus of elasticity, E
Nominal stress
Plastic strain
Properties
Reduction of area, R of A
Resistivity, ρ
Rockwell hardness, R
Significant figures
Specific heat
• Standard deviation, SD
Strain, ϵ
 elastic
 plastic
 • true, ϵ_{tr}
Strength
 breaking, BS
 tensile, TS
 yield, YS
Stress, σ
 nominal
 true, σ_{tr}
Système International, SI
Thermal diffusivity, h
Thermal expansion coefficient
 linear, α_l
 volume, α_v
Toughness
Yield point
Young's modulus, E

FOR DISCUSSION

A_1 Compare and contrast: strength, ductility, and toughness.

B_1 Why is the true stress always higher than the nominal stress? Why does the engineer use the nominal stress when he designs a bridge?

C_1 Refer to Fig. 1–2.1(c). Why is yield strength commonly defined on the basis of 0.2% permanent strain?

D_1 Loading is removed when the top point of the curve in Fig. 1–2.1(b) is reached. Will the unloading curve be vertical or sloping? Why?

• Terms encountered initially as optional topics.

E_1 Refer to D_1. After unloading, gage marks are placed on the test piece. It is now reloaded and carried to failure. In what way will the calculated elongation be affected?

F_1 Suggest a design for a testing machine to measure the toughness of a steel.

G_1 Cite a design situation where we would multiply kilograms by a number other than 9.8 to get newtons.

H_1 Provide proof for Eq. (1–3.2(b)).

I_1 Iron has discontinuities in its thermal conductivity coefficient slightly above 900°C (Fig. 1–3.1). It also changes volume sharply at that temperature. Suggest whether this abrupt volume change is an expansion or contraction.

J_1 List the kinds of carriers which can transport electrical charge through a liquid.

K_1 One ampere of current flows through a wire. How many electrons flow through the wire during each second?

L_1 Refer to Eq. (1–4.1). There are 7.1×10^{19} carriers/m^3, and the mobility is 0.2 m^2/volt · sec. The hand-calculator answer for resistivity is 0.440140 ohm · m. Express this with the appropriate number of significant figures.

M_1 The team of 5 basketball players had heights of 1.93 m, 1.98 m, 2.00 m, 1.80 m, and 1.89 meters. Their mean and median heights are ____ and ____.

STUDY PROBLEMS

1–1.1 Take the cord of a household appliance, such as a toaster or coffee-maker. List the materials used and the probable reason for their selection.

1–1.2 Examine an incandescent light bulb closely. How many different types of materials can you cite? What thermal and electrical characteristics are required of each?

1–2.1 a) A steel bar 0.5 in. (1.27 cm) in diameter supports a load of 15,400 lb (7000 kg, and therefore a force of 68,600 N). What is the stress placed on the bar? b) If the bar of part (a) has a modulus of elasticity of 30,000,000 psi (205,000 MPa), how much will the bar be strained with this load?

Answer: a) 78,000 psi (540 MPa) b) 0.0026

1–2.2 The bar in Problem 1–2.1 supports a maximum load of 11,800 kg (26,000 lb) without plastic deformation. What is its ____ strength?

1–2.3 The bar of Problem 1–2.1 breaks with a load of 25,000 lb (11,400 kg, and therefore a force of 111,000 N). Its final diameter is 0.31 in. (0.787 cm). a) What is its true breaking strength? b) What is its nominal breaking strength? •c) What is the true strain at fracture?

Answer: a) 331,000 psi (2280 MPa) b) 127,000 psi (875 MPa) c) 96%

1–2.4 A copper alloy has a modulus of elasticity of 110,000 MPa (16,000,000 psi), a yield strength of 330 MPa (48,000 psi), and a tensile strength of 350 MPa (51,000 psi). a) How much stress would be required to stretch a 3-m (118-in.) bar of this alloy 0.15 cm? b) What size round bar would be required to support a force of 22,000 N (5000 lb_f) without yielding?

1–2.5 Aluminum (6151 alloy) has a modulus of elasticity of 10^7 psi (70,000 MPa) and a yield strength of 40,000 psi (275 MPa). a) How much load can be supported by a 0.108-in. (2.75-mm) wire of this alloy without yielding? b) If a load of 97 lb (44 kg, and therefore 430 N) is supported by a 100-ft (30.5-m) wire of this size, what is the total strain?

Answer: a) 366 lb_f (1630 N force, or 167 kg mass) b) 1.27 in. (3.2 cm) elongation

1–2.6 Monel metal (70 Ni–30 Cu) has a modulus of elasticity of 180,000 MPa (26 × 10^6 psi) and a yield strength of 450 MPa (65,300 psi). a) How much load could be supported by a rod 18 mm (0.71 in.) in diameter without yielding? b) If a maximum total elongation of 2.5 mm (0.1 in.) is permissible in a 2.1-m (6.9-ft) bar, how large a load could be applied to this rod?

1–2.7 A long-used rule of thumb says that *steel* has a tensile strength of approximately 500 times its Brinell hardness number (that is, $TS_{psi} \approx 500$ BHN). a) Check this against Fig. 1–2.3. b) Establish a similar constant relating TS_{MPa} to BHN.

Answer: 3.5 (BHN) $\approx TS_{MPa}$

1–2.8 A 0.25-in. (6.4-mm) diameter 1020 steel bar 6 ft (1.83 m) long supports a weight of 1000 lb_f (4450 N). What is the difference in strain if the bar is changed to a 70–30 Monel? (See Problem 1–2.6.)

1–3.1 a) Calculate the heat transfer through a plate of copper which is substituted for the stainless steel in Example 1–3.2. b) What is the wattage per cm^2? (Joules are volt · amp · sec, or watt · sec.)

Answer: a) 42,000 joules/cm^2 · min b) 700 watts/cm^2

1–3.2 Using the data from Appendix C, determine the ratios of thermal to electrical conductivities, k/σ, for each metal.

1–3.3 The average coefficient of thermal expansion of a special steel rod is $12 \times 10^{-6}\,°C^{-1}$. a) How much temperature change is required to provide the same linear change as a stress of 620 MPa (90,000 psi)? b) What volume change does this temperature change produce?

Answer: a) 250°C (450°F) b) 0.9%

1–3.4 A welded steel rail is laid in place at 35°C (95°F) and anchored so that shrinkage cannot occur when the temperature drops to 0°C. How much stress develops in the rail?

1–4.1 There are 10^{20} electrons per m^3 in a semiconductor which has an electron mobility of 0.19 $m^2/V \cdot s$. a) What is the resistivity? b) The conductivity?

Answer: a) 0.33 $\Omega \cdot m$ b) 3.04 $ohm^{-1} \cdot m^{-1}$

1–4.2 a) If a pure copper wire (resistivity = 1.7×10^{-6} ohm · cm) and 0.1 cm in diameter is used for an electrical circuit carrying 10 amp, how many watts of heat will be lost per foot (30.48 cm)? b) How many more watts would be lost if the copper wire were replaced by a brass wire of the same size (resistivity = 3.2×10^{-6} ohm · cm)?

1–4.3 What new relative dielectric constant will be required for a capacitor to retain the same charge density if the spacing between the foil electrodes is reduced from 0.10 mm to 0.06 mm without modifying the electric field? The present insulation has a relative dielectric constant of 3.3.

Answer: 2.0

1–4.4 The dielectric constant of a glass ribbon is 6.1. Would a capacitor using such a glass ribbon 0.25 mm thick have greater or less capacitance than another similar capacitor using a 100-μm plastic with a dielectric constant of 2.1?

• **1–5.1** Obtain the scores for last year's basketball games. Determine the mean, median and standard deviation of the scores of your team, a) at home, b) away.

1–5.2 What are the a) mean and b) median of the following density data for a certain type of foam rubber: 0.26, 0.30, 0.23, 0.24, 0.21, 0.26, 0.27, 0.29, 0.22, 0.28 g/cm^3?

• **1–5.3** What is the standard deviation of the data in the previous study problem?

Answer: 0.029 g/cm^3

• Problems preceded by a bullet • are based, in part, on optional sections or examples.

CHAPTER TWO

INTRODUCTION TO MATERIALS: REVIEW OF CHEMICAL BONDING

PREVIEW

In this chapter, the concepts of bonding which are normally presented in college chemistry courses are reviewed. Additional attention is given to the strengths of various bonds and to the role of the valence electrons. This provides the basis for categorizing materials as metals, polymers (plastics), and ceramics. It also permits us to generalize about the role of bonding in those properties introduced in Chapter 1.

2–1 INDIVIDUAL ATOMS

At one time the atom was believed to be the ultimate unit in the subdivision of matter. However, subsequently it became known that the atom is composed of still smaller units. It is now possible to subdivide atoms and to explore their internal structure.

No attempt will be made here to consider all of the subatomic relationships. However, it is necessary to consider the general structure of the atom in order to understand the factors that govern the properties of materials. For example, when a material is stressed, the attractive forces between the atoms resist the stress and keep the material from deforming and pulling apart. Electrical conductivity arises from the mobility of electrons away from their "parent" atoms. Oxidation of metals is caused by the diffusion of metal atoms or of the oxygen atoms through the surface scale to form the oxide product. These and other phenomena are best explained by considering a model of an atom.

Neutrons, protons, and electrons. The atom contains a *nucleus* and surrounding *electrons*. The nucleus is composed of *protons* and *neutrons*. The electrons are charged particles with 1/1836 the mass of a proton. The charge of the electron is taken to be negative. Since electrons are components of all atoms, their electrical charge is frequently regarded as unity. In physical units, this charge is equal to 0.16×10^{-18} amp · sec (coulomb) per electron. We know that a proton carries a charge that is equal but opposite to that of an electron. The neutron is electrically neutral and has approximately the same *mass* as the proton.

Atomic weight and atomic number. Two common properties of materials, *density* and *heat capacity*, are closely related to the mass of their atoms. In addition, the melting and boiling temperatures of small molecules can be related to their molecular weights. Since atoms are extremely small in comparison to the day-to-day concepts of mass, we use the *atomic mass unit*, amu, as the basis for atomic-weight calculations. The amu is defined as one-twelfth of the atomic mass of carbon-12, the most common isotope of carbon. There are $0.6022\ldots \times 10^{24}$ amu per g. We will use this conversion factor (called *Avogadro's number* N) in numerous calculations. Since natural carbon contains approximately one percent C^{13} along with 98.9% C^{12}, the average atomic weight of carbon is 12.011... amu. This is the value presented in the *periodic table* (Fig. 2–1.1) and in tables of elements (Appendix B). Those atomic weights encountered most commonly by the reader include:

H 1.0079... amu (or 1.0079 g/(0.602×10^{24} atoms))
C 12.011...
O 15.9994...
Cl 35.453...
Fe 55.847...

These values may be rounded off to 1, 12, 16, 35.5, and 55.8 amu, respectively, for all but the most precise of our calculations.

IA	IIA	IIIB	IVB	VB	VIB	VIIB	VIII	VIII	VIII	IB	IIB	IIIA	IVA	VA	VIA	VIIA	0
1 **H** 1.00797																1 **H** 1.0079	2 **He** 4.0026
3 **Li** 6.939	4 **Be** 9.012											5 **B** 10.811	6 **C** 12.011	7 **N** 14.007	8 **O** 15.9994	9 **F** 18.998	10 **Ne** 20.183
11 **Na** 22.990	12 **Mg** 24.312											13 **Al** 26.98	14 **Si** 28.086	15 **P** 30.97	16 **S** 32.064	17 **Cl** 35.453	18 **Ar** 39.95
19 **K** 39.102	20 **Ca** 40.08	21 **Sc** 44.96	22 **Ti** 47.90	23 **V** 50.94	24 **Cr** 52.00	25 **Mn** 54.94	26 **Fe** 55.85	27 **Co** 58.93	28 **Ni** 58.71	29 **Cu** 63.54	30 **Zn** 65.37	31 **Ga** 69.72	32 **Ge** 72.59	33 **As** 74.92	34 **Se** 78.96	35 **Br** 79.91	36 **Kr** 83.80
37 **Rb** 85.47	38 **Sr** 87.62	39 **Y** 88.91	40 **Zr** 91.22	41 **Nb** 92.91	42 **Mo** 95.94	43 **Tc** 99	44 **Ru** 101.07	45 **Rh** 102.91	46 **Pd** 106.4	47 **Ag** 107.87	48 **Cd** 112.40	49 **In** 114.82	50 **Sn** 118.69	51 **Sb** 121.75	52 **Te** 127.60	53 **I** 126.90	54 **Xe** 131.30
55 **Cs** 132.91	56 **Ba** 137.34	57–71 **La** series*	72 **Hf** 178.49	73 **Ta** 180.95	74 **W** 183.85	75 **Re** 186.2	76 **Os** 190.2	77 **Ir** 192.2	78 **Pt** 195.1	79 **Au** 196.97	80 **Hg** 200.59	81 **Tl** 204.37	82 **Pb** 207.19	83 **Bi** 208.98	84 **Po** 210	85 **At** 210	86 **Rn** 222
87 **Fr** 223	88 **Ra** 226	89– **Ac** series†															

58 **Ce** 140.12	59 **Pr** 140.91	60 **Nd** 144.24	61 **Pm** 145	62 **Sm** 150.35	63 **Eu** 151.96	64 **Gd** 157.25	65 **Tb** 158.92	66 **Dy** 162.50	67 **Ho** 164.93	68 **Er** 167.26	69 **Tm** 168.93	70 **Yb** 173.04	71 **Lu** 174.97
90 **Th** 232.04	91 **Pa** 231	92 **U** 238.03	93 **Np** 237	94 **Pu** 242	95 **Am** 243	96 **Cm** 247	97 **Bk** 249	98 **Cf** 251	99 **Es** 254	100 **Fm** 253	101 **Md** 256	102 **No** 253	103 **Lw** 257

Metals — Nonmetals

Fig. 2–1.1. Periodic table of the elements, showing the atomic number and atomic weight (in amu). There are 0.6×10^{24} amu per gram; therefore the atomic weights are grams per 0.6×10^{24} atoms. Metals readily release their outermost electrons. Nonmetals readily accept or share additional electrons.

The *atomic number* indicates the number of electrons associated with each neutral atom (and the number of protons in the nucleus). Each element is unique with respect to its atomic number. Appendix B lists selected elements, from hydrogen, with an atomic number of one, to uranium (92). It is the electrons, particularly the outermost ones, which affect most of the properties of engineering interest: (1) they determine the chemical properties; (2) they establish the nature of the interatomic bonding, and therefore the mechanical and strength characteristics; (3) they control the size of the atom and affect the electrical conductivity of materials; and (4) they influence the optical characteristics. Consequently, we shall pay specific attention to the distribution and energy levels of the electrons around the nucleus of the atom.

The *periodicity of elements* is emphasized in chemistry courses. We shall not repeat those characteristics here except to observe that the periodic table (Fig. 2–1.1) arranges the atoms of sequentially higher atomic numbers so that the vertical columns, called *groups*, possess atoms of similar chemical characteristics. In brief, those elements at the far left of the periodic table are readily ionized to give positive ions, *cations*. Those in the upper right corner of the periodic table more readily share or accept electrons. They are *electronegative*.

Orbitals. The electrons which accompany an atom are subject to very rigorous rules of behavior because they have the characteristics of standing waves during their movements in the neighborhood of the atomic nuclei. Again the reader is referred to introductory chemistry texts; however, let us summarize several features here. With individual atoms, electrons have specific energy states—*orbitals*. As shown in Fig. 2–1.2, the available electron energy states around a hydrogen atom can be very definitely identified.* To us, the important consequence is that there are large ranges

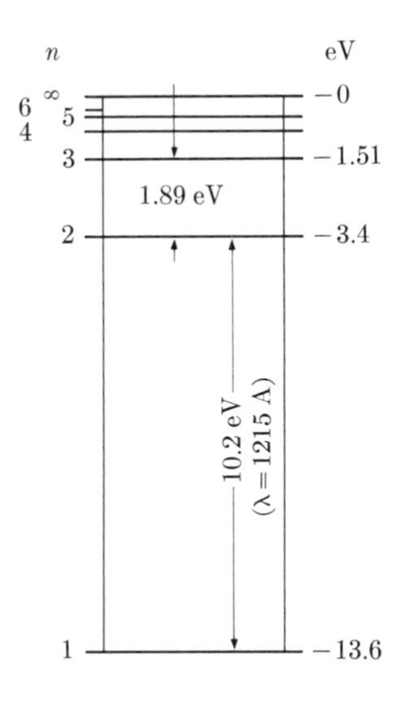

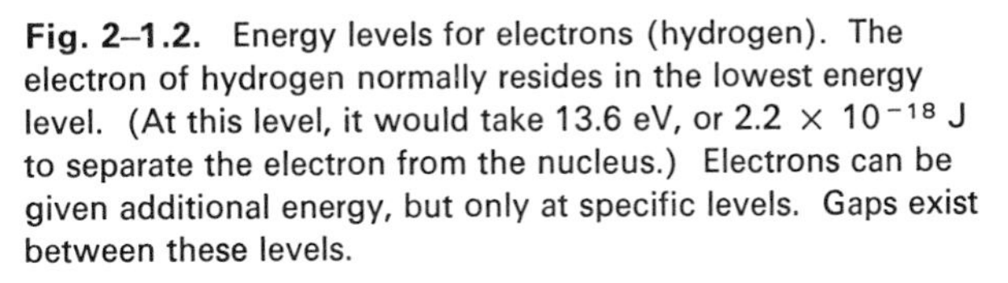
Fig. 2–1.2. Energy levels for electrons (hydrogen). The electron of hydrogen normally resides in the lowest energy level. (At this level, it would take 13.6 eV, or 2.2×10^{-18} J to separate the electron from the nucleus.) Electrons can be given additional energy, but only at specific levels. Gaps exist between these levels.

* This is done by spectrographic experiments.

of intermediate energies *not* available for the electrons. They are not available because the corresponding frequencies do not permit standing waves. Unless excited by external means, the one electron of a hydrogen atom will occupy the lowest orbital of Fig. 2–1.2.

Figure 2–1.3 shows schematically the energies of the lowest orbitals for sodium. Each orbital can contain no more than two electrons. These must be of opposite spins. Again, there are energy gaps between the orbitals which are available for electron occupancy.

In our considerations, the topmost occupied orbital will have special significance since it contains the *valence electrons*. These electrons may be removed by a relatively small electric field, to give us the cations mentioned a few paragraphs ago. The energy requirements are called the *ionization energies* (Table 2–1.1). In the next section we will see that these outermost or valence electrons are *delocalized* in metallic solids and free to move throughout the metal rather than remaining bound to individual atoms. This provides the basis for electrical and thermal conductivity.

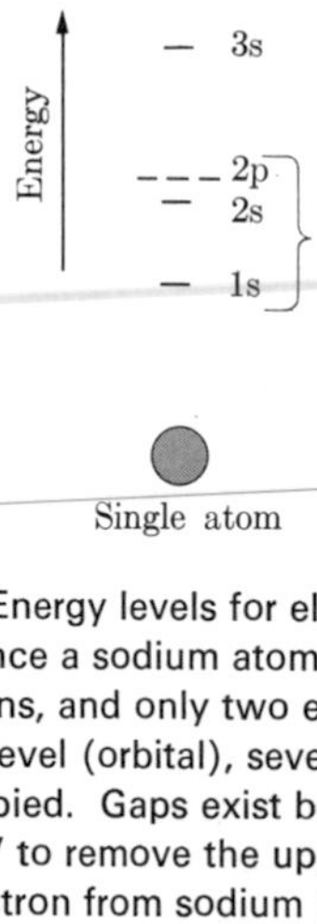

Fig. 2–1.3. Energy levels for electrons (sodium). Since a sodium atom possesses eleven electrons, and only two electrons may occupy each level (orbital), several orbitals must be occupied. Gaps exist between them. It takes 5.1 eV to remove the uppermost (valence) electron from sodium (Table 2–1.1).

Table 2–1.1

IONIZATION ENERGIES, eV*

Element		Value	Element		Value
H	I	13.6	He	I	24.5
				II	54.1
Li	I	5.4	Be	I	9.3
				II	18.2
Na	I	5.1	Mg	I	7.6
				II	15.0
K	I	4.3	Ca	I	6.1
				II	11.8
Rb	I	4.2	Sr	I	5.7
				II	11.0
Cs	I	3.9	Ba	I	5.2
				II	9.9

* 1 eV = 0.16×10^{-18} joule.
I, First electron removed.
II, Second electron removed.

Table 2–1.2
ELECTRONEGATIVITIES OF THE REPRESENTATIVE ELEMENTS
(From Mahan)

H 2.1						
Li 0.97	Be 1.5	B 2.0	C 2.5	N 3.1	O 3.5	F 4.1
Na 1.0	Mg 1.2	Al 1.5	Si 1.7	P 2.1	S 2.4	Cl 2.8
K 0.90	Ca 1.0	Ga 1.8	Ge 2.0	As 2.2	Se 2.5	Br 2.7
Rb 0.89	Sr 1.0	In 1.5	Sn 1.72	Sb 1.82	Te 2.0	I 2.2
Cs 0.86	Ba 0.97	Tl 1.4	Pb 1.5	Bi 1.7	Po 1.8	At 1.9

When the valence orbitals are not filled, the atom may accept a limited number of extra electrons within these unfilled energy states, to become a negative ion, *anion.* The *electronegativities* shown in Table 2–1.2 reveal the relative attractions of atoms for excess electrons. These electronegative atoms with unfilled valence orbitals may also share electrons. This becomes important in bonding and will be reviewed in the next section.

Example 2–1.1 Sterling silver contains approximately 7.5 w/o* copper and 92.5 w/o silver. What are the a/o* copper and a/o silver?

Solution: Basis: 10,000 amu alloy = 9250 amu Ag + 750 amu Cu.

Ag: 9250 amu Ag/(107.87 amu Ag/atom) = 85.75 atoms = 88 a/o

Cu: 750 amu Cu/(63.54 amu Cu/atom) = 11.80 atoms = 12 a/o. ◀

97.55

Example 2–1.2 From Appendix B, indicate the orbital arrangements for a single iron atom, and for Fe^{2+} and Fe^{3+} ions.

---

* Weight percent, w/o; atom percent, a/o; linear percent, l/o; volume percent, v/o; mole percent, m/o; etc. In condensed phases (*solids* and *liquids*) weight percent is implied unless specifically stated otherwise. In *gases*, v/o or m/o are implied unless specifically stated otherwise.

Solution: The argon core is $1s^2 2s^2 2p^6 3s^2 3p^6$; therefore,

$$\text{Fe:}\quad 1s^2 2s^2 2p^6 3s^2 3p^6 3d^6 4s^2. \tag{2–1.1}$$

During ionization, the $4s$ electrons are removed first; therefore:

$$Fe^{2+}:\quad 1s^2 2s^2 2p^6 3s^2 3p^6 3d^6, \tag{2–1.2}$$

$$Fe^{3+}:\quad 1s^2 2s^2 2p^6 3s^2 3p^6 3d^5. \blacktriangleleft \tag{2–1.3}$$

2–2 STRONG BONDING FORCES (PRIMARY BONDS)

Since most products are designed with solid materials, it is desirable to understand the attractions that hold the atoms together. The importance of these attractions may be illustrated with a piece of copper wire, in which each gram contains ($0.602 \times 10^{24}/63.54$) atoms. Based on the density of copper, each cubic centimeter contains 8.9 times this number, or 8.4×10^{22} atoms/cm^3. Under these conditions, the forces of attraction that bond the atoms together are strong. If this were not true, they would easily separate, the metal would deform under small loads, and atomic vibrations associated with thermal energy would gasify the atoms at low temperatures. As in the case of this wire, the engineering properties of any material depend on the interatomic forces which are present.

Interatomic attractions are caused by the electronic structure of atoms. The noble (inert or chemically inactive) gases, such as He, Ne, Ar, etc., have only limited attractions to other atoms because they have a very stable arrangement of eight electrons (2 for He) in their outer, or valence, electron orbitals, and at the same time have no net charge as a result of an unbalanced number of protons and electrons. Most other elements, unlike the noble gases, must achieve the relatively stable configuration of having eight electrons available for their outer orbitals through one of the following procedures: (1) receiving extra electrons, (2) releasing electrons, or (3) sharing electrons. The first two of these processes produce ions with a net negative or positive charge and thus provide the ions with coulombic attractions to other ions of unlike charge. The third process obviously requires an intimate association between atoms in order for the sharing of electrons to be operative. Where applicable, the above three processes produce strong or primary bonds. Energies approximating 500 kJ/mole (i.e., 500,000 joules per 0.602×10^{24} bonds) are required to rupture these bonds. Other weaker or secondary bonds (less than 40 kJ/mole) are also always present, but gain importance when they are the only forces present (Section 2–4).

Ionic bonds. The interatomic bond that is easiest to describe is the ionic bond, which results from the mutual attraction of positive and negative charges. Atoms of elements such as sodium and calcium, with one and two electrons in their valence orbitals, respectively, easily release these outer electrons and become positively charged ions. Likewise, chlorine and oxygen atoms readily add to their valence orbitals until they have eight electrons by accepting one or two electrons and thus becoming negatively charged ions. Since there is always a *coulombic attraction* between negatively and

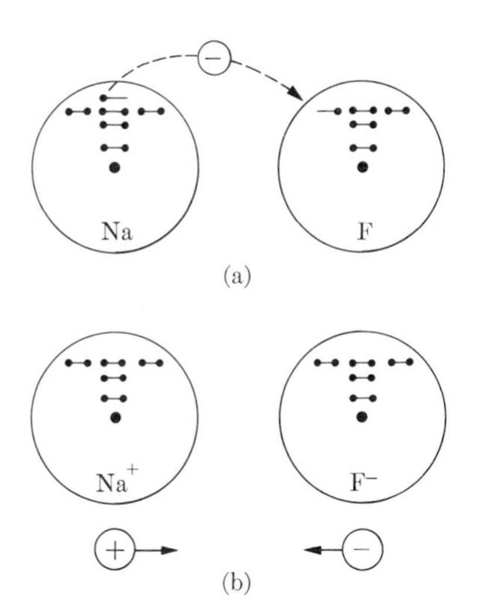

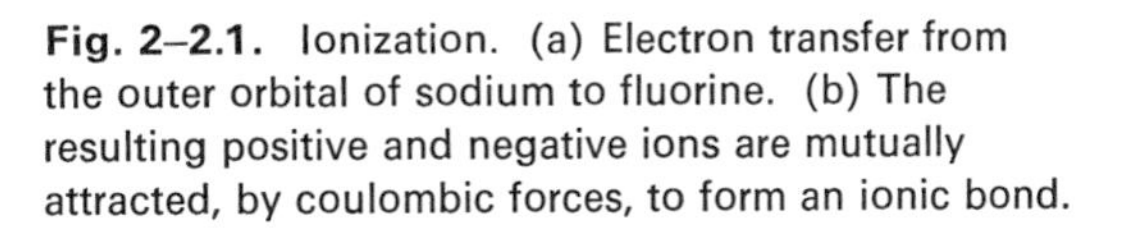
Fig. 2–2.1. Ionization. (a) Electron transfer from the outer orbital of sodium to fluorine. (b) The resulting positive and negative ions are mutually attracted, by coulombic forces, to form an ionic bond.

positively charged materials, a bond is developed between neighboring ions of unlike charges (Fig. 2–2.1).

A negative charge possesses an attraction for *all* positively charged particles and a positive charge for *all* negatively charged particles. Consequently, sodium ions surround themselves with as many negative chlorine ions as possible, and chlorine ions surround themselves with the maximum number of positive sodium ions, the attraction being equal in all directions (Fig. 2–2.2). The major requirement in an ionically bonded material is that the number of positive charges equal the number of negative charges. Thus, sodium chloride has a composition of NaCl. Magnesium chloride has a composition of $MgCl_2$, because the magnesium atom can supply two electrons from its valence shell but each chlorine atom can accept only one.

Since these coulombic attractions involve all neighbors, ionically bonded materials may be very stable, particularly if multivalent ions are involved. As an example, when magnesium and oxygen combine to form MgO, 136 kcal/mole are released, i.e., 136,000 calories, or 570 kJ per 0.6×10^{24} Mg^{2+} ions and 0.6×10^{24} O^{2-} ions in the product. Further, MgO must be raised to approximately 2800°C (~5000°F) before it melts.

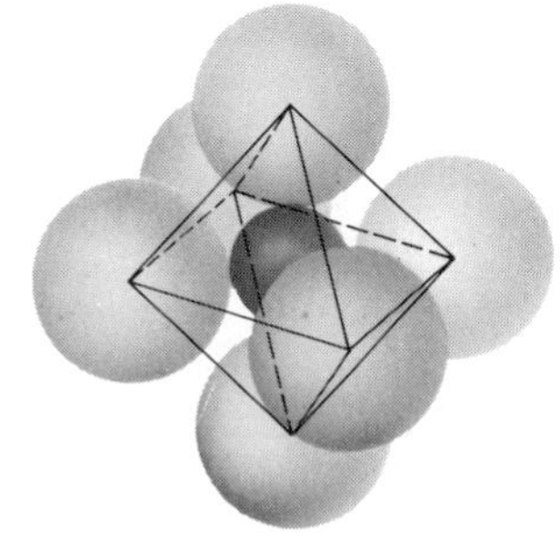

Fig. 2–2.2. Three-dimensional structure of sodium chloride. The positive sodium ion is coordinated with and has equal attraction for all six neighboring negative chlorine ions. (Compare with Fig. 3–1.1.)

Covalent bonds. Another primary force of strong attraction is the *covalent* bond in which electrons are shared. Figure 2–2.3 shows three representations of this sharing for two fluorine atoms in F_2. Commonly the first representation (electron dots or a "bond line") will suffice for our purposes; however, the reader should be aware that electrons cannot be precisely located without a degree of uncertainty. Therefore, the presentation of Fig. 2–2.3(b) is used to show the locations of greatest electron probabilities. The molecular orbitals developed for F_2 in Fig. 2–2.3(c) let us see that the *average* energy of the outer or valence electron drops when the molecule is formed from the two individual atoms. Therefore we may consider covalent bonds in terms of energy, because energy would be required to reverse the reaction sketched in Fig. 2–2.3(c).

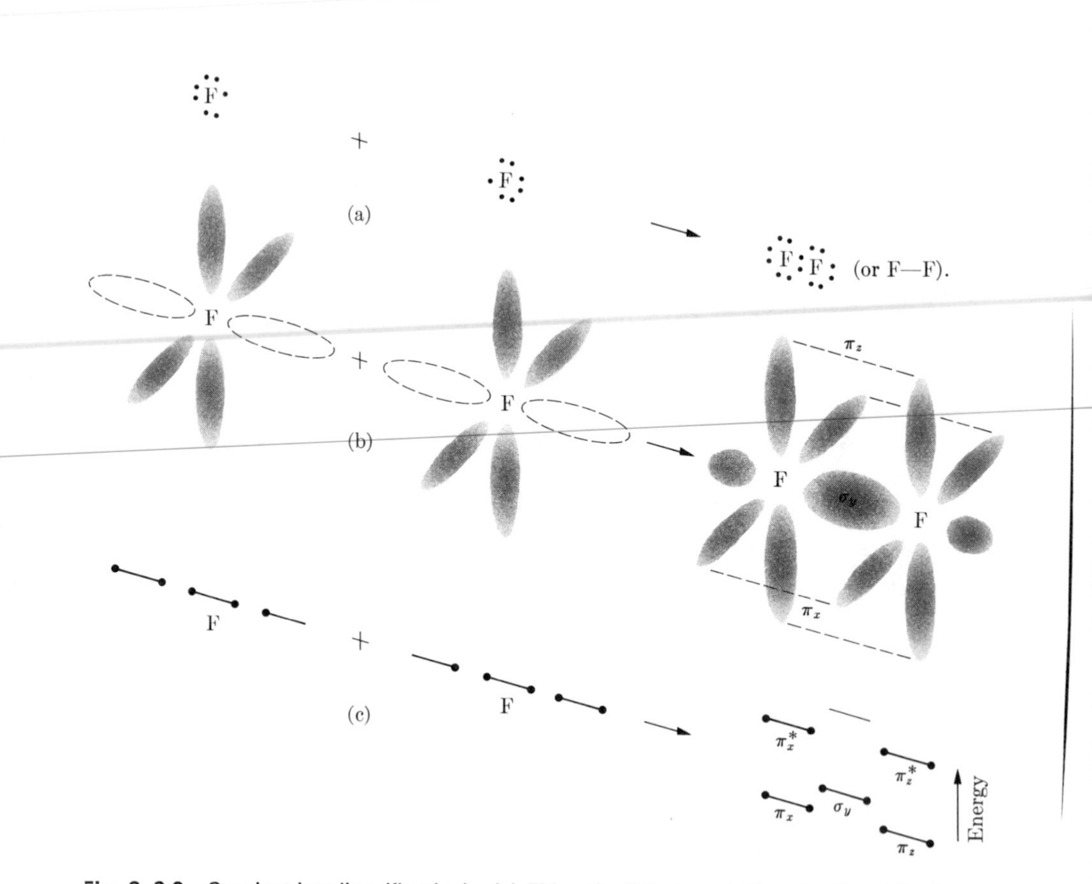

Fig. 2–2.3. Covalent bonding (fluorine). (a) Either the "electron dot" or the "bond line" are commonly used for simplicity. (b) The electron probability regions. (c) Orbital energy levels (schematic). It takes 160,000 joules (38,000 calories) to break a mole (0.6×10^{24}) of these bonds. (Only the $2p$ electrons are shown in (b) and (c).)

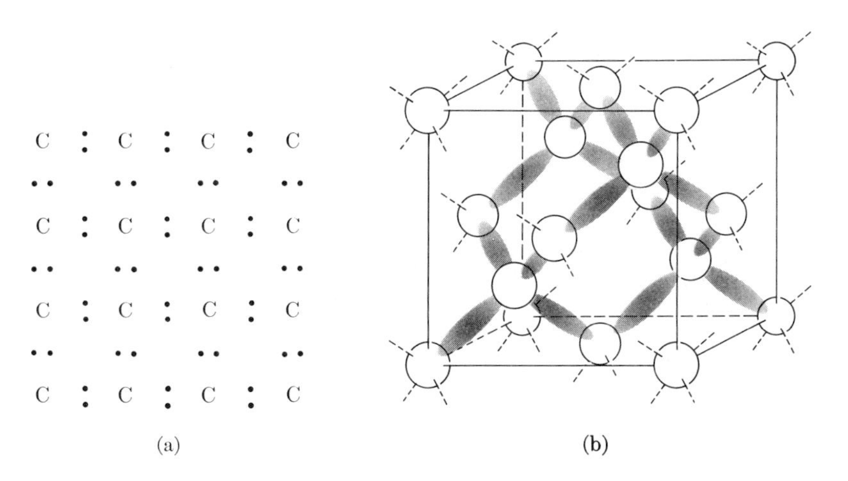

Fig. 2–2.4. Diamond structure. The strength of the covalent bonds is what accounts for the great hardness of diamond. (a) Two-dimensional representation. (b) Three-dimensional representation.

That covalent bonds provide strong attractive forces between atoms is evidenced in diamond, which is the hardest material found in nature, and which is wholly carbon. Each carbon has four valence electrons. These are shared with adjacent atoms to form a three-dimensional lattice entirely bonded by covalent pairs (Fig. 2–2.4). The strength of the covalent bond in carbon is demonstrated not only by the great hardness of diamond, but also by the extremly high temperature ($>3000°C$) to which it can be heated before the structure is disrupted (melted) by thermal energy.*

Unlike coulombic attractions which bring as many unlike ions into neighboring positions as space will allow, covalent bonds must be formed between specific atoms. In the diamond of Fig. 2–2.4(b), the number of neighbors is limited by the number of bonds and not by the available space. In Fig. 2–2.3, the two fluorines are held together with a covalent connection of 160 kJ/0.6×10^{24} bonds. However, neither of these two atoms develops strong attractions to other fluorine atoms (or molecules) which may approach them. As evidence, F_2 vaporizes to a gas at 85°K (-188°C or -306°F) with only $\sim$3 kJ/mole. When the bond involves a given pair of atoms, we apply the term *stereospecific.*

An exception to the above stereospecificity of covalent bonds occurs in compounds with a benzene ring, which is discussed in chemistry texts (Fig. 2–2.5). One electron per carbon atom (for a total of six) is *delocalized.* These six electrons have equal probability of being found anywhere around the ring.† They can respond to alternating

---

* It would take about 180 kcal, or 750 kJ, to break all the bonds in a mole (0.6×10^{24}) of carbon atoms.

† They reside in π bonds which are perpendicular to the plane of the molecule.

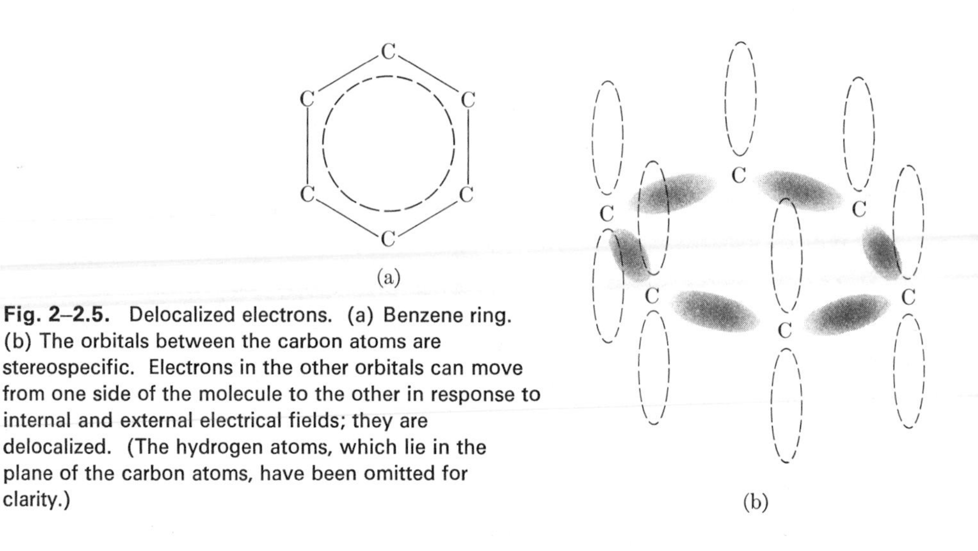

Fig. 2–2.5. Delocalized electrons. (a) Benzene ring. (b) The orbitals between the carbon atoms are stereospecific. Electrons in the other orbitals can move from one side of the molecule to the other in response to internal and external electrical fields; they are delocalized. (The hydrogen atoms, which lie in the plane of the carbon atoms, have been omitted for clarity.)

electric fields by moving from one side of the molecule to another, but cannot leave the molecule (except under unusually catastrophic conditions). There are as many wave patterns for these delocalized electrons *as there are atoms in the ring.*

Metallic bonds. In addition to ionic and covalent bonds, a third type of primary interatomic attractive mechanism is the metallic bond. The model of metallic bonding is not as simple to construct as the other two. However, we can adapt the concept of delocalized electrons from the previous paragraph to serve our purpose. First consider graphite (Fig. 2–2.6); the layers of carbon atoms in graphite possess delocalized electrons (as well as electron pairs between specific atoms). The delocalized electrons can respond to electric fields, moving within the graphite sheet in a wavelike pattern.* In fact, conductivity becomes possible if a positive electrode is present to

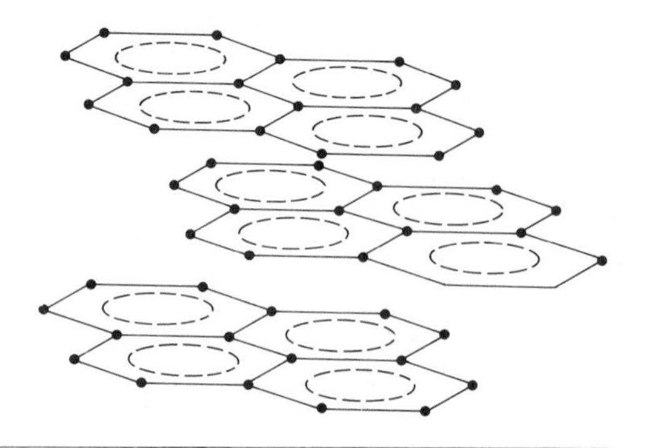

Fig. 2–2.6. Delocalized electrons in graphite layers. Each layer contains "multiple benzene rings" (Fig. 2–2.5). The conductivity is more than 100 times greater in the parallel direction than in the perpendicular direction.

* As with the benzene ring, there are as many wave patterns possible *as there are carbon atoms in the layer.* Also note that while the delocalized electrons can readily move within the layer, there is not a comparable mechanism to move from one layer to another. (See discussion question U_2 at the end of the chapter.)

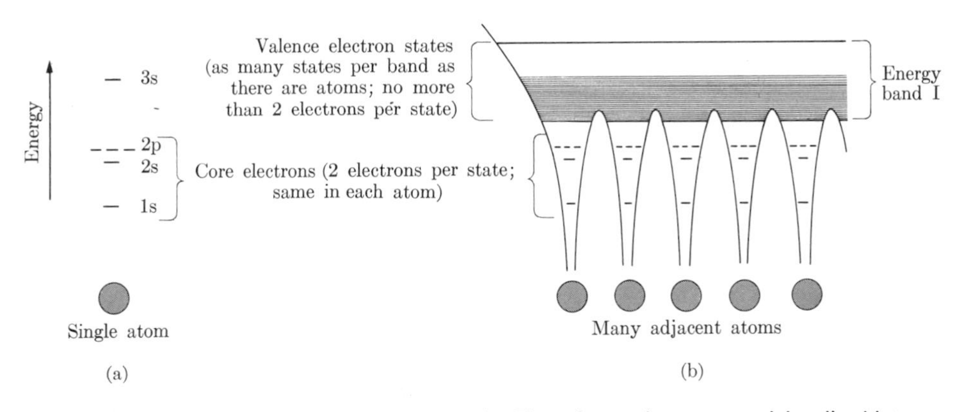

Fig. 2–2.7. Valence electrons in metal (sodium). The valence electrons are delocalized into an energy band. These electrons are able to move throughout the metal. The valence electrons fill only the bottom half of the band. Their average energy is lower than that of the 3*s* electrons with individual atoms. This energy difference provides the metallic bond.

remove electrons from one end of the layer, and a negative electrode is available to supply electrons to the other end.

Typical metals have delocalized electrons which can move in three dimensions. It is thus common to speak of an electron "cloud" or "gas" because the outer, least strongly bonded electrons are able to move throughout the metal structure. The orbitals for metals are sketched schematically in Fig. 2–2.7 for sodium. Just as the molecular orbitals of F_2 on the right side of Fig. 2–2.3(c) are modified from the atomic orbitals, the energy levels in multiatomic sodium differ from that of the single-atom orbitals that were first shown in Fig. 2–1.3. The prime change between Fig. 2–2.7(a) and 2–2.7(b) is that the upper valence orbital has split into as many levels as there are atoms in the system. Note that the average energy of the outer or valence electrons of Fig. 2–2.7(b) is less than the energy of the 3*s* orbital for the individual atoms. This accounts for the bonding in metals; in brief, energy would have to be supplied to overcome the metallic bond and separate the atoms from one another and to restablish the individual atomic orbitals. Qualitatively speaking, we find very strong bonds holding tungsten atoms together. Its melting and boiling temperatures are very high (~3400 and 5900°C, respectively); also, it has an extremely high modulus of elasticity (50,000,000 psi, or 345,000 MPa; see Appendix C for comparisons). In contrast, the bonds in sodium are low as evidenced by its melting point (97.8°C) and soft behavior. Both have delocalized electrons for electrical and thermal conduction.

Before concluding this section on primary bonds, or strong attractive forces, we should observe that while one bond type may be prevalent in a material, other bond types can be present, too; so *mixed-bond types* are widespread.

Example 2–2.1 The covalent bond between two carbon atoms, C—C, is 370 kJ/mole (88 kcal/0.6 × 10^{24} bonds). The energy of light is

$$E = h\nu \tag{2–2.1}$$

where h is Planck's constant (0.66 × 10^{-33} joule · sec) and ν is the frequency of light. What wavelength is required to break a C—C bond?

Solution

$$370{,}000 \text{ J}/0.6 \times 10^{24} = (0.66 \times 10^{-33} \text{ J} \cdot \text{s})\nu,$$

$$\nu = 9.34 \times 10^{14}/\text{s} = c/\lambda,$$

$$\lambda = (0.299 \times 10^{9} \text{ m/s})/(9.34 \times 10^{14}/\text{s})$$

$$= 0.320 \times 10^{-6} \text{ m} = (3200 \text{ Å}).$$

Comments. This is in the ultraviolet range. It is for this reason that ultraviolet light can cause deterioration in plastics that contain C—C covalent bonds. ◀

2–3 MOLECULES

A *molecule* may be defined as a group of atoms which are strongly bonded together, but whose bonds to other, similar groups of atoms are relatively weak. Our prototype for a molecule may be fluorine, F_2, which was discussed in the previous section. Recall that 160 kJ/mole (~1.65 eV/bond) would have to be present to break the covalent bond joining the two atoms (Fig. 2–2.3). In contrast, only 3 kJ/mole (0.03 eV/bond) provided the thermal agitation which is required to separate the molecules into a gas by boiling.

The more common examples of molecules include compounds such as H_2O, CO_2, CCl_4, O_2, N_2, and HNO_3. Other small molecules are shown in Fig. 2–3.1. Within each of these molecules, the atoms are held together by strong attractive forces that usually have covalent bonds, although ionic bonds are not uncommon. Unlike the forces that hold atoms together, the bonds between molecules are weak and consequently each molecule is free to act more or less independently. These observations are borne out by the following facts: (1) Each of these molecular compounds has a low melting and a low boiling temperature compared with other materials. (2) The molecular solids are soft because the molecules can slide past each other with small stress applications. (3) The molecules may remain intact in the liquid and gaseous forms.

The molecules listed above are comparatively small; other molecules have large numbers of atoms. For example, pentatriacontane (shown in Fig. 2–3.2(c)) has over 100 atoms, and some molecules contain many thousand. Whether the molecule is small like CH_4, or much larger than that shown in Fig. 2–3.2(c), the distinction between the strong intramolecular and the weaker intermolecular bonds still holds.

Other materials, such as metals, MgO, SiO_2 and phenol-formaldehyde plastics have continuing three-dimensional structures of primary bonds. The difference

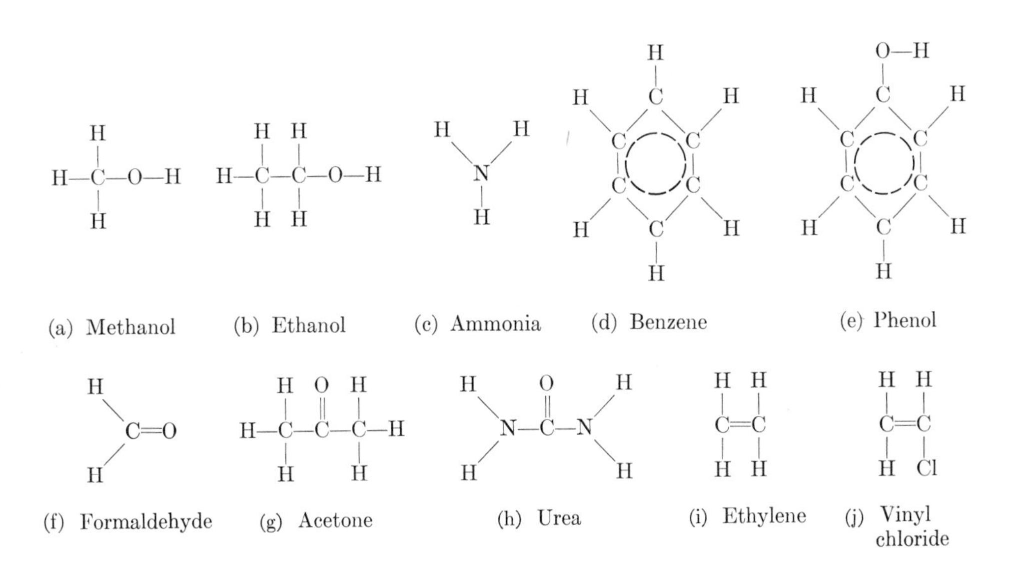

Fig. 2–3.1. Small organic molecules. Each carbon is surrounded by four bonds, each nitrogen by three, each oxygen by two, and each hydrogen and chlorine by one.

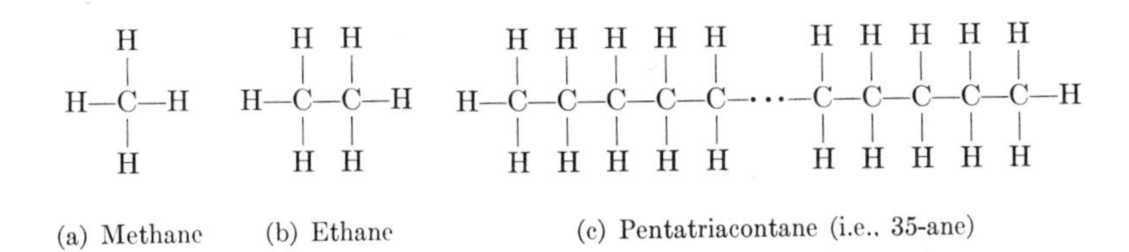

Fig. 2–3.2. Examples of molecules. Molecules are discrete groups of atoms. Primary bonds hold the atoms together within the molecule. Weaker, secondary forces attract molecules to each other.

between the structures of molecular materials and those with primary bonds continuing in all three dimensions produces major differences in properties. These differences will be considered in subsequent chapters.

Bond lengths and energies. The strength of bonds between atoms in a molecule, of course, depends on the kind of atoms and the other neighboring bonds. Table 2–3.1 is a compilation of bond lengths and energies for those atom couples most frequently encountered in molecular structures. The energy reported is the amount required to break one mole (Avogadro's number) of bonds. For example, 370,000 joules of energy are required to break 0.602×10^{24} C—C bonds, or $370{,}000/(0.602 \times 10^{24})$ joules per bond. Likewise, this same amount of energy is released (-0.61×10^{-18} J) if one of these C—C bonds is formed. Only the sign is changed.

Table 2–3.1
BOND ENERGIES AND LENGTHS

Bond	Bond energy*		Bond length, Å
	kcal/mole	kJ/mole	
C—C	88†	370†	1.54
C=C	162	680	1.3
C≡C	213	890	1.2
C—H	104	435	1.1
C—N	73	305	1.5
C—O	86	360	1.4
C=O	128	535	1.2
C—F	108	450	1.4
C—Cl	81	340	1.8
O—H	119	500	1.0
O—O	52	220	1.5
O—Si	90	375	1.8
N—H	103	430	1.0
N—O	60	250	1.2
F—F	38	160	1.4
H—H	104	435	0.74

* Approximate. The values vary with the type of neighboring bonds. For example, methane (CH_4) has the above value for its C—H bond; however the C—H bond energy is about 5% less in CH_3Cl, and 15% less in $CHCl_3$.
† All values are negative for forming bonds (energy is released), and positive for breaking bonds (energy is required).

Bond angles. The chemist recognizes hybrid orbitals in certain covalent compounds, where the *s* and *p* orbitals are amalgamated. The most important hybrid for us to review is the sp^3 orbital. We have already sketched it in Fig. 2–2.4 for the four bonds of carbon in diamond. Four equal orbitals are formed instead of having distinct 2*s* and 2*p* orbitals that occur in individual atoms (e.g., the individual sodium and fluorine atoms of Fig. 2–2.1). Methane (CH_4) and carbon tetrachloride (CCl_4), like diamond, have sp^3 orbitals which connect four identical atoms to the central carbon. Therefore, we find them equally spaced around the central carbon at 109.5° from each other.* Geometrically, this is equivalent to placing the carbon in a cube center and

* The angle 109.5° is a time-averaged value. Any particular H—C—H angle in CH_4 will vary rapidly as a result of thermal vibrations.

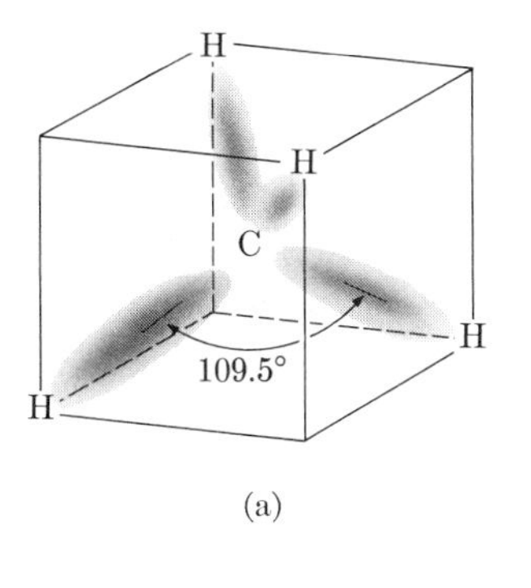

Fig. 2–3.3. Bond angles. (a) Methane, CH_4, is symmetrical with each of the six angles equal to 109.5°. (b) Chloromethane, CH_3Cl, is distorted.

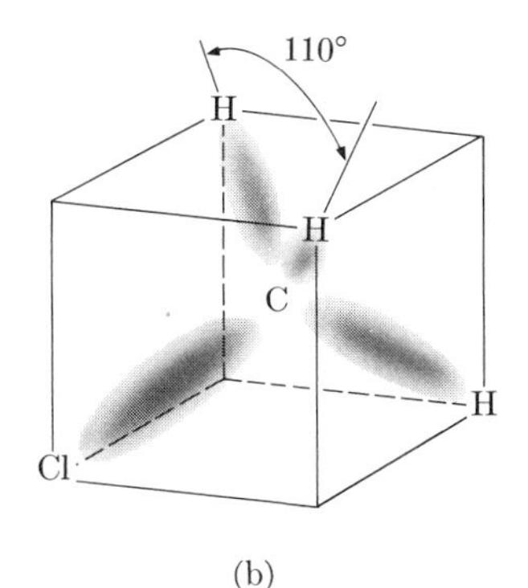

pointing the orbitals toward four of the eight corners (Fig. 2–3.3(a)). However, if the orbitals do not bond identical atoms to the central carbon, these time-averaged angles are distorted slightly, as shown for CH_3Cl (Fig. 2–3.3(b)).

Greater distortions occur in hybrid orbitals when some of the electrons occur as *lone pairs* rather than in the covalent bond. This is particularly evident in NH_3 and H_2O (Fig. 2–3.4) where 107.3° and 104.5° are the time-averaged values for H—N—H and H—O—H, respectively.

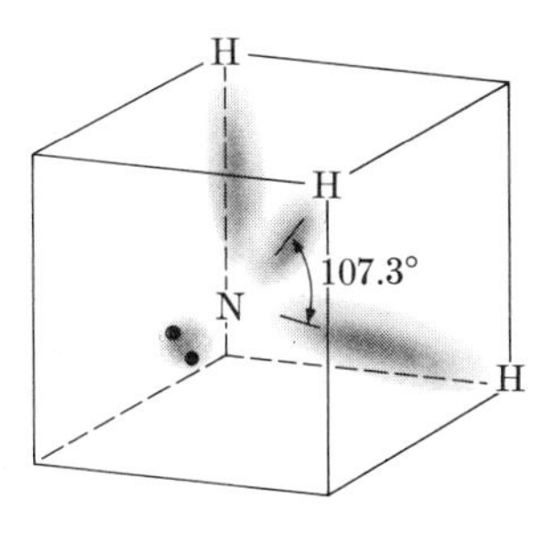

Fig. 2–3.4. Bond angles. (a) Ammonia, NH_3, and (b) water have angles between the 109.5° of Fig. 2–3.3(a) and 90°. Ammonia has one lone-pair of electrons; water, two.

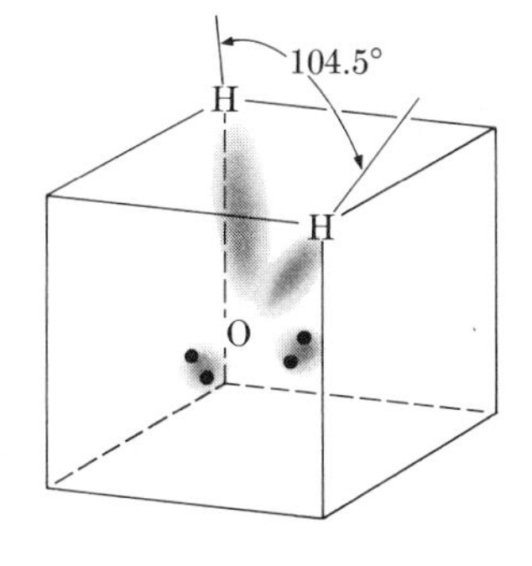

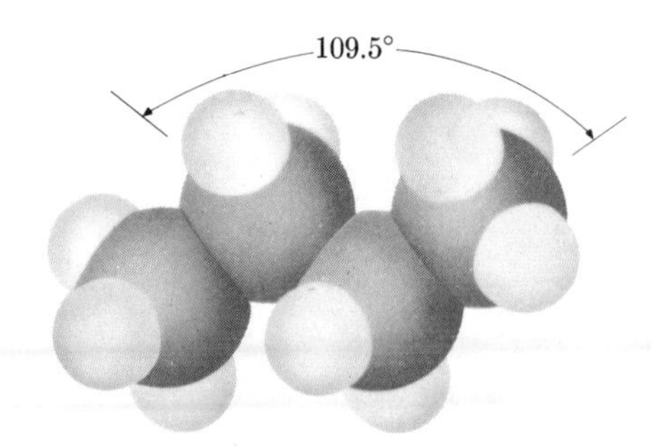

Fig. 2–3.5. Bond angles (butane). Although we commonly draw straight chains (Figs. 2–3.2(c) and 2–3.7(b)), there is a C—C—C bond angle of about 109°.

One of the bond angles most frequently encountered in the study of materials is the C—C—C angle of the hydrocarbon chains (Fig. 2–3.5). While this will differ slightly, depending upon whether hydrogen or some other side radical is present, we may assume for our purposes that the C—C—C angle is close to 109.5°.

Isomers. In molecules of the same composition, more than one atomic arrangement is usually possible. This is illustrated in Fig. 2–3.6 for propyl and isopropyl alcohol. Variations in the structure of molecules with the same composition are called *isomers.* Differences in structure affect the properties of molecules because of the resulting change in molecular polarity (Section 2–4). For example, the melting and boiling temperatures for propyl alcohol are −127°C and 97.2°C, respectively, whereas for isopropyl alcohol the corresponding temperatures are −89°C and 82.3°C.

Fig. 2–3.6. Isomers of propanol. (a) Normal propyl alcohol. (b) Isopropyl alcohol. The molecules have the same composition but different structures. Consequently, the properties are different. Compare with polymorphism of crystalline materials (Section 3–5).

```
                                  H
                                  |
  H  H  H                      H  O  H
  |  |  |                      |  |  |
H—C—C—C—O—H                 H—C—C—C—H
  |  |  |                      |  |  |
  H  H  H                      H  H  H
    (a)                          (b)
```

Example 2–3.1 How much energy is given off when 70 g of ethylene (Fig. 2–3.7(a)) reacts to give polyethylene (Fig. 2–3.7(b))?

Solution: Each added C_2H_4 molecule breaks one C═C bond and forms two C—C bonds. From Table 2–3.1:

$$\frac{+162{,}000 \text{ cal}}{0.602 \times 10^{24} \text{ molecules}} - \frac{2(88{,}000 \text{ cal})}{0.602 \times 10^{24} \text{ molecules}} = -2.32 \times 10^{-20} \text{ cal}/C_2H_4$$

$$70 \text{ g } (0.602 \times 10^{24} \text{ amu/g})/(28 \text{ amu}/C_2H_4) = 1.5 \times 10^{24}\ C_2H_4$$

$$(-2.32 \times 10^{-20} \text{ cal}/C_2H_2)(1.5 \times 10^{24}\ C_2H_4) = -35{,}000 \text{ cal}$$

$$-35{,}000 \text{ cal } (4.18 \text{ J/cal}) = -150 \text{ kJ}.$$

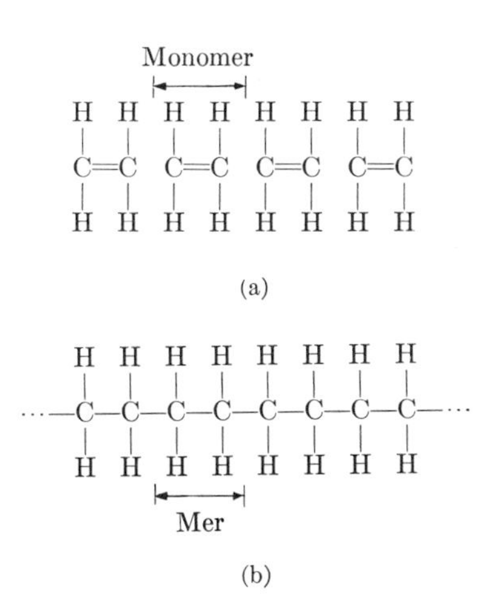

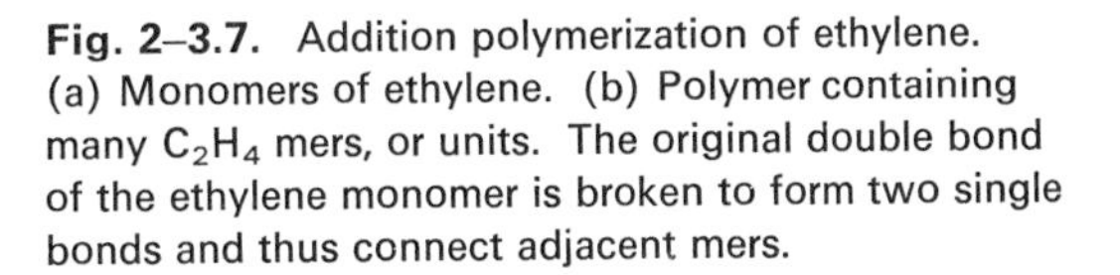

Fig. 2–3.7. Addition polymerization of ethylene. (a) Monomers of ethylene. (b) Polymer containing many C_2H_4 mers, or units. The original double bond of the ethylene monomer is broken to form two single bonds and thus connect adjacent mers.

Comments. Convention treats required energy as (+) and released energy as (−). The reaction of Fig. 2–3.7 is the basic reaction for making large vinyl-type molecules that are used in plastics (Chapter 7). ◀

2–4 SECONDARY BONDING FORCES

The three types of bonds considered in Section 2–2 are all relatively strong primary bonds which hold atoms together. Weaker, secondary bonds, which also supply interatomic attraction, are grouped here as *van der Waals forces*, although there are actually several different mechanisms involved. Were it not for the fact that sometimes they are the only forces which operate, van der Waals bonding might be overlooked.

In a noble gas like helium, the initial orbital, with its two electrons, is complete; and other noble gases, such as neon and argon, have a full complement of eight electrons in their valence orbitals. In these stable situations none of the primary bonds can be effective, since covalent, ionic, and metallic bonds all require adjustments in the valence electrons. As a result, atoms of these noble gases have little attraction for one another, and with rare exceptions they remain monatomic at ordinary temperatures. Only at extremely low temperatures, when thermal vibrations have been greatly reduced, do these gases condense (Table 2–4.1(A)). It is this condensation that makes it evident that there are weak interatomic attractions which pull the atoms together.

Similar evidence for these weak attractions is found in the molecules listed in Table 2–4.1(B). As pointed out earlier, these gases have satisfied their valence requirements by covalent bonding within the molecule. The condensation of these simple molecules occurs only when thermal vibrations are sufficiently reduced in energy to permit the weak van der Waals forces to become operative.

Induced dipoles. Most of the gases and molecules of Table 2–4.1 are symmetric; i.e., over any extended period of time, the center of positive charges from the protons in

Table 2–4.1
MELTING AND BOILING TEMPERATURES OF GASES (ABSOLUTE TEMPERATURE)

A. Noble gases			B. Simple molecules		
Gas	Melting temperature, °K	Boiling temperature, °K	Molecule	Melting temperature, °K	Boiling temperature, °K
He	0.96*	4.25	H_2	14.02	21
Ne	24.5	27	N_2	63	78
Ar	84	87.5	O_2	55	90
Kr	116	120	CH_4	88	145
Xe	161	166	CCl_4	250	349
Rn	202	211	C_4H_{10}	135	274
			NH_3	195	240
			CH_3Cl	113	259
			H_2O	273	373

* Melting point with 26 atmospheres of pressure. At one atmosphere pressure, helium remains as a liquid as 0°K (−273.16°C) is approached.

the nuclei, and the center of negative charges from the electrons, are at the center of each molecule (or noble gas atom). Momentarily, however, the electron motions and the atom vibrations may disrupt this electrical symmetry. When this happens, a small electrical *dipole* is established. In that small fraction of a second,* the centers of positive and negative charges are not coincident, so an electrical dipole is established giving the molecule a positive end and a negative end. In turn, this induces a dipole into the adjacent molecules by displacing their electrons in response to this minute electric field. However, attractive forces are established, though admittedly weak. This is demonstrated by the data of Table 2–4.1, which shows that these gases do condense; however, not until low temperatures are reached.

Polar molecules. Asymmetric molecules, such as NH_3, CH_3Cl, and H_2O, always have a noncoincidence of their positive and negative charges. This *polarity* is best illustrated in Fig. 2–3.4 for ammonia (NH_3). Three hydrogen nuclei, which are no more than bare protons (+), are exposed to the upper right. The lone pair of electrons on the other side of the molecule makes that position the negative end of the molecule. The *inter*molecular bonding forces of CH_4 and NH_3 may be compared through their melting and boiling temperatures, since each has about the same mass (16 amu and 17 amu, respectively). Table 2–4.1 shows that NH_3 must be raised to 240°K (−33°C) before thermal agitation breaks the *inter*molecular bonds completely, while the more weakly bonded, symmetrical CH_4 is vaporized at 145°K (−128°C). In further contrast, CH_3Cl (Fig. 2–3.3(b)) does not vaporize until 259°K (−14°C) because the chlorine possesses a large number of electrons which are located relatively far from

* Between 10^{-16} and 10^{-12} sec.

the center of the molecule and therefore give strong attractions to neighboring molecular dipoles.

Hydrogen bridge. This third type of van der Waals bonding force is actually a special case of a polar molecule. However, it is by far the strongest of these secondary bonding forces and is widely encountered. It therefore warrants special attention and a special name.

The exposed proton at the end of a C—H, O—H, or N—H bond is not screened by electrons. Therefore, this positive charge can be attracted to valence electrons of adjacent molecules. A coulombic-type bond is developed, called a *hydrogen bridge.* Our most common example of this is found in water, where the proton of the hydrogen in one molecule is attracted to the lone pairs of electrons on the oxygen in an adjacent molecule (Fig. 2–4.1). The maximum energy of this bond is about 30 kJ/mole (7 kcal/mole). This is in contrast to (1) a maximum of 5 kJ/mole (and usually $\ll 1$ kJ/mole) for the other types of van der Waals bonds, and (2) several hundred kilojoules per 0.6×10^{24} bonds for primary bonds (Section 2–2).

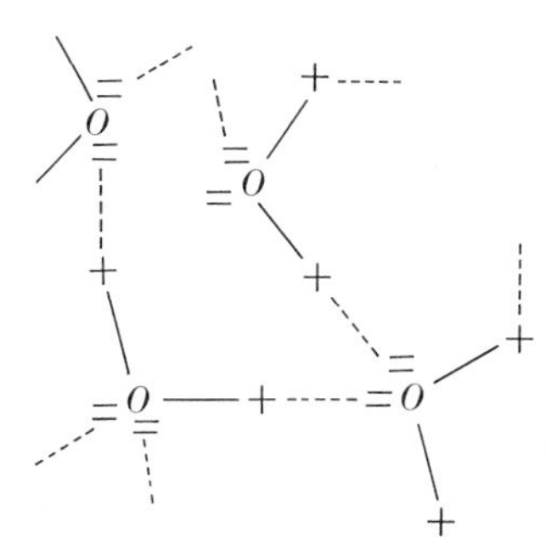

Fig. 2–4.1. Hydrogen bridge (in water). The hydrogen at the end of the orbital is an exposed proton (+). It is attracted to the electron lone-pairs of the adjacent water molecules. (Cf. Fig. 2–3.4(b).) The hydrogen bridge makes water the highest boiling of any material with a low molecular weight (18 amu).

Chemists point out that it would be difficult to overemphasize the importance of the hydrogen bridge. For example, H_2O, with a molecular weight of only 18 amu, has the highest boiling temperature of any molecule with a molecular weight of less than 100 amu. If it had a boiling point comparable to other 3- or 4-atom molecules, our oceans would be nonexistent and all biological and geological conditions would be completely altered. In our study of materials, the hydrogen bridge affects the properties and behavior of plastics (Chapter 7) and certain ceramics (Chapter 8).

Example 2–4.1 Each OH arm of a water molecule has an electric dipole of 5×10^{-30} coul · m. What is the dipole of the whole molecule?

Solution: Solve for the resultant of the two dipoles at 104.5° (Fig. 2–3.4).

$$p = 2(5 \times 10^{-30} \text{ coul} \cdot \text{m}) \cos (104.5°/2)$$
$$= 6 \times 10^{-30} \text{ C} \cdot \text{m}.$$

Comment. This is the product of the charge and the distance between the centers of positive and negative charges. ◀

2–5 INTERATOMIC DISTANCES

Although in the case of diatomic molecules there is bonding and coordination of only two atoms, most materials involve a coordination of many atoms into an integrated structure. Two main factors, interatomic distances and spatial arrangements, are of importance. Let us therefore consider them in some detail.

The forces of attraction between atoms, which we considered in the preceding sections, pull the atoms together; but what keeps the atoms from being drawn still closer together? It should be apparent from the preceding figures and the discussion that there is much vacant "space" in the volume surrounding the nucleus of an atom. The existence of this space is evidenced by the fact that neutrons can move through the fuel and the other materials of a nuclear reactor, traveling among many atoms before they are finally stopped (see Fig. 6–10.1).

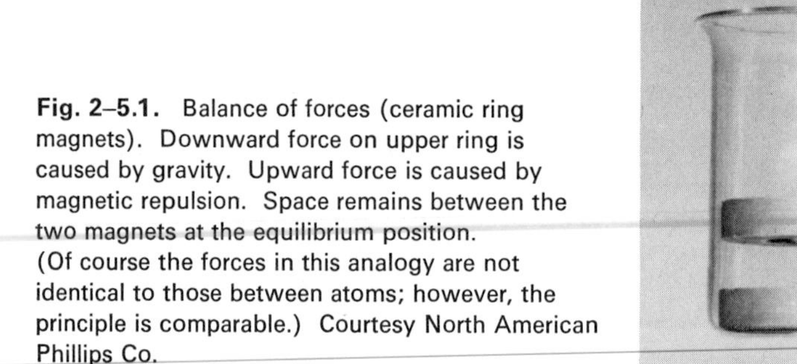

Fig. 2–5.1. Balance of forces (ceramic ring magnets). Downward force on upper ring is caused by gravity. Upward force is caused by magnetic repulsion. Space remains between the two magnets at the equilibrium position. (Of course the forces in this analogy are not identical to those between atoms; however, the principle is comparable.) Courtesy North American Phillips Co.

The space between atoms is caused by interatomic repulsive forces, which exist in addition to the interatomic attractive forces described above. Primarily because the close proximity of two atoms places too many electrons into interacting locations, mutual repulsion results. The equilibrium distance is that distance at which the repulsive and the attractive forces are equal. An analogy may be made between the interatomic distances among atoms and the spacing between the two ring magnets of Fig. 2–5.1. (In this example, the magnets are aligned to give repulsion rather than attraction.) Of course the forces in this analogy are not identical to those between atoms; however, the principle is comparable. The top ring magnet is moved by a force (gravity) toward the lower ring magnet (which in this case is fixed by the container). Since the force of gravity is essentially constant over the distance considered here, the top magnet falls to the point where it is repulsed by an equal magnetic force, of opposite direction. Because the repulsive force increases as an inverse function of the distance, equilibrium distance is achieved. Note that the magnets remain separated by space.

Coulombic forces. The ionic bond will be used to illustrate the balance between attractive and repulsive forces in materials. The force F_C developed between two point charges is related to the quantity of the two charges $Z_1 q$ and $Z_2 q$, and their separation distance a_{1-2} as follows:

$$F_C = -k_0(Z_1 q)(Z_2 q)/a_{1-2}^2. \tag{2–5.1}$$

For ions, Z is the valence (+ or −) and q is 0.16×10^{-18} coulomb. The proportionality constant k_0 depends on the units used.*

Electronic repulsive forces. The repelling force F_R between the electronic fields of two atoms or ions is also an inverse function with distance, but to a higher power:

$$F_R = -bn/a_{1-2}^{n+1}. \tag{•2–5.2}$$

Both b and n are empirical constants, with n equal to approximately 9 in ionic solids. Comparing, $F_C \propto a^{-2}$, and $F_R \propto a^{-10}$. Thus, the attractive forces predominate at greater distances of atomic separation, and the repulsive forces predominate at closer interatomic spacings (Fig. 2–5.2(a)). The equilibrium spacing, o–a', is a natural result when

$$F_C + F_R = 0. \tag{2–5.3}$$

A tension force is required to overcome the predominant forces of attraction if the spacing is to be increased. Conversely, a compressive force has to be applied to push the atoms closer together against the rapidly increasing electronic repulsion.

The equilibrium spacing is a very precise distance for a given pair of atoms, or ions. It can be measured to five significant figures by x-ray diffraction (Chapter 3), if temperature and other factors are controlled. It takes a large force/area to stretch or compress that distance as much as one percent. (Based on Young's modulus, this is a stress of 300,000 psi (2000 MPa) for steel.) It is for this reason that the *hard ball* provides a usable model for atoms for many purposes where strength or atom arrangements are considered.†

Bonding energy. The sum of the above two forces provides us with a basis for bonding energies (Fig. 2–5.2(b)). Since energy is the product of force and distance,

$$E = \int (F_C + F_R)\, da. \tag{2–5.4}$$

---

* With SI units, k_0 is 9×10^9 V · m/C, since $k_0 = 1/4\pi\epsilon_0$.

† The hard-ball model is not suitable for all explanations of atomic behavior. For example, a neutron (which doesn't have a charge), can travel through the space among the atoms without being affected by the electronic repulsive forces just described. Likewise, atomic nuclei can be vibrated vigorously by increased thermal energy, with only a small expansion of the average interatomic spacing. Finally, by a momentary distortion of their electrical fields, atoms can move past one another in a crowded solid. (*See* diffusion in Chapter 4.)

We will use infinite atomic separation as our energy reference, $E_{a=\infty} = 0$. As we bring the atoms together, energy is *released* in an amount equal to the shaded area of Fig. 2–5.2(a). The amount of energy released is shown in Fig. 2–5.2(b). Note, however, that at o–a', where $F = 0 = dE/da$, there is a minimum of energy because energy would have to be supplied to force the atoms still closer together. The depth of this *energy well*, $E_{a=\infty} - E_{\min}$, represents the bonding energy, because that much energy would be released (−) as two atoms are brought together (at 0°K). The data of Table 2–3.1 lists such values for covalent bonds.*

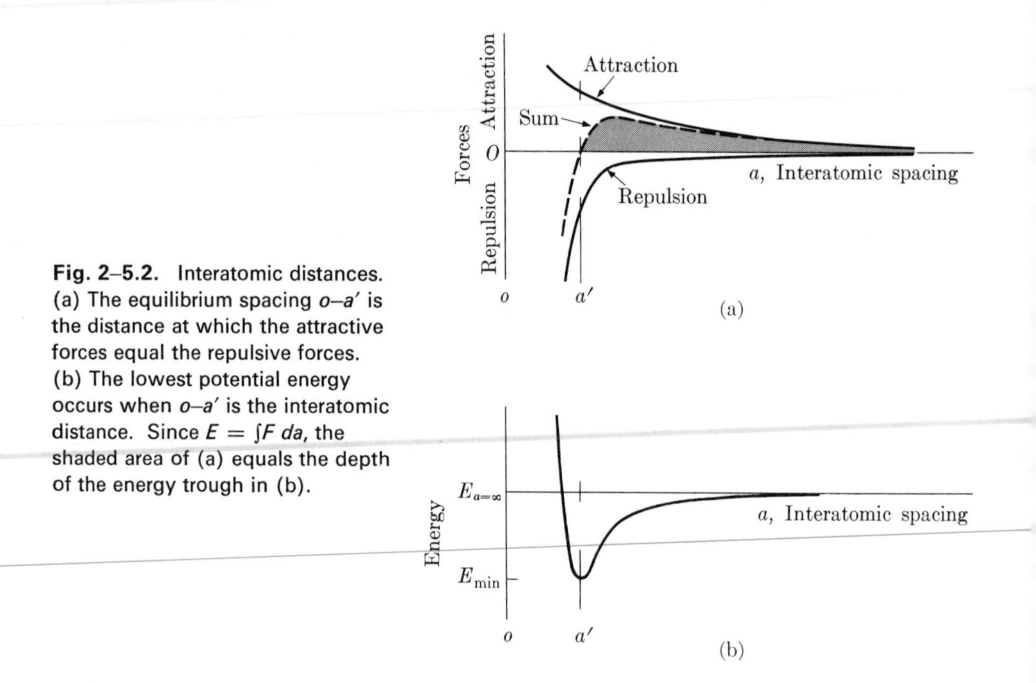

Fig. 2–5.2. Interatomic distances. (a) The equilibrium spacing o–a' is the distance at which the attractive forces equal the repulsive forces. (b) The lowest potential energy occurs when o–a' is the interatomic distance. Since $E = \int F\, da$, the shaded area of (a) equals the depth of the energy trough in (b).

The schematic representation of Fig. 2–5.2 will be useful to us on various occasions in subsequent sections, when we pay attention to elastic moduli, thermal expansion, theoretical strengths, melting and vaporization temperatures, etc.

Atomic and ionic radii. The equilibrium distance between the centers of two neighboring atoms may be considered to be the sum of their radii (Fig. 2–5.3). In metallic iron, for example, the mean distance between the centers of the atoms is 2.482 Å (angstrom units†) at room temperature. Since both atoms are the same, the radius of the iron atom is 1.241 Å.

---

* Their attractions are not the same as the coulombic forces which were presented in Eq. (2–5.1). However, an energy well of the same type may be visualized.

† The metallic radii used in this book are from the ASM *Metals Handbook*, Vol. I. The ionic radii are patterned after Ahrens.

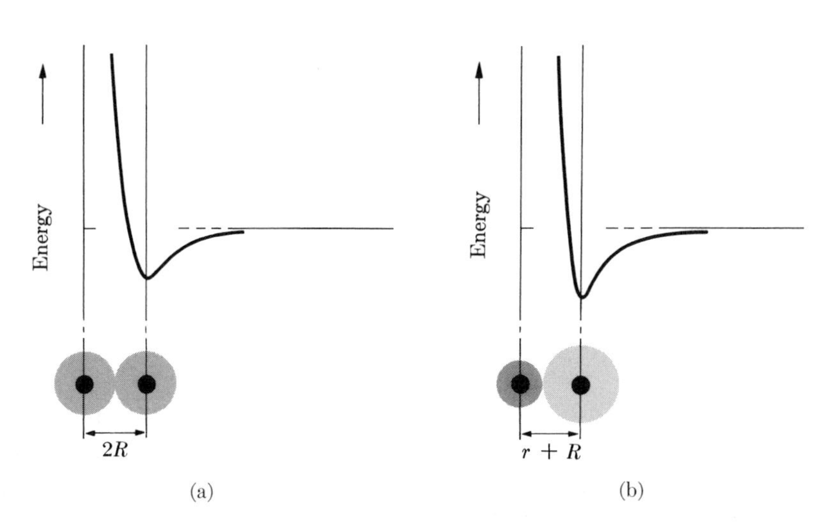

Fig. 2–5.3. Bond lengths. The distance of minimum energy between two adjacent atoms is the bond length. It is equal to the sum of the two radii. (a) In a pure metal, all atoms have the same radius. (b) In an ionic solid, the radii are different because the two adjacent ions are never identical.

Several factors can change this distance. The first is temperature. Any increase in energy above the minimum point shown in Fig. 2–5.2(b) will increase the mean distance because of the asymmetric shape of the energy trough. This increase in the mean spacing between atoms accounts for the thermal expansion of materials.

Ionic valence also influences interatomic spacing. The ferrous iron ion (Fe^{2+}) has a radius of 0.74 Å, which is smaller than that of the metallic iron atom (Appendix B). Since the two outer valence electrons have been removed (Fig. 2–5.4), the remaining 24 electrons are pulled in closer to the nucleus, which still maintains a positive charge of 26. A further reduction in interatomic spacing is observed when another electron is removed to produce the ferric ion (Fe^{3+}). The radius of this ion is 0.64 Å, or only about one-half that of metallic iron.

A negative ion is larger than its corresponding atom. Since there are more electrons surrounding the nucleus than there are protons in the nucleus, the added electrons are not as closely attracted to the nucleus as were the original electrons.

A third factor affecting the size of the atom or ion is the number of adjacent atoms. An iron atom has a radius of 1.241 Å when it is in contact with eight adjacent iron atoms, which is the normal arrangement at room temperature. If the atoms are rearranged to place this one iron atom in contact with twelve other iron atoms, the radius of each atom is increased slightly, to 1.269 Å. With more adjacent atoms, there is more electronic repulsion from neighboring atoms, and consequently the interatomic distances are increased.

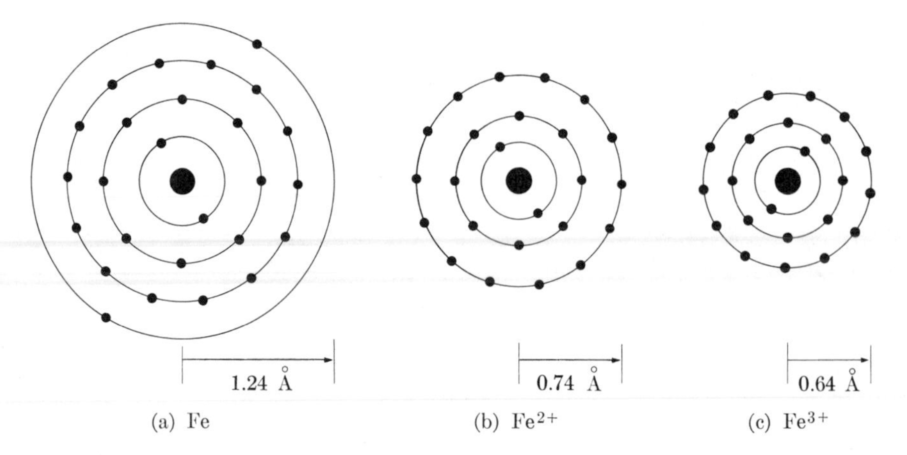

Fig. 2–5.4. Atom and ion sizes. (a) Both iron atoms and iron ions have the same number of protons (26). (b) If two electrons are removed, the remaining 24 electrons and adjacent negative ions are pulled closer to the 26-proton nucleus. (c) A ferric ion has its 23 electrons still closer to the nucleus.

We generally do not speak of atomic radii in covalently bonded materials because the electron distributions may be far from spherical (Fig. 2–3.3(a)). Furthermore, with stereospecific bonds (Section 2–2), the limiting factor in atomic coordination is not the atom size, but rather the number of electron pairs available. Even so, we may make some comparisons of interatomic distances when we look at Table 2–3.1. In ethane with a single C—C bond, the interatomic distance is 1.54 Å as compared with 1.3 Å for a C═C bond and 1.2 Å for the C≡C bond. This change is to be expected, since the bonding energies are greater with the multiple bonds.

Example 2–5.1 MgO and NaCl are comparable, except that Mg^{2+} and O^{2-} ions are divalent and Na^+ and Cl^- ions are monovalent. Therefore, the Mg—O interatomic distance is 2.1 Å while the latter is 2.8 Å. Compare the coulombic attractive forces (→ ←) that are developed at these two distances for the two pairs of ions.

Solution: From Eq. (2–5.1) and the adjacent footnote,

$$F_{\text{Mg}\rightarrow\leftarrow\text{O}} = -(9 \times 10^9 \text{ V}\cdot\text{m/C})\left(\frac{(+2)(-2)(0.16 \times 10^{-18}\text{ C})^2}{(2.1 \times 10^{-10}\text{ m})^2}\right)$$

$$= 20.9 \times 10^{-9}\text{ J/m}.$$

Similarly,

$$F_{\text{Na}\rightarrow\leftarrow\text{Cl}} = 2.9 \times 10^{-9}\text{ J/m}.$$

Comments. The opposing electronic repulsive forces (Eq. 2–5.2) will be −20.9 nJ/m and −2.9 nJ/m at these equilibrium distances. Thus, from Eq. (2–5.2), the empirical constant b is $\sim 0.4 \times 10^{-105}$ Jm^9 for MgO, and $\sim 10^{-105}$ Jm^9 for NaCl (assuming $n = 9$). ◀

• **Example 2–5.2** Compare the energy of the $Mg^{2+} \rightarrow \leftarrow O^{2-}$ bond with the energy of the $Na^{+} \rightarrow \leftarrow Cl^{-}$ bond. By combining Eqs. (2–5.1) and (2–5.2) into Eq. (2–5.4), and integrating from ∞ to a, we obtain

$$E = k_0 Z_1 Z_2 q^2/a + b/a^n. \qquad (2\text{–}5.5)$$

Solution: Using data from Example 2–5.1,

$$E_{\mathrm{Mg—O}} = \frac{(9 \times 10^9\ \mathrm{V \cdot m/C})(-4)(0.16 \times 10^{-18}\ \mathrm{C})^2}{2.1 \times 10^{-10}\ \mathrm{m}} + \frac{0.4 \times 10^{-105}\ \mathrm{Jm}^9}{(2.1 \times 10^{-10}\ \mathrm{m})^9}$$

$$= -4.4 \times 10^{-18}\ \mathrm{J} + 0.5 \times 10^{-18}\ \mathrm{J} = -3.9 \times 10^{-18}\ \mathrm{J};$$

$$E_{\mathrm{Na—Cl}} = -0.8 \times 10^{-18}\ \mathrm{J} + 0.1 \times 10^{-18}\ \mathrm{J}$$

$$= -0.7 \times 10^{-18}\ \mathrm{J}.$$

Comments. We integrated from ∞ to a, rather than from a to ∞, since our reference energy is at infinite separation. The negative values indicate that energy is given off as the ions approach each other from a distance. In this energy range, it is common to use electron volts (1 joule = 6.24×10^{18} eV); therefore, -24 eV and -4.4 eV respectively. ◀

2–6 COORDINATION NUMBER

Much of our discussion has been about diatomic combinations which involve only two atoms. However, since most engineering materials have coordinated groups of many atoms, attention must be given to polyatomic groups. Therefore, when we are analyzing the bonding of atoms within materials, we speak of a *coordination number.* The coordination number, CN, simply refers to the number of first neighbors which an atom has. Thus, in Fig. 2–3.3, the coordination number for carbon is four. In contrast, the hydrogens have only one immediate neighbor, so that their coordination numbers are only one. In Fig. 2–6.1, the Mg^{2+} ion has CN = 6.

Two factors control the coordination number of an atom. The first is covalency. Specifically, the number of covalent bonds around an atom is dependent on the number of its valence electrons. Thus the halides, which are in Group VII of the periodic table (Fig. 2–1.1), form only one bond and thus have a coordination number of one when bonded covalently. The members of the oxygen family in Group VI are held in a molecule with two bonds, and normally have a maximum coordination number of two. (Of course, oxygen may be coordinated with only one other atom through a double bond.) The nitrogen elements have a maximum coordination number of three since they are in Group V. Finally, carbon and silicon, in Group IV, have four bonds with other atoms, and a maximum coordination number of four (Fig. 2–2.4(b)).

The second factor affecting the coordination number is efficient atomic packing. Since energy is released as ions of unlike charges approach each other, ionic compounds generally have high *coordination numbers*, i.e., as many neighbors as possible without introducing the strong mutual repulsion forces between ions of like charges.

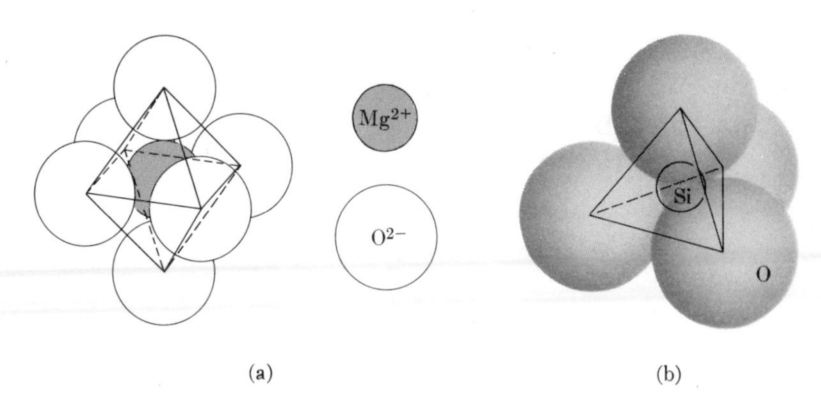

Fig. 2–6.1. Coordination numbers for ionic bonding. (a) A maximum of six oxygen ions (O^{2-}) can surround each magnesium ion (Mg^{2+}). (b) The coordination number of Si^{4+} among O^{2-} is only four because the ion-size ratio is less than 0.414 (Table 2–6.2).

This was illustrated in Fig. 2–2.2 with NaCl, and is shown again in Fig. 2–6.1(a) with Mg^{2+} ions surrounded by O^{2-} ions. The Mg^{2+} ion has a radius r of 0.66 Å (Table 2–6.1 and Appendix B). This is large enough to permit six O^{2-} ions ($R = 1.40$ Å) to surround it without direct "contact" of negative ions with one another. The minimum radius ratio (r/R), possible for six neighbors without interference, is 0.414 (Table 2–6.2). A coordination number of six (CN = 6) is encountered widely in ionic compounds.

Later, in Chapter 8, we will observe that silicon of SiO_2 has CN = 4 because an Si^{4+} ion is too small to have six coordinating oxygen ions. Since r/R for Si/O is approximately 0.3, this is consistent with the prediction of Table 2–6.2 and is shown schematically in Fig. 2–6.1(b). Another factor also favors CN = 4 for silicon among oxygens. There is considerable electron sharing between the two atoms. (Recall that the last short paragraph of Section 2–2 indicated that mixed bonding is widespread.) As with carbon, four is the maximum number of covalent bonds for silicon under normal conditions. Thus, with sharing, the probability of a CN = 4 is increased over what it would be on the basis of radius alone.

Example 2–6.1 Show the significance of 0.155 as the minimum ratio for a coordination number of three (Table 2–6.2).

Solution: The minimum ratio of sizes possible to permit a coordination number of three is shown in Fig. 2–6.2(a). In this relationship,

$$\cos 30° = \frac{R}{R + r} = 0.866, \qquad \frac{r}{R} = \frac{1 - 0.866}{0.866} = 0.155. \blacktriangleleft$$

Example 2–6.2 Show the significance of 0.414 as the minimum ratio for a coordination number of six. (Solution on page 56.)

Table 2–6.1
SELECTED ATOMIC RADII

Element	Metallic atoms		Ions			Covalent bonds	
	CN	Radius, Å	Valence	CN*	Radius, Å†	(Bond distance)/2, Å	
Carbon						Single	0.77
						Double	0.65
						Triple	0.6
Silicon			4+	6	0.42	Single	1.17
			4+	4	0.38		
Oxygen			2−	8	1.44	Single	0.75
			2−	6	1.40	Double	0.65
			2−	4	1.28		
			2−	2	~1.14		
Chlorine			1−	8	1.87	Single	0.99
			1−	6	1.81		
Sodium	8	1.857	1+	6	0.97		
Magnesium	12	1.61	2+	6	0.66		
Aluminum	12	1.431	3+	6	0.51		
			3+	4	0.46		
Iron	8	1.241	2+	6	0.74		
	12	1.27	3+	6	0.64		
Copper	12	1.278	1+	6	0.96		

* Coordination number. For ions, $1.1\, R_{CN=4} \approx R_{CN=6} \approx 0.97\, R_{CN=8}$.
† These values vary slightly with the system used. Patterned after Ahrens.

Table 2–6.2
COORDINATION NUMBERS VERSUS MINIMUM RADII RATIOS

Coordination number	Radii ratios, r/R*	Coordination geometry
3-fold	≥0.155	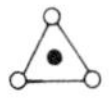
4	≥0.225	
6	≥0.414	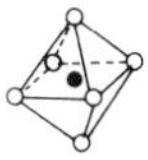
8	≥0.732	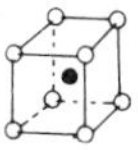
12	1.0	—

* r—smaller radius; R—larger radius.

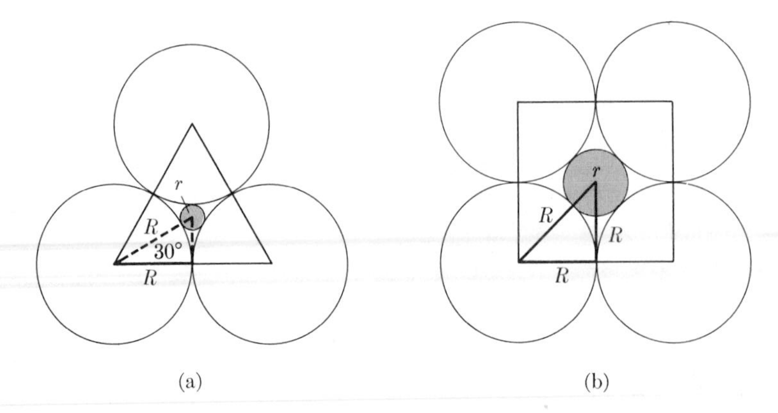

Fig. 2–6.2. Coordination calculations. (a) Three-fold coordination. (b) Six-fold coordination. (Compare with Example problems and Fig. 2–6.1 (a).)

Solution: The minimum ratio of sizes possible to permit a coordination number of six is shown in Fig. 2–6.2(b). In this relationship,

$$(r + R)^2 = R^2 + R^2, \qquad r = \sqrt{2}\,R - R, \qquad \text{and} \qquad \frac{r}{R} = 0.414.$$

Comment. From Fig. 2–6.1(a), note that the fifth and sixth ions sit above and below the center ion. ◀

2–7 TYPES OF MATERIALS

It is convenient to group materials into three main types: *metals*, *polymers* (or *plastics*), and *ceramics*. In reality, these are three idealized categories, since most materials have intermediate characteristics.

Metals. These materials are characterized by their high thermal and high electrical conductivity. They are opaque and usually may be polished to a high luster (Fig. 2–7.1). Commonly, but not always, they are relatively heavy and deformable.

What accounts for the above characteristics? The simplest answer is that metals owe their behavior to the fact that some of the electrons are delocalized and can leave their "parent" atoms (Fig. 2–2.7(b)). Conversely, electrons are not free to roam to the same extent in polymers and ceramics. Since some of the electrons are delocalized in metals, they can quickly transfer an electric charge and thermal energy. The opacity and reflectivity of a metal arises from the response of these delocalized electrons to electromagnetic vibrations at light frequencies. It is another result of the partial independence of some of the electrons from their parent atoms. In short, *metals are characterized by an ability to give up valence electrons.*

Polymers (often called plastics). Plastics are noted for their low density and their use as insulators, both thermal and electrical. They are poor reflectors of light, tending to be transparent or translucent (at least in thin sections). Finally, some of

Fig. 2–7.1. Metallic materials (stainless steel utensils for commercial kitchens). Metals possess ductility for the required processing. The luster and conductivity of metals arise from the electronic characteristics of the interatomic bonds. (Courtesy of the Vollrath Company.)

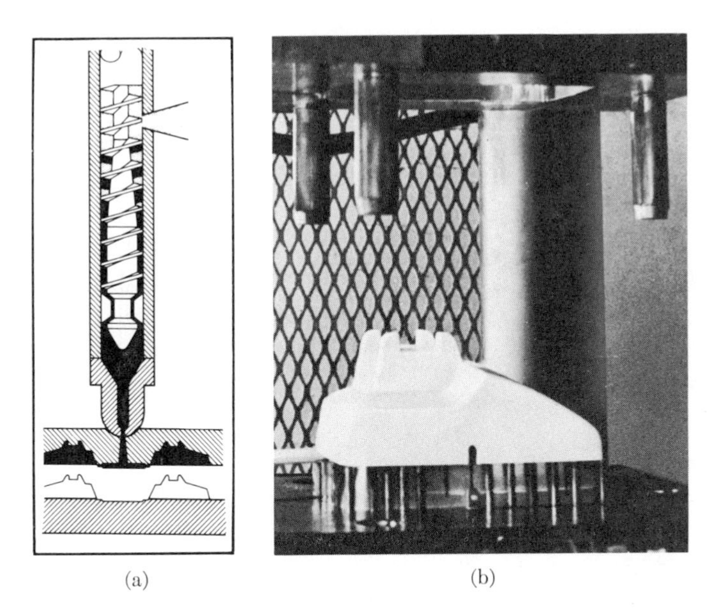

(a) (b)

Fig. 2–7.2. Plastic products (injection-molded telephone receiver). (a) The soft heated plastic was forced through the sprues into the die cavity. (b) The cooled product, which is rigid, required only one manufacturing step. (Courtesy Western Electric Company.)

them are flexible and subject to deformation. This latter characteristic is used in manufacturing (Fig. 2–7.2).

Unlike metals, which have some migrant electrons, the *nonmetallic elements* of the upper right corner of the periodic table (Fig. 2–1.1) have a great *affinity* or *attraction for additional electrons.* Each electron becomes associated with a specific atom (or pair of atoms). Thus, in plastics, we find only limited electrical and thermal conductivity because all the thermal energy must be transferred from hot to cold regions by atomic vibrations, a much slower process than the electronic transport of energy which takes place in metals. Furthermore the less mobile electrons in plastics are more able to adjust their vibrations to those of light, and therefore do not absorb light rays.

Materials which contain *only* nonmetallic elements share electrons to build up large molecules. These are often called *macromolecules.* We shall see in Chapter 7 that these large molecules contain many repeating units, or *mers*, from which we get the word *polymers.*

Ceramics. Simply stated, ceramics are *compounds which contain metallic and nonmetallic elements.* There are many examples of ceramic materials, ranging from the cement of concrete (and even the rocks themselves) to glass, to spark-plug insulators (Fig. 2–7.3), and to oxide nuclear fuel elements of UO_2, to name but a few.

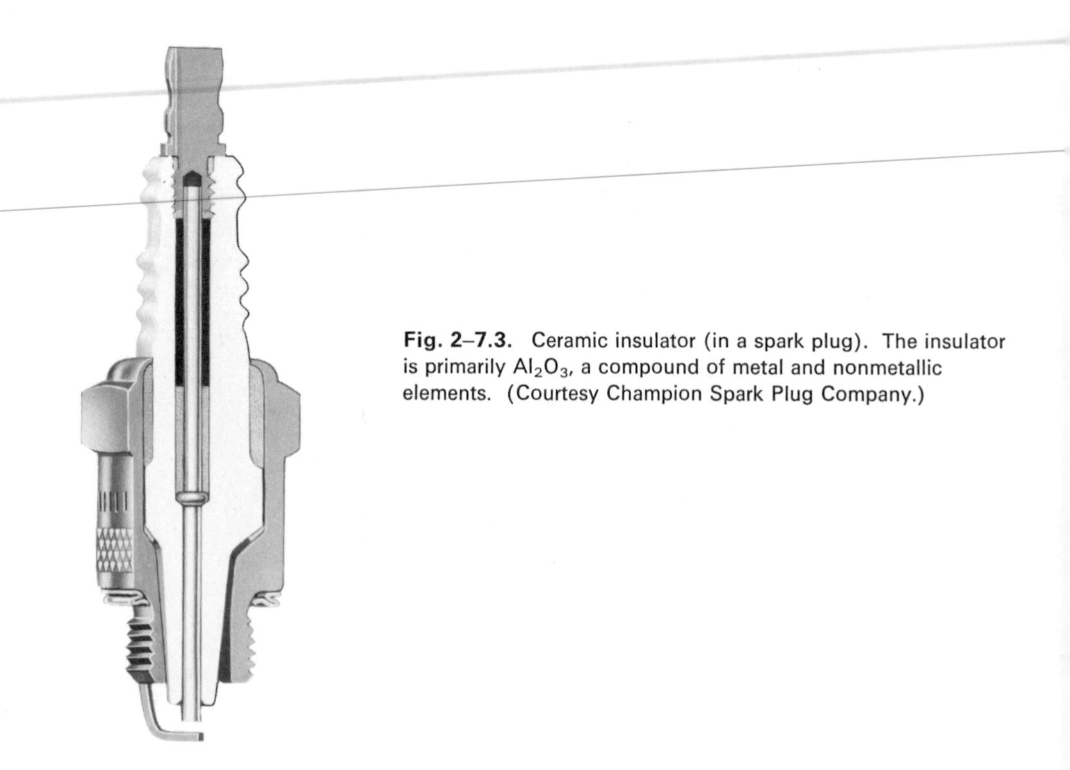

Fig. 2–7.3. Ceramic insulator (in a spark plug). The insulator is primarily Al_2O_3, a compound of metal and nonmetallic elements. (Courtesy Champion Spark Plug Company.)

Each of these materials is relatively hard and brittle. Indeed, hardness and brittleness are general attributes of ceramics, along with the fact that they tend to be more resistant than either metals or polymers to high temperatures and to severe environments. The basis for these characteristics is again the electronic behavior of the constituent atoms. Consistent with their natural tendencies, the metallic elements release their outermost electrons and give them to the nonmetallic atoms, which retain them. The result is that these electrons are immobilized, so that the typical ceramic material is a good insulator, both electrically and thermally.

Equally important, the positive metallic ions (atoms that have lost electrons) and the negative nonmetallic ions (atoms that have gained electrons) develop strong attractions for each other. Each *cation* (positive) surrounds itself with *anions* (negative). Considerable energy (and therefore considerable force) is usually required to separate them. It is not surprising that ceramic materials tend to be hard (mechanically resistant), refractory (thermally resistant), and inert (chemically resistant).

REVIEW AND STUDY

GENERALIZATIONS REGARDING PROPERTIES

Several of the engineering properties of Chapter 1 may be related *qualitatively* to the atomic bonding characteristics described in this chapter.

1. Density is controlled by atomic weight, atomic radius, and coordination number. The latter is a significant factor because it controls the packing factor.

2. Melting and boiling temperatures can be correlated with the depth of the energy trough shown in Fig. 2–5.2(b). Atoms have minimum energy (at the bottom of the trough) at a temperature of absolute zero. Increased temperatures raise the energy until the atoms are able to separate themselves one from another.

3. Strength is also correlatable with the height of the total force or sum curve of Fig. 2–5.2(a). That force, when related to the cross-sectional area, gives the stress required to separate atoms. (As we shall see in Section 6–4, materials can deform through a process other than direct separation of the atoms. However, the amount of stress required to deform them is still governed by the interatomic forces.) Also, since larger interatomic forces of attraction imply deeper energy troughs, we observe that materials with high melting points are often the harder materials: e.g., diamond, Al_2O_3, TiC, etc. In contrast, in materials with weaker bonds there is a correlation between softness and low melting point: e.g., lead, plastics, ice, and grease. Apparent exceptions to these generalizations can arise when more than one type of bond is present, as in graphite and clay.

4. The modulus of elasticity can be calculated from the slope of the sum curve of Fig. 2–5.2(a) because at the equilibrium distance, where the net force is zero, dF/da

relates stress to strain. As long as the strain or change in interatomic distance is a fraction of a percent, the modulus of elasticity remains essentially constant. Extreme compression or extreme tension respectively raise or lower the modulus of elasticity.

5. Thermal expansions of materials with comparable atomic packing factors vary inversely with their melting temperatures. This indirect relationship exists because the higher-melting-point materials have deeper and therefore more symmetrical energy troughs. Thus their mean interatomic distances increase less with a given change in thermal energy. Examples of several metals include: Hg, melting temperature equals $-39°C$, coefficient of linear expansion equals 40×10^{-6} cm/cm · °C; Pb, 327°C, 29×10^{-6} cm/cm · °C; Al, 660°C, 22×10^{-6} cm/cm · °C; Cu, 1084°C, 17×10^{-6} cm/cm · °C; Fe, 1538°C, 12×10^{-6} cm/cm · °C; W, 3387°C, 4.2×10^{-6} cm/cm · °C.

6. Electrical conductivity is very dependent on the nature of the atomic bonds. Both ionically and covalently bonded materials are extremely poor conductors, because electrons are not free to leave their host atoms. On the other hand, the delocalized electrons of metals easily move along a potential gradient. Semiconductors will be considered in Chapter 5; however, we can note here that their conductivity is controlled by the freedom of movement of their electrons.

7. Thermal conductivity is high in materials with metallic bonds, because delocalized electrons are very efficient carriers of thermal as well as electrical energy.

8. The influence of the structure of atoms on chemical properties has not been elaborated on here, since the chemical differences between elements depend primarily on the number of valence electrons. Furthermore, all chemical reactions involve the formation and the disruption of bonds. So far as engineering materials are concerned, the *corrosion* reaction (Chapter 12) is probably the most significant chemical reaction. In corrosion, the separation of a metallic ion from the metal propei involves the removal of valence electrons from the outer shell of the atom and its *ionization potential.*

In subsequent chapters, we shall develop the principles which control the properties of engineering materials by utilizing the concepts of structure.

TERMS FOR REVIEW

Atomic weight
Atomic mass unit, amu
Atomic number
Atomic radii
Avogadro's number, *N*
Bond angle
Bond energy
Bond length
Ceramics
Coordination number, CN
Coulombic forces
Covalent bonds
Delocalization
Dipole
Electronegativity
Electronic repulsion

TERMS FOR REVIEW—CONTINUED

Energy well
Hydrogen bridge
Induced dipole
Ionic bond
Ionic radii
Ionization energy
Isomer
Lone pair
Metallic bond
Metals
Molecules
  polar
Orbital
Periodic table
Polymer
Primary bond
Secondary bond
Stereospecific
Valence electrons
Van der Waals forces

FOR DISCUSSION

A_2 Distinguish between atomic number and atomic weight.

B_2 According to Table 2–1.1, it takes approximately twice as much energy to remove the second electron from Group II elements as the first electron. Why? Would the same ratio hold for Group I elements?

C_2 Why are there more metallic elements than nonmetallic elements (Fig. 2–1.1)?

D_2 Refer to Fig. 2–2.1. Make a comparable sketch for KCl.

E_2 Covalent bonds are directional; ionic bonds are not. Why?

F_2 A small diamond has 10^{20} atoms. How many covalent bonds?

G_2 Silicon tetrafluoride (SiF_4) is a stable compound, but has a low melting temperature (−77°C). Account for these facts by predicting the nature of its bonds. (Use a sketch if necessary.)

H_2 Sulfur dichloride has a molecular weight of 103 amu and a boiling point of 59°C. Draw a diagram showing the valence electron structure of this compound by using the "dot" representation (Fig. 2–2.3(a)).

I_2 Using the "dot" representation, sketch the electron structure for ClO_4^-, SO_4^{2-}, PO_4^{3-}, SiO_4^{4-} ions.

J_2 Carbon dioxide (CO_2) has no electric dipole. What does this suggest about the O—C—O bond angle?

K_2 The electric dipole decreases from HF to HCl to HBr to HI. Suggest a reason.

L_2 HBr boils at a higher temperature (−67°C) than HCl (−85°C) even though it has a lower electric dipole. Explain.

M_2 Why can a neutron move through materials, even though the atoms "touch"?

N_2 Barium is not listed in Appendix B. On the basis of the periodic table (Fig. 2–1.1) and other data from Appendix B, predict the radius of the Ba^{2+} ion; Ba atom; Ag^{2+} atom.

O_2 Why is there a lower limit, but not an upper limit, on the radii ratios for CN = 6?

P_2 List the factors which affect the coordination number for ionic compounds; covalent compounds.

Q_2 Categorize the following into the three types of materials of Section 2–7: bronze, portland cement, "rubber" cement, wood, window glass, bakelite, rust, ethylene glycol (antifreeze).

R_2 Cite materials which lie intermediate between metals and ceramics; between polymers and ceramics.

S_2 List various metals which are used to make a kitchen stove; various ceramics; various polymers.

T_2 Plot α_1 vs. T_m data of paragraph (5) in Review and Study. Add other metals to this list from data in the handbook. Explain your results in your own words.

U_2 Graphite is very anisotropic, i.e., its properties vary markedly with direction. Explain why the conductivity of graphite is more than 100 times greater in the horizontal direction of Fig. 2–2.6 than in the vertical direction.

STUDY PROBLEMS

2–1.1 a) What is the weight of an aluminum atom? b) The density of aluminum is 2.70 gm/cm^3; how many atoms per cm^3?

Answer: a) 4.48×10^{-23} g/atom b) 6.02×10^{22} atom/cm^3

2–1.2 a) How many iron atoms are there per gram? b) What is the volume of a grain of metal containing 10^{20} iron atoms?

2–1.3 a) Al_2O_3 has a density of 3.8 g/cm^3. How many atoms are present per cm^3? b) Per gram?

Answer: a) 1.12×10^{23} atoms/cm^3 b) 2.95×10^{22} atoms/g

2–1.4 A cubic volume of MgO which is 4.2 angstroms along each edge contains 4 Mg^{2+} ions and 4 O^{2-} ions. What is the density of MgO?

• **2–1.5** Give the notation for the electronic structure of (a) zirconium atoms, (b) Zr^{4+} ions.

Answer: a) $1s^2 2s^2 2p^6 3s^2 3p^6 3d^{10} 4s^2 4p^6 4d^2 5s^2$ b) $1s^2 2s^2 2p^6 3s^2 3p^6 3d^{10} 4s^2 4p^6$

• **2–1.6** Indicate the number of $3d$ electrons in each of the following:

a) Ti^{2+}	b) Ti^{4+}	c) Cr^{3+}	d) Fe^{3+}
e) Fe^{2+}	f) Mn^{2+}	g) Mn^{4+}	h) Ni^{2+}
i) Co^{2+}	j) Cu^{+}	k) Cu^{2+}.	

2–1.7 An ampere of current is supplied by a rechargeable battery. How many electrons does this provide per minute?

Answer: 3.75×10^{20} electrons/min

2–2.1 It takes approximately 5×10^{-19} joule to break the covalent bond between carbon and nitrogen. What wavelength would be required of a photon to supply this energy? (See

Appendix A for constants.)

Answer: 4000 Å

2–2.2 An electron absorbs all the energy from a photon of ultraviolet light ($\lambda = 2768$ Å) How many eV are absorbed?

2–3.1 Determine the molecular weight for each of the molecules of Fig. 2–3.1.

Answer: a) 32 b) 46 c) 17 d) 78 e) 94 f) 30 g) 58 h) 60 i) 28 j) 62.5

2–3.2 An organic compound contains 62.1 w/o carbon, 10.3 w/o hydrogen, and 27.6 w/o oxygen. Name a possible compound.

2–3.3 Refer to Fig. 2–3.3 Methyliodide (CH_3I) has a similar structure; however, the H—C—H angles equal 111.4°. What is the H—C—I angle in CH_3I? (This is a problem in trigonometry; but it will illustrate the distortion in a polar molecule.)

2–3.4 Sketch the structure of the various possible isomers for octane, C_8H_{18}.

2–4.1 Calculate the electric dipole of each H—S arm of H_2S if the electric dipole of the molecule is 3.7×10^{-30} C · m and the H—S—H bond angle is 93°.

Answer: 2.7×10^{-30} C · m

2–4.2 The H—F distance in HF is 1 Å and there are a total of 10 electrons present. The electric dipole of an HF molecule is 6.4×10^{-30} C · m. a) How far from the center of positive charge (our reference point) is the center of negative charge? b) What fraction is this of the total bond length?

2–5.1 Refer to Example 2–5.1 and its comments. a) What are the attractive and repulsive forces between Mg^{2+} and O^{2-} ions at $a = 2.0$ Å? b) At 2.2 Å?

Answer: a) 23 nJ/m, −35 nJ/m b) 19 nJ/m, −13 nJ/m

•**2–5.2** Plot the sum of the attractive and repulsive forces ($F_C + F_R$) between Mg^{2+} and O^{2-} over the distance of 1.9 Å to 2.8 Å. [*Hint:* The answers for Example 2–5.1 and Study Problem 2–5.1 provide some of the data for three of the points.]

•**2–5.3** Plot the energy versus separation distance between Mg^{2+} and O^{2-} from 1.9 Å to 2.8 Å.

Answer: 1.9 Å, -3.6×10^{-18} J; 2.1 Å, -3.9×10^{-18} J; 2.8 Å, -3.3×10^{-18} J

2–6.1 Show the origin of 0.732 in Table 2–6.2.

Answer: $2(r + R) = \sqrt{3}(2R)$

2–6.2 Show the origin of 0.225 in Table 2–6.2. [*Hint:* Place the four large ions of Fig. 2–6.1(b) at four corners of a cube and the small ion at the cube center.] (See Fig. 2–3.3(a).)

2–6.3 a) What is the radius of the smallest cation that can have a six-fold coordination with O^{2-} ions without distortion? b) eight-fold coordination?

Answer: a) 0.58 Å b) 1.05 Å

2–6.4 a) From Appendix B, cite three divalent cations which can have CN = 6 with S^{2-}, but not CN = 8. b) Cite two divalent ions which can have CN = 8 with F^-.

CHAPTER THREE

ATOMIC ORDER IN SOLIDS

PREVIEW

After considering atom-to-atom bonding, our next step along the structural scale is to look at the long-range patterns of atomic order. This is made relatively simple in crystalline solids because unit cells are formed which repeat in each of the three dimensions. Each unit cell has all of the geometric characteristics of the total crystal.

It will be the purpose of this chapter to look specifically at the atomic arrangements in a few of the more simple structures (bcc, fcc, and hcp), and establish a credibility for their existence through density calculations.

The reader should become familiar with the notations for unit cell location, crystal directions, and crystal planes, because they will be used subsequently to relate the crystal structure to the properties and behavior of materials.

An optional section of the chapter gives a basis for diffractional procedures which are used to determine crystal structures.

3–1 CRYSTALLINITY

Essentially all metals, a significant fraction of ceramics, and certain polymers crystallize when they solidify. By this, we mean that the atoms arrange themselves into an ordered, repeating, 3-dimensional pattern. Such structures are called *crystals* (Fig. 3–1.1).

This ordered pattern over a *long range* of many atomic distances is due to the atomic coordination (Section 2–6) within the material; in addition, the pattern sometimes controls the external shape of the crystal; the six-pointed outline of snowflakes is probably the most familiar example of this. The planar surfaces of gems, quartz (SiO_2) crystals, and even ordinary table salt (NaCl) are all external manifestations of internal crystalline arrangements. In each case the internal atomic arrangement persists even though the external surfaces are altered. For example, the internal structure of a quartz crystal is not altered when the surfaces of the crystal are abraded to produce round silica sand. Likewise, there is a hexagonal arrangement of water molecules in chunks of crushed ice as well as in snowflakes.

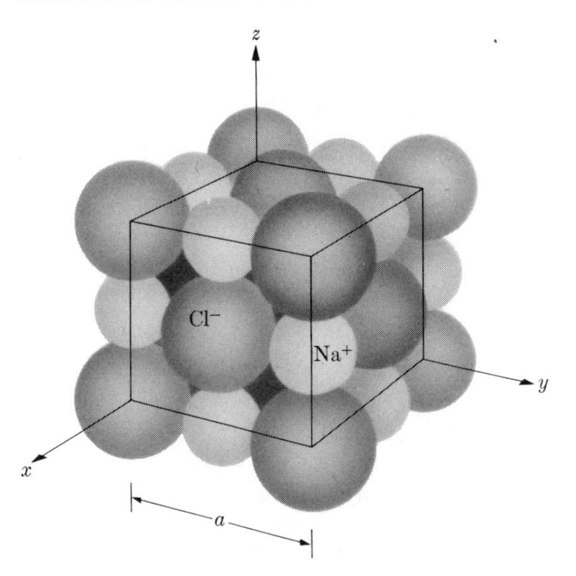

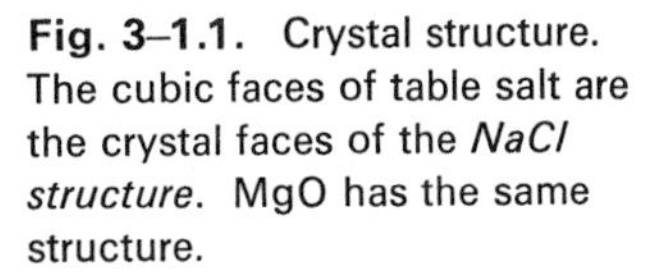
Fig. 3–1.1. Crystal structure. The cubic faces of table salt are the crystal faces of the *NaCl structure*. MgO has the same structure.

Let us use sodium chloride as an illustration of the role of atom packing on crystallinity. The ion-size ratio of Na^+ and Cl^- is 0.97/1.81, or 0.54. Recall from Table 2–6.2 that this ratio favors a coordination number of six. This was shown in Fig. 2–6.1(a) for Mg^{2+} and O^{2-}, but would also be applicable for Na^+ and Cl^-. The first-neighbor arrangement of Fig. 2–6.1(a) shows only part of the structure; a more complete pattern is shown in Fig. 3–1.1, in which we may note the following features for the NaCl structure.

1. Each Na^+ and each Cl^- ion has six neighbors (if the pattern is extended further into three dimensions).

2. There are equal numbers of Na^+ ions and Cl^- ions (if the pattern is continued).

3. A small cube is established with planar faces and an edge of length $(2r + 2R)$, where r and R are the radii of the Na^+ and Cl^- ions, respectively.

4. The pattern in this cube, called the *unit cell*, is identical to the pattern of all other cubes of NaCl. Thus if we know the structure of the repeating unit cells, we are able to describe the structure of the crystal.

5. The Na–Na and the Cl–Cl interatomic distances are both greater than the Na–Cl distances by a factor of $\sqrt{2}$. This difference is important because the coulombic attractive forces between unlike ions must be greater than the coulombic repulsive forces between like ions (Eq. (2–5.1)).

Each of the above observations will be discussed at further length. However, our immediate goal will be to consider the various possible types of crystal structures.

3–2 CRYSTAL SYSTEMS

As atoms and molecules are coordinated with neighbors to form a solid, we find specific patterns forming. The cubic pattern described in the previous section is but one of seven *crystal systems*. The cubic system has three mutually perpendicular edges or *axes*: x, y, and z. The lengths of the edges in each of the three dimensions are equal: $a_1 = a_2 = a_3$.

Not all of the seven crystal systems are so regular (Table 3–2.1). These seven represent all combinations of similar and dissimilar *lattice constants* a, b, and c, plus the alternatives of 90° and non-90° axial angles.

Table 3–2.1
GEOMETRY OF CRYSTAL SYSTEMS

System	Axes	Axial angles
Cubic	$a_1 = a_2 = a_3$	All angles = 90°
Tetragonal	$a_1 = a_2 \neq c$	All angles = 90°
Orthorhombic	$a \neq b \neq c$	All angles = 90°
Monoclinic	$a \neq b \neq c$	2 angles = 90°; 1 angle ≠ 90°
Triclinic	$a \neq b \neq c$	All angles different; none equal 90°
Hexagonal	$a_1 = a_2 = a_3 \neq c$	Angles = 90° and 120°
Rhombohedral	$a_1 = a_2 = a_3$	All angles equal, but not 90°

The majority of crystals we encounter in this book fall into the cubic system since we will generally consider the simpler materials. Examples of cubic crystals include most of the common metals (except zinc and magnesium, which are hexagonal, and tin, which is tetragonal), and a number of the simpler ceramic compounds such as MgO, TiC, and $BaTiO_3$. As a rule, polymeric crystals are not cubic: however, there will be less opportunity for us to consider the details of their crystallinity.

Unit cells. In discussing Fig. 3–1.1, we spoke of a unit cell as a small, characteristic volume which can be used to describe the structure of the total crystal. The atomic arrangement within each unit cell is identical to the arrangement in all others. Figure 3–2.1 gives us a 2-dimensional illustration. If the outlined rectangle is repeated in the two dimensions, we obtain the whole pattern. Three-dimensional unit cells are shown in the following sections.

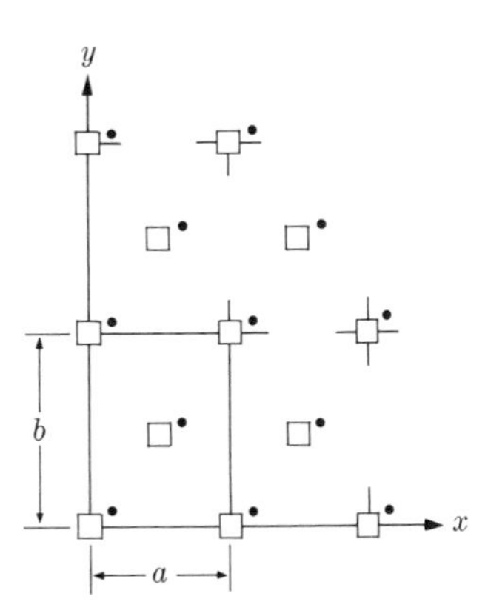

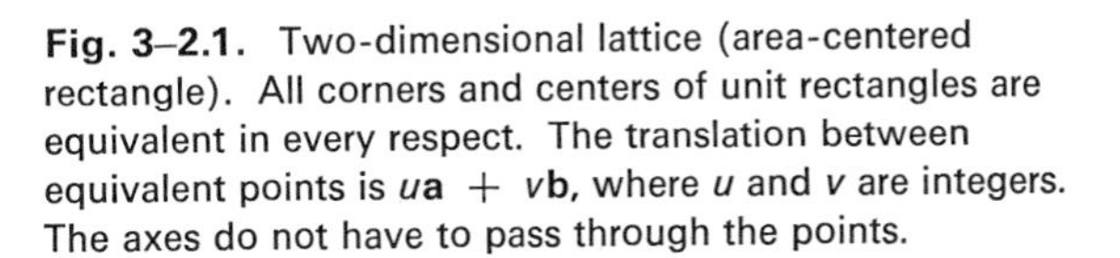
Fig. 3–2.1. Two-dimensional lattice (area-centered rectangle). All corners and centers of unit rectangles are equivalent in every respect. The translation between equivalent points is $u\mathbf{a} + v\mathbf{b}$, where u and v are integers. The axes do not have to pass through the points.

Before we leave the 2-dimensional pattern of Fig. 3–2.1, the reader should note that:

1. In *this* "unit area", $a \neq b$;
2. The axial angle is 90°;
3. While small, the outlined unit area is not the smallest repeatable area that could have been drawn to define this 2-dimensional array of points;*
4. *Any* position within the sketch may be duplicated by an equivalent location through a *translation* of $u\mathbf{a}$ and $v\mathbf{b}$, where u and v are integers. Thus, all dots of Fig. 3–2.1 are at *equivalent points* to all other dots. Whether we consider the squares or dots, the *lattice* is the same, *viz*, an "area-centered rectangle." The reference location, i.e., the *origin*, does not need to be a marked point; *it may be anywhere within the lattice.*

•**Lattices.** There are 14 *space lattices* when three dimensions are used (Fig. 3–2.2). The cubic system contains three. In each cubic lattice the unit-cell axes are 90° apart, and the unit-cell dimensions are equal. However, the various equivalent points are distributed differently in each of the three cubes. There is only one equivalent

---

* Connect the center point of the rectangle to the four corners. (Then eliminate the present rectangle.) The two lines of the resulting X can serve as axes for unit areas which are half as big as the one originally sketched. However, these are not 90° axes, and therefore are less useful in analyses for calculations.

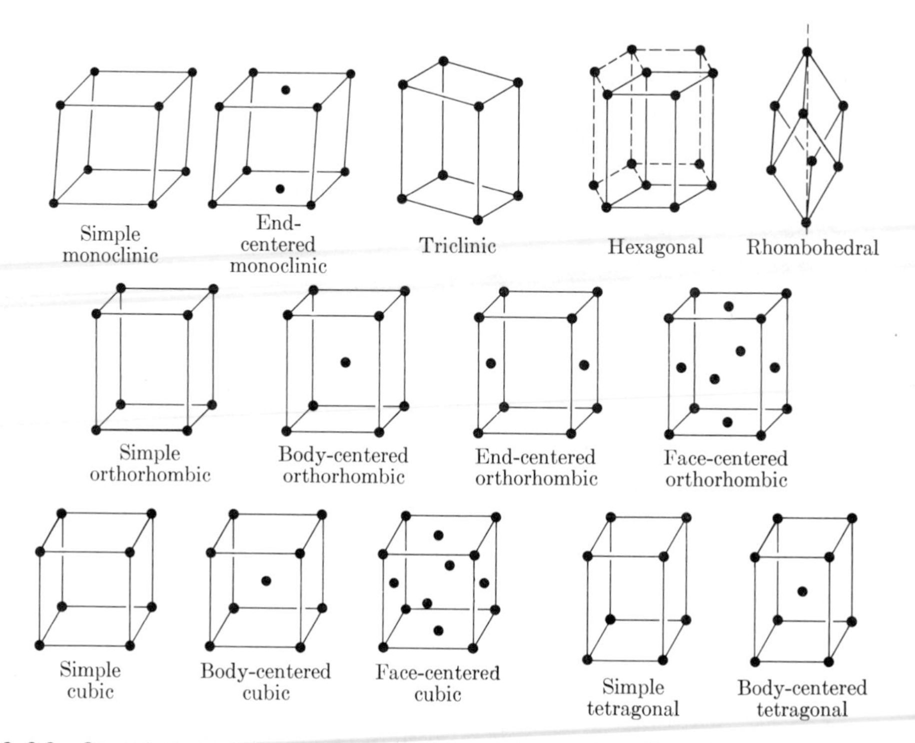

Fig. 3–2.2. Space lattices. These 14 *Bravais lattices* continue in three dimensions. Each indicated point has identical surroundings. Compare with Table 3–2.1.

point per unit cell in the *simple cubic* (sc) space lattice; two in the *body-centered cubic* (bcc) space lattice; and four in the *face-centered cubic* (fcc) lattice. No other possible cubic lattices exist.* These lattices, called *Bravais lattices*, are simply arrays of equivalent points. That is, each point in an infinite 3-dimensional lattice has (1) the same numbers of neighboring points, (2) in the same directions, and (3) at the same distances. The points would be impossible to distinguish from each other without some arbitrary reference origin and axes.

When we consider cubic metals, we commonly locate the origin and axes so that the atoms are found at these lattice points. This, however, is not necessary, because we could just as well place the origin at an interstice (or elsewhere). In either event, the translation $u\mathbf{a}$, $v\mathbf{b}$, and $w\mathbf{c}$ will produce an equivalent point.

Other lattices fall within the other crystal systems. Among these, we will give special attention in Section 3–4 to one variant of the hexagonal structure called the *hexagonal close-packed* (hcp), because it includes the unit cells of a number of common metals. We should be aware of structures other than (sc), (bcc), (fcc), and (hcp);

* Larger cubic lattices can be subdivided or reoriented into one of these three.

however, we can refer to Fig. 3–2.2, when we discuss them. The reader should be able to envision the above four without reference to Fig. 3–2.2.

3–3 CUBIC CRYSTALS

Although metals have unit cells which are relatively simple in concept, we do not find elemental metals with the *simple cubic* lattice of Fig. 3–2.2.* Let us consider briefly why this is so, by placing one atom at each lattice point of Fig. 3–3.1. Each atom has only six neighbors (CN = 6). Recall that, since energy is released as atoms come together, a higher coordination number generally gives a more stable structure by releasing more energy. We will observe shortly that CN = 8 for bcc metals and CN = 12 for fcc and for hcp metals. As a general rule, metals are more stable with these higher coordination numbers.

The concept of atomic packing factor (APF) will also help us. The simple cubic structure of Fig. 3–3.1 has an average of one atom per unit cell.† Thus, with a "hard-

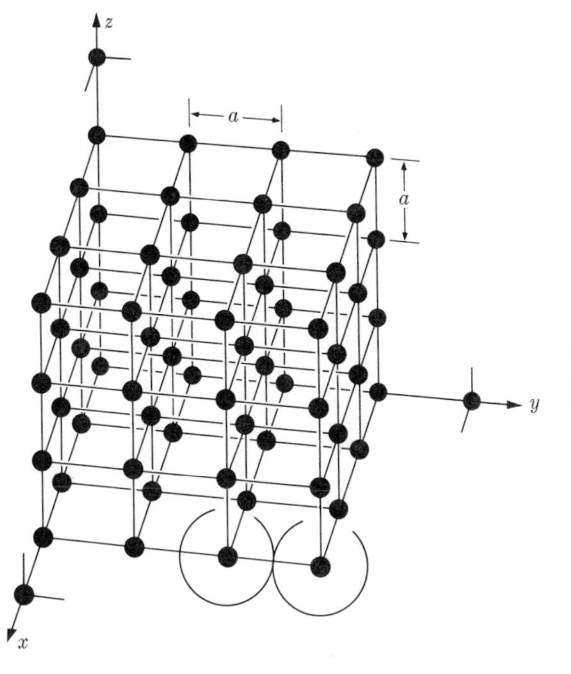

Fig. 3–3.1. Simple cubic lattice. Space is divided by three sets of equally spaced parallel planes. The x, y, and z reference axes are mutually perpendicular. Each point of intersection is equivalent.

---

* Compounds may have a simple cubic structure with only one equivalent point (of several atoms each) per unit cell (Section 3–6).

† We can check this readily by relocating the axes so that their origin is positioned among eight atoms. Now the atoms are located at the center of the resulting unit cells.

ball model" of the atoms that touch at a radius of R, only 52% of the space is occupied since $a = 2R$:

$$\text{Atomic packing factor} = \frac{\text{Volume of atoms}}{\text{Volume of unit cell}} \tag{3–3.1}$$

$$= \frac{4\pi R^3/3}{(2R)^3} = 0.52.$$

Other metallic structures give us higher packing factors.

Body-centered cubic. Iron has a cubic structure. At room temperature, the unit cell of iron has an atom at each corner of the cube, and another atom at the body center of the cube (Fig. 3–3.2).

(a) (b)

Fig. 3–3.2. Body-centered cubic structure of a metal. Part (a) is a schematic view showing the location of atom centers. (b) Model made from hard balls. (G. R. Fritterer. Reproduced by permission from B. Rogers, *The Nature of Metals,* 2nd Ed., American Society for Metals, 1964.)

Each iron atom in a body-centered cubic (bcc) structure is surrounded by eight adjacent iron atoms, whether the atom is located at a corner or at the center of the unit cell. Therefore each atom has the same geometric environment (Fig. 3–3.2(a)). There are two atoms per unit cell in a bcc metal. One atom is at the center of the cube, and eight octants are located at the eight corners (Fig. 3–3.3). The lattice constant a is related to the atomic radius R by

$$a_{\text{bcc}} = 4R/\sqrt{3}. \tag{3–3.2}$$

Therefore the atomic packing factor is 0.68, which is significantly greater than that for a simple cubic structure of a metal.

Iron is the most common material with a body-centered cubic structure, but not the only one. Chromium and tungsten, among others listed in Appendix B, also have body-centered cubic arrangements.

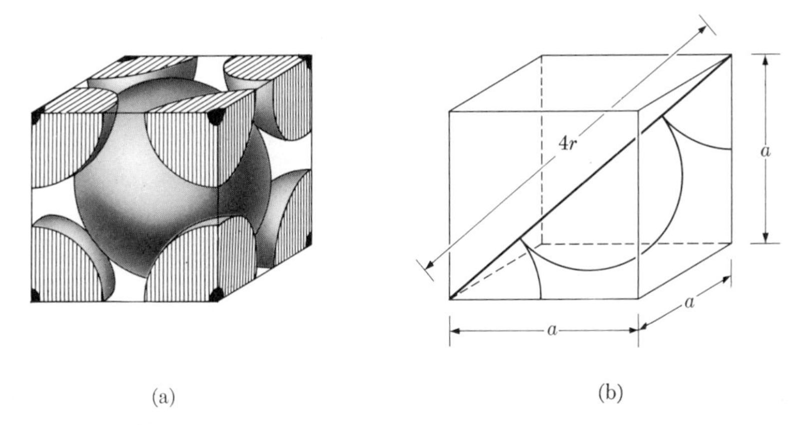

Fig. 3–3.3. Body-centered cubic unit cell. In a metal the bcc structure has two atoms per unit cell and an atomic packing factor of 0.68. The lattice constant a is related to the atomic radius as shown by Eq. (3–3.2).

Face-centered cubic. The atom arrangement in copper (Fig. 3–3.4) is not quite the same as that in iron, although it has a cubic structure. In addition to an atom at the corner of each unit cell, there is one at the center of each face, but none at the center of the cube.

Face-centered cubic (fcc) structures are somewhat more common among metals than are body-centered cubic structures. Aluminum, copper, lead, silver, and nickel possess this atomic arrangement (as does iron at elevated temperatures). Face-centered cubic structures are also found in compounds as revealed in Fig. 3–1.1, where the Cl^- ions at the cube corners and at the face centers are all equivalent.

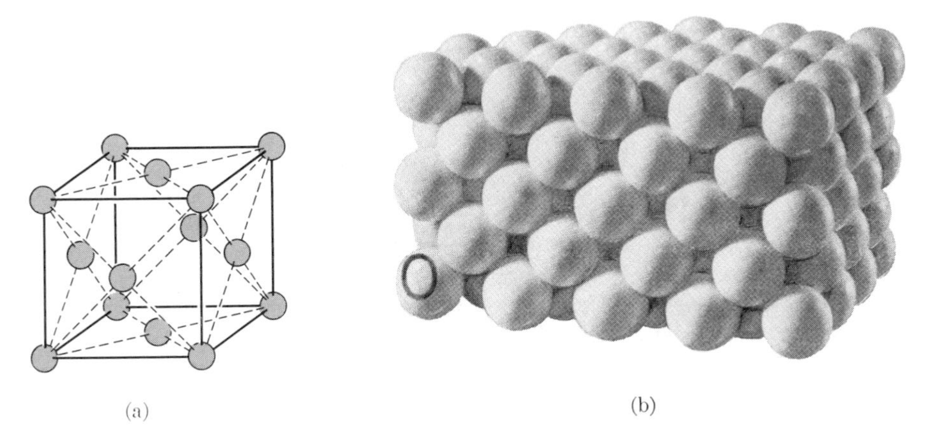

Fig. 3–3.4. Face-centered cubic structure of a metal. Part (a) is a schematic view showing location of atom centers. (b) Model made from hard balls. (G. R. Fritterer. Reproduced by permission from B. Rogers, *The Nature of Metals*, 2nd Ed., American Society for Metals, 1964).

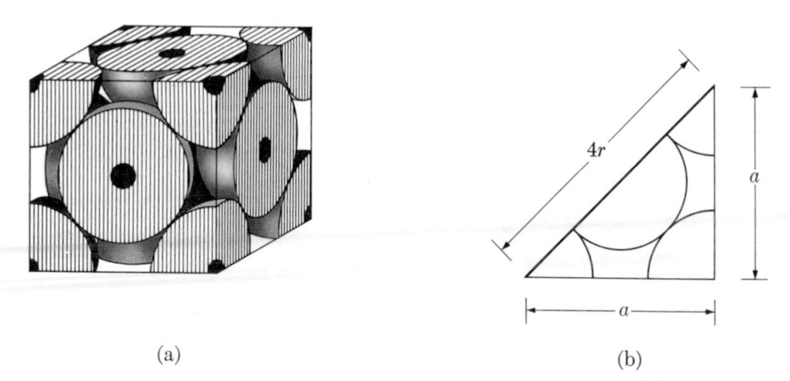

Fig. 3–3.5. Face-centered cubic unit cell. In a metal the fcc structure has four atoms per unit cell and an atomic packing factor of 0.74. In metals, the lattice constant a is related to the atomic radius as shown by Eq. (3–3.3).

A metal with an fcc structure has four times as many atoms as it has unit cells. The eight corner octants contribute a total of one atom, and the six face-centered atoms contribute a total of three atoms per unit cell (Fig. 3–3.5). The lattice constant a is related to the atomic radius R by

$$a_{\text{fcc}} = 4R/\sqrt{2}. \tag{3–3.3}$$

Example 3–3.1 Calculate (a) the atomic packing factor of an fcc metal (Fig. 3–3.5); (b) the ionic packing factor of fcc NaCl (Fig. 3–1.1).

Solution

a) Equations (3–3.1) and (3–3.3),

$$\text{PF} = \frac{4(4\pi R^3/3)}{a^3} = \frac{16\pi R^3(2\sqrt{2})}{(3)(64R^3)} = 0.74.$$

b) Equation (3–3.1) and Fig. 3–1.1,

$$\text{PF} = \frac{4(4\pi r^3/3) + 4(4\pi R^3/3)}{(2r + 2R)^3} = \frac{16\pi(0.97^3 + 1.81^3)}{3(8)(0.97 + 1.81)^3} = 0.67.$$

Comments. It is apparent, from this example, that the packing factor is independent of atom size, if only one size is present. In contrast, the relative sizes do affect the packing factor when more than one type of atom is present. The face-centered cubic structure has the highest packing factor (0.74) that is possible for a pure metal, and thus this structure could also be called a *cubic close-packed* (ccp) structure. As we might expect, many metals have this structure, although we shall see in a moment that a hexagonal close-packed structure also has a packing factor of 0.74. The coordination number in an fcc metal is 12, which accounts for the high packing factor. (This may be contrasted to the CN of 8 and the PF of 0.68 of a bcc metal.) ◀

Example 3–3.2 Copper has an fcc structure and an atom radius of 1.278 Å. Calculate its density and check this value with the density listed in Appendix B.

Solution: Equation (3–3.3),

$$a = \frac{4}{\sqrt{2}}(1.278\ \text{Å}) = 3.61\ \text{Å}.$$

From Figure 3–3.5,

$$\frac{\text{Atoms}}{\text{Unit cell}} = \frac{8}{8} + \frac{6}{2} = 4,$$

$$\text{Density} = \frac{\text{Mass/Unit cell}}{\text{Volume/Unit cell}} \tag{3–3.4(a)}$$

$$= \frac{(\text{Atoms/Unit cell})(\text{g/atom})}{(\text{Lattice constant})^3}. \tag{3–3.4(b)}$$

$$\text{Density} = \frac{4[63.5/(0.602 \times 10^{24})]}{(3.61)^3 \times 10^{-24}} = 8.97\ \text{g/cm}^3.$$

The experimental value listed in Appendix B is 8.92 g/cm³. ◀

3–4 HEXAGONAL CRYSTALS

The structures of Figs. 3–4.1(a) and 3–4.1(b) are two representations of *hexagonal* unit cells. These cells have no internal positions which are equivalent to the corner positions. Although the volume of the cell is three times as great in Fig. 3–4.1(a) as in Fig. 3–4.1(b), there are three times as many atoms (3 versus 1); therefore the

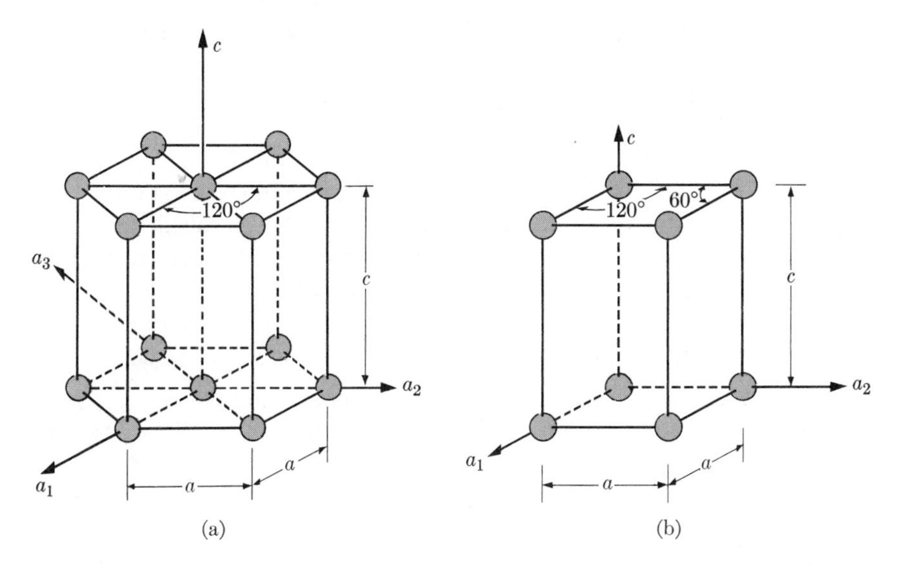

Fig. 3–4.1. Simple hexagonal unit cells. (a) Hexagonal representation. (b) Rhombic representation. The two are equivalent, with a ≠ c, a basal angle of 120°, and vertical angle of 90°.

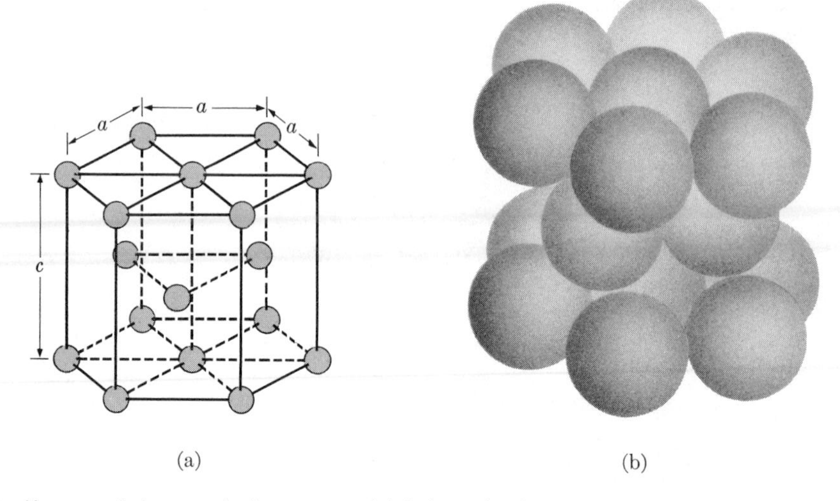

Fig. 3–4.2. Hexagonal close-packed structure. (a) Schematic view showing the location of atom centers. (b) Model made from hard balls.

number of atoms per unit volume remains the same. Metals do not crystallize with atoms arranged according to Fig. 3–4.1 because the packing factor is too low.

Hexagonal close-packed. The specific hexagonal structure formed by magnesium is shown in Fig. 3–4.2. Such a structure, which is more dense than that represented by Fig. 3–4.1, is called a *hexagonal close-packed* (hcp) structure. It is characterized by the fact that each atom in one layer is located directly above or below interstices among three atoms in the adjacent layers. Consequently, each atom touches three atoms in the layer below its plane, six atoms in its own plane, and three atoms in the layer above. There is an average of six atoms per unit cell in the hcp structure of Fig. 3–4.2 (or two per unit cell if we use the related rhombic representation).

The atomic packing factor for an hcp metal may be readily calculated, and is found to equal 0.74. This is identical to the packing factor of an fcc metal, which is predictable because each has a coordination number of 12.

• **Example 3–4.1** Assume spherical atoms. What is the c/a ratio of an hcp metal?

Solution: Refer to Fig. 3–4.2(a) and consider the three central atoms plus the one at the center of the top. This is an equilateral tetrahedron with edges $a = 2R$. From geometry,

$$h = a\sqrt{2/3}.$$

$$c = 2h = 2a\sqrt{2/3} = 1.63a. \tag{3–4.1}$$

Comments. The c/a ratios of hcp metals depart somewhat from this figure: Mg, 1.62; Ti, 1.59; Zn, 1.85. This means we must envision magnesium and titanium atoms as slightly compressed spheres, and zinc atoms as prolate spheroids. ◀

3–5 POLYMORPHISM

We recall from Section 2–3 that two molecules may possess different structures even though their compositions are identical. We called those molecules *isomers*. An analogous situation occurs in crystalline solids which will be extremely important to us. *Polymorphs* are two or more distinct types of crystals which have the same composition.* The most familiar example is the dual existence of graphite and diamond as two polymorphs of carbon.

The prime example of polymorphism in metals will be iron, since our whole ability to heat-treat steel and modify its properties stems from the fact that, as iron is heated, it changes from the bcc to an fcc lattice. Furthermore, the change is reversible as iron cools. At room temperature bcc iron has a coordination number of 8, an atomic packing factor of 0.68, and an atomic radius of 1.241 Å. Pure iron changes to fcc at 912°C, at which point its coordination number is 12, its atomic packing factor is 0.74, and its atomic radius is 1.29 Å. [At 912°C (1673°F) the atomic radius of bcc iron, due to thermal expansion, is 1.26 Å.]

Many other compositions have two or more polymorphic forms. In fact some, such as SiC, have as many as 20 crystalline modifications; however, this is unusual. Invariably, polymorphs have differences in density and other properties. In succeeding chapters we shall be interested in the property variations and in the time required to change from one crystal modification (phase) to another.

Example 3–5.1 Iron changes from bcc to fcc at 912°C (1673°F). At this temperature the atomic radii of the iron atoms in the two structures are 1.26 Å and 1.29 Å, respectively.

a) What is the percent of volume change, v/o, as the structure changes?

b) Of linear change, l/o?

[*Note:* As indicated in Section 2–6 and Table 2–6.1, the higher the coordination number the larger the radius.]

Solution: Basis: 4 iron atoms, or *two* unit cells of bcc iron, and *one* unit cell of fcc iron.

a) In bcc, Eq. (3–3.2):

$$\text{Volume} = 2a_{\text{bcc}}^3 = 2\left[\frac{4(1.26)}{\sqrt{3}}\right]^3 = 49.3\ \text{Å}^3.$$

In fcc, Eq. (3–3.3):

$$\text{Volume} = a_{\text{fcc}}^3 = \left[\frac{4(1.29)}{\sqrt{2}}\right]^3 = 48.6\ \text{Å}^3;$$

$$\frac{48.6 - 49.3}{49.3} = -1.4\ \text{v/o change.}$$

b)

$$(1 + \Delta L/L)^3 = 1 + \Delta V/V,$$

$$\Delta L/L = \sqrt[3]{1 - 0.014} - 1 = -0.47\ \text{l/o change.}$$

Iron expands by thermal expansion until it reaches 912°C, where there is an abrupt shrinkage; further heating continues the expansion (Fig. 10–1.1(a)). ◀

---

* When this is found in elemental solids such as metals, the term *allotropes* is sometimes used.

3–6 MULTIPLE-ATOM LATTICE SITES

Our examples in the previous sections presented crystals with only *one* atom per lattice point. These could have included aluminum for fcc, titanium for hcp, or chromium for bcc. It is also possible to have more than one atom per lattice point. First, we may find molecules such as CH_4 occupying lattice sites (Fig. 3–6.1). Of course, each methane molecule has five atoms.

The sodium chloride of Fig. 3–1.1 is fcc. As drawn, chloride ions are at the cube corners and at the center of each face. An equal number of sodium ions are also present, so there are two atoms, or ions, per fcc lattice point.* The structures of CsCl and β'-brass are shown in Fig. 3–6.2. At first glance, this would appear to be bcc; however, a quick check shows that the cube center and the cube corners are *not* equivalent, since they possess different types of atoms (or ions). This is a simple cubic structure with two atoms per lattice point.

Finally, let us consider diamond, in Fig. 3–6.3. It is identical with Fig. 2–2.4(b), except that we have shaded alternate atoms of the structure for identification purposes.† The unshaded atoms reveal an fcc lattice. The shaded atoms have the same number of neighbors (four) as the unshaded atoms. Note, however, that they are not strictly equivalent because the neighbors are not in the same directions from the unshaded atoms as they are from the shaded atoms. There are eight carbon atoms per fcc unit cell, which has four equivalent sites.

Example 3–6.1 Calculate the volume of the unit cell of LiF, which has the same structure as NaCl (Fig. 3–1.1).

Solution: Although LiF is fcc, we cannot use the geometry of Fig. 3–3.5(b), since the fluoride ions do not touch, as did the metal atoms. (*Check* Fig. 3–1.1 again.) From Appendix B:

$$a = 2(0.68 + 1.33)\ \text{Å},$$

$$a^3 = 65\ \text{Å}^3 \quad (= 65 \times 10^{-24}\ \text{cm}^3). \blacktriangleleft$$

Example 3–6.2 Cesium iodide (CsI) has the structures shown in Fig. 3–6.2. What is its packing factor if the radii are 1.72 Å and 2.3 Å, respectively?

Solution: The body diagonal of the unit cell is equal to $(2r + 2R)$. Therefore,

$$a = [2(1.72 + 2.3)]/\sqrt{3} = 4.64\ \text{Å}.$$

$$\text{Packing factor} = (4\pi/3)(1.72^3 + 2.3^3)/(4.64)^3 = 0.72.$$

Comments. The packing factor is greater than that obtained in the calculation following Eq. (3–3.1) because that simple cubic lattice did not contain a second (different) atom in the interstice.

The radii will differ from the radii of Appendix B, since its data are for CN = 6. Here CN = 8. ◀

---

* Since all Na^+ ions are in comparable surroundings, they also possess an fcc lattice (with the intervening Cl^- ions). However, the Na^+ and Cl^- sites are *not* equivalent to each other.

† They will also serve as a basis for the ZnS structures of Chapters 5 and 8.

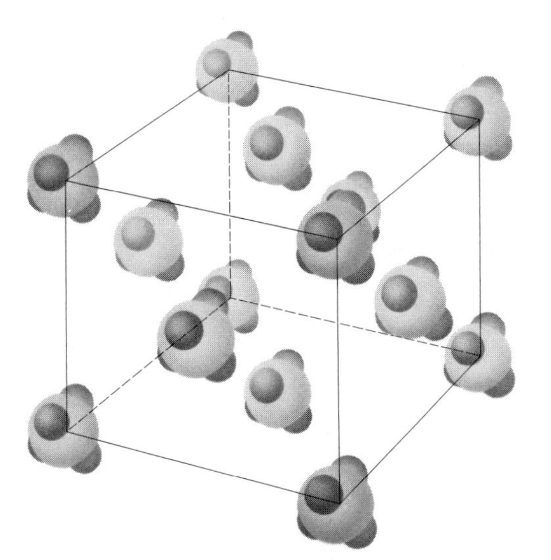

Fig. 3–6.1. Fcc compound (methane, CH_4). Each fcc lattice point contains a molecule of five atoms.

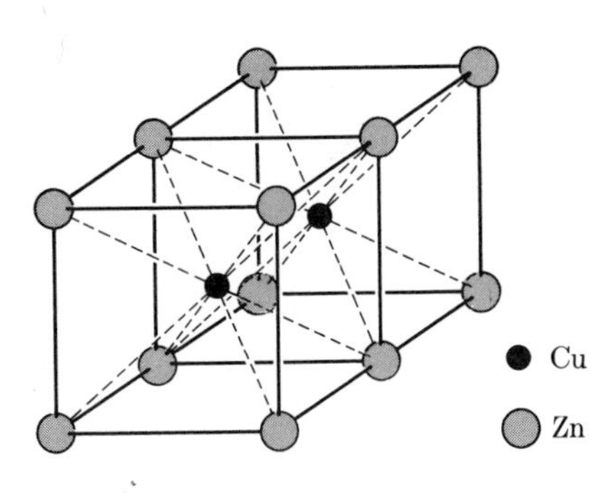

Fig. 3–6.2. Simple cubic compound (β' brass). Each copper atom is coordinated with eight zinc atoms; and each zinc atom is coordinated with eight copper atoms. The lattice is simple cubic, because the center site is not equivalent to the corner sites. The prototype for this structure is CsCl (Fig. 8–2.1(a)).

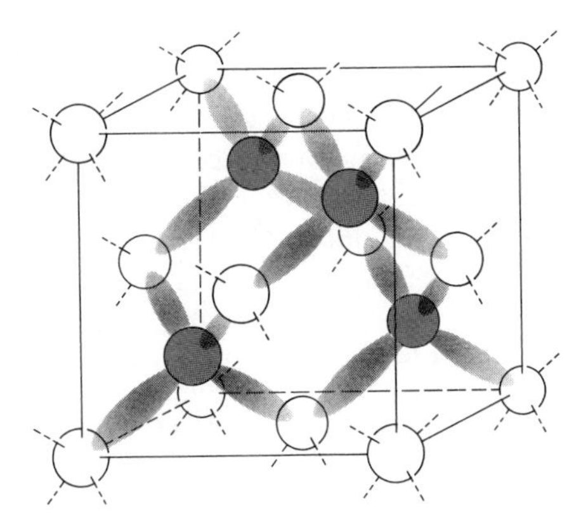

Fig. 3–6.3. Diamond (fcc). There are eight atoms and four equivalent points per unit cell. The four shaded atoms do not have neighbors in the same direction as the unshaded atoms do. Therefore the two sets are not fully equivalent.

3–7 CRYSTAL DIRECTIONS

When we correlate various properties with crystal structures in subsequent chapters, it will be necessary to identify specific crystal directions, because many properties are directional. Directions can be indexed relatively simply if the unit cell is used as a basis. The reference axes for the crystal are commonly placed so that a corner of the unit cell is at the origin.* The coefficients for the origin are 0,0,0. The coefficients for other *locations* are always expressed in unit-cell dimensions. Thus, the far corner of the unit cell is *always* 1,1,1, whether we are considering cubic cells, tetragonal cells, or one of the other systems of Table 3–2.1, or Fig. 3–2.2. Likewise, the center of the unit cell is always $\frac{1}{2},\frac{1}{2},\frac{1}{2}$ regardless of the system (Fig. 3–7.1(a)).

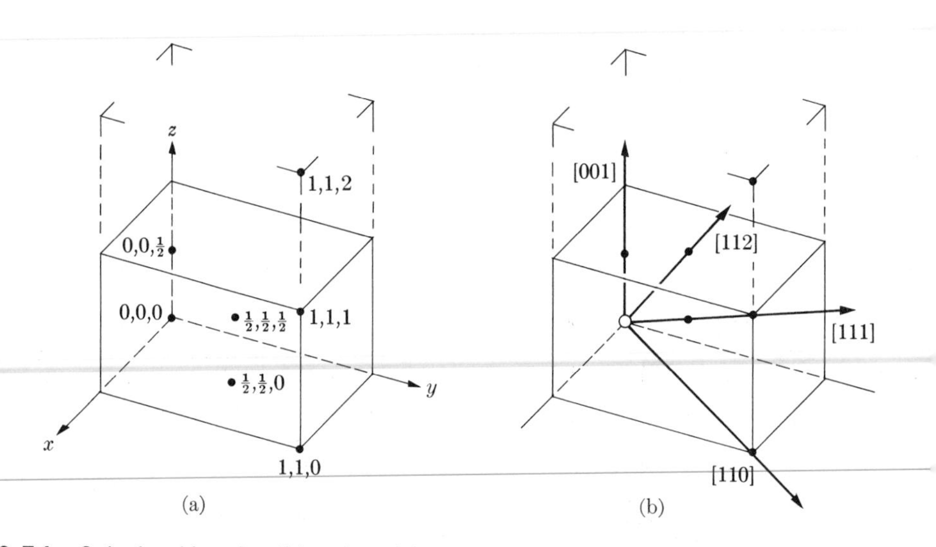

Fig. 3–7.1. Orthorhombic unit cell locations (a), and crystal directions (b). The location of the origin is commonly, but not necessarily, at the lower left rear corner. By convention we use square brackets [*hkl*] to indicate specific directions, and carets ⟨*hkl*⟩ to indicate a family of directions. This is to avoid confusion with indices for planes (*hkl*). Locations are indexed without closures: *u, v, w*.

The coefficients for *unit-cell locations* are expressed in sequence according to the *x*-, *y*-, and *z*-axes. Thus, a $\frac{1}{2},\frac{1}{2},0$ location is at the center of the lower face of the unit cell (if we use the orientation of axes which is generally accepted as standard for crystal geometry discussions). Likewise, a $0,0,\frac{1}{2}$ location is halfway along the *z*-axis.

---

* This is not necessary, but it is convenient. Furthermore, it is not critical which corner is used as the origin. The figures in this text generally use the left, lower, rear corner as the origin, simply because the sketch falls in the $+x$, $+y$, $+z$ octant, as we commonly view the axes. This procedure may be overruled if some other location is a more convenient origin for our purposes.

Before we continue to the main topic of this section, which is crystal directions, note that

0,0,0	0,0,1	0,1,0	1,0,0
1,1,1	1,1,0	1,0,1	0,1,1

are all equivalent, since each is a corner of the lattice (Fig. 3–2.2). Thus it is redundant to list them all, if we wish to simply list various unit-cell locations.* (See Example 3–7.1.)

Crystal directions are indexed simply as a *ray* extending from the origin through the locations with the lowest integer index. Thus the [111] direction passes from 0,0,0 through 1,1,1. Note, however, that this direction also passes through $\frac{1}{2},\frac{1}{2},\frac{1}{2}$ (and 2,2,2). Likewise the [112] passes through $\frac{1}{2},\frac{1}{2},1$; but for simplicity's sake, we use the integer notation. Observe that we enclose the direction indices in square brackets $[hkl]$, and use the letters h, k, and l for the indices arising from the three principal directions, x, y, and z, respectively. Parallel directions always have the same indices. Finally, note that we may have negative coefficients; a $[11\bar{1}]$ direction will have a vector in the minus z direction.

Angles between directions. In certain calculations (e.g., resolved shear stresses), it will be necessary to calculate the angle between two different crystal directions. For most of the calculations we will encounter, this may be performed by simple inspection. Thus, in Fig. 3–7.1, the angle between [110] and [112] directions (that is, [110] $\measuredangle$ [112]) is arctan $2c/\sqrt{a^2 + b^2}$. If that unit cell had been cubic rather than orthorhombic, so that $a = b = c$, the angle would have been arctan $2a/a\sqrt{2}$, or arccos $a\sqrt{2}/a\sqrt{6}$. In fact, with cubic crystals (*only*), we can determine cos $[hkl] \measuredangle [h'k'l']$ by the *dot product*. This latter procedure will be useful to us since most of our calculations in this introductory text will involve the simpler cubic crystals.

• **Linear densities.** Deformation occurs most readily in those directions with the shortest distance **b** between *equivalent* sites. In bcc crystals, $\mathbf{b}_{110}$ will be equal to $a\sqrt{2}$; in fcc crystals, $\mathbf{b}_{110}$ equals $a/\sqrt{2}$. (Check these out with Fig. 3–2.2.) It is often convenient to refer to the reciprocal of these distances, $\mathbf{b}_{hkl}^{-1}$, as the *linear density* of equivalent sites. We can also calculate the linear density of atoms. In most metals, the two values are the same. However, Example 3–7.3 shows that the two may differ in crystals with multiatom lattice sites.

• **Families of directions.** In the cubic crystal, the following directions are identical except for our arbitrary choice of the x-, y-, and z-labels on the axes:

$[111]$	$[11\bar{1}]$	$[1\bar{1}1]$	$[\bar{1}11]$
$[\bar{1}\bar{1}\bar{1}]$	$[\bar{1}\bar{1}1]$	$[\bar{1}1\bar{1}]$	$[1\bar{1}\bar{1}]$

* The 0,0,1 location of Fig. 3–7.1(a) would have been the 0,0,0 location, had we arbitrarily considered the next higher unit cell rather than the one sketched.

Any directional property* will be identical in these eight directions. Therefore, it is convenient to identify this *family* of directions as ⟨111⟩ rather than writing the eight individual directions. Note that the closure symbols are pointed brackets ⟨ ⟩.

Example 3–7.1 Index the unit-cell locations of atoms in fcc copper, if we place an atom at the origin.

Solution

$$0,0,0 \qquad 0,\tfrac{1}{2},\tfrac{1}{2} \qquad \tfrac{1}{2},0,\tfrac{1}{2} \qquad \tfrac{1}{2},\tfrac{1}{2},0.$$

Comments. There are only 4 atoms per unit cell (Example 3–3.2). The atoms at the other corners are simply translations (and serve as origins for adjacent unit cells). Likewise, $\tfrac{1}{2},\tfrac{1}{2},1$ is redundant with $\tfrac{1}{2},\tfrac{1}{2},0$. ◀

Example 3–7.2 a) What is the angle between the [111] and [001] directions in a cubic crystal? b) $[111] \measuredangle [\bar{1}\bar{1}1]$?

Solution: From the *cubic* analog of Fig. 3–7.1(b),

a)
$$\cos [111] \measuredangle [001] = a/a\sqrt{3},$$
$$[111] \measuredangle [001] = 54.75°.$$

b) Observe that [001] bisects $[111] \measuredangle [\bar{1}\bar{1}1]$; therefore

$$[111] \measuredangle [\bar{1}\bar{1}1] = 2(54.75°) = 109.5°;$$

or, by the dot product (*since the crystal is cubic*):

$$\cos [111] \measuredangle [\bar{1}\bar{1}1] = -\tfrac{1}{3},$$
$$[111] \measuredangle [\bar{1}\bar{1}1] = 109.5°.$$

Comments. Compare the results here with Fig. 2–3.3(a) where all four of the bond angles are 109.5°. ◀

• **Example 3–7.3** The lattice constant a is 3.57 Å for diamond, which is cubic. a) What is the linear density of equivalent sites in the $[11\bar{1}]$ direction? b) Of atoms?

Solution: Refer to Fig. 3–6.3. Use the *upper*, rear, left corner as the origin.

a) In the $[11\bar{1}]$ direction, the repeating distance 0,0,0 to 1,1,−1 is $a\sqrt{3}$, or 6.18 Å; therefore,

$$\mathbf{b}^{-1}_{[11\bar{1}]} = 16.1 \times 10^6/\text{cm}.$$

b) There are atoms at 0,0,0 and $\tfrac{1}{4},\tfrac{1}{4},-\tfrac{1}{4}$; therefore,

$$2 \text{ atoms}/a\sqrt{3} = 32 \times 10^6/\text{cm}.$$

---

* For example, Young's modulus, magnetic permeability, index of refraction, etc.

Comment. The location of the $\frac{1}{4},\frac{1}{4},-\frac{1}{4}$ atom can be checked by relating it to its four neighbors at

$$0,0,0 \qquad \tfrac{1}{2},\tfrac{1}{2},0 \qquad \tfrac{1}{2},0,-\tfrac{1}{2} \qquad 0,\tfrac{1}{2},-\tfrac{1}{2}.$$ ◀

3–8 CRYSTAL PLANES

A crystal contains planes of atoms; these influence the properties and behavior of a material. Thus it will be advantageous to identify various planes within crystals.

The lattice planes most readily visualized are those which outline the unit cell, but there are many other planes. The more important planes for our purposes are those sketched in Figs. 3–8.1, 3–8.2, and 3–8.3. These are labeled (010), (110), and $(\bar{1}11)$ respectively, where the numbers within the parentheses (hkl) are called the *Miller indices.*

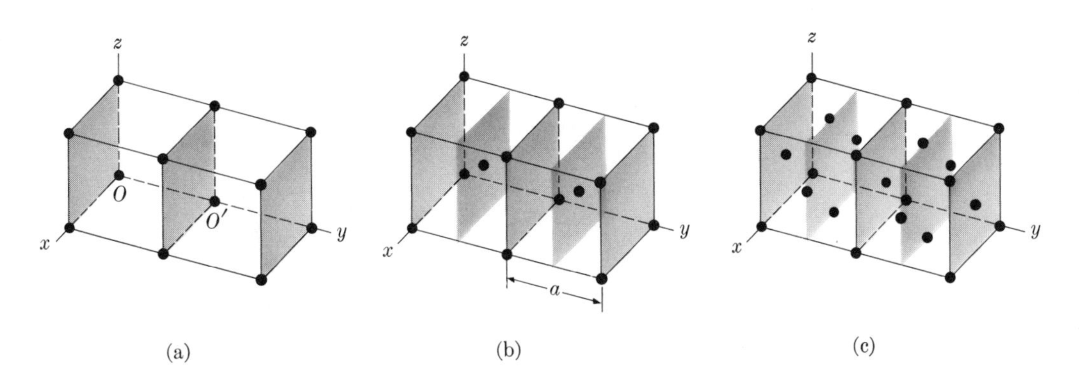

Fig. 3–8.1. (010) planes in cubic structures. (a) Simple cubic. (b) Bcc. (c) Fcc. (Note that the (020) planes included for bcc and fcc are equivalent to (010) planes.)

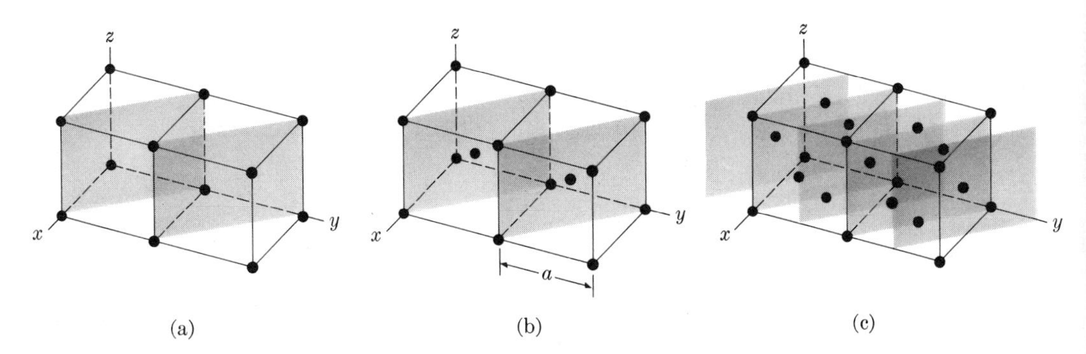

Fig. 3–8.2. (110) planes in cubic structures. (a) Simple cubic. (b) Bcc. (c) Fcc. (The (220) planes included for fcc are equivalent to (110) planes.)

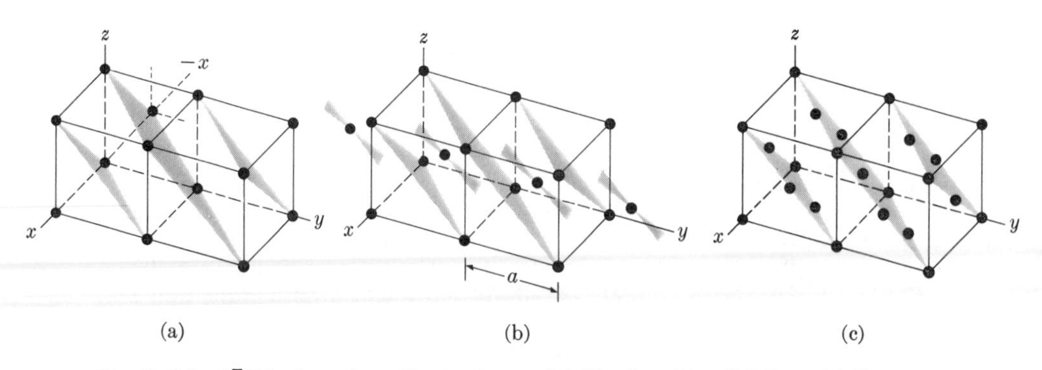

Fig. 3–8.3. ($\bar{1}11$) planes in cubic structures. (a) Simple cubic. (b) Bcc. (c) Fcc. Negative intercepts are indicated by bars above the index. (The ($\bar{2}22$) planes included for bcc are equivalent to ($\bar{1}11$) planes.)

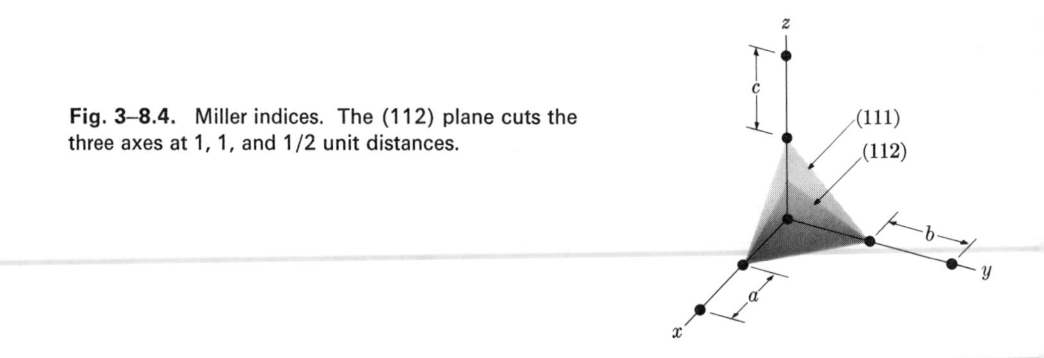

Fig. 3–8.4. Miller indices. The (112) plane cuts the three axes at 1, 1, and 1/2 unit distances.

We may use the plane with the darker tint in Fig. 3–8.4 to explain how (*hkl*) numbers are obtained. The plane intercepts the *x*-, *y*-, and *z*-axes at **1***a*, **1***b*, and **0.5***c*. The Miller indices are simply the reciprocals of these intercepts: (112). The plane of lighter tint in Fig. 3–8.4 is the (111) plane, since it intercepts the axes at 1*a*, 1*b*, and 1*c*. Returning to the earlier figures, we have:

Figure	Plane	Intercepts	Miller indices
3–8.1(a)	Middle	∞a, $1b$, ∞c	(010)
3–8.2(a)	Left	$1a$, $1b$, ∞c	(110)
3–8.3(a)	Middle	$-1a$, $1b$, $1c$	($\bar{1}11$)

Note that a *minus intercept* is handled readily with an overbar. Furthermore, observe that we use parentheses (*hkl*) to denote planes (and no commas), in order to avoid confusion with individual directions which were denoted in Section 3–7 with square brackets: [*hkl*].

All parallel planes are identified with the same indices. Figure 3–8.1(a) will explain this:

Plane	Origin	Intercepts	Miller indices
Right	0	$\infty a, 2b, \infty c$	$(0\frac{1}{2}0)$
Right	0′	$\infty a, 1b, \infty c$	(010)
Left	0′	$\infty a, -1b, \infty c$	$(0\bar{1}0)$

Since the choice of the origin, O or O', is arbitrary, we can use either $(0\frac{1}{2}0)$ or (010) to index the right plane. The (010) indices are simpler to say and write; therefore we use them. This is entirely permissible since these three shaded planes are geometrically identical and behave the same during plastic deformation (Chapter 6) and under other circumstances. We can carry this one step further in Fig. 3–8.1(c):

Plane	Intercepts	Miller indices
3rd from left	$\infty a, 1b, \infty c$	(010)
2nd from left	$\infty a, 0.5b, \infty c$	(020) = (010)

Again these planes are parallel *and identical* (but offset $\frac{1}{2}$ in the x-direction and $\frac{1}{2}$ in the z-direction). Ordinarily, we do not need to distinguish between the (010), $(0\frac{1}{2}0)$, and (020) planes. Furthermore, in Fig. 3–8.2(c), the (220) plane is identical to the (110) plane, and the $(\bar{2}22)$ plane is the same as the $(\bar{1}11)$ plane (Fig. 3–8.3(b)). As a result, we use the indices with the lowest set of integers.*

• **Crystal forms.** Depending on the crystal system, two or more planes may possess the same geometry, just as two or more directions may belong to the same family of directions. In the cubic system, an example of multiple planes includes the following, which constitute a *form*:

$$\begin{vmatrix} (100) & (010) & (001) \\ (\bar{1}00) & (0\bar{1}0) & (00\bar{1}) \end{vmatrix} = \{100\}.$$

The collective notation for a crystal form is $\{hkl\}$. Figure 3–1.1 provides us with an illustration of the {100} form which has the six planes listed above. Each face is identical except for the consequences of our arbitrary choice of axis labels and directions. The reader is asked to verify that the {111} form includes eight planes, and the {110} form includes 12 planes, when all of the permutations and combinations of individual planes are included.

* However, in Section 3–9 we will distinguish between d_{010} and d_{020} when we examine *interplanar spacings*.

• **Indices for hexagonal crystals (*hkil*).** The three Miller indices (*hkl*) can describe all possible planes through any crystal. In hexagonal systems, however, it is frequently useful to establish four axes, three of them coplanar (Fig. 3–4.1(a)). This leads to four intercepts and (*hkil*) indices.* The fourth index i is an additional index which mathematically is related to the sum of the first two:

$$h + k = -i. \tag{3–8.1}$$

These optional (*hkil*) indices are generally favored because they reveal hexagonal symmetry more clearly. Although partially redundant, they are used almost exclusively, in preference to the equivalent (*hkl*) indices, in scientific papers which discuss the behavior of hexagonal crystals.

Planar densities of atoms. When we consider plastic deformation later, we will need to know the density of atoms on a crystal plane. Example 3–8.3 shows how we may calculate these by means of the relationship:

$$\text{Planar density} = \frac{\text{Atoms}}{\text{Unit area}}. \tag{3–8.2}$$

As we did in dealing with linear density, we may also calculate the planar density of equivalent points.

Example 3–8.1 Sketch a (111) plane through a unit cell of a simple tetragonal crystal having a c/a ratio of 0.62.

Solution: Figure 3–8.5 shows this plane. The (111) plane cuts the three axes at unit distances. However, the unit distance along the z-axis is shorter than the unit distances on the x- and y-axes. ◀

Example 3–8.2 Sketch two planes with (122) indices on the axes of Fig. 3–8.5.

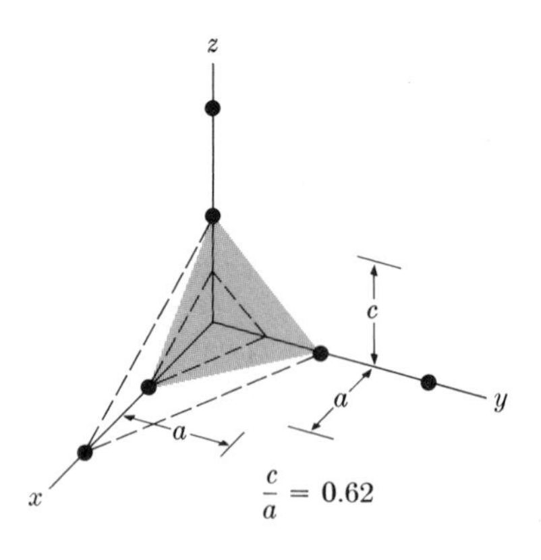

Fig. 3–8.5. Noncubic intercepts (tetragonal structure). The shaded (111) plane cuts the three axes of any crystal at equal unit distances. However, since *c* may not equal *a*, the actual intercepting distances are different. (The dashed lines refer to Example 3–8.2.)

---

* Called Miller–Bravais indices.

Solution: The intercepts will be the reciprocals of (122)—$1a$, $0.5b$, and $0.5c$. This is shown by the inner set of dashed lines. A plane with intercepts—$2a$, $1b$, and $1c$—is parallel and therefore has the same indices (outer set of dashed lines). ◀

Example 3–8.3 How many atoms per mm² are there on the (100) and (111) planes of lead (fcc)?

Solution

$$\text{Pb radius} = 1.750\ \text{Å} \qquad \text{(from Appendix B)},$$

$$a_{\text{Pb}} = \frac{4R}{\sqrt{2}} = \frac{4(1.750\ \text{Å})}{1.414} = 4.95\ \text{Å}.$$

Figure 3–8.6 shows that the (100) plane contains two atoms per unit-cell face.

$$(100):\quad \text{atoms/mm}^2 = \frac{2\ \text{atoms}}{(4.95 \times 10^{-7}\ \text{mm})^2} = 8.2 \times 10^{12}\ \text{atoms/mm}^2.$$

Figure 3–8.7(b) shows that the (111) plane contains three one-sixth atoms in the triangular shaded area.

$$(111):\quad \text{atoms/Å}^2 = \frac{\frac{3}{6}}{\frac{1}{2}bh} = \frac{\frac{3}{6}}{\frac{1}{2}(2)(1.750\ \text{Å})(\sqrt{3})(1.750\ \text{Å})}$$

$$= 0.095\ \text{atoms/Å}^2 = 9.5 \times 10^{12}\ \text{atoms/mm}^2. \quad ◀$$

Fig. 3–8.6. (100) atom concentration (fcc). A (100) plane in an fcc structure has two atoms per a^2.

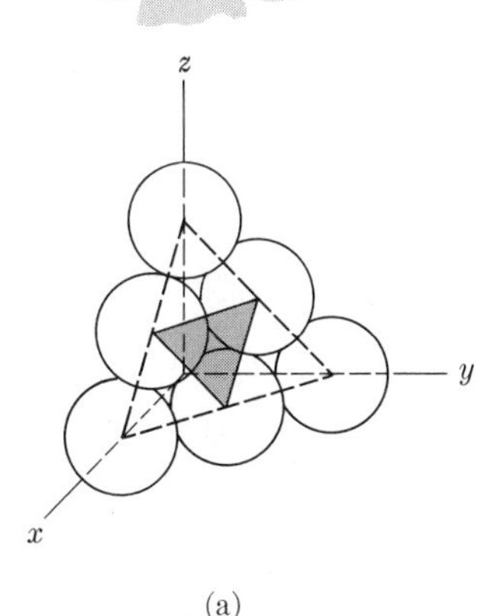

(a)

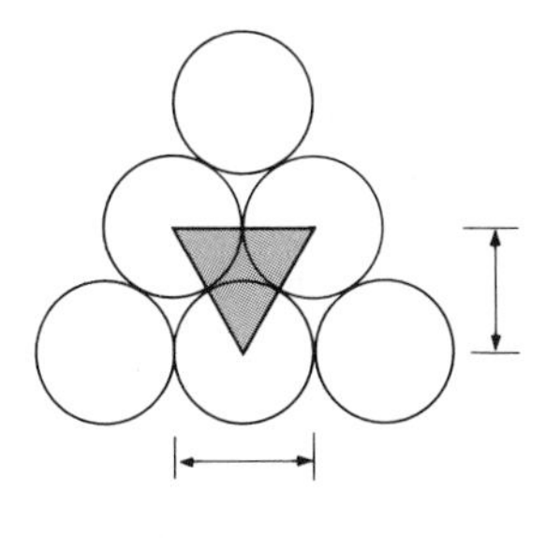

(b)

Fig. 3–8.7. (111) atom concentration (fcc). A (111) plane has one-half atom per $\sqrt{3}\,R^2$.

•3–9 X-RAY DIFFRACTION

Excellent experimental verification for the crystal structures which we have been discussing is available through x-ray diffraction. When these high-frequency electromagnetic waves are selected to have a wavelength slightly greater than the *interplanar* spacings of crystals, they are diffracted according to very exacting physical laws. The angles of diffraction let us decipher crystal structures with a high degree of accuracy. In turn, one can readily determine the interplanar spacings (and therefore atomic radii) in metals to four significant figures, and even more precisely, if necessary. Let us first examine the spacings between planes. Then we will turn to diffraction.

Interplanar spacings. Recall, from Section 3–8, that all parallel planes bear the same (hkl) notation. Thus the several (110) planes of Fig. 3–8.2 have still another (110) plane which passes directly through the origin. As a result, if we measure a *perpendicular* distance from the origin to the next adjacent (110) plane, we have measured the interplanar distance, d. We will observe that in the simple cubic structures of Figs. 3–8.1(a), 3–8.2(a), and 3–8.3(a), the interplanar distances are a, $(a\sqrt{2})/2$, and $(a\sqrt{3})/3$ for d_{010}, d_{110}, and $d_{\bar{1}11}$, respectively. That is, there is one spacing per cell edge a for d_{010}; two spacings per face diagonal, $a\sqrt{2}$, for d_{110}; and three spacings per body diagonal, $a\sqrt{3}$, for $d_{\bar{1}11}$. We may formulate a general rule for d-spacings in *cubic* crystals:

$$d_{hkl} = \frac{a}{\sqrt{h^2 + k^2 + l^2}}, \tag{3–9.1}$$

where a is the lattice constant and h, k, and l are the indices of the planes.* The interplanar spacings for noncubic crystals may be expressed with an equation that is related to Eq. (3–9.1), but takes into account the variables of Table 3–2.1.

Bragg's law. When x-rays encounter a crystalline material, they are diffracted by the planes of the atoms (or ions) within the crystal. The *diffraction angle* θ depends upon the wavelength λ of the x-rays and the distance d between the planes:

$$n\lambda = 2d \sin \theta. \tag{3–9.2}$$

Consider the parallel planes of atoms in Fig. 3–9.1, from which the wave is diffracted. The waves may be "reflected" from the atom at H or H', and remain in phase at K. However, x-rays are reflected not only from the surface plane, but also from the adjacent subsurface planes. If these reflections are to remain in phase and be *coherent*, the distance $MH''P$ must equal one or more integer wavelengths of the rays. The value n of Eq. (3–9.2) is the integer number of waves which occur in the distance $MH''P$.

---

* The reciprocal nature of Miller indices permits this type of simplified calculation. Likewise, the cross-product of the Miller indices for two intersecting planes gives the direction indices for the line of intersection. Finally, in order for a direction to lie in a plane, the dot product of the indices for that direction and for the plane must equal zero. In brief, there is purpose in using the "down-side up" Miller indices.

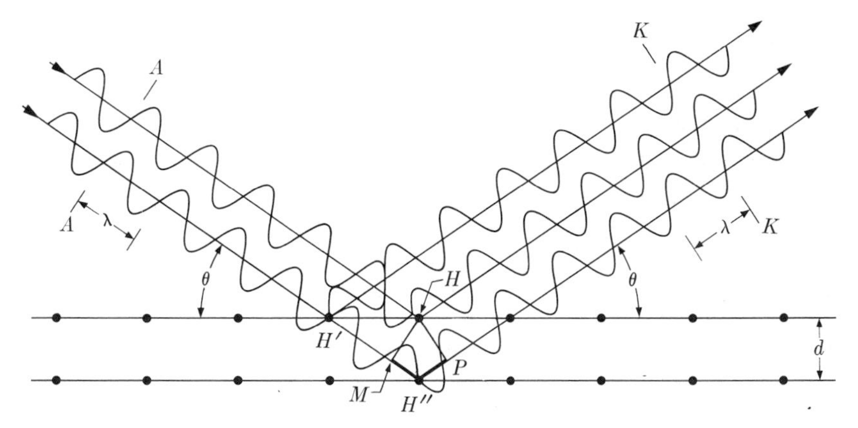

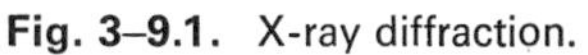

Fig. 3–9.1. X-ray diffraction.

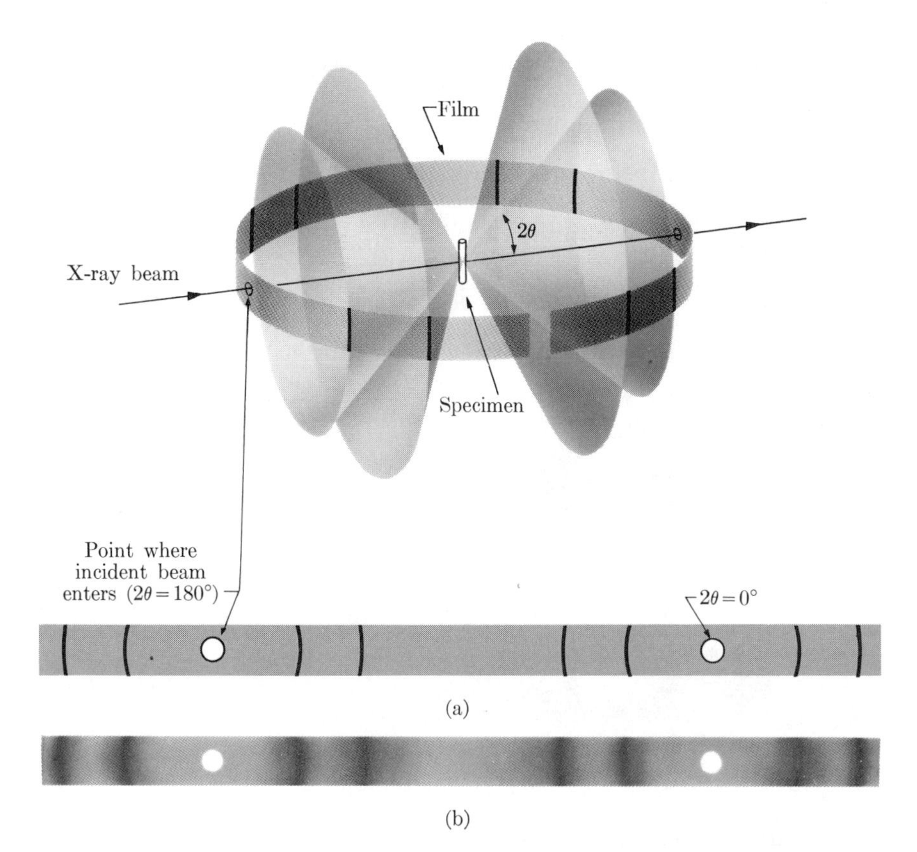

Fig. 3–9.2. The exposure of x-ray diffraction patterns. Angle 2θ is precisely fixed by the lattice spacing d and the wavelength λ as shown in Eq. (3–9.2). Every cone of reflection is recorded in two places on the strip of film. (B. D. Cullity, *Elements of X-ray Diffraction.*)

Diffraction analyses. The most common procedure for making x-ray diffraction analyses utilizes very fine powder of the material in question. It is mixed with a plastic cement and formed into a very thin filament which is placed at the center of a circular camera (Fig. 3–9.2). A collimated beam of x-rays is directed at the powder. Since there is a very large number of powder particles with all possible orientations, the diffracted beam emerges as a cone of radiation at an angle 2θ from the initial beam. (Observe in Fig. 3–9.1 that the diffracted beam is 2θ away from the initial beam.)

The diffraction cone exposes the film strip in the camera at two places; each is 2θ from the straight-through exit port. There is a separate cone (or pair of *diffraction lines*) for each d_{hkl} value of interplanar spacings. Thus, the diffraction lines may be measured and the *d*-spacings calculated from Eq. (3–9.2). All fcc metals will have a similar set of diffraction lines, but with differing 2θ values, since they have different lattice constants; for example, $a_{\text{Cu}} = 3.615$ Å; $a_{\text{Al}} = 4.049$ Å; $a_{\text{Pb}} = 4.95$ Å, etc. Thus, we can differentiate between various fcc metals.

In Fig. 3–9.3, we see x-ray diffraction films for copper (fcc), tungsten (bcc) and zinc (hcp). It is immediately apparent that the sequences of diffraction lines are

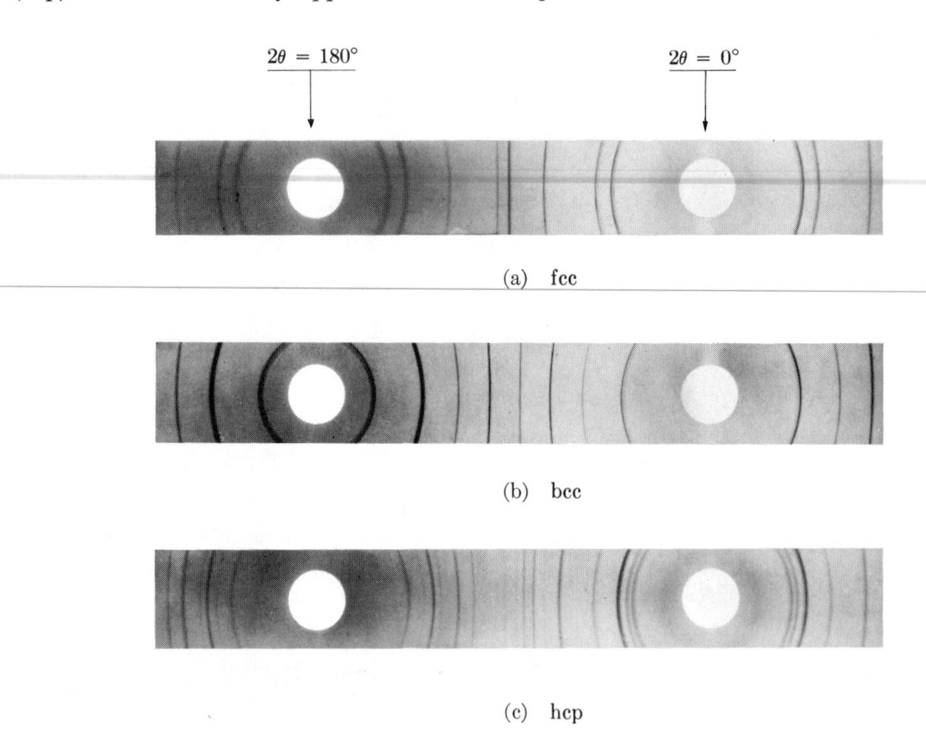

Fig. 3–9.3. X-ray diffraction patterns for (a) copper, fcc, (b) tungsten, bcc, and (c) zinc, hcp. The crystal structure and the lattice constants may be calculated from patterns such as these. (B. D. Cullity, *Elements of X-ray Diffraction*.)

different for the three types of crystals. Since we lack space here to explain these differences, we will have to simply observe that, with different "fingerprints," it is possible not only to determine the size of the lattice constants, with utmost precision, but also to identify the crystal lattice. X-ray diffraction is an extremely powerful tool in the study of the internal structure of materials.

Example 3–9.1 X-rays of an unknown wavelength are diffracted 43.4° by copper (fcc) whose lattice constant a is 3.615 Å. Separate determinations indicate that this diffraction line for copper is the first-order ($n = 1$) line for d_{111}.

a) What is the wavelength of the x-rays?

b) The same x-rays are used to analyze tungsten (bcc). What is the angle, 2θ, for the second-order ($n = 2$) diffraction lines of the d_{010} spacings?

Solution

a) Since $2\theta = 43.4°$,

$$(1)\lambda = 2\left[\frac{3.615 \text{ Å}}{\sqrt{1^2 + 1^2 + 1^2}}\right] \sin 21.7°$$

$$= 1.54 \text{ Å}.$$

b) From Appendix B,

$$R_W = 1.367 \text{ Å}.$$

From Eq. (3–3.2),

$$a_W = 4(1.367 \text{ Å})/\sqrt{3} = 3.157 \text{ Å}.$$

From Eq. (3–9.1),

$$\sin\theta = 2(1.54 \text{ Å})(\sqrt{0 + 1 + 0})/(2)(3.157 \text{ Å}),$$

$$2\theta = 58.4°.$$

Comments. The second-order diffraction of d_{010} is equivalent to the first-order diffraction of a d_{020} spacing; i.e., the perpendicular distance from the origin to a plane which cuts the x-, y-, and z-axes at ∞, $\frac{1}{2}$, and ∞, respectively. This may be checked with Eqs. (3–9.1) and (3–9.2).

The d_{hkl}-spacings for all planes of the same form are equal. For example, $d_{100} = d_{010} = d_{001}$ in the cubic system. ◀

REVIEW AND STUDY

In Chapter 2 we considered the bonding of atoms with their immediate neighbors and the resulting coordination. This chapter carries our understanding of internal structure one step further by looking at longer-range geometric order. Since all atoms of one kind have similar coordination requirements, it is not surprising that we find the same patterns repeated time and time again throughout the material.

SUMMARY

We may divide crystalline solids into unit cells with each having all the characteristics of atomic coordination found throughout the crystal. Comparable positions in each unit cell have the *same neighbors* in the *same directions* and at the *same distances.* When we consider properties, this simplifies our description of the internal structure, because we only have to describe the unit cell.

There are seven crystal systems based on the geometry of the unit cell (unit-cell dimensions and axial angles). These seven systems possess a total of 14 lattices, based on the internal arrangement of the equivalent sites within the unit cell. Our attention was directed primarily toward the body-centered cubic (bcc), face-centered cubic (fcc) and a common structure of the hexagonal lattice called the hexagonal close-packed (hcp). On occasion, we will also consider body-centered tetragonal (bct) and simple cubic (sc) lattices. Cubic metals commonly have only one atom per lattice site. In contrast, noncubic metals and compounds have more than one atom per lattice site.

Crystal directions are indexed on the basis of the unit cell dimensions. Thus, a [111] direction is a ray from the origin, 0,0,0, through the center of the unit cell and out to the far corner, regardless of the crystal system. Families of directions $\langle hkl \rangle$ include all directions which are identical except for our arbitrary choice of coordinate references.

Crystal planes are indexed by the reciprocal of the axial intercepts (followed by the elimination of fractions). We will give the greatest, but not exclusive, attention to the (100), (110), and (111) planes; therefore, the reader should be familiar with the following facts:

1. Any plane of the {100} form parallels two of the coordinate axes and cuts the third;
2. Any plane of the {110} form parallels one axis and cuts the other two with equal intercept coefficients; and
3. The planes of the {111} form cut all three axes with equal intercept coefficients.

Since this is not a course in crystallography, we did not discuss in detail how crystal structures are established and verified in the laboratory. However, we did note that use is made of x-ray diffraction by the crystal lattice. The diffraction angles are interpreted by Bragg's law. There are few if any structural characteristics which can be measured with a greater degree of certainty than these crystal parameters.

TERMS FOR REVIEW

Atomic packing factor, APF

• Bragg's law

Body-centered cubic, bcc

Crystal

Crystal direction $[hkl]$

Crystal plane (hkl)

Crystal system

• Diffraction lines

TERMS FOR REVIEW—CONTINUED

Equivalent points
Face-centered cubic, fcc
• Family of directions $\langle hkl \rangle$
• Form $\{hkl\}$
Hexagonal close-packed, hcp
• Interplanar spacing, d_{hkl}
Lattice
Lattice constant
Long-range order
Miller indices
NaCl structure
Polymorphism
Simple cubic, sc
Unit cell
• X-ray diffraction

FOR DISCUSSION

A_3 What is meant by long-range order?

B_3 Cite the several crystal systems and their distinguishing characteristics.

C_3 How many equivalent sites are there per bcc unit cell? Fcc unit cell?

D_3 The corner site of a cubic unit cell is associated with how many unit cells? The corner site of a rectangular area of a plane belongs to how many "unit areas"?

E_3 From Fig. 3–3.5, show that CN = 12 for fcc.

F_3 A rhombohedral lattice has ____ equivalent sites per unit cell.

G_3 What is the ratio of volumes for the two presentations of the hexagonal unit cell in Fig. 3–4.1? The ratio of equivalent sites?

H_3 Diamond requires a higher pressure, to be stable, than does graphite. What does this indicate about their densities?

I_3 From Fig. 3–3.5, show that fcc iron could be categorized as a body-centered tetragonal (bct) with a c/a ratio of 1.414. Why do we use fcc rather than bct?

J_3 Gray tin has the structure of diamond (Figs. 2–2.4 and 3–6.3). White tin has a body-centered tetragonal structure with atoms at:

$$0,0,0 \qquad \tfrac{1}{2},\tfrac{1}{2},\tfrac{1}{2} \qquad \tfrac{1}{2},0,\tfrac{1}{4} \qquad 0,\tfrac{1}{2},\tfrac{3}{4}$$

Sketch the unit cell. Show that the second and third sites listed above are not equivalent, but the third and fourth sites are equivalent.

•K_3 Show that the hexagonal close-packed unit cell is not a separate space lattice, but rather a special case of the hexagonal lattice. (Compare the equivalency, or lack thereof, for an atom in the middle layer with one in the upper or lower layers.)

•L_3 How many directions are in the $\langle 012 \rangle$ family of a cubic crystal? (We usually consider $[hkl]$ and $[\overline{hkl}]$ to be the same, since a reversal of all indices simply follows the same ray but in the opposite sense; that is, $[\overline{hkl}] = -[hkl]$.)

•M_3 Repeat L_3 but for a tetragonal crystal. (Recall $c \neq a$ in tetragonal; therefore we cannot permutate the 2 with the 0 and 1 here, as we can in the cubic system.)

•N_3 List all the (hkl) planes in the $\{110\}$ form of a cubic crystal; a tetragonal crystal.

O_3 The [111] direction is normal to the (111) plane in a *cubic* crystal, but not in a tetragonal crystal. Why?

P_3 Which of the following directions lie in the (110) plane:

$$[112] \quad [1\bar{1}0] \quad [001] \quad [1\bar{1}2] \quad [8\bar{8}9]?$$

•Q_3 Place another plane midway between the two traces shown in Fig. 3–9.1. Is the resulting diffraction increased (or eliminated)?

•R_3 With Fig. 3–8.2(c), prove to yourself that second-order ($n = 2$) diffraction for d_{110} equals first-order ($n = 1$) diffraction for d_{220}.

STUDY PROBLEMS

3–2.1 The unit cell of aluminum is a cube with $a = 4.049$ Å. a) How many unit cells are in an aluminum foil 0.005 cm thick and 25 cm square? b) It weighs 8.44 g. How many atoms are present? c) How many per unit cell?

Answer: a) 47×10^{21} unit cells b) 188×10^{21} atoms c) 4

3–2.2 Sketch a "face-centered tetragonal" lattice of four equivalent points per unit cell. This lattice is not in Fig. 3–2.2 because the same points are presented under one of the basic 14 lattices. a) Which one? b) How many equivalent points per unit cell?

3–2.3 An fcc lattice may be presented alternately as a rhombohedral lattice. a) How do the two lattices compare? b) What is the axial angle? c) What is the ratio of unit cell volumes? •d) Why do we prefer the fcc over the rhombohedral?

Answer: a) $a_{\text{fcc}} = a_{\text{rh}}\sqrt{2}$ b) 60° c) $V_{\text{fcc}} = 4V_{\text{rh}}$ (compare the number of atoms per unit cell.) •d) The cubic lattice is more symmetrical

3–2.4 There are fourteen 3-dimensional lattices (Fig. 3–2.2). a) List the five 2-dimensional lattices. b) Why do we *not* consider the "area-centered square" to be a separate 2-dimensional lattice?

3–3.1 Show in tabular form the relationship between atom radii and unit-cell dimensions for face-centered, body-centered, and simple cubic metals. (Assume one atom per lattice site.)

	FCC	BCC	SC
Side of unit cell	$a = 4R/\sqrt{2}$		
Face diagonal			
Body diagonal			

3–3.2 Lead is fcc and its atomic radius is 1.750×10^{-8} cm. What is the volume of its unit cell?

3–3.3 Silver is fcc and its atomic radius is 1.444 Å. How large is the side of its unit cell?

Answer: 4.084 Å

3–3.4 Gold is face-centered cubic. Its lattice constant is 4.076 Å and its atomic weight is 197.0. a) Calculate its density. b) Check the value in a handbook.

3–3.5 Calculate the density of chromium from the data in Appendix B.

Answer: 7.198 g/cm^3

3–3.6 Calculate the density of calcium from data in Appendix B.

3–4.1 Zinc has an hcp structure. The height of the unit cell is 4.94 Å. The centers of the atoms in the base of the unit cell are 2.665 Å apart. a) How many atoms are there per hexagonal unit cell? (Show reasoning.) b) What is the volume of the hexagonal unit cell? c) Would the calculated density be greater or less than the actual density of 7.135 g/cm^3? (Justify your answer.)

Answer: a) 6 b) 9.1×10^{-23} cm^3 c) 7.15 g/cm^3. (This makes no allowance for imperfections.)

3–4.2 Magnesium is hcp and has nearly spherical atoms with a radius of $\sim$1.61 Å. What is its density? [*Hint:* $c/a = 1.62$.]

• **3–4.3** Refer to the data of Study Problem 3–4.1. a) What is the c/a ratio for zinc? b) What are the *two* interatomic distances between adjacent atoms?

Answer: a) 1.85 b) 2.665 Å, 2.91 Å

3–5.1 Titanium is bcc in its high-temperature form. The radius increases 2% when the bcc changes to hcp during cooling. What is the percentage volume change? (*Recall* that there will be a change in APF.)

Answer: -2.5 v/o ($= -0.8$ l/o)

3–5.2 Graphite has a density of 2.25 g/cm^3. How much expansion will there be if 1 mg of diamond changes to graphite? (*Hint:* The interatomic distance in Fig. 3–6.3 is 1.54 Å.)

3–6.1 Sodium and chlorine ions weigh 22.99 and 35.45 amu, respectively. If the density is 2.165 g/cm^3, calculate the dimensions of the cubic unit cell of NaCl.

Answer: 5.64 Å

• **3–6.2** The hcp structure is not among the 14 lattices of Fig. 3–2.2. a) Why? b) To which lattice does it belong? c) How many atoms of magnesium (hcp) are there per lattice site? [*Hint:* Extend a line from the center, top site of Fig. 3–4.2(a) through the front site in the middle layer.]

3–6.3 What is the distance between centers of adjacent molecules of solid methane (fcc and $\rho = 0.415$ g/cm^3)?

Answer: 4.5 Å

3–7.1 Draw a line from $\frac{1}{2},\frac{1}{2},0$ to the center of the *next* unit cell, which could be indexed $\frac{1}{2},\frac{3}{2},\frac{1}{2}$. What is the direction?

Answer: [021]

3–7.2 What is the value of [110] ∢ [011] in a cubic crystal?

• 3–7.3 a) What is the linear density of atoms along the [112] direction of iron? b) Of nickel?

Answer: a) 1.42×10^7 atom/cm b) 2.32×10^7 atom/cm

• 3–7.4 List the directions of the ⟨110⟩ family (a) in a cubic crystal; (b) in a tetragonal crystal.

3–8.1 a) How many atoms are there per square millimeter on a (100) plane of copper; (b) on a (110) plane? (c) on a (111) plane?

Answer: a) 15.3×10^{12} Cu/mm^2 b) 10.8×10^{12} Cu/mm^2 c) 17.7×10^{12} Cu/mm^2

3–8.2 Sketch a unit cell of silver (fcc). a) Shade in the (012) plane. b) What is the planar density of atoms?

3–8.3 a) Sketch a unit cell of chromium (bcc). a) Shade the (102) plane. b) What is the planar density of atoms?

Answer: b) 5.4×10^{14}/cm^2

• 3–8.4 Sketch those ⟨110⟩ directions which lie in the (111) plane.

3–8.5 What is the line of intersection of the (111) and (101) planes?

Answer: $[10\bar{1}]$, or its inverse

• 3–8.6 List the planes which belong to the {012} form of a cubic crystal.

• 3–8.7 What are the (*hkil*) planes which make up the hexagonal prism of Fig. 3–4.1(a)?

Answer: $(10\bar{1}0)$, $(1\bar{1}00)$, $(01\bar{1}0)$, and their negative parallels.

• 3–9.1 The distance between (110) planes in a body-centered cubic structure is 2.03 Å. a) What is the size of the unit cell? b) What is the radius of the atoms? c) What might the metal be?

Answer: a) 2.87 Å b) 1.24 Å c) bcc iron or Cr (not Ni)

• 3–9.2 Nickel is face-centered cubic with an atom radius of 1.246 Å. a) What is the d_{200} spacing? b) The d_{220} spacing? c) The d_{111} spacing?

• 3–9.3 The lattice constant for a unit cell of aluminum is 4.049 Å. a) What is d_{220}? b) d_{111}? c) d_{200}?

Answer: a) 1.431 Å b) 2.337 Å c) 2.024 Å

• 3–9.4 Sodium has a bcc unit cell with a = 4.29 Å. Show, by an appropriate diagram (drawn approximately to scale), the arrangement of atoms in the crystal planes having Miller indices (110). Calculate the spacing between these planes.

• 3–9.5 X-rays with a wavelength of 0.58 Å are used for calculating d_{200} in nickel. The reflection angle θ is 9.5°. What is the size of the unit cell?

Answer: 3.5 Å

•**3–9.6** A sodium chloride crystal is used to measure the wavelength of some x-rays. The diffraction angle is 5.1° for the d_{111} spacing of the chloride ions. What is the wavelength? (The lattice constant is 5.63 Å.)

3–10.1 Assuming atoms are spherical, calculate the atomic packing factor of diamond. (See Fig. 2–2.4(b).)

Answer: 0.34

3–10.2 Assuming that ions are spherical (a) calculate the atomic packing factor of MgO; (b) of LiF. (They both have the structure shown in Fig. 3–1.1.)

CHAPTER FOUR

ATOMIC DISORDER IN SOLIDS

PREVIEW

Nature is not perfect: inherently there is some disorder present. Thus the structures just described in Chapter 3 are subject to exceptions. Often these exceptions are minor: maybe one atom out of 10^{10} is out of place. Even so, they can become important. Imperfections account for the behavior of semi-conductors, for the ductility of metals, and for the strengthening of alloys. Imperfections account for the color of sapphire. They also permit the movement of atoms during heat treating, so that new structures and enhanced properties may be realized.

This chapter considers impurities first, then crystalline imperfections. Some solids are so disordered that we do not detect crystallinity. Finally we will look at atomic vibrations and movements within solids.

4–1 IMPURITIES IN SOLIDS

We admire "the real thing"; thus we commonly prefer pure wool, refined sugar, and like to think of 24-carat gold. Although these ideals may be noble, there are instances where, because of cost, availability, or properties, it is more desirable to have impurities present. An example is *sterling silver*, which contains 7.5% copper and only 92.5% silver (Example 2–1.1). This material, which we rate highly, could be refined to well over 99% purity.* It would cost more; however, it would be of inferior quality. Without altering its appearance, the 7.5% Cu makes the silver stronger, harder, and therefore more durable—at a lower cost!

Of course, we must pay attention to the properties pertinent to our design. Zinc added to copper produces *brass*, again at a lower cost than the pure copper. Brass is harder, stronger, and more ductile than copper. On the other hand, brass has lower electrical conductivity than copper, and so we use the more expensive pure copper for electrical wiring and similar applications where conductivity is important.

Alloys are combinations of two or more metals into one material. These combinations may be *mixtures* of two kinds of crystalline structures (e.g., bcc iron and Fe_3C in a steel bridge beam). Alternatively, alloys may involve *solid solutions*, which will be exemplified in the next section by brass. Although the term of alloy is generally not used specifically, various combinations of two or more oxide components may be incorporated advantageously into ceramic products (e.g., in the sparkplug insulator of Fig. 2–7.3). Likewise, the plastic telephone casing of Fig. 2–7.2 contains a combination of several types of molecules.

4–2 SOLID SOLUTIONS IN METALS

Solid solutions form most readily when the solvent and solute atoms have similar sizes and comparable electron structures. For example, the individual metals of brass—copper and zinc—have atomic radii of 1.278 Å and 1.39 Å, respectively. They both have 28 subvalent electrons and they each form crystal structures of their own with a coordination number of 12. Thus, when zinc is added to copper, it substitutes readily for the copper within the fcc lattice, until a maximum of nearly 40 percent of the copper atoms has been replaced. In this solid solution of copper and zinc, the distribution of zinc is entirely random (see Fig. 4–2.1).

Substitutional solid solutions. The solid solution described above is called a *substitutional* solid solution because the zinc atoms substitute for copper atoms in the crystal structure. This type of solid solution is quite common among various metal systems. The solution of copper and nickel to form *monel* is another example. In monel, any fraction of the atoms in the original copper structure may be replaced by nickel. Copper–nickel solid solutions may range from practically no nickel and almost

---

* As stated in the footnote with Example 2–1.1, weight percent (w/o) is implied in liquids and solids *unless specifically stated otherwise.*

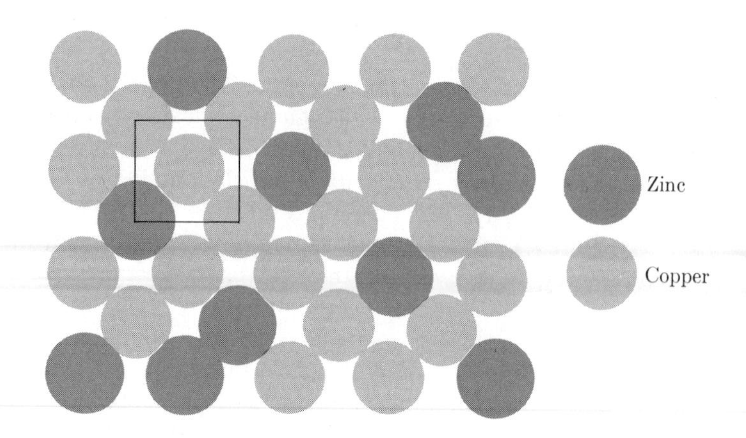

Fig. 4–2.1. Random substitutional solid solution (zinc in copper, i.e., brass). The crystal pattern is not altered.

100 percent copper, to almost 100 percent nickel and practically no copper. All copper–nickel alloys are face-centered cubic.

On the other hand, there is a very definite limit to the amount of tin which may replace copper to form *bronze*, and retain the face-centered cubic structure of the copper. Tin in excess of the maximum amount of *solid solubility* must form another phase. This *solubility limit* will be considered in more detail in Chapter 9.

If there is to be extensive replacement in a substitutional type of solid solution, the atoms must be nearly the same size. Nickel and copper have a complete range of solutions because their structures are the same and their radii are 1.246 Å and 1.278 Å, respectively. As the difference in size increases, less substitution can occur. Only 20 percent of copper atoms can be replaced by aluminum, because the latter has a radius of 1.431 Å as compared with only 1.278 Å for copper. Extensive solid solubility rarely occurs if there is more than about 15 percent difference in radius between the two kinds of atoms. There is further restriction on solubility when the two components have different structures or valences.

The limiting factor is the *number* of substituted atoms rather than the *weight* of the atoms which are substituted. However, engineers ordinarily express composition as weight percent. It is therefore necessary to know how to express weight percent in terms of atomic percent, and vice versa (see Example 2–1.1).

Ordered solid solutions. Figure 4–2.1 showed a *random substitution* of one atom for another in a crystal structure. In such a process, the chance of one element occupying any particular atomic site in the crystal is equal to the atomic percent of that element in the alloy. In this case, there is no *order* in the substitution of the two elements.

However, it is not unusual to find an *ordering* of the two types of atoms into a specific arrangement. Figure 4–2.2 shows an ordered structure in which most dark "atoms" are surrounded by light "atoms." Such ordering is more common at lower temperatures, since greater thermal agitation tends to destroy the orderly arrangement.

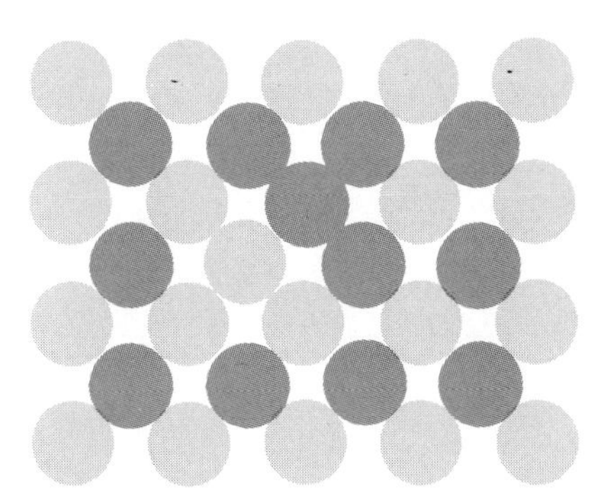

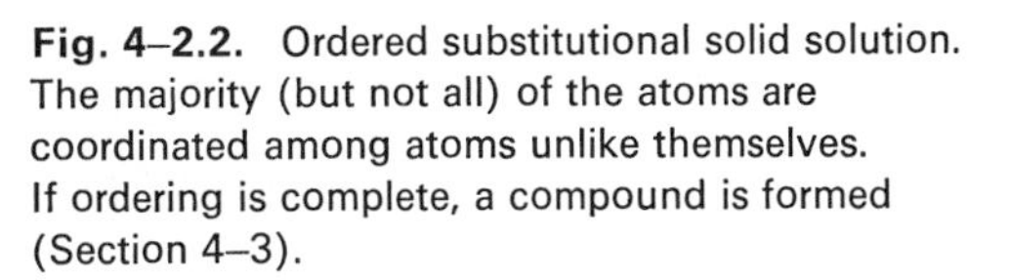

Fig. 4–2.2. Ordered substitutional solid solution. The majority (but not all) of the atoms are coordinated among atoms unlike themselves. If ordering is complete, a compound is formed (Section 4–3).

Interstitial solid solutions. In another type of solid solution, illustrated in Fig. 4–2.3, a small atom may be located in the interstices between larger atoms. Carbon in iron is an example. At temperatures below 912°C (1673°F), pure iron occurs as a body-centered cubic structure. Above 912°C (1673°F), there is a temperature range in which iron has a face-centered cubic structure. In the face-centered cubic lattice, a relatively large *interstice*, or "hole," exists in the center of the unit cell. Carbon, being an extremely small atom, can move into this hole to produce a solid solution of iron and carbon. At lower temperatures, where the iron has a body-centered cubic structure, the interstices between the iron atoms are much smaller. Consequently the solubility of carbon in body-centered cubic iron is relatively low.

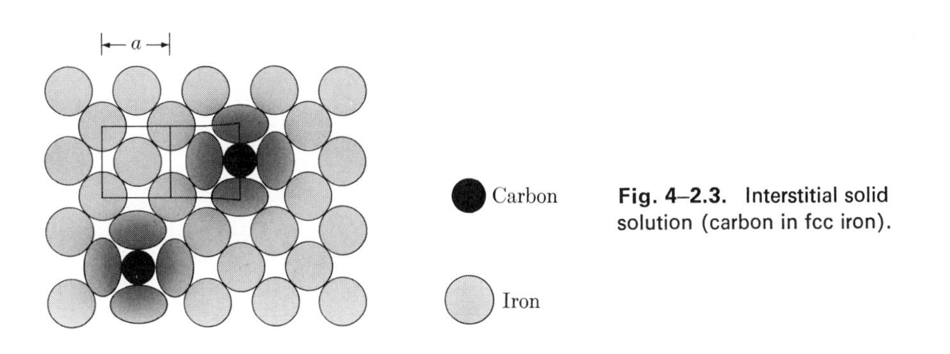

Fig. 4–2.3. Interstitial solid solution (carbon in fcc iron).

Example 4–2.1 Bronze is a solid-solution alloy of copper and tin in which 3%, more or less, of the copper atoms are replaced by tin atoms. The fcc unit cell of copper is retained, but expanded a bit because the tin atoms have a radius of approximately 1.51 Å. a) What is the weight percent in a 3 a/o tin bronze? b) Assuming the lattice constant increases linearly with the atomic fraction of tin, what is the density of this bronze?

Solution: Basis = 100 atoms = 25 fcc unit cells.

a) Mass of copper: 97(63.54 amu) = 6163 amu = 94.5 w/o;

mass of tin: 3(118.69 amu) = 356 amu = 5.4 w/o.

6519

b) Average atom size = 0.97(1.278 Å) + 0.03(1.51 Å) = 1.285 Å;

$$a = (4)(1.285)/\sqrt{2} = 3.634\ \text{Å};$$

$$\rho = \frac{6519\ \text{amu}/(0.602 \times 10^{24}\ \text{amu/g})}{25(3.634 \times 10^{-8}\ \text{cm})^3} = 9.0\ \text{g/cm}^3. \blacktriangleleft$$

Example 4–2.2 The maximum solubility limit of tin in bronze is 15.8 w/o at 586°C. What is the atom percent tin?

Solution: Basis = 100,000 amu = 15,800 amu Sn + 84,200 amu Cu.

Number of Cu atoms: 84,200/63.54 = 1325 = 90.9 a/o Cu;

Number of Sn atoms: 15,800/118.69 = 133 = 9.1 a/o Sn. ◀

1458

4–3 SOLID SOLUTIONS IN COMPOUNDS

Substitutional solid solutions can occur in ionic phases as well as in metals. In ionic phases, just as in the case of solid metals, atom or ion size is important. A simple example of an ionic solid solution is shown in Fig. 4–3.1. The structure is that of MgO (Fig. 3–1.1) in which the Mg^{2+} ions have been replaced by Fe^{2+} ions. Inasmuch as the radii of the two ions are 0.66 Å and 0.74 Å, complete substitution is possible. On the other hand, Ca^{2+} ions cannot be similarly substituted for Mg^{2+} because their radius of 0.99 Å is comparatively large.*

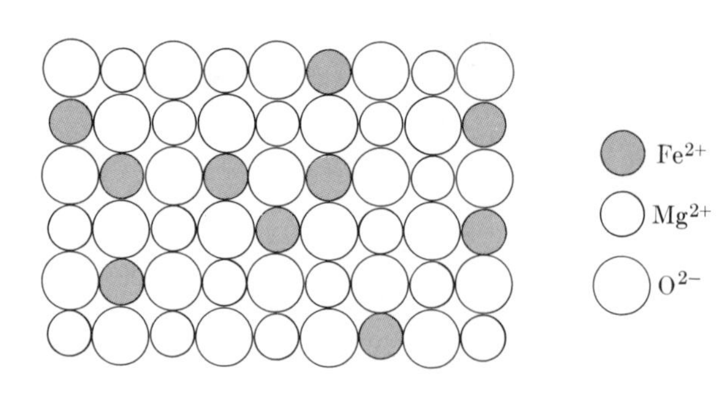

Fig. 4–3.1. Substitutional solid solution in a compound, Fe^{2+} is substituted for Mg^{2+} in the MgO structure.

* See Appendix B for ionic radii.

An additional requirement, which is more stringent for solid solutions of ceramic compounds than for similar solid solutions of metàls, is that the valence charges on the replaced ion and the new ion must be identical. For example, it would be difficult to replace the Mg^{2+} in MgO with an Li^{+}, although the two have identical radii, because there would be a net deficiency of charges. Such a substitution could be accomplished only if there were other compensating changes in charge. (See Section 8–4.)

Nonstoichiometric compounds. Many compounds have exact ratios of elements (e.g., H_2O, CH_4, MgO, Al_2O_3, Fe_3C, to name but a few). They have a regularity that is comparable to the ordered structure of Fig. 4–2.2. In that figure there is an exact 1-for-1 ratio of the two types of atoms. The above five compounds are *stoichiometric.*

Other compounds deviate from specific integer ratios for the two (or more) elements which are present. Thus we find that "Cu_2Al" ranges from 31 a/o to 37 a/o Al (16 to 20 w/o Al) rather than being exactly $33\frac{1}{3}$ a/o Al. Likewise, at 1000°C "FeO" ranges from 51 to 53 a/o oxygen, rather than being exactly 50 a/o. We call these compounds *nonstoichiometric* because they do not have a fixed ratio of atoms.

Nonstoichiometric compounds always involve some solid solution. In the Cu_2Al example cited above, the atoms are nearly enough the same size and sufficiently comparable electronically so that with excess aluminum, some of the copper atoms are replaced by aluminum atoms (up to the maximum of 37 a/o Al). Conversely, in the presence of excess copper, the Cu/Al atom ratio reaches 69/31, by substituting a few copper atoms into the aluminum sites of Cu_2Al.

The nonstoichiometry of $Fe_{1-x}O$ arises from a different origin. The iron ions and the oxygen ions are too different to permit any measurable substitution. However, iron compounds always include some ferric (Fe^{3+}) ions along with the ferrous (Fe^{2+}) ions. Thus in order to balance charges there always must be more than 50 a/o oxygen. In fact, each two Fe^{3+} ions require an extra O^{2-} ion; or conversely, each pair of Fe^{3+} ions must be accompanied by a cation vacancy (Fig. 4–3.2). This is called a

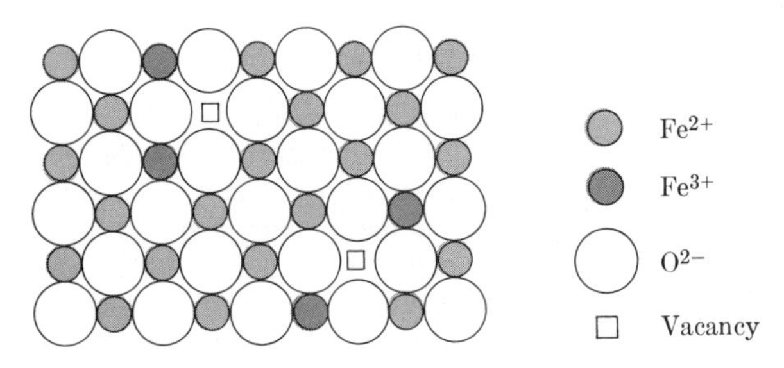

Fig. 4–3.2. Defect structure ($Fe_{1-x}O$). This structure is the same as NaCl (Fig. 3–1.1) except for some iron ion vacancies. Since a fraction of the iron ions are Fe^{3+} rather than Fe^{2+}, the vacancies are necessary to balance the charge. The value of x ranges from 0.04 to 0.16, depending on temperature and the amount of available oxygen.

defect structure since there are irregularities in the atom packing.* At 1000°C and with iron saturation, the composition is $Fe_{0.96}O$; with oxygen saturation, the composition is $Fe_{0.88}O$.

These defect structures will be important to us in the next chapter when we discuss electrical conduction. Also, we will observe later in this chapter that the atoms or ions can diffuse more readily through the crystalline solid when vacancies are present.

Example 4–3.1 An iron oxide (Fig. 4–3.2) contains 52 a/o oxygen, and has a lattice constant of 4.29 Å. a) What is the Fe^{2+}/Fe^{3+} ion ratio? b) What is the density? (This structure is like the NaCl's except for the vacancies, □.)

Solution: Basis $=$ 100 ions $= 52\ O^{2-} + 48Fe + 4\square$.

a) Charge balance: $52(2-) + (y)(3+) + (48 - y)(2+) = 0$,

$$y = 8\ Fe^{3+}; \qquad 48 - y = 40\ Fe^{2+};$$

$$Fe^{2+}/Fe^{3+} = 5.$$

b) With 4 anions per unit cell (*see* Fig. 3–1.1): 13 unit cells.

$$\rho = \frac{[48(55.8) + 52(16)]\ \text{amu/13 u.c.}}{(0.602 \times 10^{24}\ \text{amu/g})(4.29 \times 10^{-8}\ \text{cm})^3/\text{u.c.}} = 5.7\ \text{g/cm}^3. \blacktriangleleft$$

• **Example 4–3.2** A β'-brass is nominally an intermetallic compound, CuZn, with the simple cubic structure shown in Fig. 3–6.2. It may also be called a *partially ordered solid solution,* particularly since it is nonstoichiometric with a range of 46 to 50 a/o zinc at 450°C. Assume 90% of the $\frac{1}{2},\frac{1}{2},\frac{1}{2}$ sites of Fig. 3–6.2 are occupied by copper atoms in a 46 a/o Zn–54 a/o Cu alloy. What per cent of the 0,0,0 sites are occupied by copper atoms?

Solution: Basis $=$ 50 unit cells $=$ 50 0,0,0 sites (and 50 $\frac{1}{2},\frac{1}{2},\frac{1}{2}$ sites)

$=$ 100 atoms $=$ 54 Cu + 46 Zn.

$\frac{1}{2},\frac{1}{2},\frac{1}{2}$ sites: 45 Cu + 5 Zn;

0,0,0 sites: 9 Cu + 41 Zn.

Therefore, 9 of the 50, or 18% of the 0,0,0 sites, are occupied by copper.

Comments. At low temperatures almost all of the neighbors of zinc atoms are copper atoms, and *vice versa.* However, as the temperature is increased, the atoms start to disorder. In this problem, at 450°C, the 0,0,0 sites are 82 Zn–18 Cu, while the $\frac{1}{2},\frac{1}{2},\frac{1}{2}$ sites are 10 Zn–90 Cu (all atom percents). Thus the two sites are not equivalent, and the structure is simple cubic.

Above 470°C, this alloy becomes fully random, with no preference for copper atoms to be surrounded by zinc atoms. Under those conditions, the substitutional solid solution (called β-brass rather than β'-brass) is bcc because the unit-cell centers and corners have equal probability for the same average composition. ◀

---

* If all atoms were present, this structure would be the same as the NaCl structure in Fig. 3–1.1.

4–4 IMPERFECTIONS IN CRYSTALS

We have just seen a type of imperfection in crystals where a vacancy is required to produce a charge balance (Fig. 4–3.2). When imperfections such as vacancies involve one or a few atoms, we call them *point defects.* Other imperfections may be lineal through the crystal; hence, the term *line defects.* They become particularly significant when crystals are plastically deformed by shear stresses. In fact, a small number of these cause metal crystals to be 1000 times more ductile than would be possible in their absence. When present in large numbers, these lineal imperfections increase the strength of the metal. Finally, other imperfections may be two-dimensional in concept and involve external *surfaces* or internal *boundaries.*

Point defects. The simplest point defect is a *vacancy*, which involves a missing atom (Fig. 4–4.1(a)) within a metal. Such defects can be a result of imperfect packing during the original crystallization, or they may arise from thermal vibrations of the atoms at elevated temperatures (Section 4–6), because as thermal energy is increased there is an increased probability that individual atoms will jump out of their position of lowest energy. Vacancies may be single, as shown in Fig. 4–4.1(a), or two or more of them may condense into a di-vacancy (Fig. 4–4.1(b)) or a tri-vacancy.

Schottky imperfections are closely related to vacancies but are found in compounds which must maintain a charge balance (Fig. 4–4.1(c)). They involve vacancies

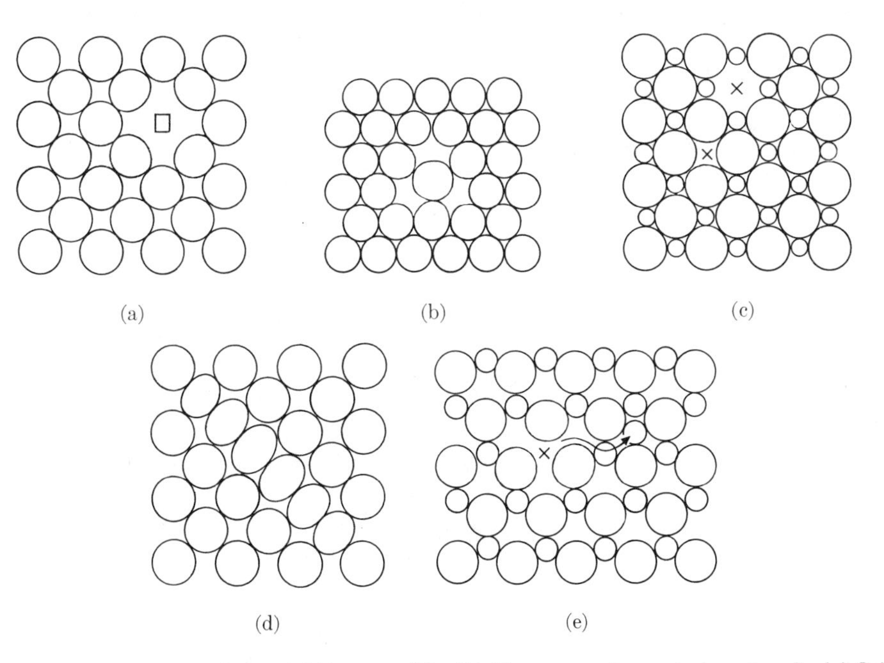

Fig. 4–4.1. Point defects. (a) Vacancy, □. (b) Di-vacancy (two missing atoms). (c) Schottky defect (ion pair vacancies). (d) Interstitialcy. (e) Frenkel defect (ion displacement).

of pairs of ions of opposite charges. Both vacancies and Schottky defects facilitate atomic diffusion (Section 4–7).

An extra atom may be lodged within a crystal structure, particularly if the atomic packing factor is low. Such an imperfection, called an *interstitialcy*, produces atomic distortion (Fig. 4–4.1(d)).

An ion displaced from the lattice into an interstitial site (Fig. 4–4.1(e)) is called a *Frenkel defect*. Close-packed structures have fewer interstitialcies and Frenkel defects than vacancies and Schottky defects, because additional energy is required to force the atoms into the new positions.

Line defects (dislocations). The most common type of line defect within a crystal is a dislocation. An *edge dislocation* is shown in Fig. 4–4.2. It may be described as an edge of an extra plane of atoms within a crystal structure. Zones of compression and of tension accompany an edge dislocation (Fig. 4–4.3) so that there is a net increase in energy along the dislocation. The displacement distance for atoms around the dislocation is called the *Burgers vector*. This vector is at right angles to the edge dislocation line.

A *screw dislocation* has its displacement, or Burgers vector, parallel to the linear defect (Fig. 4–4.4). Shear stresses are associated with adjacent atoms; therefore, extra energy is involved here as in the previously cited edge dislocations.

Dislocations of both types may originate during crystallization. Edge dislocations, for example, arise when there is a slight mismatch in the orientation of adjacent parts of the growing crystal so that an extra row of atoms is introduced or eliminated. As shown in Fig. 4–4.4, a screw dislocation provides for easy crystal growth because additional atoms and unit cells can be added to the "step" of the screw. Thus the term screw is very apt, because the step swings around the axis as growth proceeds.

Dislocations more commonly originate during deformation. We see this in Fig. 4–4.5, where shear is seen to introduce both edge dislocations and screw dislocations.

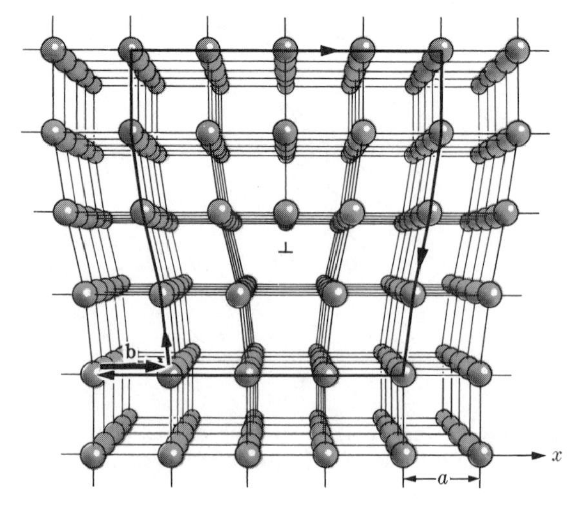

Fig. 4–4.2. Edge dislocation. A linear defect occurs at the edge of an extra plane of atoms. (Guy, A. G., *Elements of Physical Metallurgy*.)

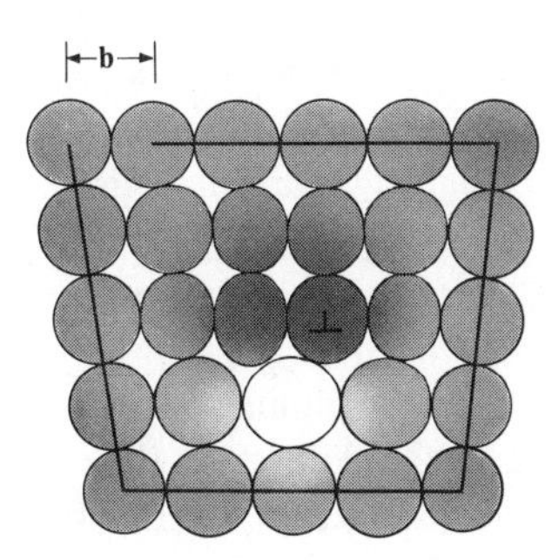

Fig. 4–4.3. Dislocation energy. Atoms are under compression (darker) and tension (lighter) adjacent to the dislocation. The displacement vector (Burgers vector). is perpendicular to the dislocation line.

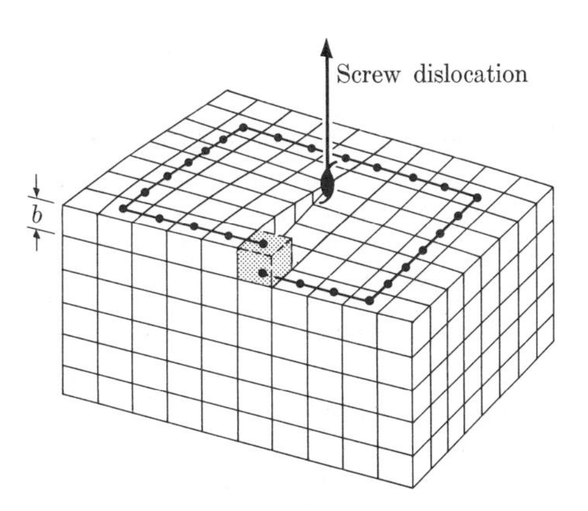

Fig. 4–4.4 Screw dislocation (unit cells shown). The displacement vector (Burgers vector) is parallel to the linear defect.

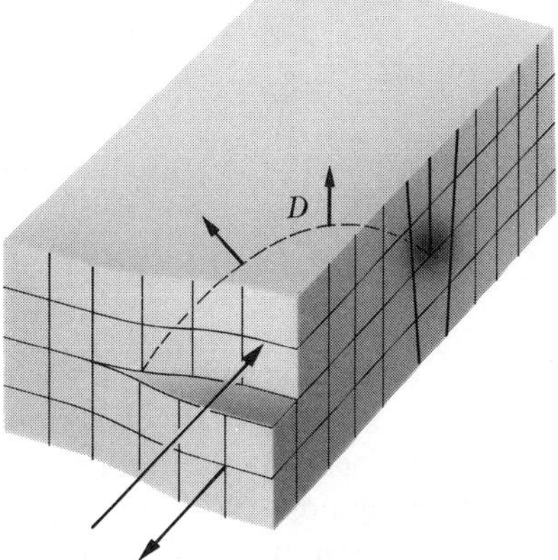

(a)

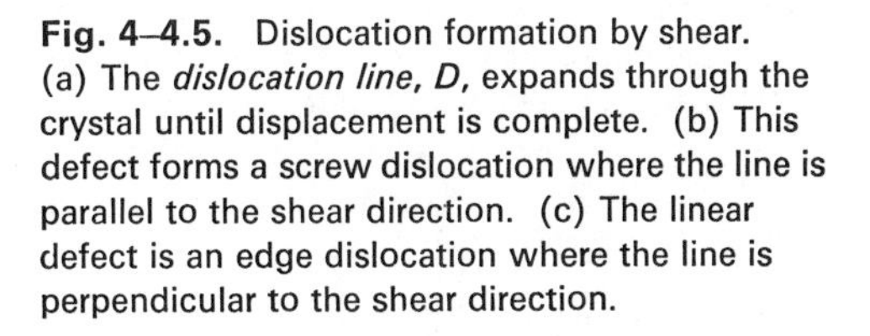

Fig. 4–4.5. Dislocation formation by shear. (a) The *dislocation line, D,* expands through the crystal until displacement is complete. (b) This defect forms a screw dislocation where the line is parallel to the shear direction. (c) The linear defect is an edge dislocation where the line is perpendicular to the shear direction.

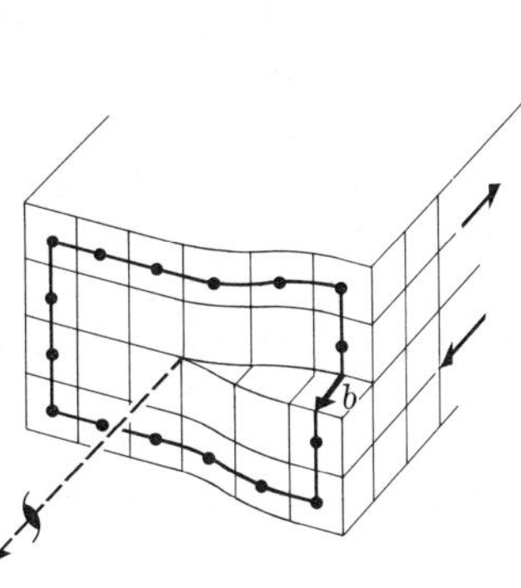

(b)

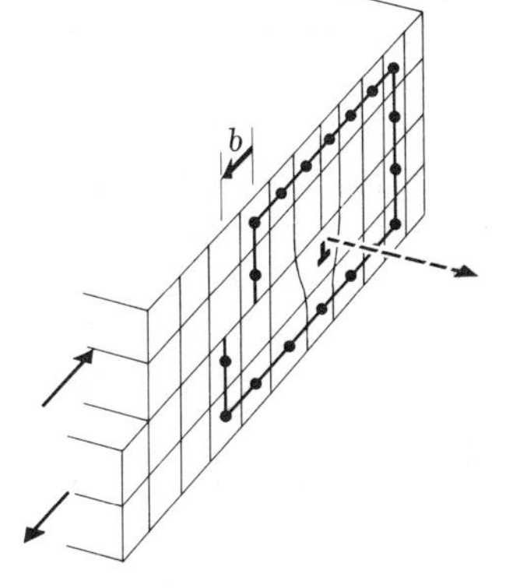

(c)

Both lead to the same final displacement and are in fact related through the dislocation line which forms.

Surfaces. Crystalline imperfections may extend in two dimensions as a boundary. The most obvious boundary is the external *surface.* Although we may visualize a surface as simply a terminus of the crystal structure, we should quickly appreciate the fact that the atoms on the surface are not fully comparable to the atoms within a crystal. The surface atoms have neighbors on only one side (Fig. 4–4.6); therefore they have higher energy than do the internal atoms. This energy may be rationalized with Fig. 2–5.2 by noting that if additional atoms were to be deposited onto the surface atoms, energy would be released just as it was for the combination of two individual atoms. We find our best visible evidence of this surface energy in the case of liquid drops which have spherical shape to minimize the surface area (and therefore the surface energy) per unit volume. Surface adsorption provides additional evidence of the energy differential at the surface.

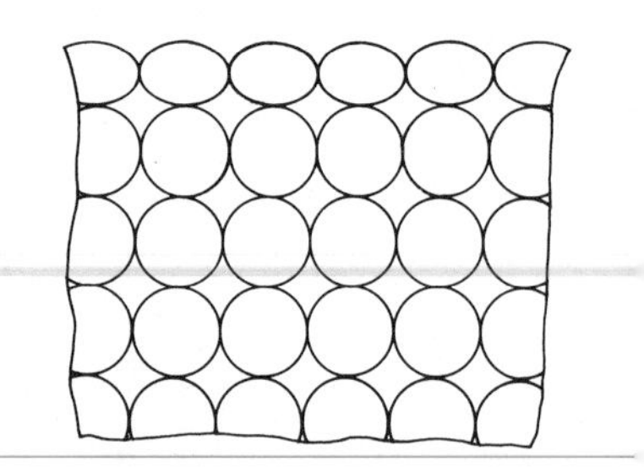

Fig. 4–4.6. Surface atoms (schematic). Since these atoms are not entirely surrounded by others, they possess more energy than internal atoms.

Grain boundaries. Although a material such as copper in an electric wire may contain only one phase, i.e., only one structure (fcc), it contains many crystals of various orientations. These individual crystals are called *grains.* The shape of a grain in a solid is usually controlled by the presence of surrounding grains. Within any particular grain, all of the unit cells are arranged with one orientation and one pattern. However, at the *grain boundary* between two adjacent grains there is a transition zone which is not aligned with either grain (Fig. 4–4.7).

When a metal is observed under a microscope, although we cannot see the individual atoms illustrated in Fig. 4–4.7, we can quite readily locate the grain boundaries if the metal has been treated by *etching.* First the metal is smoothly polished so that a plane, mirrorlike surface is obtained, and then it is chemically attacked for a short period of time. The atoms in the area of transition between one grain and the next will dissolve more readily than other atoms and will leave a line which can be seen with the microscope (Fig. 4–4.8); the etched grain boundary does not act as a perfect mirror as does the remainder of the grain (Fig. 4–4.9).

We may consider that the grain boundary is two-dimensional, although it may be curved, and actually has a finite thickness of 2 or 3 atomic distances. The mismatch

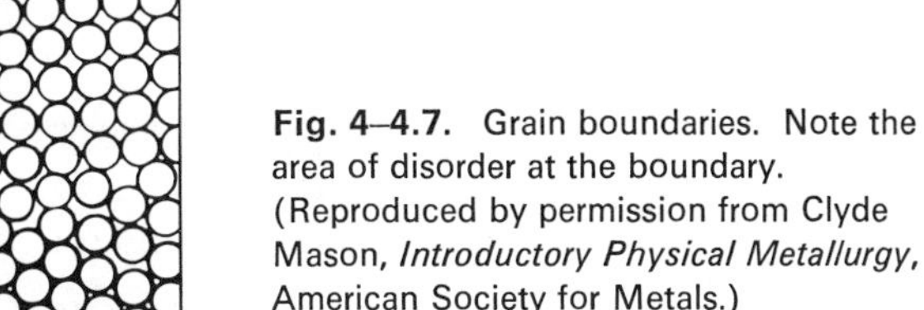

Fig. 4–4.7. Grain boundaries. Note the area of disorder at the boundary. (Reproduced by permission from Clyde Mason, *Introductory Physical Metallurgy*, American Society for Metals.)

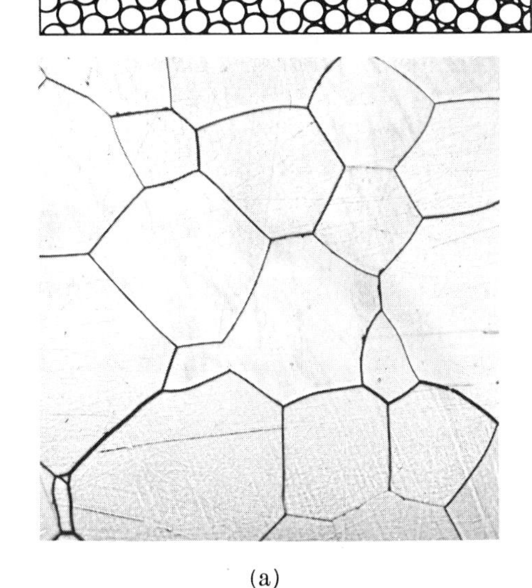

(a)

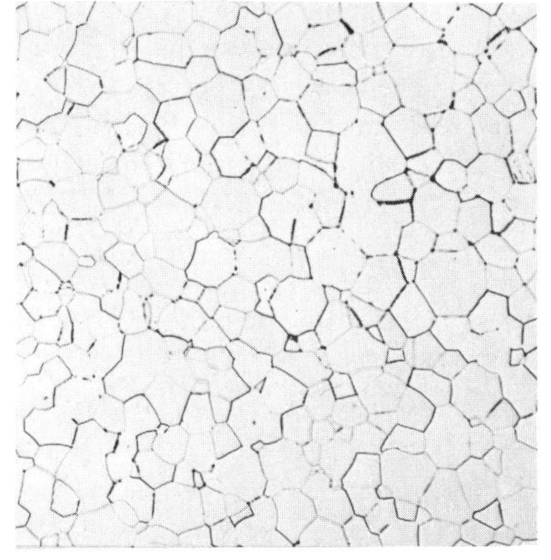

(b)

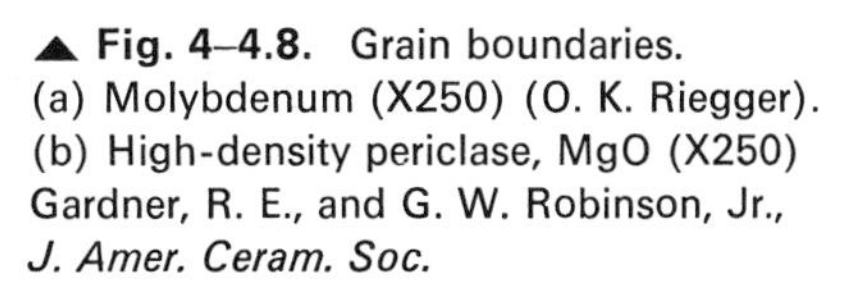

▲ **Fig. 4–4.8.** Grain boundaries. (a) Molybdenum (X250) (O. K. Riegger). (b) High-density periclase, MgO (X250) Gardner, R. E., and G. W. Robinson, Jr., *J. Amer. Ceram. Soc.*

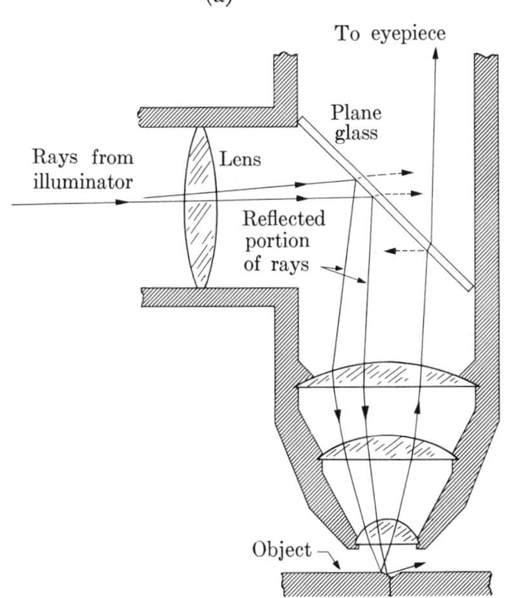

Fig. 4–4.9. Grain boundary observation. The metal has been polished and etched. The corroded boundary does not reflect light through the microscope. (Reproduced by permission from B. Rogers, *The Nature of Metals*, 2nd ed., American Society for Metals, 1964.)

of the orientation of adjacent grains produces a less efficient packing of the atoms along the boundary. Thus the atoms along the boundary have a higher energy than those within the grains. This accounts for the more rapid etching along the boundaries described above. The higher energy of the boundary atoms is also important for the nucleation of polymorphic phase changes (Section 3–5). The lower atomic packing along the boundary favors atomic diffusion (Section 4–7), and the mismatch between adjacent grains modifies the progression of dislocation movements (Fig. 4–4.5). Thus the grain boundary modifies plastic strain of a material.*

Grain size and grain boundary area. It is obvious that the two *microstructures* of Fig. 4–4.8 are different. There is more grain-boundary area in the microstructure which is finer grained. Since grain size and grain boundary area are important with respect to the behavior of a material, we will want to have some quantitative measure of them. The American Society for Testing and Materials (ASTM) has established the reference nets shown in Fig. 4–4.10 with various *grain-size numbers* (GS #) for comparison with observations through a microscope at ×100 (i.e., when the linear dimension is magnified 100 times). The number of grains N per in.^{2} at ×100 is:

$$N = 2^{n-1}, \tag{4–4.1}$$

where n is the grain-size number. We shall refer to these grain-size numbers in subsequent chapters.

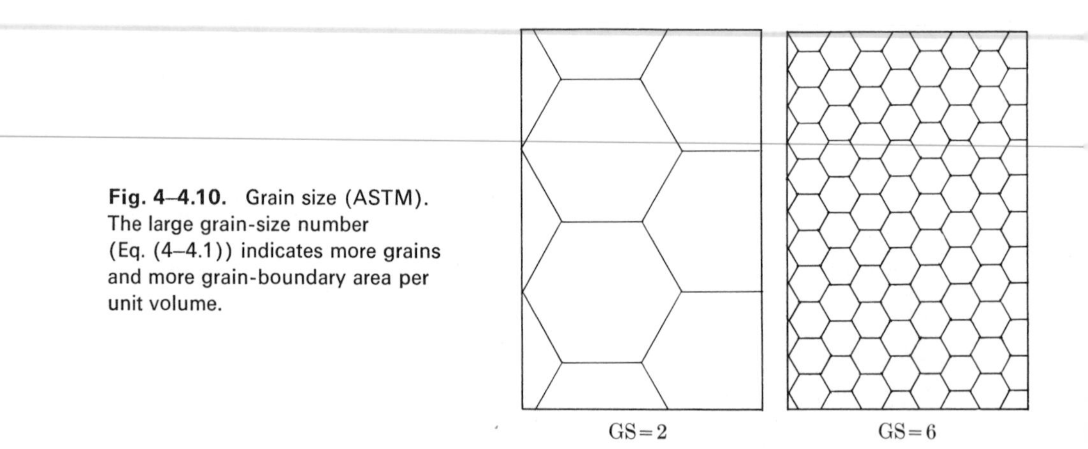

Fig. 4–4.10. Grain size (ASTM). The large grain-size number (Eq. (4–4.1)) indicates more grains and more grain-boundary area per unit volume.

The *grain boundary area* can be readily calculated. Consider the circle of Fig. 4–4.11, which is shown at ×250 and therefore is (2.0π in.)/250, or 0.025 in. in circumference (0.06 cm). If that circle is placed on a photomicrograph, the number of

* At normal temperatures the grain boundaries interfere with slip. Therefore a *fine-grained* material is stronger than a *coarse-grained* material. At elevated temperatures, the boundaries can accommodate the dislocations. As a result the situation is reversed, and creep results (Section 6–8).

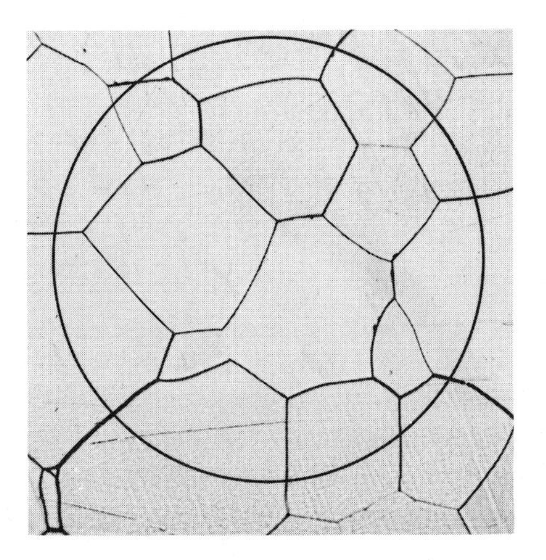

Fig. 4–4.11. Grain boundary area calculation. Since the magnification is X250, the length of the circle is π(5 cm)/250, or 0.063 cm (0.025 in.). It intersects 11 boundaries in that distance. Therefore, there are 2(11/0.063)cm, or 350 cm^2 of boundary area per cm^3. (See Eq. (4–4.2) and Example 4–4.1.)

intercepts of grain boundary, per unit length P_L, is simply half the grain boundary area per unit volume, S_V, or

$$S_V = 2P_L. \tag{4–4.2}$$

We shall not prove this, but will use it in the following example.

Example 4–4.1 a) Determine the grain boundary area per cubic inch in the molybdenum of Figs. 4–4.8 and 4–4.11. b) What is the ASTM grain-size number?

Solution: a) The circumference of the 2.00-in. circle intercepts 11 boundaries at $\times$250. Therefore P_L equals 11/(2.00π in./250) = 440/in.:

$$S_V = 2(440/\text{in.}) = 880 \text{ in.}^2/\text{in.}^3 \qquad (= 350 \text{ cm}^2/\text{cm}^3).$$

b) At $\times$250, the 2.3 × 2.3 in.^2 area contains 17 grains, i.e., (corner grains × $\frac{1}{4}$) + (edge grains × $\frac{1}{2}$) + internal grains. At $\times$100, these 17 grains would occupy 0.92 × 0.92 in.^2, giving approximately 20 grains/in.^2, or $2^{4.3}$. Thus, from Eq. (4–4.1), GS = 5^+.

Comments. The circle of Fig. 4–4.11 must be at least three times as large as the largest grain in order to get a good statistical sample of P_L. Otherwise, a random sampling procedure is required.

Often we speak of grain size in terms of a typical "diameter." Thus the grains in Fig. 4–4.11 could be said to have an average grain-size diameter of ~0.06 mm (after the magnification is taken into account). Note, however: (a) the grains are not spherical; and (b) the grains do not all have the same "diameter." This latter difference may arise in part because we are "catching only the top corner" of some of the grains, and in part because the grains do vary in size. In general, grain "diameters" do not offer a fully satisfactory measure of either the average size of the grains or the grain boundary area. ◀

Example 4–4.2 A very accurate measurement was made to four significant figures of the density of aluminum. When cooled rapidly from 650°C, ρ = 2.698 g/cm^3. Compare that value with the theoretical density obtained from diffraction analyses where a was determined to be 4.049 Å.

Solution: Since the atomic weight is 26.98 amu and aluminum is fcc,

$$\rho_{\text{theor.}} = \frac{4(26.98 \text{ amu})/(0.6022 \times 10^{24} \text{ amu/g})}{(4.049 \times 10^{-8} \text{ cm})^3} = 2.700 \text{ g/cm}^3;$$

$$\rho/\rho_{\text{theor.}} = 2.698/2.700 = 0.999 \quad \text{or} \sim 1 \text{ vacancy per 1000 atoms.}$$ ◀

4–5 NONCRYSTALLINE MATERIALS

Long-range order is absent in some materials of major engineering and scientific importance. Included are all liquids, glass, the majority of plastics, and a few metals if the latter are cooled extremely rapidly from their liquid. In principle, we can view this lack of repetitive structure as a volume, or 3-dimensional, disorder, and as a continuation of our sequence: point defects, lineal defects, and 2-dimensional boundaries. We call these materials *amorphous* (literally "without form") in contrast to crystalline materials.

Liquids. For the most part, liquids are *fluids* (i.e., they *flow* under their own mass). However, just as "molasses in January" gets semirigid, various liquids of technical importance can become very viscous and even solid, without crystallizing.

First let us look at the disorder which occurs in a single-component metal as it approaches melting, then transforms into a liquid. We can use aluminum for our example. As was implied in the last section, the greater thermal energy at higher temperatures introduces not only greater thermal vibrations, but also some vacancies. Just short of the melting point, crystalline aluminum may contain up to 0.1% vacancies in its lattice. When the vacancies approach one percent in a close-packed structure, "turmoil reigns." The regular 12-fold coordination is destroyed and the long-range order of the crystal structure disappears (Fig. 4–5.1).

This disorder of melting increases the volume of most materials* (Fig. 4–5.2). With this disorder, the number of nearest neighbors drops from 12 to only 11 or 10;

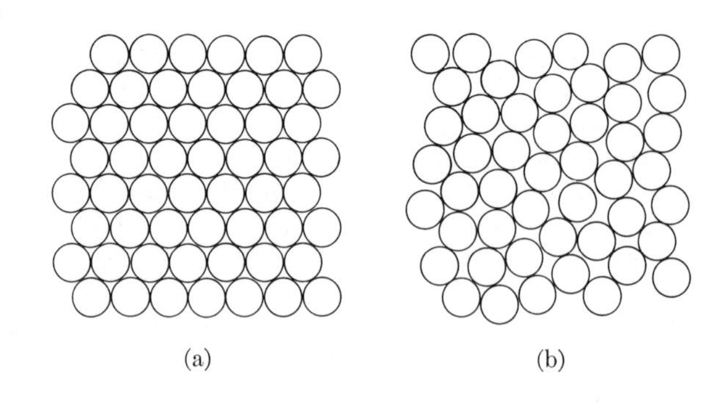

Fig. 4–5.1. Melting (metal). (a) Crystalline metal with CN = 12. (The 6 in the plane plus 3 above and 3 below.) (b) Liquid metal. Long-range order is lost; CN < 12, and the average interatomic distance increases slightly.

---

* It increases the volume of all those materials which are close-packed and which do not have directional bonds. A *few* materials with low packing factors and stereospecific bonds (Chapter 2) collapse into denser structures when they are thermally excited. Water is the prime example of this exception. (And are we lucky!!)

Table 4–5.1
HEATS OF FUSION OF METALS

Metal	Melting temperature, °C (°F)	Heat of fusion, joule/mole*
Tungsten, W	3387 (6129)	32,000
Molybdenum, Mo	2623 (4753)	28,000
Chromium, Cr	1863 (3385)	21,000
Titanium, Ti	1672 (3042)	21,000
Iron, Fe	1538 (2800)	15,300
Nickel, Ni	1455 (2651)	17,900
Copper, Cu	1084 (1984)	13,500
Aluminum, Al	660 (1221)	10,500
Magnesium, Mg	649 (1200)	9,000
Zinc, Zn	419 (787)	6,600
Lead, Pb	327 (621)	5,400
Mercury, Hg	−38.9 (−38)	2,340

* joule/0.6 × 10^{24} atoms; 4.18 J = 1 cal.

however, these are not in a regular pattern, so the space per atom and the average interatomic distance is increased a few percent. Each material requires a characteristic amount of energy to be melted. We call this energy the *heat of fusion.* Since the heat of fusion, ΔH_f, is the energy required to disorganize a mole of atoms, and the melting temperature, T_m, is a measure of the atomic bond strength, we find a general correlation between the two (Table 4–5.1).*

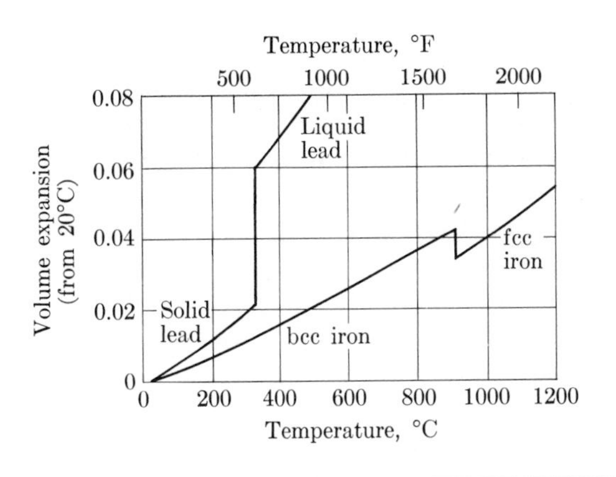

Fig. 4–5.2. Volume changes. (a) Lead with ~4 v/o expansion between fcc and liquid. (b) Iron has a volume change between bcc and fcc (Example 3–5.1).

* For the correlation to be good, we must have comparable materials, e.g. among metals.

Glasses. As indicated earlier, glasses are sometimes considered to be very viscous liquids, inasmuch as they are noncrystalline. However, only a few liquids can actually form glasses. Therefore, in order to make a distinction, we must look at the structure of glass more critically.

At high temperatures glasses form true liquids. The atoms have freedom to move around and respond to shear stresses. When a commercial glass at its liquid temperature is supercooled, there is thermal contraction caused by atomic rearrangements which produce more efficient packing of the atoms. This contraction (Fig. 4-5.3) is typical of all liquid phases; however, with more extensive cooling, there is an abrupt change in the expansion coefficient of glasses. Below a certain temperature, called the *glass temperature*, T_g, there are no further rearrangements of the atoms and the only contraction is a result of smaller thermal vibrations. This lower coefficient is comparable to the thermal coefficient in crystals where thermal vibrations are the only factor causing contraction, and no rearrangement occurs.

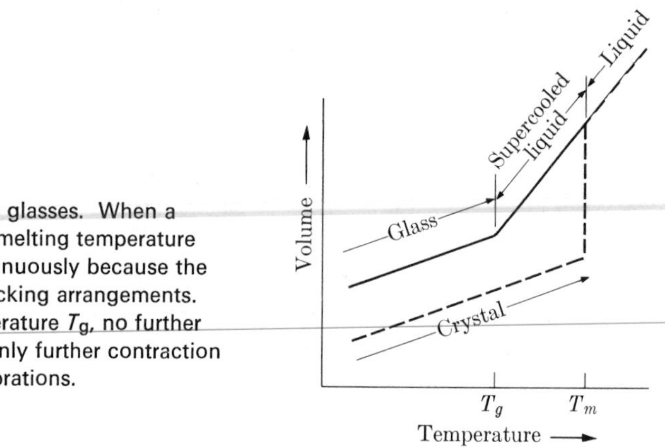

Fig. 4-5.3. Volume changes in glasses. When a liquid is supercooled below the melting temperature T_m, it contracts rapidly and continuously because the atoms develop more efficient packing arrangements. Below the glass transition temperature T_g, no further rearrangement occurs, and the only further contraction is caused by reduced thermal vibrations.

The term *glass* applies to those materials which have the expansion characteristics of Fig. 4-5.3. Glasses may be either inorganic or organic, and are characterized by a short-range order (and an absence of long-range order). Figure 4-5.4 presents one of the simplest glasses (B_2O_3), in which each small boron atom fits among three larger oxygen atoms. Since boron has a valence of three and oxygen a valence of two, electrical balance is maintained if each oxygen atom is located between two boron atoms. As a result, a continuous structure of strongly bonded atoms is developed. Below the glass temperature, where the atoms are not readily rearranged, the fluid characteristics are lost and a noncrystalline solid exists. Such a solid has a significant resistance to shear stresses and therefore cannot be considered a true liquid.

The temperature-volume characteristics of Fig. 4-5.3 were first observed in silicate glasses (Section 8-6). However, it soon became apparent that these charac-

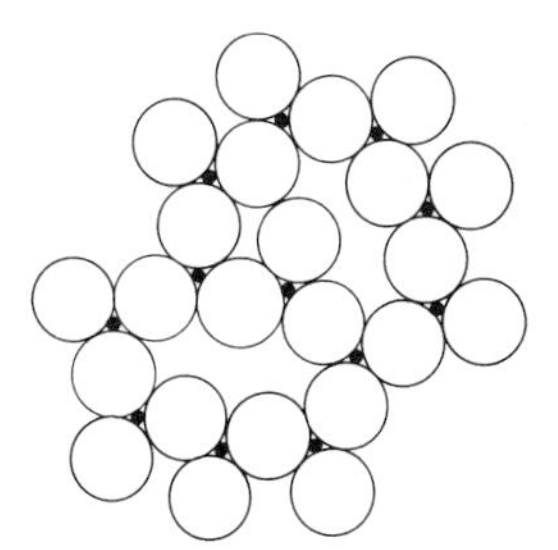

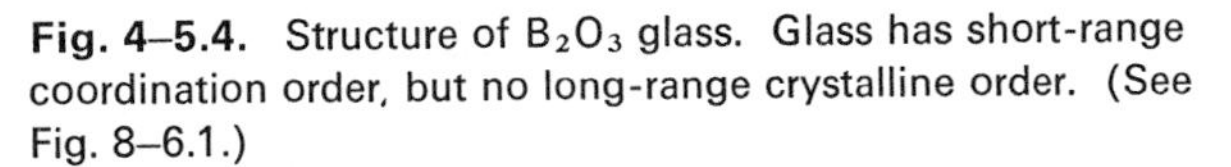

Fig. 4–5.4. Structure of B_2O_3 glass. Glass has short-range coordination order, but no long-range crystalline order. (See Fig. 8–6.1.)

teristics have major significance in polymeric materials. Thus in Chapter 7, we will look at the glass temperature, T_g, closely. Below the T_g, polymers are hard and brittle, and have low dielectric constants. Above the T_g, a plastic becomes flexible and even rubbery. Concurrently, it changes its dielectric behavior.

Phases. Reference has already been made to phases, and will be made again numerous times throughout this text. We are now in a position to define a phase as that part of a material which is *distinct from others in structure and/or composition.* Consider "ice–water." While of the same composition, the ice is a crystalline solid with a hexagonal lattice; the water is a liquid. The *phase boundary* between the two locates a discontinuity in structure: they are *separate phases.* Or consider silver-plated copper. Both silver and copper are fcc; however, the silver atoms are sufficiently larger than the copper atoms* so that there is nearly complete composition discontinuity at room temperature. Thus, they form two separate phases.

Commonly, two phases of a material have distinct differences in both composition and structure, e.g., a plastic with a fiberglass reinforcement. In contrast, some phases lose their distinctiveness and dissolve, e.g., after dissolving in coffee, sugar is no longer a separate phase. The same is true of zinc, which, by itself hcp, dissolves in copper (fcc) to produce brass, a single-phase solid solution.

In terms of the above discussion, a *solution* (liquid or solid) is a phase with more than one *component.* A *mixture* is a material with more than one *phase.*

There are many crystalline phases because there are innumerable permutations and combinations of atoms, or groups of atoms. There are relatively few amorphous phases because, lacking long-range order, their atomic arrangements are less definite and permit a greater range of solution than do crystals.† There is only one gaseous phase. The atoms or molecules are far apart and randomly distributed; as a result additional vapor components may be introduced into one "structure." No discontinuities are observed in a gas other than at the atomic or molecular level.

---

* $R_{Cu} = 1.278$ Å; $R_{Ag} = 1.444$ Å.

† The fiberglass and plastic just cited are both amorphous. While their structures are not sufficiently similar to produce a single phase, each can be a solvent for large quantities of solutes.

Example 4–5.1 From Fig. 4–5.2, calculate the packing factor (a) of solid lead at 326°C; (b) of liquid lead at 328°C. (Assume the hard-ball radius of 1.750 Å is retained.)

Solution: Lead is fcc with a calculated APF of 0.74 (see Example 3–3.1(a)); thus an increase in volume decreases the APF accordingly.

a) $$\text{APF}_{326°\text{C}} = 0.74/1.021 = 0.725;$$

b) $$\text{APF}_{328°\text{C}} = 0.74/1.06 = 0.698.$$

Comment. One could suggest that the radius increases by 0.7 l/o from 1.750 Å at 20°C to 1.762 Å at 326°C to give the 2.1 v/o expansion. If so, the packing factor at 326°C would remain at 74%, and the packing factor of the liquid becomes 0.713. One cannot argue against that suggestion, since one definition of radius is one-half of the closest interatomic distance. In either event, there is an abrupt discontinuity in packing efficiency at the melting temperature. ◀

4–6 ATOMIC VIBRATIONS

The atoms of a material become static only at 0°K (−273°C or −460°F). Under that condition the atoms settle down to their lowest energy positions among their neighbors (Fig. 2–5.2(b)). As the temperature is increased, the increased energy permits the atoms to vibrate into greater and shorter interatomic distances. However, as can be noted by the energy well of that earlier figure, the displacements in the two directions are not symmetric; i.e., for a given energy level (temperature), the atoms can move farther apart more readily than they can be pushed together. This produces a *thermal expansion* because the mean interatomic distance is increased.

Even at a given temperature, not all atoms (or molecules) have identical energies at any specific *instant* of time. Rather there is a spectrum of energy among the atoms which extends from near zero to very high values. Of course the majority of the atoms have energies somewhere near the mean value. Conversely, over a period of time, a specific *atom* will experience a range of energies which extends from near zero to very high values. The majority of the time, however, its energy is somewhere near the mean value. We will need to look at this *energy spectrum* so we can predict the probability of an atom having enough energy to break its bond and jump to new locations, a process we call *diffusion*.

Thermal expansion. In the vicinity of our ambient temperature, the heat capacity of many solids is essentially constant. Therefore, we can assign a succession of equal temperature intervals to the energy trough (Fig. 4–6.1(a)). With successive energy increases, the median interatomic spacing increases; hence, the thermal expansion we spoke of two paragraphs back. Note two features, however. There is less dimensional change per ΔT in a strongly-bonded (high-melting) material characterized by a deep energy trough, than in a weakly-bonded material (Fig. 4–6.1(b)). This is illustrated in Fig. 4–6.2, a graphical presentation of the data given in generalization (5) on page 60. Figure 4–6.2 emphasizes that our generalization is limited to comparable materials.

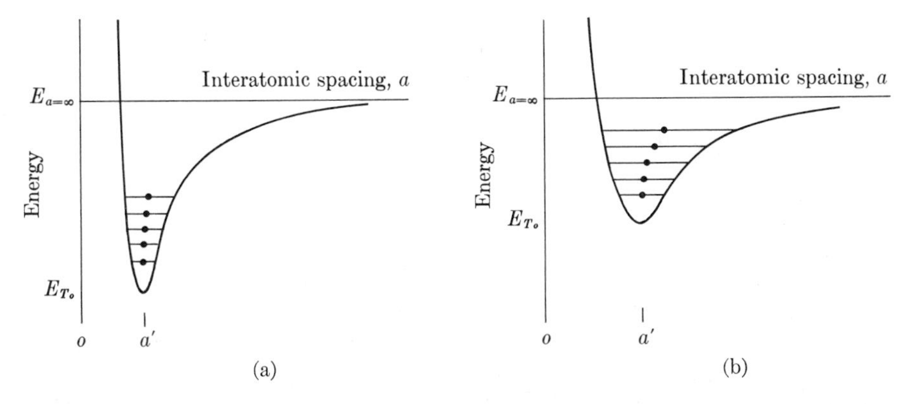

Fig. 4–6.1. Energy and expansion. (a) Strongly bonded solid. (b) Weakly bonded solid. With equal additions of thermal energy, the mean interatomic spacing changes less in a material with a deeper energy trough (cf. Fig. 2–5.2(b)). The expansion becomes more pronounced at higher temperatures (higher energy).

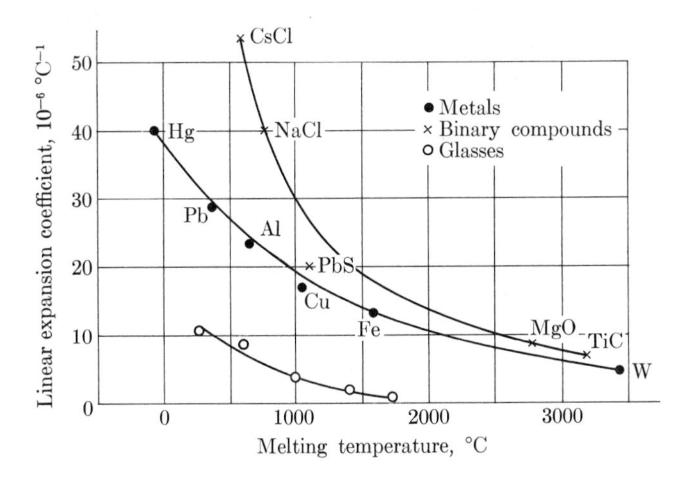

Fig. 4–6.2. Melting temperatures and expansion coefficients (20°C). The more strongly bonded, higher melting materials have lower expansion coefficients (cf. Fig. 4–6.1). Comparisons must be made between comparable materials.

The second feature of Fig. 4–6.1 is that, as the temperature increases, the change in dimension becomes relatively more pronounced. The median curves of Fig. 4–6.1 bend to the right. This explains the data of Fig. 1–3.1 where the expansion coefficients of copper and iron increased with temperature.

Thermal expansion is *isotropic* only in cubic and amorphous materials. In other crystals, the amount of expansion varies with orientation. Consider graphite (Fig. 2–2.6); the structure is markedly *anisotropic*. As a result, the thermal agitation expands the lattice more in the vertical direction than in the two horizontal directions.* This is not surprising because the bonds between layers are weaker than those within the layers.

Thermal energy distribution. The total kinetic energy, K.E., of a mole of gas increases in proportion to the temperature T, so that the equation

$$\text{K.E.} = \tfrac{3}{2}RT \tag{4–6.1}$$

is appropriate. The R of this equation is the same gas constant encountered in introductory chemistry courses, where its units are commonly reported as 1.987 cal/mole · °K. If we switch to joules of the SI units and pay attention to the individuals instead of moles (or 0.602×10^{24}), the value becomes 13.8×10^{-24} J/°K. This is called *Boltzmann's constant* and is identified by k. Thus, the average K.E. of an individual molecule of gas is

$$\text{K.E.} = \tfrac{3}{2}kT. \tag{4–6.2}$$

However, as discussed in the second paragraph of this section, this does not imply that all molecules of air in this room have this same energy. Rather, there will be a statistical distribution of energies as indicated in Fig. 4–6.3. At any particular instant of time, a very few molecules will have nearly zero energy; many molecules will have energies near to the average energy, and some molecules will have extremely high energies. As the temperature increases, there is (1) an increase in the average energy of the molecules, and (2) an increase in the number of molecules with energies in excess of any specified value.

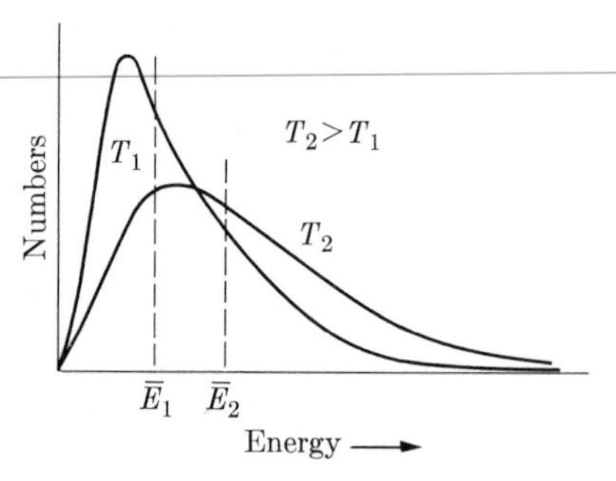

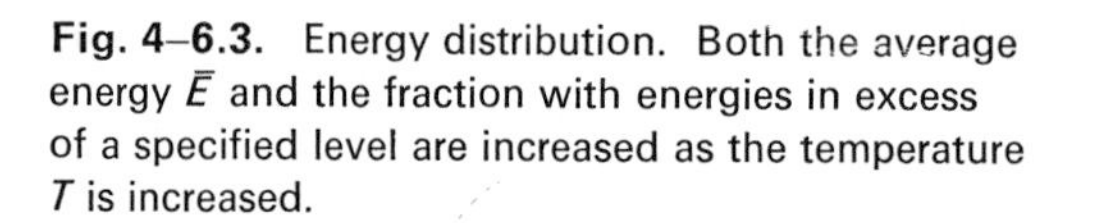

Fig. 4–6.3. Energy distribution. Both the average energy $\bar{E}$ and the fraction with energies in excess of a specified level are increased as the temperature T is increased.

The above applies to the kinetic-energy distribution of molecules in a gas. However, the same principle applies to the distribution of vibrational energy of atoms in a

* At 700°C, for example, the thermal-expansion coefficient of graphite is only 0.8×10^{-6}/°C within the plane of the layers, but 29×10^{-6}/°C perpendicular to the layers. Both of these values decrease at lower temperatures, and in fact, the first coefficient passes through zero at 400°C, and becomes slightly negative at room temperature. The second value ($\perp$ to layers) is 26×10^{-6}/°C at 20°C, so that the overall volume coefficient $\alpha_V \simeq \alpha_x + \alpha_y + \alpha_z$ is positive. (See Example 4–6.1.)

liquid or solid. Specifically, at any particular instant of time, a negligible number of atoms will have zero energy; many atoms will have energies near the average energy; and some atoms will have extremely high energies.

Our interest will be directed toward those atoms which have high energies. Very often we should like to know the probability of atoms possessing more than a specified amount of energy, e.g., what fraction of the atoms has energy greater than E of Fig. 4–6.4. The statistical solution to this problem has been worked out by Boltzmann as follows:

$$\frac{n}{N_{\text{tot}}} \propto e^{-(E-\bar{E})/kT}, \tag{4–6.3(a)}$$

where k is the previously described Boltzmann's constant of 13.8×10^{-24} J/°K. The number n of atoms with an energy greater than E, out of the total number N_{tot} present, is a function of the temperature T. This equation is applicable when E is considerably in excess of the average energy $\bar{E}$, so that the equation reduces to

$$\frac{n}{N_{\text{tot}}} = Me^{-E/kT}, \tag{4–6.3(b)}$$

where M is the proportionality constant. As presented, the value of E must be expressed in joules/atom; however, conversions may be made from other units by means of Appendix A.

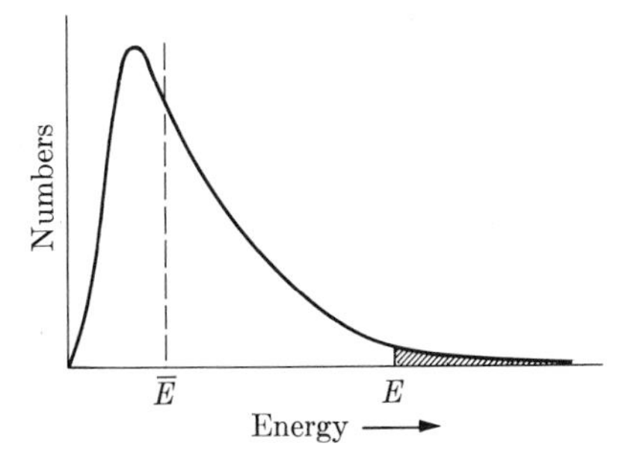

Fig. 4–6.4. Energies. The ratio of the number of high-energy atoms (shaded) to total number of atoms is an exponential function $(-E/kT)$ when $E \gg \bar{E}$.

Example 4–6.1 At 700°C, the linear-expansion coefficients, α, for graphite are as follows:

$$\alpha_{\perp} = 29 \times 10^{-6}/°\text{C},$$

$$\alpha_{\parallel} = \alpha_{\parallel} = 0.8 \times 10^{-6}/°\text{C}$$

where $\alpha_{\perp}$ is the expansion coefficient perpendicular to the graphite layers of Fig. 2–2.6, and $\alpha_{\parallel}$ is the coefficient in the two directions parallel to the layers. What is the volume increase in graphite between 600°C and 800°C?

Solution: Since $V = L^3$, and $V + \Delta V = (L + \Delta L_\perp)(L + \Delta L_\parallel)^2$,

$$\Delta V = L^2(\Delta L_\perp + 2\Delta L_\parallel + \cdots),$$

$$\Delta V/V \simeq \Delta L_\perp/L + 2\Delta L_\parallel/L,$$

$$\alpha_V\,\Delta T \simeq (\alpha_\perp + 2\alpha_\parallel)\,\Delta T = (29 + 1.6)(10^{-6}/°C)(200°C)$$

$$= 0.6\ \text{v/o}.$$

Comments. In general,

$$\alpha_V \simeq \alpha_x + \alpha_y + \alpha_z, \tag{4-6.4}$$

where the three subscripts at the right refer to the linear-expansion coefficient of the three coordinate directions. ◀

Example 4-6.2 At 500°C (773°K), one out of 10^{10} atoms has the energy required to jump out of its lattice site into an interstitial position. At 600°C (873°K), this fraction is increased to 10^{-9}. a) What is the energy required for this jump? b) What fraction has enough energy at 700°C (973°K)?

Solution: From Eq. (4-6.3(b)),

$$\ln\left(\frac{n}{N_{\text{tot}}}\right) = \ln M - \frac{E}{kT} \tag{4-6.5(a)}$$

or, with base 10 logarithms,

$$\log\left(\frac{n}{N_{\text{tot}}}\right) = C - \frac{E}{2.3kT}. \tag{4-6.5(b)}$$

The value of C is a constant.

a)

$$-10 = C - E/2.3(13.8 \times 10^{-24}\ \text{J/°K})(773°\text{K}),$$

and

$$-9 = C - E/2.3(13.8 \times 10^{-24}\ \text{J/°K})(873°\text{K}).$$

Solving simultaneously,

$$C = -1.27,$$

and

$$E = 0.214 \times 10^{-18}\ \text{J};$$

or, in terms of a mole,

$$E = 128{,}900\ \text{J/mole} \qquad (= 30{,}800\ \text{cal/mole}).$$

b)

$$\log\left(\frac{n}{N_{\text{tot}}}\right) = -1.27 - (0.214 \times 10^{-18}\ \text{J})/2.3(13.8 \times 10^{-24}\ \text{J/°K})(973°\text{K}),$$

$$n/N_{\text{tot}} = 6.3 \times 10^{-9}.$$

Comments. The relationship shown in Eq. (4–6.5) has a logarithmic value which is lineal with reciprocal temperature, $1/T$,

$$\log x = C + \frac{B}{T}. \tag{4–6.6}$$

Equation (4–6.6) is called an *Arrhenius* equation. It is widely encountered in any situation where temperature is the driving force. Examples include the intrinsic conductivity of semiconductors, the catalytic reaction of emission control, the diffusion processes in metals, the creep in plastics, and the viscosity of fluids. We will encounter this relationship again. ◀

4–7 ATOMIC REARRANGEMENTS

As the temperature is increased and the atoms vibrate more energetically, a small fraction of the atoms will relocate themselves in the lattice. Example 4–6.2 related this fraction to temperature. Of course, the fraction depends not only on temperature, but also on how tightly the atoms are bonded in position. The energy requirement for an atom to change position is called the *activation energy*. We will identify this energy later as E per atom, or Q per mole (Table 4–7.2).

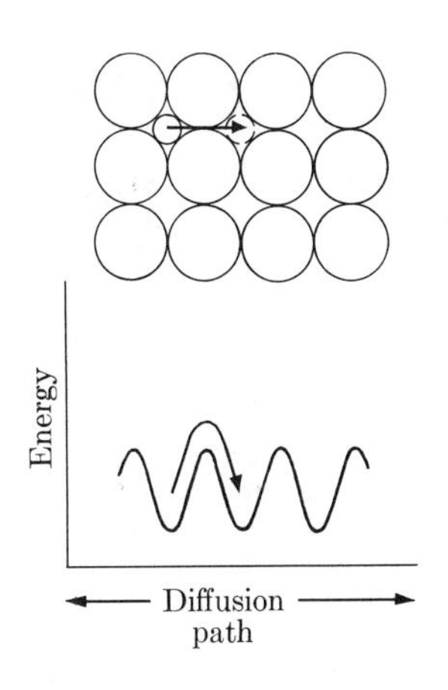

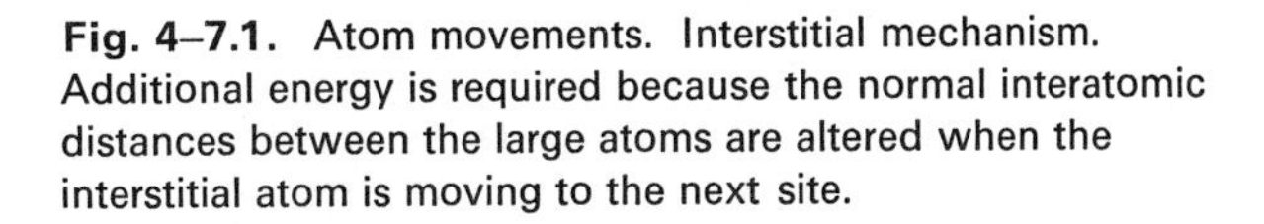

Fig. 4–7.1. Atom movements. Interstitial mechanism. Additional energy is required because the normal interatomic distances between the large atoms are altered when the interstitial atom is moving to the next site.

Let us use Fig. 4–7.1 to illustrate activation energy schematically. A carbon atom is small ($r \simeq 0.7$ Å) and can sit interstitially among a number of fcc iron atoms. If it has enough energy,* it can squeeze between the iron atoms to the next interstice when it vibrates in that direction. At 20°C there is only a small probability that it will have that much energy. At higher temperatures, the probability increases (cf. Example 4–6.2).

---

* About 34,000 cal/mole. In other units, this is about 0.24×10^{-18} J/atom, or about 1.5 eV/atom.

Other diffusion mechanisms are sketched in Fig. 4–7.2. When all the atoms are the same size, or nearly so, the vacancy mechanism becomes predominant. The vacancies may be present, either as part of a defect structure (Fig. 4–3.2), or because of extensive thermal agitation (e.g., Example 4–4.2, where aluminum was just 10° below its melting point).

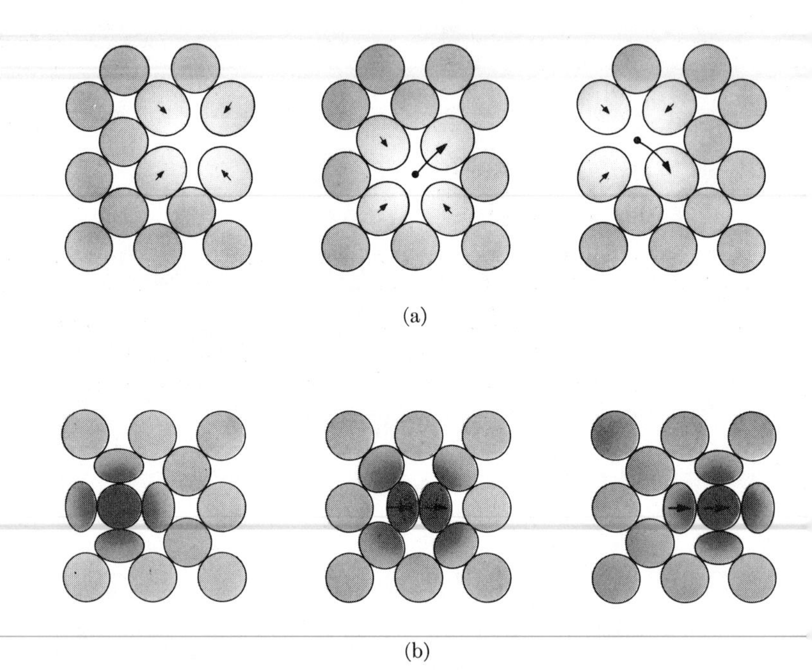

Fig. 4–7.2. Diffusion mechanisms. (a) By vacancies. (b) By interstitialcies. The vacancies move in the opposite direction from the diffusing atoms. (In the next chapter we shall observe an analogy for the movements of electron holes.) The movements follow "random-walk" statistics.

Diffusion coefficients. When an atom moves into a vacancy, a new hole is opened. In turn, this may receive an atom from *any* of the neighboring sites. As a result, a vacancy makes a "random walk" through a crystal. The same random-walk mechanism may be described for a carbon atom moving among iron atoms from interstice to interstice. However, we frequently encounter *concentration gradients.* As an example, assume there is one carbon atom per 20 unit cells of fcc iron at point (1), and only one carbon atom per 30 unit cells at point (2), which is one millimeter away. Now, since there are random movements of carbon atoms at *each* point, we will find a net flux of carbon atoms from point (1) to point (2), simply because there are half again as many atoms jumping in the vicinity of point (1). (See Example 4–7.1.) This net flow of atoms (or molecules) is called *diffusion.*

Table 4–7.1
DIFFUSION COEFFICIENTS*

Solute	Solvent (host structure)	Diffusion coefficient, cm^2/sec	
		500°C (930°F)	1000°C (1830°F)
1. Carbon	fcc iron	(5×10^{-11})**	3×10^{-7}
2. Carbon	bcc iron	10^{-8}	(2×10^{-5})
3. Iron	fcc iron	(2×10^{-19})	2×10^{-12}
4. Iron	bcc iron	10^{-16}	(3×10^{-10})
5. Nickel	fcc iron	(10^{-19})	2×10^{-12}
6. Manganese	fcc iron	(3×10^{-20})	10^{-12}
7. Zinc	Copper	4×10^{-14}	5×10^{-9}
8. Copper	Aluminum	4×10^{-10}	10^{-6} M†
9. Copper	Copper	10^{-14}	2×10^{-9}
10. Silver	Silver (crystal)	10^{-13}	10^{-8} M
11. Silver	Silver (grain boundary)	10^{-7}	—
12. Carbon	hcp titanium	3×10^{-12}	(2×10^{-7})

* Calculated from data in Table 4–7.2.
** Parentheses indicate that the phase is metastable.
† M—Calculated, although temperature is above melting point.

The *flux* J of atoms (expressed in atoms/$cm^2 \cdot$ sec) is proportional to the concentration gradient, $(C_2 - C_1)/(x_2 - x_1)$. In mathematical terms:

$$J = -D\frac{dC}{dx}. \tag{4–7.1*}$$

The proportionality constant D is called the *diffusion coefficient.* The negative sign indicates that the flux is in the down-hill gradient direction.

Diffusion coefficients vary with the nature of the solute atoms, with the nature of the solid structure, and with changes in temperature. Several examples are given in Table 4–7.1. Some reasons for the various values of Table 4–7.1 are:

1. Higher temperatures provide higher diffusion coefficients, because the atoms have higher thermal energies and therefore greater probabilities of being activated over the energy barrier between atoms (Fig. 4–7.1).

---

* This is called *Fick's first law.* There is also a Fick's second law,

$$\frac{\partial C}{\partial t} = D\left(\frac{\partial^2 C}{\partial x^2}\right), \tag{4–7.2}$$

which shows the rate at which the concentration will change with time. In fact, the values of $\partial C/\partial t$ and $\partial^2 C/\partial x^2$ were determined experimentally in the laboratory to calculate the values of D found in Table 4–7.1. [See Van Vlack, *Materials Science for Engineers*, Reading, Mass.: Addison-Wesley (1970), p. 171f.]

2. Carbon has a higher diffusion coefficient in iron than does nickel in iron because the carbon atom is a small one (Appendix B).

3. Copper diffuses more readily in aluminum than in copper because the Cu–Cu bonds are stronger than the Al–Al bonds (as evidenced by their melting temperatures).

4. Atoms have higher diffusion coefficients in bcc iron than in fcc iron because the former has a lower atomic packing factor (0.68 versus 0.74). (We shall observe later that the fcc structure has larger interstitial holes; however, the passageways between the holes are smaller than in the bcc structure.)

5. The diffusion proceeds more rapidly along the grain boundaries because this is a zone of crystal imperfections (Fig. 4–4.7).

• **Diffusion coefficients versus temperature.** The discussion of Section 4–6 related the distributions of thermal energy to temperature. Boltzmann was able to quantify this with Eq. (4–6.3), which showed that the number of atoms that have more than a specified amount of energy increases in proportion to an exponential function that includes that energy and the reciprocal of temperature. With diffusion, the *activation energy* for atom movements corresponds to the energy E of Boltzmann's equation. Thus,

$$D = D_0 e^{-E/kT}, \tag{4–7.3(a)}$$

where D_0 is the proportionality constant that relates the logarithm of the diffusion coefficient to the reciprocal of the temperature, $1/T$:

$$\ln D = \ln D_0 - \frac{E}{kT}. \tag{4–7.3(b)}$$

The other term is the same Boltzmann constant k as in Eq. (4–6.3); that is, 13.8×10^{-24} J/°K.

The data of Table 4–7.2 provide us with the values of D_0 and the activation energy for a number of diffusion reactions. Since we think in terms of base-10 logarithms more commonly than base e, and the chemist prefers molar and calorie units, we can rewrite the above equation

$$\log D = \log D_0 - \frac{Q}{4.575T}. \tag{4–7.4}$$

Energy is expressed as Q (cal/mole), while the value 4.575 includes the gas constant R (1.987 cal/mole · °K) and 2.30. . ., the conversion from base-e to base-10 logarithms.

The twelve sets of data in Tables 4–7.1 and 4–7.2 are plotted in Fig. 4–7.3. The Arrhenius-type plots* may be established from either Eq. (4–7.3(a)) or Eq. (4–7.4).

* See the comments after Example 4–6.2.

Table 4–7.2

DIFFUSION COEFFICIENTS* ($\log D = \log D_0 - Q/2.3RT = \log D_0 - E/2.3\ kT$)†

Solute	Solvent (host structure)	D_0, cm²/sec	Q, cal/mole	E, J/atom
1. Carbon	fcc iron	0.2	34,000	0.236×10^{-18}
2. Carbon	bcc iron	2.2	29,300	0.204
3. Iron	fcc iron	0.22	64,000	0.445
4. Iron	bcc iron	2.0	57,500	0.400
5. Nickel	fcc iron	0.77	67,000	0.465
6. Manganese	fcc iron	0.35	67,500	0.469
7. Zinc	Copper	0.34	45,600	0.317
8. Copper	Aluminum	0.15	30,200	0.210
9. Copper	Copper	0.2	47,100	0.327
10. Silver	Silver (crystal)	0.4	44,100	0.306
11. Silver	Silver (grain boundary)	0.14	21,500	0.149
12. Carbon	hcp titanium	5.1	43,500	0.302

* See Askill, J., *Tracer Diffusion Data for Metals, Alloys, and Simple Oxides*, New York: Plenum, (1970), for a more complete listing of diffusion data.

† $R = 1.987$ cal/mole · °K; $k = 13.8 \times 10^{-24}$ J/atom · °K.

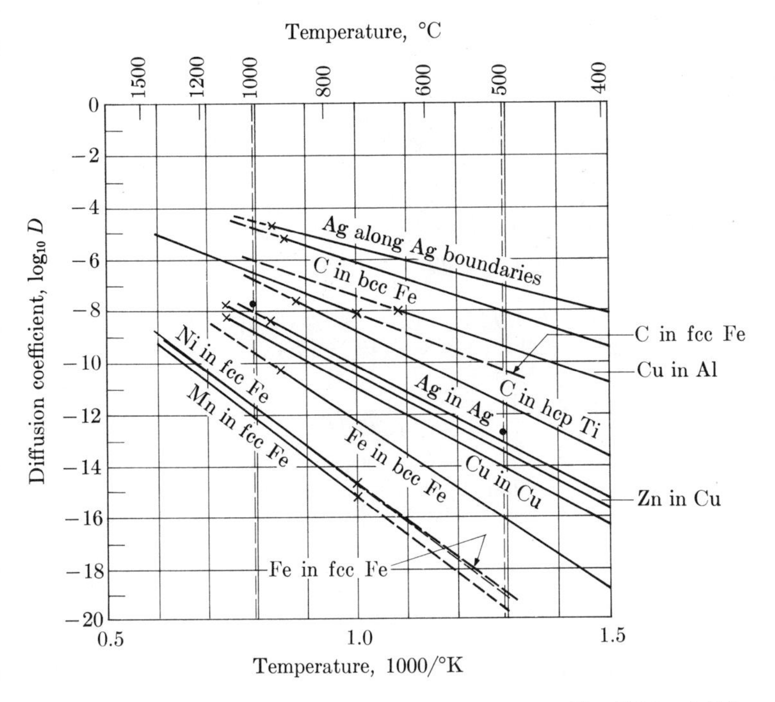

Fig. 4–7.3. Diffusion coefficients versus temperature. (See Tables 4–7.1 and 4–7.2.)

Example 4–7.1 At the surface of a steel bar there is one carbon atom per 20 unit cells of iron. One millimeter behind the surface, there is one carbon atom per 30 unit cells. The diffusion coefficient at 1000°C is 3×10^{-7} cm²/sec. The structure is fcc (a = 3.65 Å at 1000°C). How many carbon atoms diffuse through each unit cell per minute?

Solution: Carbon concentrations are:

$$C_2 = 1/(30(3.65 \times 10^{-8} \text{ cm})^3) = 0.68 \times 10^{21}/\text{cm}^3;$$

$$C_1 = 1/(20(3.65 \times 10^{-8} \text{ cm})^3) = 1.03 \times 10^{21}/\text{cm}^3.$$

From Eq. (4–7.1),

$$J = -(3 \times 10^{-7} \text{ cm}^2/\text{sec})(0.68 - 1.03)(10^{21}/\text{cm}^3)/(0.1 \text{ cm}) = 1.05 \times 10^{15}/\text{cm}^2 \cdot \text{sec}.$$

Each unit cell has an area of $(3.65 \times 10^{-8} \text{ cm})^2$. Therefore,

$$J_{\text{u.c.}} = (1.05 \times 10^{15}/\text{cm}^2 \cdot \text{sec})(3.65 \times 10^{-8} \text{ cm})^2(60 \text{ sec/min}) = 84 \text{ atoms/min}.$$

Comments. It is apparent that a piece of steel is not a static material, but that, in fact, changes can occur within it.

We use the above process to *carburize* steel. In Chapter 13 we will see how this can be used to advantage to modify the surface hardness of a steel. ◀

• **Example 4–7.2** The diffusion coefficient of aluminum in copper is 2.6×10^{-13} cm²/sec at 500°C and 1.6×10^{-8} cm²/sec at 1000°C. a) What is the diffusion coefficient at 750°C? b) Determine the values of D_0, Q, and E for this diffusion couple.

Solution: Using the *Arrhenius plot* (log vs. $1/T$) of Fig. 4–7.3, we can interpolate a value of $\sim 10^{-9.6}$ cm²/sec for $D_{750°\text{C}}$; or, using mathematics with Eq. (4–7.4),

$$\log 2.6 \times 10^{-13} = \log D_0 - \frac{Q}{4.575(773)},$$

$$\log 1.6 \times 10^{-8} = \log D_0 - \frac{Q}{4.575(1273)}.$$

Solving simultaneously, we get:

$$D_0 = 0.39 \text{ cm}^2/\text{sec},$$

and

$$Q = 43{,}000 \text{ cal/mole} \qquad (= 0.3 \times 10^{-18} \text{ J/atom}).$$

From Eq. (4–7.4),

$$\log D = \log 0.39 - (43{,}000\ \text{cal/mole})/(4.575\ \text{cal/mole}\cdot{}^\circ\text{K})(1023^\circ\text{K})$$
$$D = 2.5 \times 10^{-10}\ \text{cm}^2/\text{sec};$$

or, from Eq. (4–7.3),

$$\ln D = \ln 0.39 - (0.3 \times 10^{-18}\ \text{J})/(13.8 \times 10^{-24}\ \text{J}/^\circ\text{K})(1023^\circ\text{K}),$$

and therefore,

$$D = 2.5 \times 10^{-10}\ \text{cm}^2/\text{sec}.$$

Comments. Observe that the diffusion coefficient of copper through aluminum is higher than for aluminum through copper. This is to be expected from our knowledge of the bond strength of the host metal:

1. (T_m of Cu) > (T_m of Al);
2. $R_{Cu} < R_{Al}$. ◀

REVIEW AND STUDY

In Chapter 3 attention was given to the preciseness and regularity of the atomic structure of crystals. They will serve as a basis for considering many properties and characteristics of materials, such as density, anisotropy, slip planes, phase stability, piezoelectricity, semiconducting compounds, etc. At the same time, crystals are not perfect, and many important properties and behaviors of materials arise from the irregularities. We cited some of these in the preview of the chapter. We must consider the *disorder* of materials as well as their order.

SUMMARY

Impurities are found in all materials unless special means are used to reduce them to a low level. An alloy is a metal with impurities which are intentionally present. Solid solutions contain atoms (or molecules) of a second component as a solute. These may be present either interstitially among the solvent atoms, or as a substitute replacement within the crystal lattice. In order for solid solution to be extensive in crystals, the solute and solvent must be comparable in size and electronic behavior.

Imperfections are found in all crystals unless special means are used to reduce them to a low level. It is convenient to categorize them by geometry:

1. Point defects, that include vacancies and/or interstitials;
2. Lineal defects, commonly called dislocations; and
3. Boundaries, which may be external surfaces or internal separations between grains or phases. The latter may be treated as two-dimensional defects.

Liquids generally lack the long-range order that characterizes crystals. For some materials, it is possible to avoid crystallization and retain the amorphous character of a liquid into a solid, (1) if the anticipated crystal structure is complex, or (2) by rapid cooling. We label these amorphous solids by the term *glass*. Although lacking a freezing temperature, amorphous materials possess a glass temperature T_g, which will be very important to us when we study the behavior of plastics.

A *phase* was defined as that part of a material which is distinct from others in structure and/or composition.

At any instant of time, most atoms or molecules possess near-average energy. However, some will possess very little energy, and some will possess abnormally high energies. We are interested in the high end of this statistical distribution, because those atoms may be activated to break bonds, and to diffuse to new locations within the material. It is only through this mechanism that the internal structure (and hence the properties) of the material may be modified. Diffusion occurs more readily (1) at high temperatures, (2) when the diffusing atom is small (e.g., carbon in iron), (3) when the packing factor of the host structure is low (e.g., bcc versus fcc), (4) when the bonds of the host structure are weak (e.g., low-melting materials), and (5) when there are defects in the material (e.g., vacancies or grain boundaries).

TERMS FOR REVIEW

- Activation energy, E or Q

Alloy
Amorphous
Anisotropic
Arrhenius equation
ASTM G.S.#
Boltzmann's constant, k
Brass
Bronze
Component
Concentration gradient
Defect structures
Diffusion
Diffusion coefficient, D
Dislocation
Energy distribution
Fick's first law
Flux (diffusion), J
Glass
Grain

TERMS FOR REVIEW—CONTINUED

Grain boundary
Heat of fusion
Imperfection
Interstice
Mixture
Nonstoichiometric compounds
Phase
Phase boundary
Point defects
Solid solutions,
  interstitial
  ordered
  substitutional
Solubility limit
Stoichiometric
Thermal expansion
Vacancy

FOR DISCUSSION

A_4 What is the difference between brass and bronze?

B_4 Distinguish between an alloy which is a mixture and one which is a solid solution.

C_4 Distinguish between solvent and solute, as applied to solid solutions.

D_4 Distinguish between interstitial and substitutional solid solutions.

E_4 From Appendix B, select elements which have a favorable size for substitutional solid solution with metallic iron. Select those ions which might substitute for Fe^{2+} ions.

F_4 There is a good argument for considering that an ordered solid solution is a compound. Discuss.

G_4 Nonstoichiometric AlMg does not possess a significant number of vacancies; nonstoichiometric FeO does. Explain the difference.

H_4 The ordered solid solution, $AuCu_3$, is cubic and at low temperatures preferentially holds gold atoms at 0,0,0 and copper atoms at:

$$\tfrac{1}{2},\tfrac{1}{2},0 \qquad \tfrac{1}{2},0,\tfrac{1}{2} \qquad 0,\tfrac{1}{2},\tfrac{1}{2}.$$

To what space lattice does it belong?

I_4 The $AuCu_3$ of question H_4 randomizes above 380°C, so that each of the above cubic sites has a 25% probability of possessing a gold atom (and 75% for Cu). To what space lattice does it now belong?

J_4 Why do Schottky defects occur in pairs?

K_4 Explain the origin of surface energy; of grain-boundary energy.

L_4 Cite various ways grain boundaries affect the behavior of materials.

M_4 Magnification is always expressed on the basis of the lineal multiplication factor. With ×100, what is the area multiplication factor?

N_4 G.S. #5 has ____ times as many grains per square inch in a two-dimensional section as G.S. #3. The lineal dimensions are ____ times as large. The grain boundary area is ____ times as much per grain, but ____ times as much per unit volume.

O_4 Compare and contrast the change that occurs at 380°C in $AuCu_3$ (Review question I_4) with melting.

P_4 Why is energy given off during the freezing of water (80 cal/g, or 6 kJ/mole)?

Q_4 The heat of fusion of copper is higher than that of zinc. Why?

R_4 The volume thermal-expansion coefficient of a liquid is almost always higher than that of the corresponding solid. Why?

S_4 Explain the glass temperature, T_g, to a friend who is not in this course.

•T_4 Which will be higher, the *mean energy* or the *median energy* of the gas molecules in this room?

U_4 Sketch an energy-vs.-distance curve for an atom adjacent to a vacancy. (Cf. Fig. 4–7.1.)

•V_4 What is meant by activation energy? By phase?

W_4 Refer to Table 4–7.1. Why are the values higher for couple #2 than for couple #1? For couple #2 than for couple #4? For couple #11 than for couple #10? For couple #8 than for couple #9?

•X_4 Check out the units in Eq. (4–7.4).

•Y_4 Fluidity (the reciprocal of viscosity) varies with temperature according to Arrhenius-type equations (Eqs. 4–6.6 and 4–7.3). Compare that property with diffusion.

Z_4 Self-diffusion is the movement of atoms within their own structure. Explain how this can occur. Suggest a way to measure it in the absence of a concentration gradient, through the use of radioactive isotopes.

STUDY PROBLEMS

4–2.1 An alloy contains 85 w/o copper and 15 w/o tin. Calculate the atomic percent (a/o) of each element.

Answer: 8.6 a/o Sn; 91.4 a/o Cu

4–2.2 There is 5 a/o magnesium in an Al–Mg alloy. Calculate the w/o magnesium.

4–2.3 Consider Fig. 4–2.3 to be an interstitial solution of carbon in fcc iron. What is the w/o carbon present?

Answer: 1.3% carbon

4–2.4 Consider Fig. 4–2.1 to be a substitutional solid solution of cadmium and magnesium. What is the w/o Cd present if (a) Cd is the more prevalent atom? (b) Mg is the more prevalent atom? (Cd = 112.4 amu.)

4–2.5 a) An alloy containing 75 w/o Cu and 25 w/o Zn has ____ a/o Cu and ____ a/o Zn. b) How much would each unit cell of this alloy weigh? c) The density of this brass is 8.5 gm/cm^3. What is the volume, and d) what is the mean lattice constant of each unit cell?

Answer: a) 75.5 a/o Cu, 24.5 a/o Zn b) 4.25×10^{-22} g/uc c) 5×10^{-23} cm^3/uc d) 3.69 Å

4–2.6 An alloy contains 80 w/o Ni and 20 w/o Cu in substitutional fcc solid solution, with a = 3.54 Å. Calculate the density of this alloy.

4–2.7 If 1.0% carbon, by weight, is present in an fcc iron, what percent of the unit cells have carbon atoms?

Answer: 19% of the unit cells have carbon.

4–3.1 a) What is the w/o FeO in the solid solution of Fig. 4–3.1? b) The w/o Fe^{2+}? c) Of O^{2-}?

Answer: a) 51 w/o b) 39.8 w/o c) 30.8 w/o

4–3.2 If all the iron ions of Fig. 4–3.1 were changed to Ni ions, what would be the w/o MgO?

4–3.3 What is the density of $Fe_{<1}O$, if the Fe^{3+}/Fe^{2+} ratio is 0.14? ($Fe_{<1}O$ has the structure of NaCl; and ($r_{Fe} + R_O$) averages 2.15 Å.)

Answer: 5.73 g/cm^3

4–4.1 Calculate the radius of the largest atom which can exist interstitially in fcc iron without crowding. [*Hint:* Sketch the (100) face of several adjacent unit cells.]

Answer: 0.53 Å

4–4.2 Determine the radius of the largest atom which can be located in the interstices of bcc iron without crowding. [*Hint:* The center of the largest hole is located at $\frac{1}{2},\frac{1}{4},0$.]

4–4.3 The number of vacancies increases at higher temperature. Between 20°C and 1020°C, the lattice constant of a bcc metal increased 0.5 l/o from thermal expansion. In the same temperature range, the density decreased 2.0%. Assuming there was one vacancy per 1000 unit cells in this metal at 20°C, estimate how many vacancies there are per 1000 unit cells at 1020°C.

Answer: 11/1000 unit cells

4–4.4 In copper at 1000°C, one out of every 473 lattice sites is vacant. If these vacancies remain in the copper when it is cooled to 20°C, what will be the density of the copper?

4–4.5 a) What is the coordination number for the interstitial site in Study Problem 4–4.2? b) How many of these sites are there per unit cell?

Answer: a) 4 b) 12

4–4.6 a) What is the coordination number of the interstitial site in Study Problem 4–4.1? b) What structure would result if *every* such site were occupied by a smaller atom or ion?

4–4.7 a) Assume that the G.S. #6 of Fig. 4–4.10 represents a two-dimensional cut through a polycrystalline solid. Estimate the corresponding grain-boundary area. b) Repeat for G.S. #2.

Answer: a) 1500 in.^{2}/in.^{3} (or 600 cm^2/cm^3) b) 150 cm^2/cm^3 (or 380 in.^{2}/in.^{3})

4–4.8 How many grains are observed in a microscope per square inch (at $\times$100) for (a) G.S. #8? (b) G.S. #5? (c) Repeat for $\times$200.

4–4.9 Assume Fig. 4–11 is at $\times$100 rather than $\times$250. a) Estimate the boundary area per cm^3. b) Per in.^{3}. c) Estimate the grain size number.

Answer: a) 140 cm^2/cm^3 b) 350 in.^{2}/in.^{3} c) Slightly less than 3

4–5.1 Based on Fig. 4–5.2, estimate the volume expansion coefficient, α_V, of (a) solid lead and (b) liquid lead at the melting point of lead.

Answer: a) 100×10^{-6}/°C b) 120×10^{-6}/°C

4–5.2 a) Estimate the heat of fusion of sodium from Table 4–5.1. b) Of silver.

4–6.1 An aluminum wire is stretched between two rigid supports at 35°C. It cools to 15°C. What stress is developed?

Answer: 31 MPa (or 4500 psi).

4–6.2 Estimate the linear exp. coef. of bcc iron at 900°C from the data in Fig. 4–5.2.

4–6.3 At 800°C, 1 out of 10^{10} atoms, and at 900°C, 1 out of 10^9 atoms, has appropriate energy for movements within a solid. a) What is the activation energy in cal/mole? b) At what temperature will 1 out of 10^8 atoms have the required amount of energy?

Answer: a) 57,500 cal/mole b) 1020°C

4–6.4 An activation energy of 2.0 eV (0.32×10^{-18} J) is required to form a vacancy in a metal. At 800°C there is one vacancy for every 10^4 atoms. At what temperature will there be one vacancy for every 1000 atoms?

4–7.1 A solid solution of copper in aluminum has 10^{20} atoms of copper per cm^3 at point X, and 10^{18} copper atoms per cm^3 at point Y. Points X and Y are 10 micrometers apart. What will be the diffusion flux of copper atoms from X to Y at 500°C?

Answer: 4×10^{13} atoms/cm$^2 \cdot$ sec

4–7.2 a) What is the ratio of diffusion coefficients for carbon in bcc iron to carbon in fcc iron at 500°C? b) Of carbon in fcc iron to nickel in fcc iron at 1000°C? c) Of carbon in fcc iron at 1000°C to carbon in fcc iron at 500°C?

4–7.3 A zinc gradient in a copper alloy is 10 times greater than the aluminum gradient in a copper alloy. Compare the flux of solute atoms/cm$^2 \cdot$ sec in the two alloys at 500°C. (The data for $D_{\text{Al in Cu}}$ are in Example 4–7.2.)

Answer: $J_{\text{Zn in Cu}}/J_{\text{Al in Cu}} = 1.5$

•**4–7.4** Aluminum is to be diffused into a silicon single crystal. At what temperature will the diffusion coefficient be 10^{-10} cm^2/sec? (Q = 73,000 cal/mole and D_0 = 1.55 cm^2/sec.)

•**4–7.5** Refer to Study Problem 4–7.1. a) What is the diffusion coefficient of copper in aluminum at 100°C? b) What will be the diffusion flux of copper atoms from X to Y at 100°C?

Answer: a) 3×10^{-19} cm^2/sec b) 30,000 atoms/cm^2 · sec

•**4–7.6** How much should the concentration gradient be for nickel in iron if a flux of 10^6 nickel atoms/cm^2 · sec is to be realized (a) at 1000°C? (b) at 1400°C?

CHAPTER FIVE

ELECTRON TRANSPORT IN SOLIDS*

* If desired, the instructor may choose to defer this chapter until later in the course with a minimum of adjustment in the intervening chapters. However, in view of the major technological role which electronic materials play in our society, the instructor is urged not to let his class skip this chapter.

PREVIEW

Atoms and their arrangements have received the bulk of our attention to date. In this chapter we will focus that attention on the electrons and their freedom to move among the atoms. Thus when we turn to metals, polymers, and ceramics in Chapters 6, 7, and 8, we will be able to consider electrical properties as well as mechanical and thermal behavior.

Our basic charge carrier is the electron with its 0.16×10^{-18} coul. We will also observe that a missing electron (either in a positive ion, or in delocalized bonds) may serve as a positive carrier of equal charge. Metals, with their weak hold on valence electrons, are good conductors both of electricity and of heat. We will observe that this conductivity occurs because very little energy is required to place delocalized electrons in a conduction level. In contrast, electrons must be raised across an energy gap before they can conduct in an insulator. If this gap is big, conduction cannot occur.

We will see that semiconductors have small energy gaps so that a useful number of electrons may be activated to conductivity levels. Also impurities may be added to modify the energy-gap "barrier" to conduction. Semiconducting materials represent an area of extended technical cooperation between the physicists, the materials scientists and engineers, and the design engineers, be they circuit designers or other types of engineers who use automatic controls and modern computers.

5–1 CHARGE CARRIERS

Various materials which are available to the engineer or scientist span a wide spectrum of conductivity (or resistivity, since $\sigma = 1/\rho$). As shown in Fig. 5–1.1, we commonly divide materials into three categories, *conductors*, *semiconductors*, and *insulators*. Metals fall in the first category, since they have delocalized electrons which are free to move throughout the structure (Sections 2–2 and 2–7). Insulators include those ceramics and polymeric materials with strongly held electrons and nondiffusing ions. Their function is to isolate neighboring conductors. It was not very long ago that only the two ends of the spectrum of Fig. 5–1.1 were considered to be useful. Today, however, the middle, semiconducting category has become exceedingly important and will, in fact, be the chief subject of this chapter.

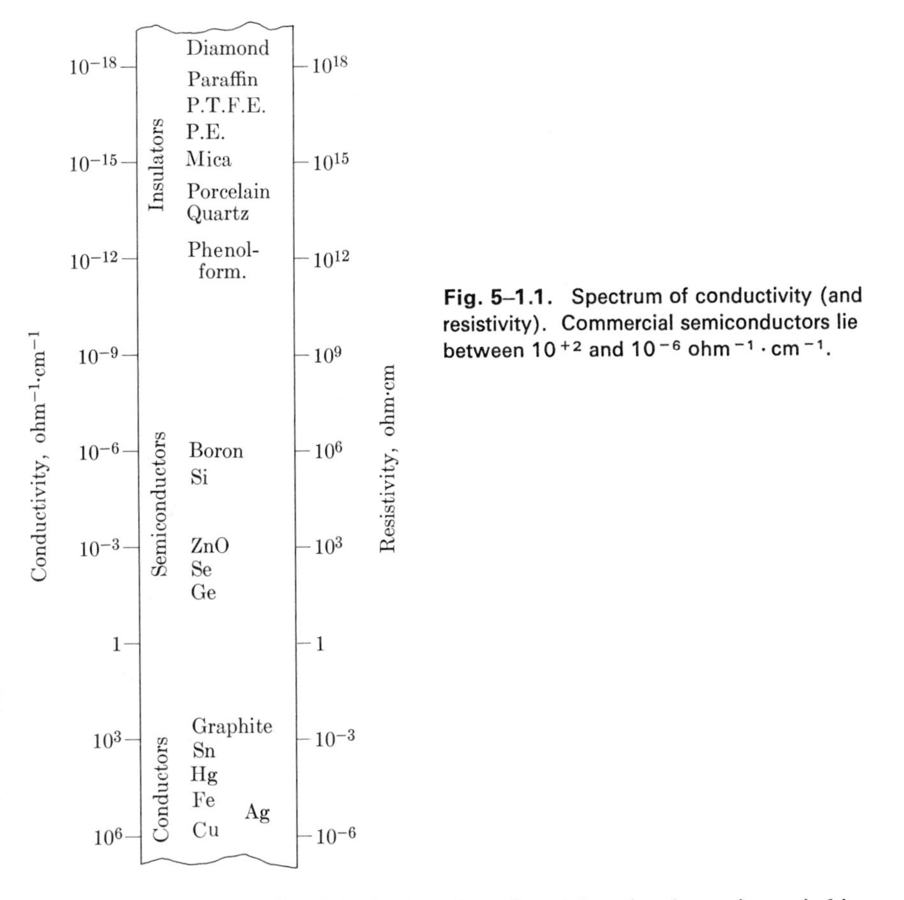

Fig. 5–1.1. Spectrum of conductivity (and resistivity). Commercial semiconductors lie between 10^{+2} and 10^{-6} ohm$^{-1}\cdot$cm^{-1}.

In those types of materials that conduct electricity, the charge is carried in modules of 0.16×10^{-18} coul, this being the charge on an individual electron. In metals, it is the individual electron that moves. In ionic materials, charge can be carried by diffusing ions. However, their charge is simply an integer number of electron charges

(− or +, for anions and cations, respectively). Thus, an SO_4^{2-} ion carries 0.32×10^{-18} coul of charge within a storage battery, and all Pb^{2+} ions have an absence of two electron charges as they move in the opposite direction.

Electrons and anions are *negative charge carriers.* In contrast, a cation such as Pb^{2+} is a *positive charge carrier* because, as we have just seen, it has an *absence* of electrons. There is another positive charge carrier which becomes important in semiconductors, *viz.*, an *electron hole.* It is an absence of an electron within the energy band for delocalized electrons discussed in Section 2–2. We will come back to these in Section 5–4.

Conductivity σ and *resistivity* ρ values for a material depend upon the number n of charge carriers per cm^3, the charge q on each, and their *mobility* μ according to Eq. (1–4.1(a)):

$$\sigma = \frac{1}{\rho} = nq\mu. \tag{5–1.1}$$

In this chapter, we shall look at atomic and structural factors which affect n and μ. This will let us anticipate factors which influence our choice of material for design and service behavior.

We can calculate the *drift velocity*, $\bar{v}$, of a charge carrier from Eq. (1–4.2),

$$\bar{v} = \mu\mathscr{E}. \tag{5–1.2}$$

Note that the drift velocity is proportional to the *electric field* $\mathscr{E}$, or volts/cm. Thus, in the absence of a voltage gradient, there is no net, or drift, velocity. This does not indicate that the charged carriers do not move; rather the movements are random, and so charge carried in one direction is balanced by charge being carried in the opposite direction.

Example 5–1.1 A semiconductor with 10^{15} charge carriers/cm^3 has a resistivity of 10 ohm · cm at 20°C. What is the drift velocity of the electrons if one ampere of current is carried across a gradient of 1.5 volts/cm?

Solution: From Eqs. (5–1.1) and (5–1.2),

$$\bar{v} = \mathscr{E}/\rho nq$$
$$= \frac{(1.5\ \text{V/cm})}{(10\ \text{ohm}\cdot\text{cm})(10^{15}/\text{cm}^3)(0.16\times 10^{-18}\ \text{amp}\cdot\text{sec})} = 940\ \text{cm/sec}. \tag{5–1.3}$$

Comments. From Eqs. (1–4.2) and (5–1.2), the mobility of the charge carriers is 625 (cm/sec)/(volt/cm). ◀

5–2 METALLIC CONDUCTIVITY

The metallic bond was described in Section 2–2 in terms of *delocalized electrons.* Specifically, the valence electrons are able to move throughout the metal. They act like standing waves. Thus, they move within the metal with as much charge carried in one direction as in the opposite direction. There is no net charge transport in the absence of the electronic field.

If the metal is placed in an electrical circuit, the electrons moving toward the positive electrode receive additional momentum, and therefore *require more energy*. Conversely, those electrons moving toward the negative electrode lose momentum and must *reduce their energy*. As a result, the *drift velocity* of Eq. (5–1.2) is developed.

Mean free path. Waves move through periodic structures without interruption. A well-ordered crystal (Chapter 3) provides one of the most regular of the periodic structures available. Thus a metallic crystal lattice provides an excellent medium for electron movements. However, any irregularity in the repetitive structures through which a wave travels deflects the wave. Thus, if an electron had been traveling toward the positive electrode and was then deflected, it would no longer continue to gain momentum. The net effect is to *reduce the drift velocity* just cited, even though we have not altered the electric field. In brief, irregularities in the lattice *decrease* the mobility of Eq. (5–1.2); therefore, they *decrease* the conductivity and *increase* the resistivity (Eq. 5–1.1).

The average distance which an electron can travel in its wavelike pattern without deflection is called the *mean free path*. We will want to identify irregularities which deflect electron movements, because that will help us understand why resistivities of metals are not all the same. We can identify two effects on the basis of Chapter 4.

Resistivity versus temperature. A metal has increasing resistivity with temperature (Fig. 5–2.1). To a first approximation it is linear (except near absolute zero). We have no basis on which to conclude that the n of Eq. (5–1.1) significantly decreases with increased temperature in a metal;* rather we must look at the mobility μ. Thermal agitation (Section 4–6) increases in intensity in proportion to increased temperature (except at very low temperatures). This increased agitation decreases

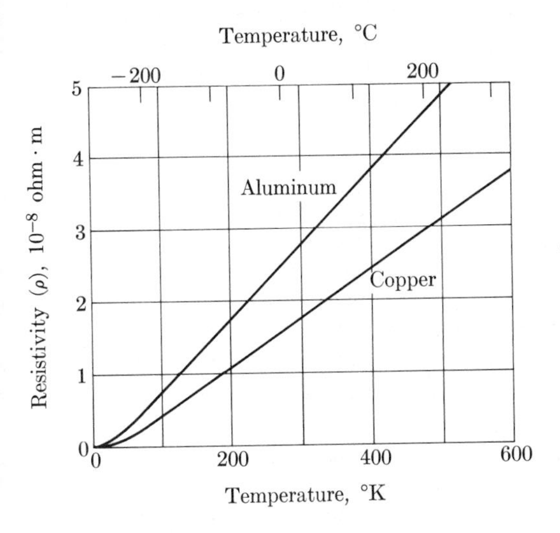

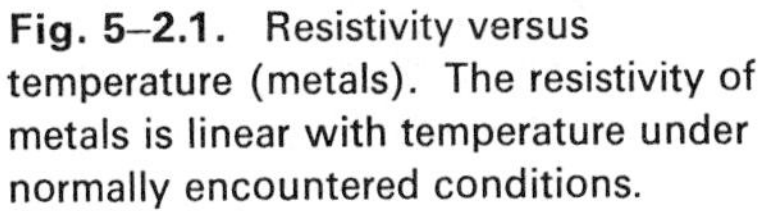
Fig. 5–2.1. Resistivity versus temperature (metals). The resistivity of metals is linear with temperature under normally encountered conditions.

* The number of charge carriers increases with increased temperature in a semiconductor (Section 5–4).

Table 5–2.1
TEMPERATURE RESISTIVITY COEFFICIENTS

Metal		Resistivity at 0°C*, microohm · cm	Temperature resistivity coefficient, y_T, °C^{-1}
Aluminum		2.7	0.0039
Copper		1.6	0.0039
Gold		2.3	0.0034
Iron		9	0.0045
Lead		19	0.0039
Magnesium		4.2	0.004
Nickel		6.9	0.006
Silver		1.7	0.0038
Tungsten		5	0.0045
Zinc		5.3	0.0037
Brass	(Cu–Zn)	6	0.002
Bronze	(Cu–Sn)	9	0.001
Constantan	(Cu–Ni)	49	0.00001
Monel	(Ni–Cu)	46	0.002
Nichrome	(Ni–Cr)	99	0.0004

* These values will not agree with those in Appendix C, since they are based on different reference temperatures.

the mean free path and, therefore, the mobility of electrons in a metal. The consequent change in resistivity is important to the engineer who is designing electrical equipment. In some cases, compensation must be introduced into a circuit to avoid an unwanted temperature sensitivity. In other cases, this temperature sensitivity provides a useful "brake." Example 5–2.1 will point this out in a familiar application (the toaster).

We may express the ρ-versus-T relationship with a *temperature resistivity coefficient* y_T. Thus,

$$\rho_T = \rho_{0°C}(1 + y_T T), \tag{5–2.1}*$$

where $\rho_{0°C}$ is the resistivity at 0°C, or 273°K. The value of this coefficient is approximately 0.004/°C for pure metals (Table 5–2.1). This suggests that the mean free path of electrons is reduced by a factor of two between 0°C and 250°C.

Resistivity in solid solutions. Another factor which can reduce the mean free path of electrons in a metal is the presence of solute atoms. A solid-solution alloy always has a lower conductivity than do its pure component metals.†

* Temperature T is expressed in °C in this equation because absolute zero, 0°K, is not in the linear part of the resistivity curves of Fig. 5–2.1.

Furthermore, the decreases in resistivity (increase in conductivity) which we are describing are not related to *superconductivity* which appears near absolute zero. That involves another phenomenon (quantum mechanical) which is beyond the scope of this book.

† Examine the data for metals and alloys in either Table 5–2.1 or Appendix C.

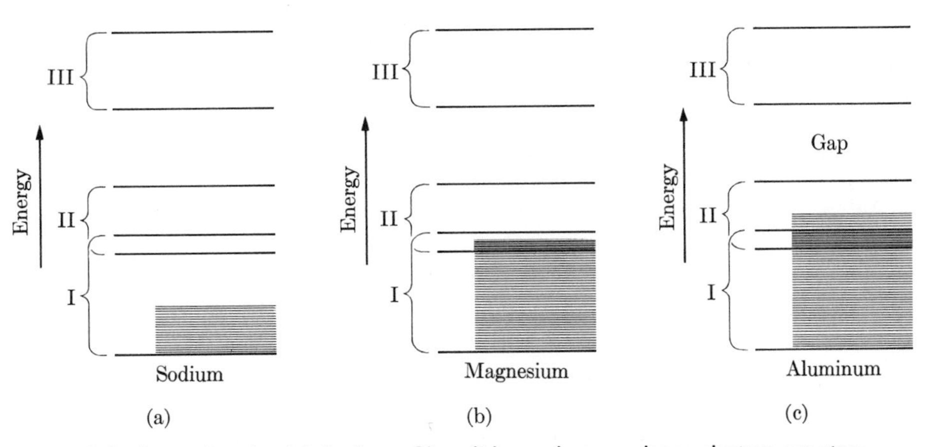

Fig. 5–2.2 Energy bands. (a) Sodium. Since it has only one valence electron per atom, its first valence band is only half filled. (b) Magnesium. Its first band would be full, except that its second band overlaps, to contain a few electrons. (c) Aluminum. With three valence electrons, its first band is filled, and its second band is half full. All of these metals have empty levels in the upper part of their valence bands.

The reason for this generalization is that an electron encounters an irregularity in the electric field when it approaches an impurity atom in the crystal lattice. In the first place, the lattice is slightly distorted in an alloy such as brass, because the atomic radii differ a few percent; in addition, a zinc atom has 30 protons rather than the 29 in copper. Although these differences seem small, they deflect additional electrons and reduce the mean free path. Since brass (70Cu–30Zn) has a resistivity 3 or 4 times as great as that of pure copper, we can assume that the mean free path for electrons is only 25–30% as long in brass as in pure copper. If reduced resistivity is paramount in a design, the engineer will turn to pure metals (Section 6–1).

Energy bands. Recall from Fig. 2–2.7, that electrons of isolated atoms occupy only specific orbitals or energy levels, and that gaps exist between these levels. In effect, the electrons establish standing waves around an atom. This pattern is also found in the inner or subvalence electrons of metals; however, the outer or valence electrons are delocalized. As a result, the valence orbitals form a band (Fig. 2–2.7(b)), and the standing wave is influenced by every atom which is involved. A consequence of this fact is that a band possesses as many discrete energy *levels* as there are atoms in the system. Since the number is exceedingly great, and the energy bands are usually only a few electron volts wide, it follows that the energy levels within a band are so infinitesimally separated that we may pretend the band forms a continuum.

A physical principle* states that only two electrons may occupy the same level (and these two must be of opposite magnetic spin). Thus, *a band may contain twice as many electrons as there are atoms.* As a result, a monovalent metal such as sodium has its first valence band only half-filled (Fig. 5–2.2(a)). Since aluminum has three

* Pauli exclusion principle.

valence electrons per atom, its first valence band is filled and its second band is half full (Fig. 5–2.2(c)). Naturally the lower energy levels of the band fill first.

Definition of a metallic conductor. We characterized metals in Section 2–7 by their "ability to give up valence electrons," and pointed out that they thus were conductors. We now have a better definition of metallic conductors, *viz.*, they have *unfilled valence bands.* Figures 5–2.2(a) and (c) show this schematically for sodium and aluminum.

The empty energy levels are important for conduction because they permit an electron to rise to a higher level when it gains momentum in its move toward the positive electrode. This would not be possible if the energy band were completely filled and an overlying energy gap were present.

Magnesium, with its two valence electrons per atom, is expected to fill the first valence band. It so happens, however, that the first and second bands overlap (Fig. 5–2.2(b)). Thus some of the $2N$ electrons (where N is the number of atoms) spill over into the second band, where there are plenty of vacant levels to receive the accelerating electrons.

Silicon, however, presents another story because its four valence electrons per atom completely fill the first two valence bands (Fig. 5–2.3). Furthermore, there is an energy gap above the second band. Thus, electrons cannot be energized within these valence bands. Silicon is not a metallic conductor (with pure materials, at 20°C,

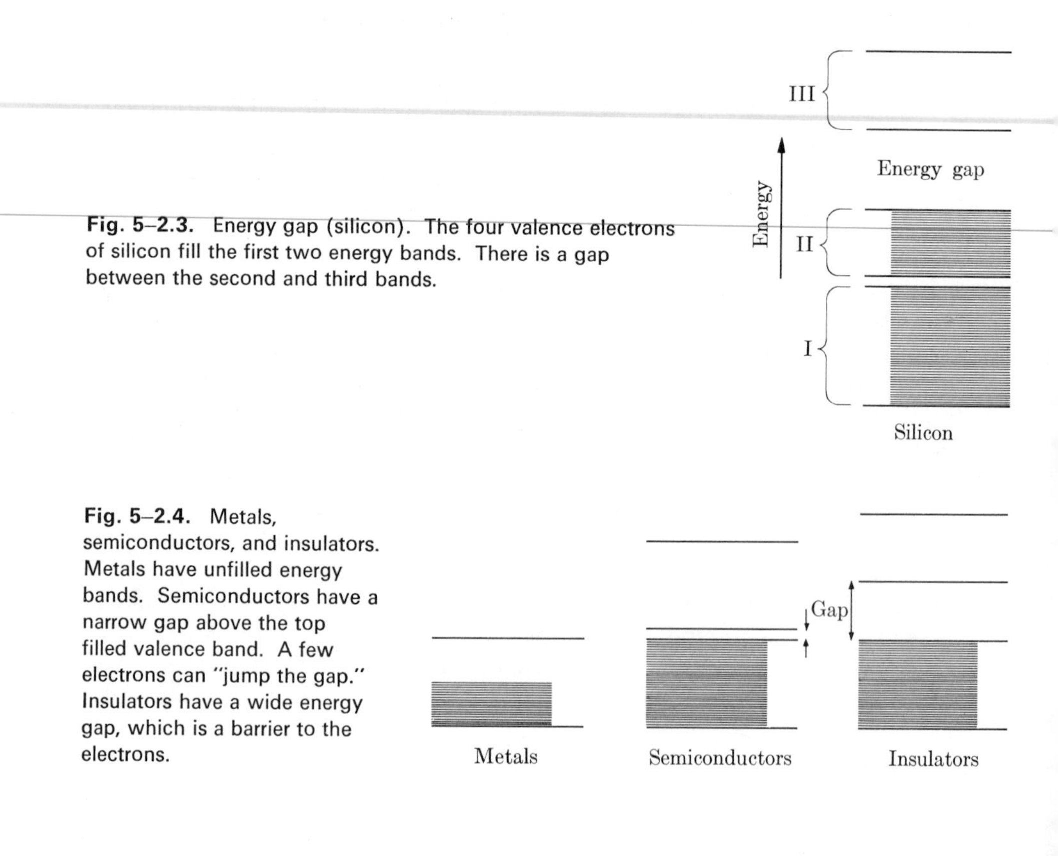

Fig. 5–2.3. Energy gap (silicon). The four valence electrons of silicon fill the first two energy bands. There is a gap between the second and third bands.

Fig. 5–2.4. Metals, semiconductors, and insulators. Metals have unfilled energy bands. Semiconductors have a narrow gap above the top filled valence band. A few electrons can "jump the gap." Insulators have a wide energy gap, which is a barrier to the electrons.

$\rho_{Si} = 0.2 \times 10^6$ ohm · cm; $\rho_{Cu} = 1.7 \times 10^{-6}$ ohm · cm—a ratio of about 10^{11}). We shall see, in the next two sections, that the difference between insulators and semiconductors is related to the size of the *energy gap* E_g which overlies the filled *valence band* (Fig. 5–2.4). Silicon is a semiconductor (Fig. 5–1.1) because its energy gap is of such size that a *few* highly energized electrons are able to "jump the gap" and gain momentum in the *conduction band* (Section 5–4).

Example 5–2.1 A toaster uses 300 watts when it is in operation and the nichrome element is at 870°C. It operates off a 110-V line. a) How many amperes does it draw when it is hot? b) When the switch is first snapped on?

Solution

a) $$I = 300 \text{ W}/110 \text{ V} = 2.7 \text{ amp.}$$

b) $$R_{870°C} = 110 \text{ V}/2.7 \text{ A} = 40 \text{ ohm.}$$

Since dimension changes are minor (and partially compensating),

$$R_{20°C}/R_{870°C} = \rho_{20°C}/\rho_{870°C}.$$

From Eq. (5–2.1) and Table 5–2.1,

$$\begin{aligned} R_{20} &= R_{870}\left[\frac{\rho_0(1 + y_T 20°\text{C})}{\rho_0(1 + y_T 870°\text{C})}\right] \\ &= 40 \text{ ohm } [1 + 0.0004(20)]/[1 + 0.0004(870)] \\ &= 40 \text{ ohm } [1.008/1.35] = 30 \text{ ohms;} \\ I_{20} &= 110 \text{ V}/30\Omega = 3.7 \text{ amp.} \end{aligned}$$

Comments. Had the element continued to draw 3.7 amperes, the temperature would continue to rise beyond 870°C, subjecting it to faster oxidation and related service deterioration.

Note that the temperature coefficients of resistivity of alloys are less than for pure metals. This is due in part to the fact that the mean free path for the electron is already short and the resistivity is initially higher. ◀

5–3 INSULATORS

In terms of energy bands, an insulator is a material with a large energy gap between the highest filled valence band and the next empty band (Fig. 5–2.4). The gap is so large that for all intents and purposes we can state that electrons are trapped in the lower band. There is no way they can become activated and gain momentum in an electric field. Their number n for Eq. (5–1.1) is insignificantly low.

We commonly describe the valence electrons as being bound within the negative ions (or in a covalent bond). Approximately 7 eV (1.1×10^{-18} J) of energy would be required to break an electron loose from the Cl^- ions in NaCl, and about 6 eV of energy to separate an electron from the covalent bond of diamond. These activation energies of 7 eV and 6 eV are also the dimensions of the energy gaps, and may be compared to 1.1 eV and 0.72 eV for silicon and germanium, respectively. Since the physicist considers an energy gap of about 4 eV as an arbitrary distinction between semiconductors and insulators, NaCl and diamond are electronic insulators, and silicon and germanium are called semiconductors* (Fig. 5–1.1).

Example 5–3.1 An 0.01-cm film of polyethylene (PE) is used as a dielectric to separate two electrodes at 110 V. Based on Fig. 5–1.1, what is the electron flux through the film?

Solution: Basis = 1 cm² and 1 sec.

$$R = \frac{(10^{16}\ \text{ohm}\cdot\text{cm})(0.01\ \text{cm})}{(1\ \text{cm}^2)} = 10^{14}\ \text{ohm};$$

$$I = \frac{110\ \text{V}}{10^{14}\ \text{ohm}} = 1.1 \times 10^{-12}\ \text{amp.}$$

$$\frac{(1.1 \times 10^{-12}\ \text{A})}{(0.16 \times 10^{-18}\ \text{A}\cdot\text{s/el})} = 7 \times 10^{6}\ \text{electrons/sec.}$$

Comments. This picoampere current is small, but measurable. Therefore, extraneous factors such as impurities, pinhole porosity, surface leakage, etc., must be considered when making measurements. ◀

5–4 INTRINSIC SEMICONDUCTORS

Semiconductors and insulators are differentiated on the basis of the size of their energy gaps (see Fig. 5–2.4). In a semiconductor, the energy gap is such that a useful number of electrons are able to jump the gap from the filled valence band to the empty conduction band (Fig. 5–4.1). Those energized electrons can now gain momentum as they

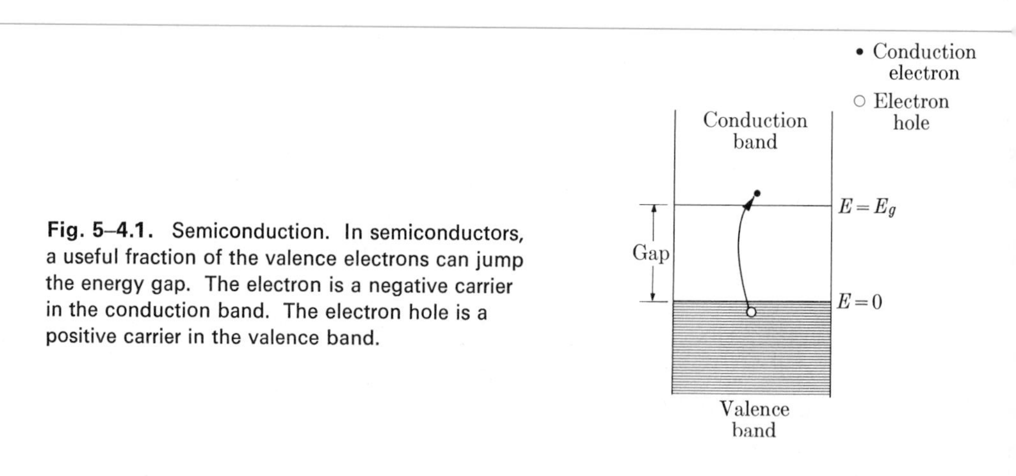

Fig. 5–4.1. Semiconduction. In semiconductors, a useful fraction of the valence electrons can jump the energy gap. The electron is a negative carrier in the conduction band. The electron hole is a positive carrier in the valence band.

---

* Diamond can be an electronic semiconductor if impurities are present; and NaCl can be an ionic semiconductor if conditions are favorable for sodium *ion* diffusion.

Table 5–4.1
ENERGY GAPS IN SEMICONDUCTING ELEMENTS

Element	Energy gap E_g, eV	At 20°C (68°F) Fraction of valence electrons with energy > E_g	Conductivity σ, $\text{ohm}^{-1} \cdot \text{cm}^{-1}$
C(diamond)	~6	$\sim 1/30 \times 10^{21}$	$< 10^{-18}$
Si	1.1	$\sim 1/10^{13}$	5×10^{-6}
Ge	0.72	$\sim 1/10^{10}$	0.02
Sn(gray)	0.08	~1/5000	10^4

move toward the positive electrode; furthermore, the resulting electron holes in the valence band become available for conduction because electrons deeper in the band can move up into those levels as they gain momentum.

Figure 5–4.2 shows the energy gap schematically for C(diamond), Si, Ge, and Sn(gray). The gap is too large in diamond to provide a usable number of charge carriers, so diamond is categorized as an insulator (Table 5–4.1). The number of carriers increases as we move down through Group IV of the periodic table to silicon, germanium, and tin; as a result, the conductivity increases, as shown in the accompanying table. This conductivity is an inherent property of these materials and does not arise from impurities. Therefore, it is called *intrinsic semiconductivity.*

The crystal structure of diamond is repeated (from Chapter 2) in Fig. 5–4.3(a). Each carbon atom has a coordination number of 4, and each neighboring pair of atoms shares a pair of electrons (Section 2–2). Silicon, germanium, and gray tin have the same structure.* Figure 5–4.4 uses germanium to represent schematically the mechanism of semiconductivity in these elements.

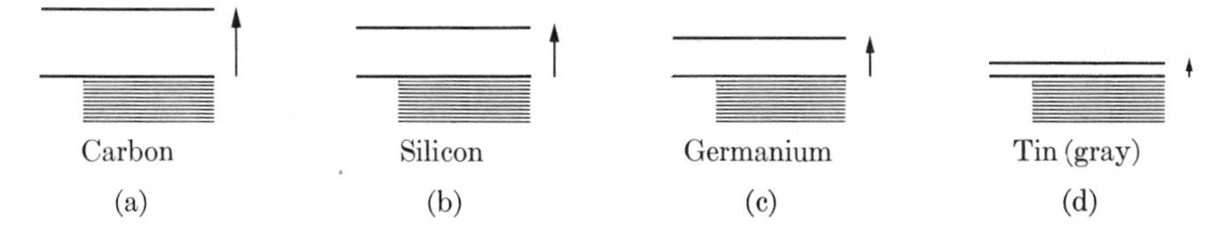

Fig. 5–4.2. Energy gaps in Group IV elements. All these elements can have the same structure since they all have filled bands. Because tin has the smallest energy gap, it requires a smaller electric field to raise its electrons to a conducting level in the next band, and its resistivity is low. (Cf. Table 5–4.1.)

* White tin is the more familiar polymorph. It is stable above 13°C (but may be supercooled to lower temperatures). White tin (bct) is denser than gray tin ($\rho_w = 7.3 \text{ g/cm}^3$, while $\rho_g = 5.7 \text{ g/cm}^3$); therefore, the energy bands of white tin overlap, and this phase is a metallic conductor (cf. Fig. 5–2.2(b)).

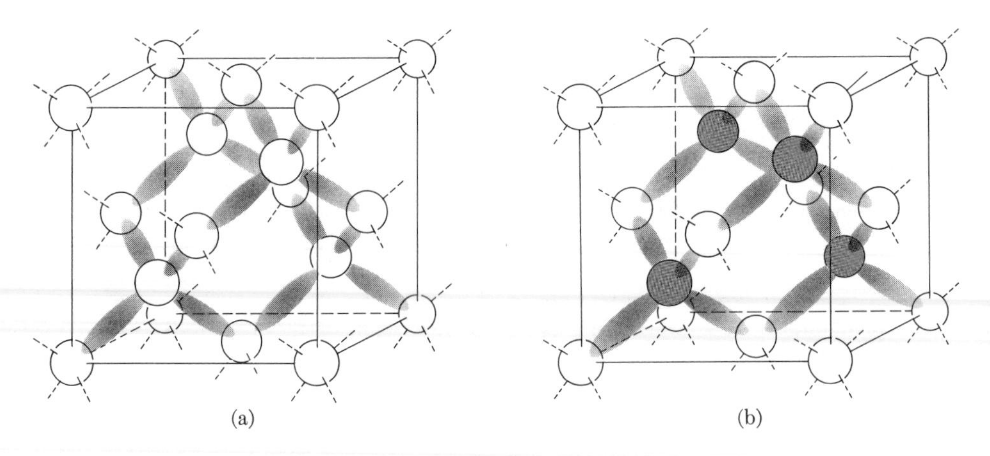

Fig. 5–4.3. Crystal structures of familiar semiconductors. (a) Diamond, silicon, germanium, gray tin. (b) ZnS, GaP, GaAs, InP, etc. (Cf. Fig. 8–2.1(c).) The two structures are similar, except that two types of atoms are in alternate positions in the semi-conducting compounds. All atoms have CN = 4; each material has an average of four valence electrons per atom, and two electrons per bond.

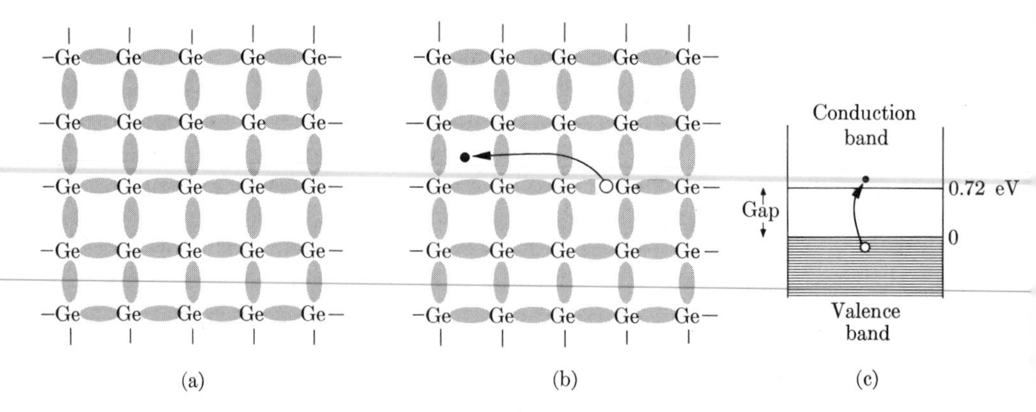

Fig. 5–4.4. Intrinsic semiconductor (germanium). (a) Schematic presentation showing electrons in their covalent bonds (and their valence bands). (b) Electron-hole pair. (Positive electrode at the left.) (c) Energy gap, across which an electron must be raised to provide conduction. For each conduction electron, there is a hole produced among the valence electrons.

The above four Group IV elements are the only elements which obtain semi-conductivity from the structure of Fig. 5–4.3(a). However, a number of III–V *compounds* are based on the same structure (Fig. 5–4.3(b)). Atoms of elements from Group III of the periodic table (B, Al, Ga, In) may alternate with atoms of elements from Group V of the periodic table (N, P, As, Sb). Most of the 16 III–V compounds which can form from these elements are semiconductors because every atom has four neighbors, and the average number of shared valence electrons is *four*. This matches exactly the situation for silicon and germanium, our predominant semiconductors.

Table 5–4.2
PROPERTIES OF COMMON SEMICONDUCTORS (20°C)

Material	Energy gap E_g, eV	Mobilities, cm^2/volt · sec		Intrinsic conductivity, $ohm^{-1} \cdot cm^{-1}$	Lattice constant, a
		Electron, μ_n	Hole, μ_p		
Elements					
C(diamond)	~6	1,700	1,200	$< 10^{-18}$	3.57 Å
Silicon	1.1	1,400	500	5×10^{-6}	5.43
Germanium	0.72	3,900	1,900	0.02	5.66
Tin (gray)	0.08	2,000	1,000	10^4	6.49
Compounds					
AlSb	1.6	1,000	400	—	6.13
GaP	2.2	500	20	—	5.45
GaAs	1.4	4,000	300	10^{-8}	5.65
GaSb	0.7	5,000	1,000	—	6.12
InP	1.3	4,000	100	5	5.87
InAs	0.33	30,000	500	100	6.04
InSb	0.17	80,000	500	—	6.48
ZnS	3.6	100	10	—	—
SiC (hex)	3	100	20	—	—

Charge mobility. Our introductory equation on conductivity (Eq. 1–4.1) must now be modified to match Fig. 5–4.1, since an intrinsic semiconductor has two types of carriers:

$$\sigma = n_n q \mu_n + n_p q \mu_p. \tag{5–4.1}$$

The *electrons* which jump into the conduction band are *negative*-type carriers. The conductivity they produce depends on their mobility μ_n through the semiconductor. The *electron holes* which are formed in the valence band are *positive*-type carriers.* The conductivity they produce depends on their mobility μ_p, through the semiconductor. The total conductivity arises from both contributors. Of course, both the hole and the electron carry the same basic charge unit of 0.16×10^{-18} coul. In an intrinsic semiconductor, where there is a one-for-one formation of conduction electrons and electron holes, $n_n = n_p$; thus we could simplify Eq. (5–4.1). However, let's leave it in its present form because n_n does not equal n_p for *extrinsic* semiconductors (Section 5–5).

Table 5–4.2 summarizes the properties of a number of semiconductors. Note that we can make two generalizations.

* Comparably, anions, with extra electrons, are negative-type, and cations, which are deficient in electrons, are positive-type.

1. The size of the energy gap commonly decreases as we move down in the periodic table (C → Si → Ge → Sn), or (GaP → GaAs → GaSb), or (AlSb → GaSb → InSb).

2. The mobility of electrons within a given semiconductor is greater than the mobility of electron holes in the same semiconductor. The latter difference will be important when considering the use of n-type semiconductors in contrast to p-type semiconductors.

Semiconductivity (intrinsic) versus temperature. Unlike metals, which have increased resistivity and decreased conductivity at higher temperatures (Fig. 5–2.1), the conductivity of intrinsic semiconductors *increases* at higher temperatures. The explanation is straightforward when one considers that the number of charge carriers, n, increases directly with the number of electrons which jump the gap (Fig. 5–4.1). At 0°K, *no* electron would have the necessary energy to do this; however, as the temperature rises, the electrons receive energy, just as the atoms do. At 20°C, a significant fraction of the valence electrons in silicon, germanium, and tin have energy in excess of E_g, the energy gap (Table 5–4.1). The same is true for compound semiconductors.

Since the thermal-energy distribution of electrons is similar to the thermal-energy distribution of atoms* (Fig. 4–6.3), we can use the equation:

$$n_i = \mathcal{N} e^{-E_g/2kT}. \tag{5–4.2}$$

The similarity of this equation to Eq. (4–6.3(b)) is apparent; n_i is the number of electrons with enough energy to jump the gap, and $\mathcal{N}$ is the proportionality constant appropriate to the material. We use $E_g/2$ in Eq. (5–4.2), rather than just E, since n_n and n_p are symmetric about the middle of the energy gap. As in Chapter 4, T is the absolute temperature (°K) and k is Boltzmann's constant, but usually expressed in electron volts (86.1×10^{-6} eV/°K) rather than 13.8×10^{-24} J/°K.

While it is true that increased temperatures shorten the mean free path of the carriers in a semiconductor, and therefore decrease their mobility, the *number* of carriers increases exponentially. As a result, n far overshadows μ in Eq. (5–4.1). Thus, to a first approximation,

$$\sigma = \sigma_0 e^{-E_g/2kT}, \tag{5–4.3(a)}$$

or

$$\ln \sigma = \ln \sigma_0 - \frac{E_g}{2kT}. \tag{5–4.3(b)}$$

This is shown in Fig. 5–4.5 for an Arrhenius-type plot.† We use a different proportionality constant σ_0 than in Eq. (5–4.2) because q and μ have been incorporated.

---

* The distribution must be modified since no more than two electrons have the same energy (Pauli exclusion principle). However, that modification is significant only near the mean energy value. At the high end of the energy spectrum (Fig. 4–6.4), the thermal-energy distribution for electrons and for atoms becomes comparable.

† See the comments in Example 4–6.2.

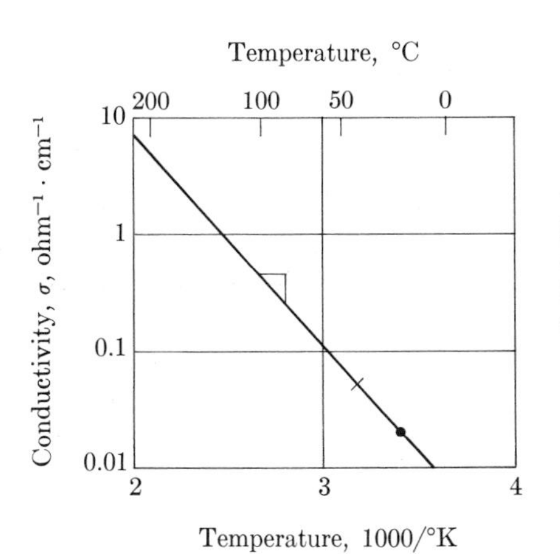

Fig. 5–4.5. Semiconduction versus temperature (intrinsic germanium). The slope is $E_g/2k$ of Eq. (5–4.3) when the ordinate is $\ln \sigma$, or $E/4.6k$ when the ordinate is $\log_{10} \sigma$. See Example 5–4.2.

If we measure the conductivity (or resistivity) in the laboratory at several different temperatures, the slope of the $\ln \sigma$-versus-$1/T$ plot in Fig. 5–4.5 provides the value of E_g, the energy gap.

Photoconduction. There is only a small probability that an electron in the valence band of silicon may be raised across the energy gap into the conduction band (1 out of 10^{13}, according to Table 5–4.1). In contrast, if an electron is hit by a photon of light, it may readily be energized to the conduction level (Fig. 5–4.6). As an example, a photon of red light (wavelength = 6.6×10^{-5} cm) has 1.9 eV of energy, more than enough to cause an electron to jump the 1.1 eV energy gap in silicon. Thus the conductivity of silicon increases markedly when it is exposed to light.

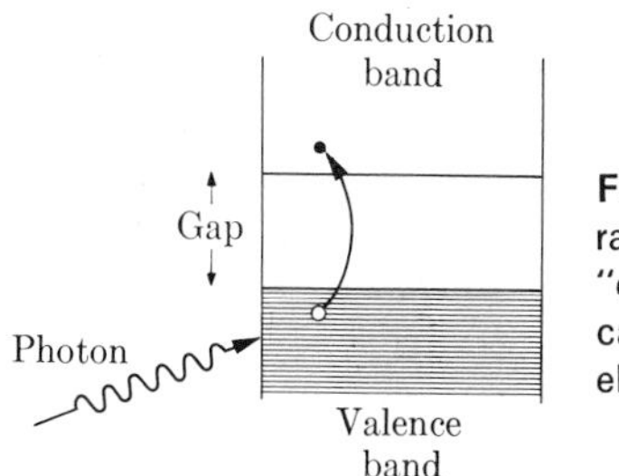

Fig. 5–4.6. Photoconduction. A photon (i.e., light energy) raises the electron across the energy gap, producing a "conduction electron + valence hole" pair, forming charge carriers. Recombination (Eq. (5–4.4(b))) occurs when the electron drops back to the valence band.

Recombination. The reaction that produces an *electron–hole pair*, as shown in Fig. 5–4.6, may be written as

$$E \rightarrow n + p, \tag{5–4.4(a)}$$

where E is energy, n is the conduction electron, and p is the hole in the valence band. In this case the energy came from light.

Since all materials are more stable when they reduce their energies, electron–hole pairs recombine sooner or later:

$$n + p \rightarrow E. \tag{5–4.4(b)}$$

In effect, the electron drops from the conduction band back to the valence band, just the reverse of Fig. 5–4.4(c). Were it not for the fact that light or some other energy source continually produces additional electron–hole pairs, the conduction band would soon become depleted.

The time required for recombination varies from material to material. However, it follows a regular pattern because, within a specific material, every conduction electron has the same probability of recombining within the next second (or minute). This leads to the relationship

$$N = N_0 e^{-t/\lambda}, \tag{5–4.5(a)}$$

which we usually rearrange to

$$2.3 \log_{10} (N_0/N) = t/\lambda. \tag{5–4.5(b))*}$$

In these equations, N_0 is the number of electrons in the conduction band at a particular moment of time (say, when the source of light is turned off). After an additional time t, the number of remaining conduction electrons is N. The term λ is called the *relaxation time*, and is characteristic of the material.

• **Luminescence.** The energy released in Eq. (5–4.4(b)) may appear as heat. It may also appear as light. When it does, we speak of *luminescence* (Fig. 5–4.7). Sometimes we subdivide luminescence into several categories. *Photoluminescence* is the light emitted after electrons have been activated to the conduction band by light photons.

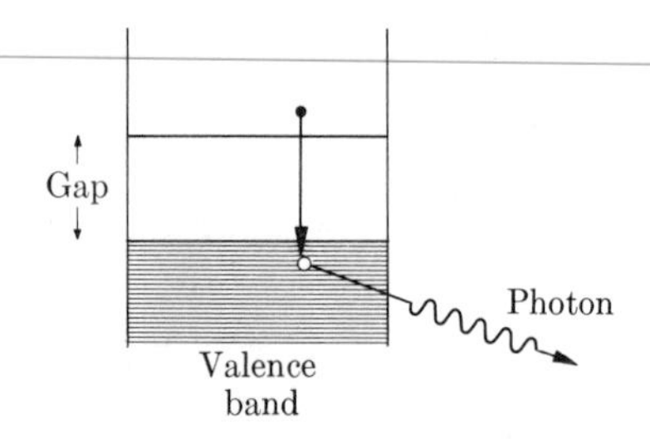

Fig. 5–4.7. Luminescence. Each millisecond, a fraction of the electrons energized to the conduction band return to the valence band. As the electron drops across the gap, the energy may be released as a photon of light.

* Equation (5–4.5) can be derived through calculus (by those who wish to do so) from the information stated above:

$$dN/dt = -N/\lambda. \tag{5–4.6}$$

Rearranging,

$$dN/N = -dt/\lambda;$$

then integrating we get,

$$\ln N/N_0 = -t/\lambda. \tag{5–4.5(c)}$$

Chemoluminescence is the word used when the initial activation is due to chemical reactions. Probably *electroluminescence* is best known, because this is what occurs in a TV tube, in which a stream of electrons scans the screen, activating the electrons in the phosphor to their conduction band. Almost immediately, however, the electrons recombine, emitting energy as visible light.

Since the recombination rate is proportional to the number of activated electrons, the intensity I of luminescence also follows Eq. (5–4.5(b)):

$$2.3 \log_{10} (I_0/I) = t/\lambda. \tag{5–4.7}$$

For a TV tube, the engineer chooses a phosphor with a relaxation time such that light continues to be emitted as the next scan comes across. Thus our eyes do not see a light–dark flickering. However, the light intensity from the previous trace should be weak enough so that it does not compete with the new scan that follows one-thirtieth of a second later. (See Example 5–4.3.)

Example 5–4.1 Each gray tin atom has four valence electrons. The unit cell size (Fig. 5–4.3(a)) is 6.49 Å. Separate calculations indicate that there are 2×10^{19} conduction electrons per cm^3. What fraction of the electrons have been activated to the conduction band?

Solution: Basing our calculations on Fig. 5–4.3, we find that there are 8 tin atoms per unit cell.

$$\text{Valence electrons/cm}^3 = \frac{(8 \text{ atoms/uc})(4 \text{ el/atom})}{(6.49 \times 10^{-8} \text{ cm})^3/\text{uc}} = 1.17 \times 10^{23}/\text{cm}^3.$$

$$\text{Fraction activated} = \frac{2 \times 10^{19}}{1.17 \times 10^{23}} \simeq 0.0002.$$ ◀

Example 5–4.2 The resistivity of germanium at 20°C (68°F) is 50 ohm · cm. What is its resistivity at 40°C (104°F)?

Solution: Based on Eq. (5–4.3) and an energy gap of 0.72 eV (Table 5–4.2):

$$\frac{\sigma_2}{\sigma_1} = \frac{\rho_1}{\rho_2} = \frac{\sigma_0 e^{-E_g/2kT_2}}{\sigma_0 e^{-E_g/2kT_1}},$$

$$2.3 \log_{10} \rho_1/\rho_2 = (E_g/2k)\left[\frac{1}{T_1} - \frac{1}{T_2}\right], \tag{5–4.8}$$

$$\log_{10} \rho_1/\rho_2 = \frac{0.72 \text{ eV}}{4.6(86.1 \times 10^{-6} \text{ eV/°K})}\left[\frac{313 - 293\text{°K}}{(293\text{°K})(313\text{°K})}\right]$$

$$= 0.396.$$

Thus, if $\rho_{20°} = 50$ ohm · cm, $\rho_{40°} = 50/10^{0.396} = 20$ ohm · cm.

Comments. By rearranging $(1/T_1 - 1/T_2)$ to $(T_2 - T_1)/T_1T_2$, it is possible to handle the above calculation with a slide rule.

It is possible to measure resistance changes (and therefore resistivity changes) of $<0.1\%$. Therefore one can measure temperature changes of a small fraction of a degree. (See Study Problem 5–6.1.) ◀

• **Example 5–4.3** The scanning beam of a television tube covers the screen with 30 frames per second. What must the relaxation time for the activated electrons of the phosphor be, if only 20% of the intensity is to remain when the following frame is scanned?

Solution: Refer to Eq. (5–4.7):

$$2.3 \log (1.00/0.20) = (0.033 \text{ sec})/\lambda,$$

$$\lambda = 0.02 \text{ sec}.$$

Comments. We use the term *fluorescence* when the relaxation time is short compared to the time of our visual perception. If the luminescence has a noticeable afterglow, we use the term *phosphorescence.* ◀

5–5 EXTRINSIC SEMICONDUCTORS

***N*-type semiconductors.** Impurities alter the semiconducting characteristics of materials by introducing excess electrons or excess electron holes. Consider, for example, some silicon containing an atom of phosphorus. Phosphorus has five valence electrons rather than the four which are found with silicon. In Fig. 5–5.1(a), the extra electron

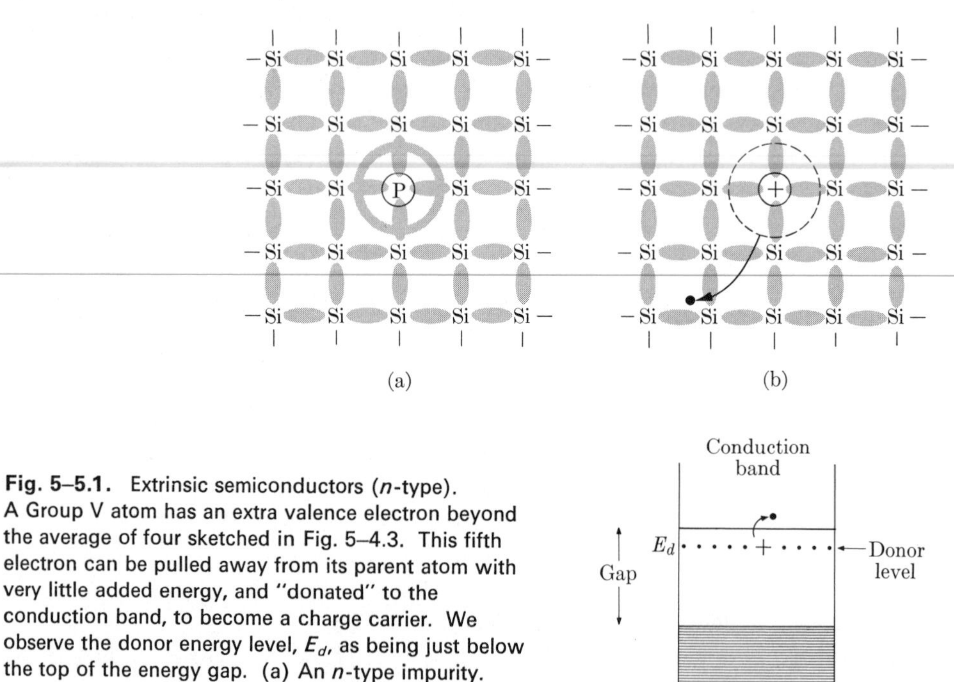

Fig. 5–5.1. Extrinsic semiconductors (n-type). A Group V atom has an extra valence electron beyond the average of four sketched in Fig. 5–4.3. This fifth electron can be pulled away from its parent atom with very little added energy, and "donated" to the conduction band, to become a charge carrier. We observe the donor energy level, E_d, as being just below the top of the energy gap. (a) An n-type impurity. (b) Ionized phosphorus atom. (Positive electrode at left.) (c) Band model.

is present independently of the electron pairs which serve as bonds between neighboring atoms. This electron can carry a charge toward the positive electrode (Fig. 5–5.1(b)). Alternatively, in Fig. 5–5.1(c), the extra electron—which cannot reside in the valence band because that is already full—is located near the top of the energy gap. From this position—called a *donor* level E_d—the extra electron can easily be activated into the conduction band. Regardless of which model is used, Fig. 5–5.1(b) or 5–5.1(c), we can see that atoms from Group V (N, P, As, and Sb) of the periodic table (Fig. 2–1.1) can supply negative, or *n*-type, charge carriers to semiconductors.

***P*-type semiconductors.** Group III elements (B, Al, Ga, and In) have only three valence electrons. Therefore, when such elements are added to silicon as impurities, electron holes come into being. As shown in Fig. 5–5.2(a) and (b), each aluminum atom can accept one electron. In the process a positive charge moves toward the negative electrode. Using the band model (Fig. 5–5.2(c)), we note that the energy difference for electrons from the valence band to the *acceptor level*, E_a, is much less than the full energy gap. The electron holes remaining in the valence band are available as positive carriers for *p*-type semiconduction.

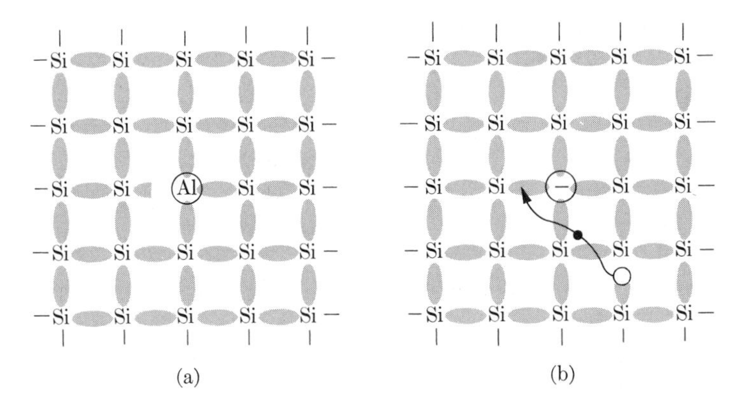

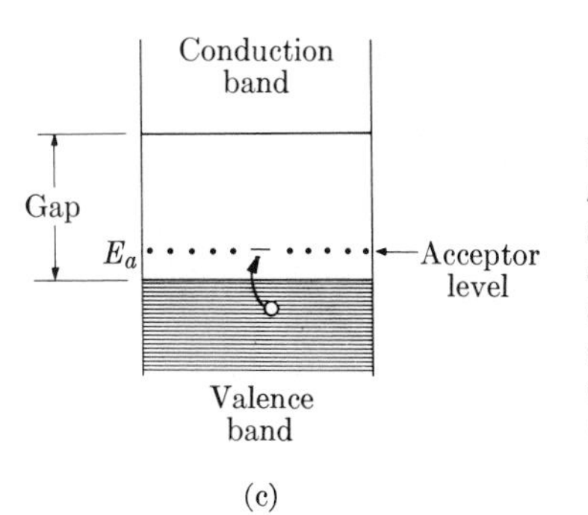

Fig. 5–5.2. Extrinsic semiconductors (*p*-type). A Group III atom has one less valence electron than the average of four sketched in Fig. 5–4.3. This atom can accept an electron from the valence band, thus leaving an electron hole as a charge carrier. The acceptor energy level, E_a, is just above the bottom of the energy gap. (a) A *p*-type impurity. (b) Ionized aluminum atom. (Negative electrode at right.) (c) Band model.

Donor exhaustion (and acceptor saturation). If the energy difference for electrons from the donor level to the conduction band is small in comparison with the size of the energy gap, all the electrons in the donor level can be raised to the conduction band, even at room temperatures. This is illustrated in Fig. 5–5.3. At absolute zero, beyond the right edge of the figure, all the electrons are either in the valence band or in the donor levels, and Fig. 5–5.1(a) applies. As the temperature is raised, more and more of the donor electrons jump to the conduction band and carry charge. When the thermal energy of the electrons is approximately equal to this small energy difference ($E_g - E_d$), essentially all the donor electrons have jumped to the conduction band and this reservoir of charge carriers arising from impurities is exhausted. As a result, there is a conductivity plateau. When the temperature is increased appreciably, the thermal energy becomes sufficient to cause electrons to jump across the full energy gap from the valence band to the conduction band.

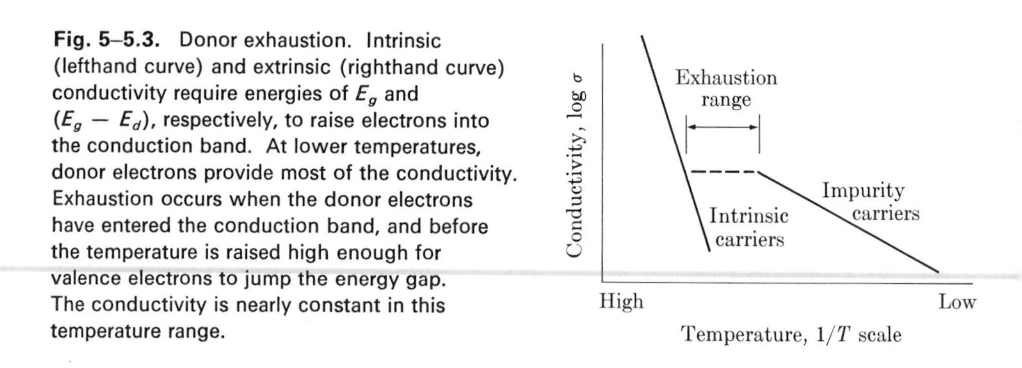

Fig. 5–5.3. Donor exhaustion. Intrinsic (lefthand curve) and extrinsic (righthand curve) conductivity require energies of E_g and ($E_g - E_d$), respectively, to raise electrons into the conduction band. At lower temperatures, donor electrons provide most of the conductivity. Exhaustion occurs when the donor electrons have entered the conduction band, and before the temperature is raised high enough for valence electrons to jump the energy gap. The conductivity is nearly constant in this temperature range.

Donor exhaustion of *n*-type semiconductors has its parallel in acceptor saturation of *p*-type semiconductors. The reader is asked to make the various comparisons. Donor exhaustion and acceptor saturation are important to materials and electrical engineers, since they provide a region of essentially constant conductivity. This means that it is less necessary to compensate for temperature changes in electrical circuits than it would be if the log σ-versus-$1/T$ characteristics followed an ever-ascending line.

• **Defect semiconductors.** The iron oxide of Fig. 4–3.2 possessed Fe^{3+} ions in addition to the regular Fe^{2+} ions. A similar situation occurs in Fig. 5–5.4(a) when NiO is oxidized to give some Ni^{3+} ions, and in fact is relatively common among transition-metal oxides which have multiple valences. In nickel oxide, three Ni^{2+} are replaced by 2 Ni^{3+} and a vacancy, □. This maintains the charge balance; it also permits easier diffusion and therefore some ionic conductivity. More important, however, is the fact that electrons can hop from an Ni^{2+} ion into acceptor sites in Ni^{3+} ions. Conversely, an electron hole moves from one nickel ion to another as it migrates toward the negative electrode. Nickel oxide and other oxides with $M_{1-x}O$ defect structures are *p*-type semiconductors.

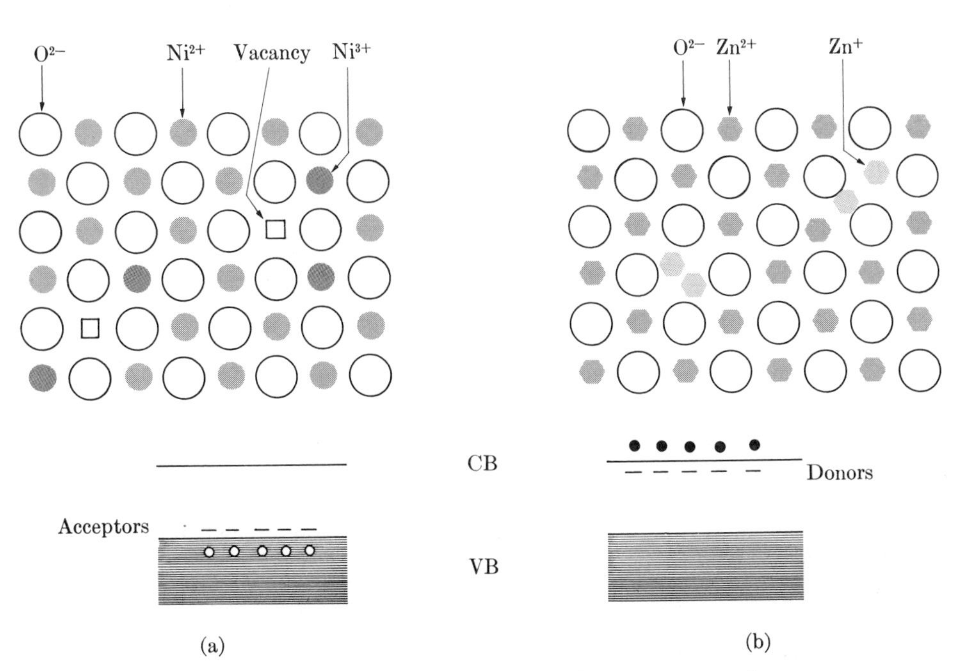

Fig. 5–5.4. Defect semiconductors. (a) $Ni_{1-x}O$. The Ni^{3+} ions serve as electron acceptors, so that holes, ○, form in the valence band. (b) $Zn_{1+y}O$. The Zn^{+} ions are donors of electrons, ●, to the conduction band for *n*-type semiconduction.

There are also *n*-type oxides. Zinc oxide, when exposed to a reducing atmosphere, produces $Zn_{1+y}O$ by removing some oxygen. However, in this case, an oxygen vacancy does not develop. Rather, a zinc ion moves into an interstitial position (Fig. 5–5.4(b)). The Zn^{+} ions which arise to balance the charge have one electron more than the bulk of the Zn^{2+} ions. These can donate electrons to the conduction band for *n*-type semiconductivity.

Example 5–5.1 Silicon, according to Table 5–4.2, has a conductivity of only 5×10^{-6} $\text{ohm}^{-1} \cdot \text{cm}^{-1}$ when pure. An engineer wants it to have a conductivity of 2 $\text{ohm}^{-1} \cdot \text{cm}^{-1}$ when it contains aluminum as an impurity. How many aluminum atoms are required per cm^3?

Solution: Since the intrinsic conductivity is negligible compared with 2 $\text{ohm}^{-1} \cdot \text{cm}^{-1}$, assume that all the conductivity comes from the holes:

$$n_p = (2\ \text{ohm}^{-1} \cdot \text{cm}^{-1})/(0.16 \times 10^{-18}\ \text{amp} \cdot \text{sec})(500\ \text{cm}^2/\text{volt} \cdot \text{sec})$$
$$= 2.5 \times 10^{16}/\text{cm}^3.$$

Comments. Each aluminum atom contributes one acceptor site, i.e., electron hole. Therefore 2.5×10^{16} aluminum atoms are required per cm^3. This is, of course, a large number;

however, it is still small (0.5 ppm) when compared with the number of silicon atoms per cm^3. (See Study Problem 5–5.1(a).) ◀

Example 5–5.2 Germanium used for an early transistor had an extrinsic resistivity of 2 ohm · cm and a conduction electron concentration of $0.8 \times 10^{15}/cm^3$. a) What is the mobility of the electrons in the germanium? b) What impurity elements could be added to the germanium to donate the conduction electrons?

Solution: Since we are considering extrinsic conductivity from electrons, i.e., *n*-type:

a)
$$\mu_n = 1/(2 \text{ ohm} \cdot \text{cm})(0.8 \times 10^{15}/\text{cm}^3)(0.16 \times 10^{-18} \text{ amp} \cdot \text{sec})$$
$$= 3900 \text{ cm}^2/\text{volt} \cdot \text{sec};$$

b) Group V elements: N, P, As, Sb.

Comments. Note that the electron mobility does not depend on which of these Group V elements is added, since the electron, once in the conduction band, moves through the silicon lattice independent of its donor.

Group VI elements could also be added. Since they have a second additional electron (beyond the four necessary for bonding), it would take only $0.4 \times 10^{15}/cm^3$ of these atoms to supply $0.8 \times 10^{15}/cm^3$ conduction electrons. ◀

•5–6 SEMICONDUCTING DEVICES

There are many electronic devices that use semiconductors. We shall consider but a few.

We have already seen (Example 5–4.2 and Study Problem 5–4.5) that the resistivity, and therefore the resistance, of semiconductors varies markedly with temperature changes (Figs. 5–4.5 and 5–5.3). Some ceramic semiconductors are available which can detect temperature changes of approximately 10^{-6} °C! Such a resistance device acts as a thermometer, and is called a *thermistor*.

Because many semiconducting materials have low packing factors, they have a high compressibility. Experiments show that as the volume contracts, the size of the energy gap is measurably reduced; this of course increases the number of electrons which can jump the energy gap. Thus pressure can be calibrated against resistance for *pressure gages*.

Rectifying junctions. Suppose that a *p*-type semiconductor and an *n*-type semiconductor are joined. Their junction can serve as a *rectifier*; i.e., it is an electrical valve that lets current pass one way and not the other. Figure 5–6.1 presents a brief description, somewhat oversimplified. In part (a) of the figure, we see positive carriers in the *p*-half and negative carriers in the *n*-half. With the voltage in one direction (Fig. 5–6.1(b)), the current feeds through the device from right to left (electrons from left to right). However, with a reverse voltage (Fig. 5–6.1(c)), the conduction electrons are attracted to the positive electrode and the holes are displaced in the opposite direction, leaving a carrier-free zone at the junction. The conductivity of this zone is low. If a greater voltage is applied, it simply widens the "insulating" zone. Current passes in only one direction, as shown in Fig. 5–6.2.

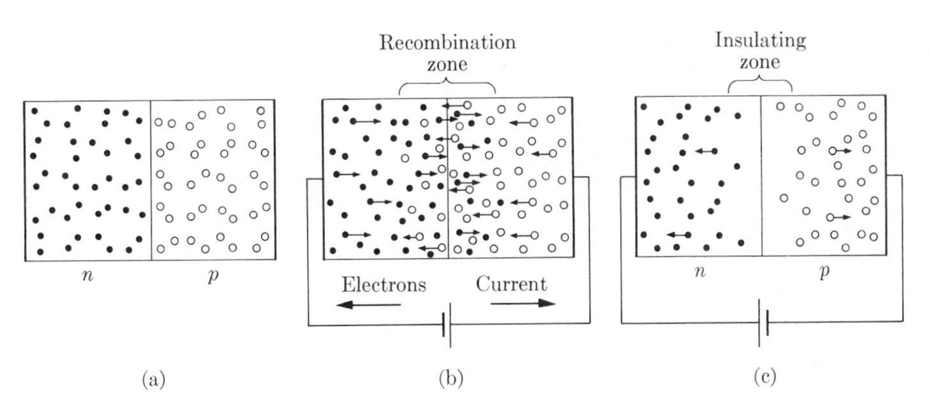

Fig. 5–6.1. An *n–p* junction (rectifier). (a) No voltage. (b) Forward bias. Charge is transported across the junction. Electrons and holes recombine beyond the junction. (c) Reverse bias. The carriers do not carry charge across the junction. The holes (in the *p*-type) and the conduction electrons (in the *n*-type) are pulled away from the junction by the electric field.

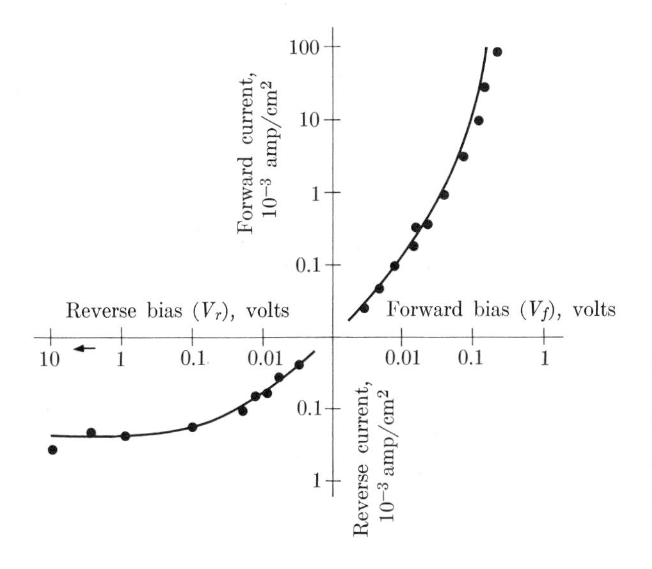

Fig. 5–6.2. Characteristics of a rectifier junction. Forward bias gives a high current density, amp/cm². A reverse bias forms a high resistance. As the reverse voltage is increased from 0.1 to 10 volts, there is only a slight increase in reverse current.

As the holes move across the junction with a forward bias (Fig. 5–6.1(b)), they recombine with the electrons in the *n*-type material according to Eq. (5–4.4(b). Likewise, the electrons combine with the holes as the electrons move into *p*-type material. These reactions do not occur immediately, however. In fact, an excess number of positive and negative carriers may move measurable distances beyond the junction. The numbers of excess, unrecombined carriers are an exponential function of the voltage V. These facts do not affect the rectifier significantly, but they are important to transistors.

Transistors. Transistors have revolutionized engineering design in the field of communications. Let us explore their operation by means of a simplified model. A transistor has two junctions in series. They may be *p-n-p* or *n-p-n*. The former is somewhat more common; however, we shall consider the *n-p-n* transistor, since it's a little easier for us to envision the movements of electrons than the movements of holes. The principles behind each type are the same, though.

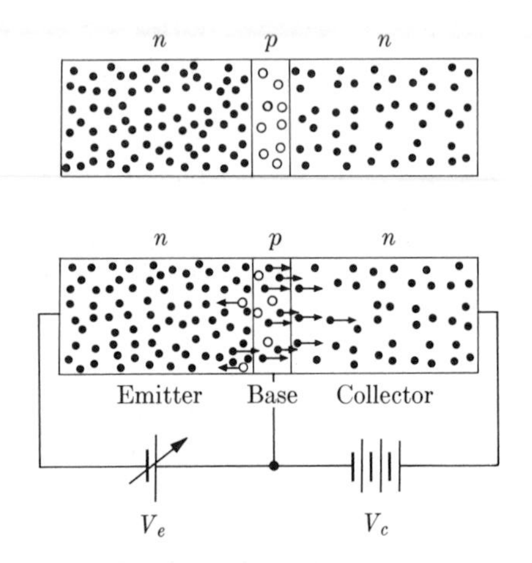

Fig. 5–6.3. Transistor (*n–p–n*). The number of electrons crossing from the emitter-base junction is highly sensitive to the emitter voltage. If the base is narrow, these carriers move to the base–collector junction, and beyond, before recombination. The total current flux, emitter to base, is highly magnified, or amplified, by fluctuations in the voltage of the emitter.

A transistor consists of an *emitter*, a *base*, and a *collector* (Fig. 5–6.3). For the moment, consider only the *emitter junction*, which is biased so that electrons move into the base (and toward the collector). As discussed a moment ago, the number of electrons that cross this junction and move into the *p*-type material is an exponential function of the emitter voltage, V_e. Of course, these electrons at once start to combine with the holes in the base; however, if the base is narrow, or if the relaxation time is long (λ of Eq. 5–4.5), the electrons keep on moving through the thickness of the base. Once they are at the second or *collector junction*, the electrons have free sailing, because the collector is an *n*-type semiconductor. The total current that moves through the collector is controlled by the emitter voltage, V_e. As the emitter voltage fluctuates, the collector current, I_c, changes exponentially; written logarithmically:

$$2.3 \log_{10} I_c \simeq A + V_e/B, \tag{5–6.1}$$

where A and B are constants for a given temperature. Thus, if the voltage in the emitter is increased even slightly, the amount of current is increased markedly. It is because of these relationships that a transistor serves as an amplifier.

Example 5–6.1 A transistor has a collector current of 4.7 milliamperes when the emitter voltage is 17 millivolts. At 28 millivolts, the current is 27.5 milliamperes. Given that the emitter voltage is 39 millivolts, estimate the current.

Solution: Based on Eq. (5–6.1),

$$2.3 \log_{10} 4.7 \simeq A + 17/B = 1.55,$$

$$2.3 \log_{10} 27.5 \simeq A + 28/B = 3.31.$$

Solving simultaneously, using *milli*units, we have

$$A \simeq -1.17 \quad \text{and} \quad B \simeq 6.25,$$

At 39 millivolts,

$$2.3 \log_{10} I_c \simeq -1.17 + 39/6.25 \simeq 5.07,$$

$$I_c \simeq 160 \text{ milliamp.}$$

Comment. The electrical engineer modifies Eq. (5–6.1) to take care of added current effects. These, however, do not change the basic relationship: The variation of the collector current is much greater than the variation of the signal voltage. ◀

REVIEW AND STUDY

SUMMARY

In contrast to atomic order found in crystals (Chapter 3) and the atomic disorder of crystalline imperfections (Chapter 4), this chapter focused on electrons and their freedom to move among the atoms. In metals the charge is carried by electrons ($q = 0.16 \times 10^{-18}$ coul); in ionic materials it is carried by either anions, which are atoms with extra electrons, or cations, which have an electron deficiency; in semiconductors the charge carriers are both electrons and electron holes. Comparable to cations, electron holes carry a positive charge.

Metals have unfilled valence energy bands. In both insulators and semiconductors, the valence bands are filled. In semiconductors, however, the overlying gap is small enough so that electrons can be energized into the conduction band.

Since many electrons are available for conductivity in metals, their conductivity is established by the mobility of the electrons. Any factors, such as thermal vibrations or impurities which introduce crystal imperfections, will reduce the mean free path of the electrons and thereby decrease the conductivity (increase the resistivity).

Semiconductors include two main categories, intrinsic and extrinsic. The former gain their conductivity by electron activation across the energy gap, independent of impurities. Thermal activation is most common. Since the numbers of charge carriers increase with temperature in the exponent, the changing values of n in Eq. (5–4.1) overshadow the effects of temperature on carrier mobility.

It is also possible to activate the electrons into the conduction band by light and other types of radiation. Each activated electron in an intrinsic semiconductor leaves an electron hole, so that each member of the electron–hole pair contributes to the conductivity. In general the electrons in the conduction band have greater mobility than the holes in the valence band.

Extrinsic semiconductors gain their conductivity from (1) impurities which donate electrons to the conduction band (*n*-type), or (2) impurities which accept electrons from the valence band (*p*-type). For silicon and similar Group IV elements, *n*-type conductivity requires Group V (or VI) elements; *p*-type conductivity requires Group III (or II) elements. Extrinsic semiconductors have permitted the materials scientist and electrical engineer to design junction devices such as rectifiers and transistors.

Defect semiconduction represents a special class of extrinsic semiconductors that originate from nonstoichiometry in multivalent ions. Thus, numerous metallic oxides can possess either *n*-type or *p*-type conductivity.

TERMS FOR REVIEW

Acceptor level
Acceptor saturation
Charge carrier, n
Conductivity, σ
Conduction band
Conductors
Donor
Donor exhaustion
Drift velocity, $\bar{v}$
Electric field, $\mathscr{E}$
Electron charge, q
Electron hole
Electron–hole pair
Energy band
Energy gap, E_g
• Fluorescence
Insulator
• Junction
• Luminescence
Mean free path
Metallic conductor
Mobility, μ
• Phosphorescence
Photoconductors
Recombination
• Rectifier
Relaxation time, λ
Resistivity, ρ
  vs. solid solution
  vs. temperature
Semiconductors
  compound
  • defect
  extrinsic
  intrinsic
  n-type
  p-type
Temperature resistivity coefficient, y_T
• Thermistor
• Transistor
Valence band

FOR DISCUSSION

A_5 Cite the various types of charge carriers.

B_5 Discuss various factors which affect the drift velocity of electrons in solids.

C_5 Why does the conductivity of metals decrease at elevated temperatures?

D_5 Why does the conductivity of semiconductors increase at elevated temperatures?

E_5 Differentiate between metallic conductors, insulators, and semiconductors.

F_5 Differentiate between intrinsic and extrinsic semiconductors.

G_5 There are II–IV, e.g., ZnS, and III_2VI semiconducting compounds. Relate these compounds to the compounds of Table 5–4.2.

H_5 Why is the slope for intrinsic carriers in Fig. 5–5.3 steeper than for impurity carriers?

I_5 Predict whether boron nitride, with the structure of Fig. 5–4.3(b), will be an insulator or semiconductor.

J_5 Explain how an electric eye works.

K_5 What fraction of electrons remain in the conduction band after $t = \lambda$? After $t = 2\lambda$?

•L_5 Distinguish between fluorescence and phosphorescence.

M_5 Differentiate between *n*-type and *p*-type semiconductivity.

N_5 Differentiate between acceptor and donor impurities.

O_5 There are 10^{14} magnesium atoms, which replace the same number of silicon atoms, in a cm^3 of silicon. How will this affect the conductivity compared with an impurity of 10^{14} aluminum atoms?

P_5 The mean free paths of electrons in semiconductors are shortened at higher temperatures. However, this effect of temperature on conductivity can only be detected in the exhaustion (or saturation) range. Why?

•Q_5 Cu_2O contains predominantly Cu^+ ions but some Cu^{2+} ions. Will it be *n*-type or *p*-type?

•R_5 Refer to Fig. 5–6.1(b). The junction can serve as a light-emitting diode. Explain, on the basis of Fig. 5–4.7.

•S_5 Explain how a rectifier works.

•T_5 The upper end of the curve in Fig. 5–6.2 becomes horizontal at exhaustion (or saturation). Why?

STUDY PROBLEMS

5–1.1 Laboratory measurements indicate that the drift velocity of electrons in a semiconductor is 1.49×10^4 cm/sec when the voltage gradient is 15 V/mm. The resistivity is 7 ohm · cm. What is the carrier concentration?

Answer: $9 \times 10^{15}/cm^3$

5–1.2 A flashlight bulb has a resistance of 5 ohms when it is used in a 2-cell flashlight battery. Assume 3 volts. How many electrons move through the filament per minute?

5–2.1 At what temperature does silver have the same resistivity as gold does at 50°C?

Answer: 153°C

5–2.2 A 6% variation is permitted in resistance between 0° and 25°C. Which metals of Table 5–2.1 meet the specification?

5–4.1 A silicon chip is 1 mm × 1 mm × 0.1 mm. How fast do electrons drift through its short dimension if 37 millivolts are applied?

Answer: 5200 cm/sec

5–4.2 Pure germanium has a conductivity of 0.02 $ohm^{-1} \cdot cm^{-1}$ with equal numbers of negative carriers, n_n, and positive carriers, n_p. What fraction of the conductivity is due to electrons and what fraction is due to electron holes?

5–4.3 How many electron carriers (and electron holes) does intrinsic silicon require to provide a conductivity of 0.011 $ohm^{-1} \cdot cm^{-1}$?

Answer: $n_n = 3.6 \times 10^{13}/cm^3 = n_p$

5–4.4 Pure silicon has 32 valence electrons per unit cell (8 atoms with 4 electrons each). Its resistivity is 2×10^5 ohm · cm. What fraction of the valence electrons are conductors?

5–4.5 The conductivity of silicon is 5×10^{-6} $ohm^{-1} \cdot cm^{-1}$ at 20°C (68°F). Estimate the conductivity at 30°C.

Answer: 10^{-5} $ohm^{-1} \cdot cm^{-1}$

5–4.6 To what temperature must germanium be cooled in order for its conductivity to be reduced by a factor of two below its 20°C (68°F) value?

5–4.7 An intrinsic semiconductor has a conductivity of 1.11 $ohm^{-1} \cdot cm^{-1}$ at 10°C and 1.72 $ohm^{-1} \cdot cm^{-1}$ at 17°C. What is the energy gap?

Answer: 0.88 eV.

•**5–4.8** Refer to Example 5–4.3. Assume a phosphor is used with a relaxation time of 0.04 sec. What fraction of the light intensity will remain when the next scan is made?

•**5–4.9** A phosphorescent material is exposed to ultraviolet light. The intensity of emitted light decreased 20% in the first 37 min after the ultraviolet light was removed. How long will it be after the uv light is removed before the emitted light has only 20% of the original intensity (a decrease of 80%)?

Answer: 267 minutes

5–5.1 Silicon has a density of 2.33 g/cm^3. a) What is the concentration of silicon atoms per cm^3? b) Phosphorus is added to silicon to make it an *n*-type semiconductor with a conductivity of 1 mho/cm and an electron mobility of 1400 $cm^2/volt \cdot sec$. What is the concentration of donor electrons per cm^3?

Answer: a) 5×10^{22} Si/cm^3 b) $4.5 \times 10^{15}/cm^3$

5–5.2 How many silicon atoms are there for each aluminum atom in Example 5–5.1?

5–5.3 Extrinsic germanium is formed by melting 3.22×10^{-6} g of antimony (Sb) with 100 g of germanium. a) Will the semiconductor be *n*-type or *p*-type? b) Calculate the concentration of antimony (in atoms/cm^3) in germanium. (The density of germanium is 5.35 g/cm^3.)

Answer: a) *n*-type b) 8.5×10^{14}/cm^3

5–5.4 Gallium arsenide is made extrinsic by adding 0.000001 a/o phosphorus (and keeping a stoichiometric Ga/As ratio). Calculate the extrinsic conductivity at exhaustion.

5–5.5 Three grams of *n*-type silicon which had been doped with phosphorus to produce a conductivity of 6 ohm^{-1} · cm^{-1} are melted with three grams of *p*-type silicon which had been doped with aluminum to produce a conductivity of 6 ohm^{-1} · cm^{-1}. What is the resulting conductivity? Will it be *p*-type or *n*-type?

Answer: 4 ohm^{-1} · cm^{-1}; *p*-type

5–5.6 Silicon has eight atoms per unit cell. Each unit cell is cubic, with a dimension of 5.43 Å. An impurity has been added to it to make it *n*-type. Only one impurity atom is added per every billion (10^9) silicon atoms; however, each impurity atom provides one negative carrier. a) What is the concentration of impurities? b) What is the conductivity?

5–5.7 Some $Fe_{<1}O$ has an Fe^{3+}/Fe^{2+} ratio of 0.1; what is the mobility of the electron holes if this oxide has a conductivity of 1 (ohm · cm)$^{-1}$ and 99% of the charge is carried by the electron holes (a = 4.3 A)?

Answer: 1.4×10^{-3} cm^2/volt · sec.

•**5–5.8** How many charge carriers are there per cubic centimeter in the previous problem? a) Electron holes? b) Cation vacancies?

5–6.1 The resistance of a certain silicon wafer is 1031 ohms at 25.1°C. With no change in measurement procedure, the resistance decreases to 1029 ohms. What is the temperature change? The energy gap of silicon is 1.1 eV. [*Hint:* $(1/T_1 - 1/T_2) = (T_2 - T_1)/T_2T_1 \simeq (T_2 - T_1)/T_1^2$ when $T_2 \approx T_1$.]

Answer: +0.03°C

5–6.2 To what temperature must you raise InP in order to make its resistivity half what it is at 0°C? Its energy gap is 1.3 eV. (See hint for Study Problem 5–6.1.)

•**5–6.3** Refer to Example 5–6.1. If the emitter voltage is doubled from 17 to 34 millivolts, by what factor is the collector current increased?

Answer: 15 (= 72 milliamps/4.7 milliamps)

•**5–6.4** A transistor operates between 10 millivolts and 100 millivolts across the emitter. At the lower voltage, the collector current is 6 milliamps; at the higher voltage, 600 milliamps. Estimate the current when the emitter voltage is 50 mV.

Answer: 46 mA

CHAPTER SIX

SINGLE-PHASE METALS

PREVIEW

This chapter is the first of three which direct our attention to single-phase materials. Metals, polymers (i.e., "plastics"), and ceramics will be discussed in sequence. Although a distinction is often made among these three major categories of materials, the boundaries are not sharp. Significant numbers of materials have characteristics intermediate between two or even three of these categories.

Our goal in this chapter is to learn how the properties and behavior of single-phase metals can be modified. This approach serves two purposes: (1) It permits the designer to understand the limitations of metals in his designs, and (2) it provides the engineer with procedures for adjusting properties to meet design specifications.

We will see that while single-phase metals have only one crystal structure, (1) their composition can be varied by solid solution, (2) their grains can have various microstructures, and (3) imperfections can be introduced by plastic deformation and by radiation

exposure, or removed by thermal treatments. Each of these affects properties.

6–1 SINGLE-PHASE ALLOYS

Single-phase metals used commercially may be pure metals with only one component. Examples of such metals include copper for electrical wiring, zinc for the coating on galvanized steel, and the aluminum used for housewares. However, in many cases a second component is intentionally added to the metal in order to improve the properties. Any such combination of metals is called an *alloy*.

Alloys are single-phase metals so long as the solid solubility limit is not exceeded. Brass (a single-phase alloy of copper and zinc), bronze (a similar alloy of copper and tin), and copper-nickel alloys are typical of metals we shall study in this chapter. Multiphase, or polyphase, alloys contain additional phases because the solid solubility limit is exceeded. The majority of our steels, as well as many other metals, are multiphase alloys. They will be discussed in later chapters.

Properties of single-phase alloys. The properties of alloys are different from those of pure metals, as shown by Figs. 6–1.2 and 6–1.3 (pp. 170–171) for brass and Cu–Ni solid solutions. The increase in strength and hardness is due to the presence of solute atoms that interfere with the movements of atoms in the crystals during plastic deformation. We shall observe later (Section 6–4) that this interference exists because dislocation movements (Fig. 4–4.5) are restricted by alloying elements.

Very small amounts of impurities reduce the electrical conductivity of a metal, because the foreign atom introduces nonuniformities in the electrical field within the crystal lattice. Therefore, the electrons encounter more deflections and reflections, with a consequent reduction in the length of the mean free path (Section 5–2).

In a metal, electrons carry the majority of the energy for thermal conduction. Thus there is a correspondence between the thermal and electrical conductivity. (Compare (e) and (f) of Figs. 6–1.2 and 6–1.3.) In fact, it was pointed out some time ago that k/σ in most metals is about 7.5×10^{-6} at normal temperatures when thermal conductivity k and electrical conductivity σ are expressed in $\text{J} \cdot \text{cm/cm}^2 \cdot \text{s} \cdot {}^\circ\text{C}$ and $\text{ohm}^{-1} \cdot \text{cm}^{-1}$, respectively (Fig. 6–1.1). This is called the *Wiedemann–Franz* (W–F) ratio. Since data are readily available on the electrical resistivity values for metals at

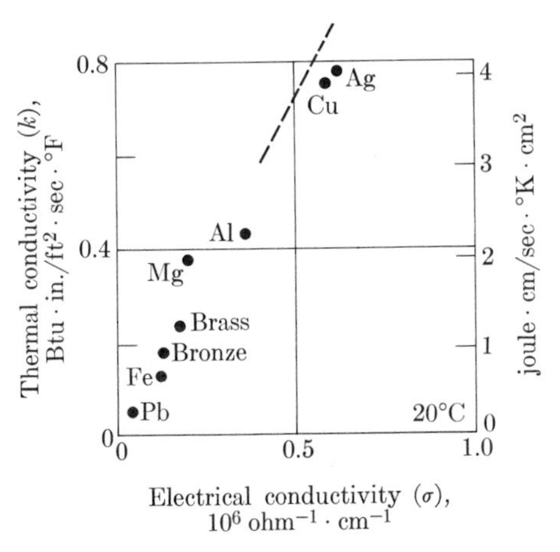

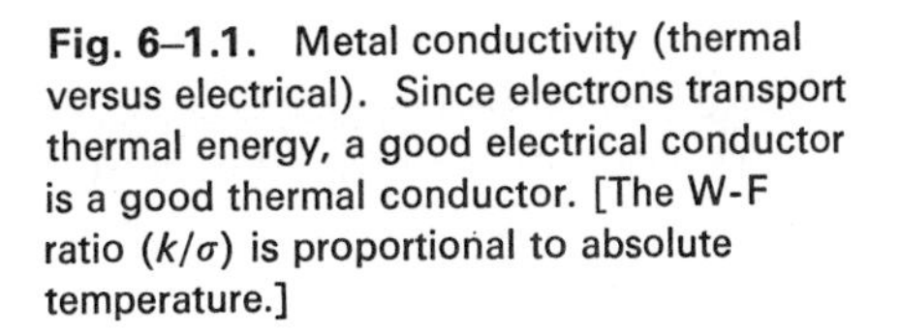
Fig. 6–1.1. Metal conductivity (thermal versus electrical). Since electrons transport thermal energy, a good electrical conductor is a good thermal conductor. [The W-F ratio (k/σ) is proportional to absolute temperature.]

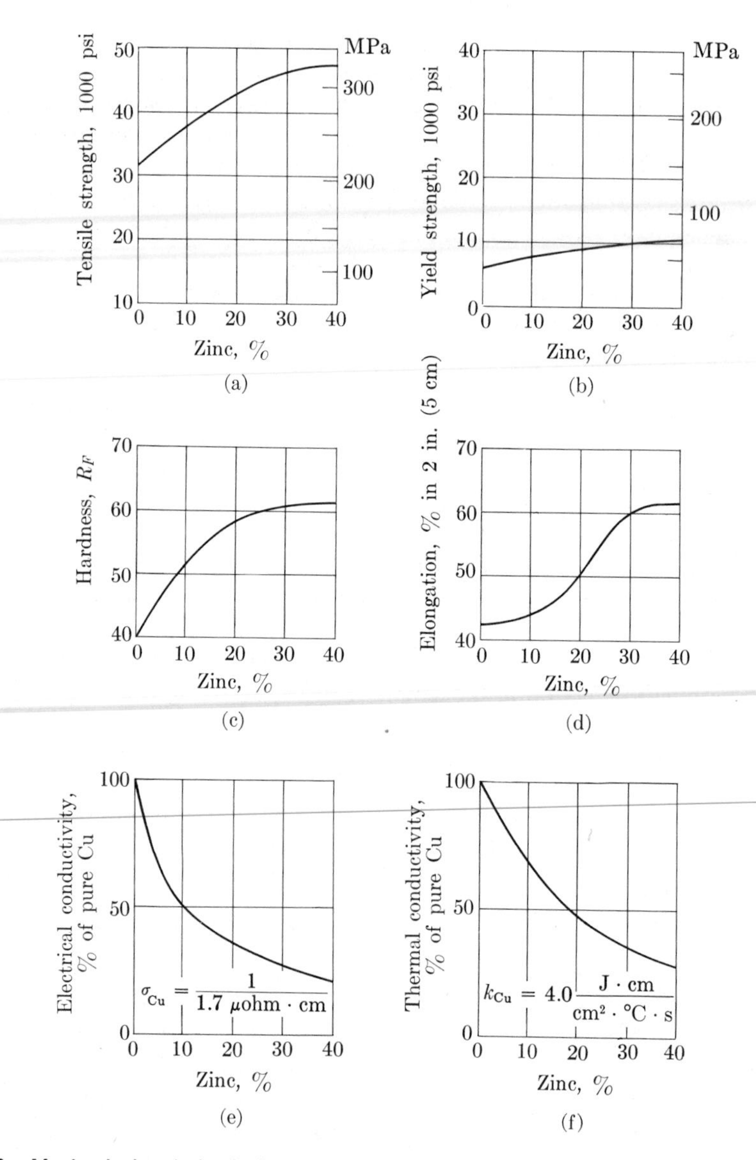

Fig. 6–1.2. Mechanical and physical properties of annealed brasses. The solubility limit of zinc in fcc copper is near 40%. (Adapted from ASM data.)

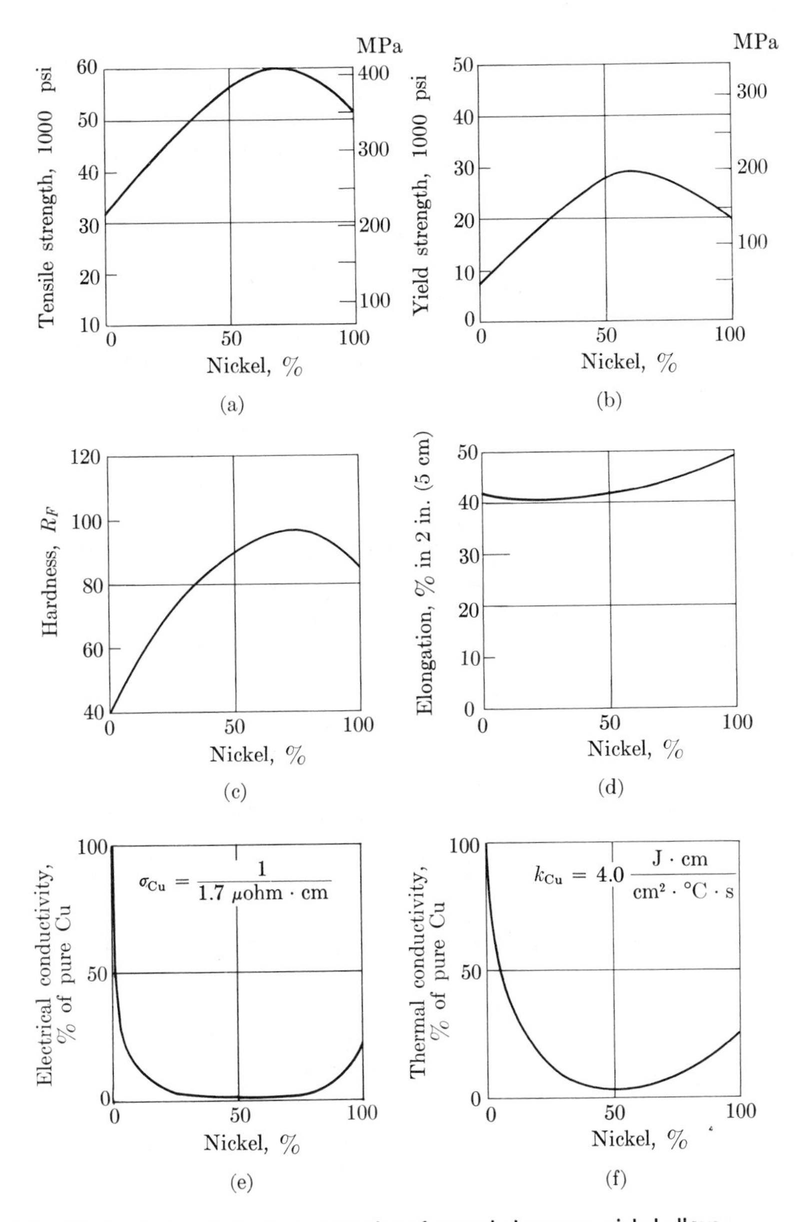

Fig. 6–1.3. Mechanical and physical properties of annealed copper–nickel alloys. Copper and nickel form a complete series of solid solutions. (Adapted from ASM data.)

various temperatures, the W–F ratio provides a convenient rule of thumb for less available thermal conductivity values.

Example 6–1.1 A copper–nickel alloy must have a tensile strength greater than 42,000 psi (>290 MPa), and a thermal conductivity greater than 0.56 J · cm/°C · cm² · s. Select an alloy from Fig. 6–1.3.

Solution: From Appendix C, $k_{Cu} = 4.0$ J · cm/°C · cm² · s; therefore, $k_x \geq 14\%\ k_{Cu}$.

Tensile strength, TS	$19\% < x < 100\%$ Ni
Thermal conductivity, k	$0\% < x < 25\%$ Ni
	and $85\% < x < 100\%$ Ni
Specification windows	19%–25% Ni
	and 85%–100% Ni.

Use 80 Cu–20 Ni, since nickel is more expensive than copper.

Comments. If the coins in one's pocket are examined, the reader will recall that nickel is more expensive than copper.

Specifications generally allow a slight margin at the edge of the window. ◀

Example 6–1.2 Iron is to be used as a thermal conductor at 300°C. Its conductivity, in Appendix C, is for 20°C. Estimate $k_{300°C}$ if k/σ is proportional to °K.

Solution: From Table 5–2.1 and Eq. (5–2.1) which require °C,

$$\sigma_{300°C} = 1/\rho_{300°C} = 1/([9\ \mu\text{ohm} \cdot \text{cm}][1 + (0.0045/°\text{C})(300°\text{C})])$$

$$= 5 \times 10^4\ \text{ohm}^{-1} \cdot \text{cm}^{-1}.$$

$$k_{300°C} \approx (5 \times 10^4\ \text{ohm}^{-1} \cdot \text{cm}^{-1})(7 \times 10^{-6}\ \text{J} \cdot \text{ohm}/°\text{C} \cdot \text{s})(573°\text{K}/293°\text{K})$$

$$= \sim 0.7\ \text{J} \cdot \text{cm/cm}^2 \cdot \text{s} \cdot °\text{C}.$$

Comments. The experimental value found in metals handbooks is 0.6 J · cm/cm² · s · °C at 300°C. The W–F ratio does not apply to nonmetallic materials where electrons are not available to transfer thermal energy. ◀

6–2 POLYCRYSTALLINE MICROSTRUCTURES

Grains were described in Section 4–4 as individual crystals. Materials with many grains are called polygranular or, more commonly, *polycrystalline*. Adjacent crystals have dissimilar orientations so that a grain boundary is present (Fig. 4–4.7). The microstructures of single-phase metals can be varied by changes in *size*, *shape*, and *orientation* of the grains (Fig. 6–2.1). These aspects are not wholly independent,

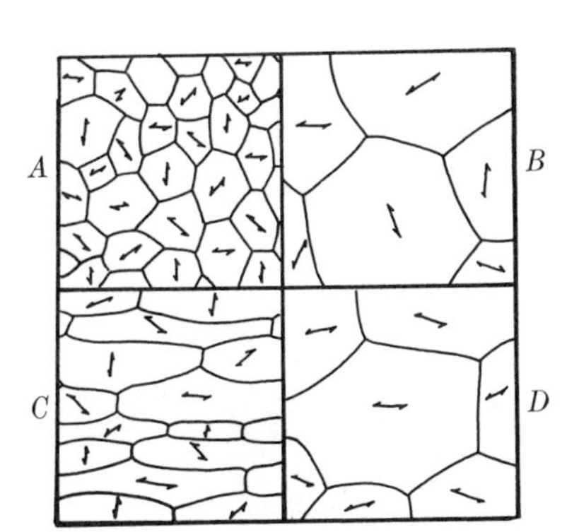

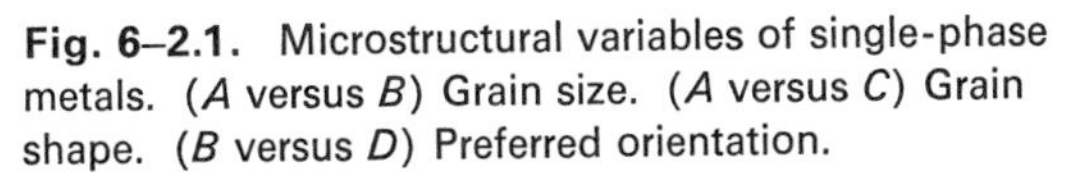

Fig. 6–2.1. Microstructural variables of single-phase metals. (*A* versus *B*) Grain size. (*A* versus *C*) Grain shape. (*B* versus *D*) Preferred orientation.

because the shape and size of grains are both consequences of grain growth. Likewise, grain shape is usually dependent on the crystalline orientation of grains during growth.

Grain growth in metals. The average grain size of a single-phase metal increases with time if the temperature is such as to produce significant atom movements (Section 4–7). The driving force for grain growth is the energy released as an atom moves across the boundary from the convex to the concave grain, where the atom is coordinated with a larger number of neighbors at equilibrium interatomic spacings (Fig. 6–2.2). As a result, the boundary moves toward the center of curvature. Since small grains tend to have surfaces of sharper convexity than large grains, they disappear

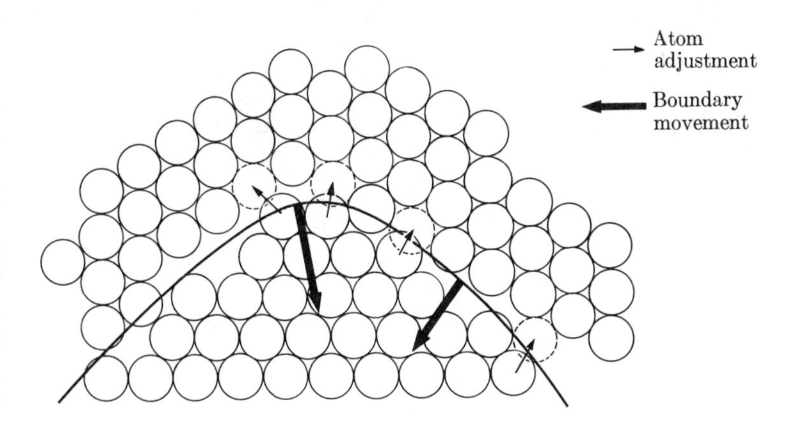

Fig. 6–2.2. Grain-boundary movement. The atoms move to the concave grain, where they are more stable. As a result the boundary is shifted toward the center of curvature.

Fig. 6–2.3. Grain growth. The boundaries move toward the center of curvature (arrows). As a result the small grains eventually disappear.

(a) 15 min at 580°C (b) 1 hr at 580°C (1076°F)

(c) 10 min at 700°C (d) 1 hr at 700°C (1292°F)

Fig. 6–2.4. Grain growth (brass at X40). (Courtesy J. E. Burke, General Electric Co.)

because they feed the larger grains (Fig. 6–2.3). The net effect is grain growth (Fig. 6–2.4).

An interesting example of grain growth can be seen in the ice of a snow bank. Snowflakes start out as numerous small ice crystals, lose their identity with time, and are replaced by larger granular ice crystals. A few of the crystals have grown at the expense of the many smaller crystals.

All crystalline materials, metals and nonmetals, exhibit this characteristic of grain growth. Its engineering importance will be treated in later sections, but the effect of temperature on grain growth must be considered first. An increase in temperature increases the thermal vibrational energy, which in turn *accelerates* the net diffusion of atoms across the boundary from small to large grains. A subsequent decrease in temperature slows down the boundary movement, but *does not reverse it.* The only way to reduce (refine) the grain size in single-phase alloys is to plastically deform the grains and start new grains by recrystallization (Section 6–7).

• **Growth rates.** The *rate* of grain growth depends on the curvature of the boundary, and is therefore inversely proportional to a power $(n - 1)$ of the dimension δ of the grains:*

$$d\delta/dt \propto 1/\delta^{n-1}. \tag{6–2.1}$$

Integrating,

$$\delta^n - \delta_i^n = kt, \tag{6–2.2}$$

or if the initial dimension, δ_i, is enough smaller than the subsequent grain size at time t to be ignored,

$$\log \delta = C + (1/n) \log t. \tag{6–2.3}$$

The value of the *grain growth exponent*, n^{-1}, is commonly between one-third and one-sixth in commercial metals (Example 6–2.1).

Grain shape. Although it is common to speak of grain size in terms of diameter, it is obvious that all grains of single-phase metals are nonspherical. Rather, they must completely fill space and also maintain a minimum of total boundary area. This was shown in Figs. 4–4.8 and 6–2.1(*A*), where the term *equiaxed* is appropriate because the grains have approximately equal dimensions in the three coordinate directions.

Nonregular shapes of grains may include shapes that are platelike, columnar crystals, and dendritic (or treelike). No attempt will be made to systematize them in this book.

Grain orientation. The orientation of grains within a metal is typically quite random (Fig. 6–2.1(*A*)). However, there are exceptions, which can be important from the standpoint of engineering properties. For example, the [100] directions of iron have a higher magnetic permeability than other directions. Therefore, if the grains within a polycrystalline transformer sheet are not random, but are processed to have a *preferred orientation* so that the [100] direction is in the direction of the magnetic field, a significantly more efficient performance may be obtained from the transformer.

* Often called "diameter," even though the grain is not spherical.

• **Example 6–2.1** Determine the grain-growth exponent n^{-1} for the brass of Fig. 6–2.5.

Solution: Since n^{-1} is the slope between two different points on a curve,

$$n^{-1} = (\log \delta_2 - \log \delta_1)/(\log t_2 - \log t_1).$$

At 500°C,

$$n^{-1} = \frac{\log (0.04/0.01)}{\log (2500/2.5)}$$

$$= \tfrac{1}{5}.$$

Comments. The grain-growth exponent n^{-1} becomes a smaller fraction when pores or minor impurity phases inhibit grain growth. The growth is inhibited still further when the grains approach the dimensions of the material. ◀

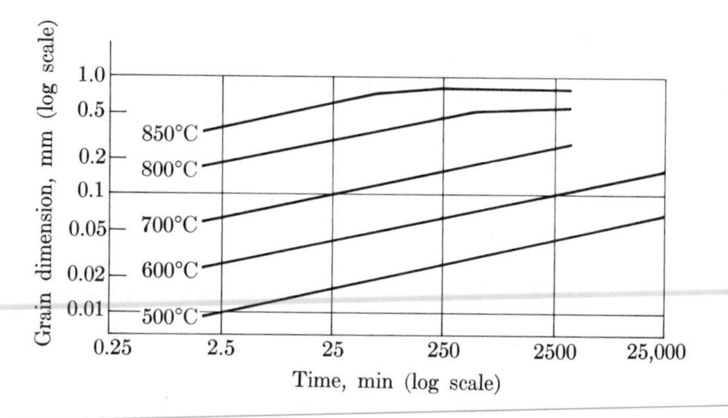

Fig. 6–2.5. Grain growth (brass). See Example 6–2.1. The logarithmic growth proceeds until the grains approach the dimensions of the metal sample. (After J. E. Burke, A.I.M.E.)

6–3 ELASTIC DEFORMATION

Elastic deformation precedes plastic deformation. It occurs when a stress is placed on a piece of metal or, for that matter, on any solid material (Section 1–2). When the load is applied in tension, the piece becomes slightly longer; removal of the load permits the specimen to return to its original dimension. Conversely, when a load is applied in compression, the piece becomes slightly shorter. Within the elastic range, the strain is a result of a slight elongation of the unit cell in the direction of the tensile load, or a slight contraction in the direction of the compressive load (Fig. 6–3.1).

When elastic deformation occurs, the strain is nearly proportional to the stress. This ratio between stress and strain is the *modulus of elasticity* (Young's modulus) and is a characteristic of the type of metal. The greater the forces of attraction between atoms in a metal, the higher the modulus of elasticity.

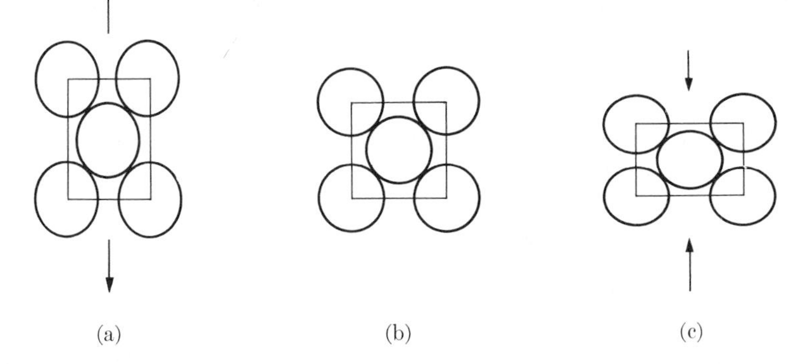

Fig. 6–3.1. Elastic normal strain (greatly exaggerated). Atoms are not permanently displaced from their original neighbors. (a) Tension. (b) No strain. (c) Compression.

Any lengthening or compression of the crystal structure in one direction, due to a uniaxial force, produces an adjustment in the dimensions at right angles to the force. In Fig. 6–3.1(a), for example, a small contraction is indicated at right angles to the tensile force. The negative ratio between the lateral strain ϵ_x and the direct tensile strain ϵ_y is called *Poisson's ratio* ν:

$$\nu = -\frac{\epsilon_x}{\epsilon_y}. \tag{6–3.1}$$

In engineering applications, *shear stresses* are also set up in crystal structures (Fig. 6–3.2). These produce a displacement of one plane of atoms relative to the adjacent plane of atoms. The elastic shear strain γ is defined as the tangent of the shear angle α:

$$\gamma = \tan \alpha; \tag{6–3.2}$$

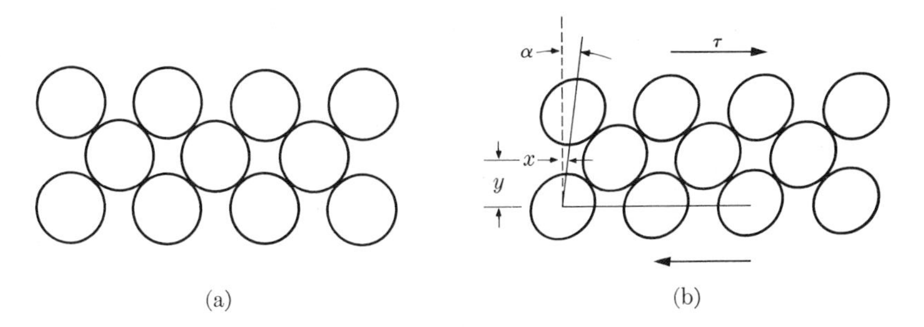

Fig. 6–3.2. Elastic shear strain. Shear couples produce a relative displacement of one plane of atoms past the next. This strain is elastic so long as atoms keep their original neighbors. (a) No strain. (b) Shear strain.

and the shear modulus G is the ratio of the shear stress τ to shear elastic strain γ:

$$G = \frac{\tau}{\gamma}. \tag{6–3.3}$$

This *shear modulus* (also called the modulus of rigidity) is different from the modulus of elasticity E; however, the two are related, a relationship which may be expressed by

$$E = 2G(1 + \nu). \tag{6–3.4}$$

Since Poisson's ratio ν is normally between 0.25 and 0.5, the value of G is approximately 35% of E.

A third elastic modulus, the *bulk modulus K*, is encountered in materials. It is the reciprocal of the compressibility β of the material, and is equal to the hydrostatic pressure σ_h per unit of volume compression, $\Delta V/V$:

$$K = \frac{\sigma_h V}{\Delta V} = \frac{1}{\beta}. \tag{6–3.5}$$

The bulk modulus is related to the modulus of elasticity as follows:

$$K = \frac{E}{3(1 - 2\nu)}. \tag{6–3.6}$$

Elastic moduli versus temperature. Elastic moduli decrease as temperature increases, as shown in Fig. 6–3.3 for four common metals. In terms of Fig. 2–5.2(a), a thermal expansion reduces the value of dF/da and thereby decreases the modulus of elasticity. The discontinuity in the curve for iron in Fig. 6–3.3 is due to the change from bcc to fcc at 912°C (1673°F).

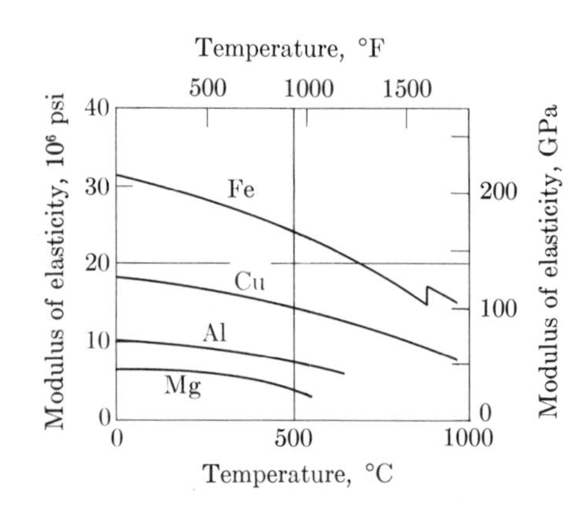

Fig. 6–3.3. Modulus of elasticity versus temperature. (Adapted from A. G. Guy, *Elements of Physical Metallurgy*.)

Elastic moduli versus crystal direction. Elastic moduli are not isotropic within materials; that is, they vary with crystallographic orientation. As an example, iron has an average modulus of elasticity of about 30,000,000 psi (205 GPa); however,

Table 6–3.1
MODULI OF ELASTICITY (YOUNG'S MODULUS)*

Metal	Maximum		Minimum		Random	
	10^6 psi	GPa	10^6 psi	GPa	10^6 psi	GPa
Aluminum	11	75	9	60	10	70
Gold	16	110	6	40	12	80
Copper	28	195	10	70	16	110
Iron (bcc)	41	280	18	125	30	205
Tungsten	50	345	50	345	50	345

* Adapted from E. Schmid and W. Boas, *Plasticity in Crystals*. English translation, London: Hughes and Co.

the actual modulus of a crystal of iron varies from 41,000,000 psi (280 GPa) in the [111] direction to only 18,000,000 psi (125 GPa) in the [100] direction (Table 6–3.1). The consequence of any such anisotropy becomes significant in polycrystalline materials. Assume, for example, that Fig. 6–3.4(a) represents the cross section of a steel wire in which the average stress is 30,000 psi (205 MPa). If the grains are randomly oriented, the elastic strain will be 0.001, because the average modulus of elasticity is 30,000,000 psi (205 GPa). However, in reality, the stress will vary from 18,000 psi (125 MPa) to 41,000 psi (280 MPa) as shown in Fig. 6–3.4(b), because grains have different orientations. Of course, this means that some grains will exceed their elastic limit before other grains reach their elastic limit.

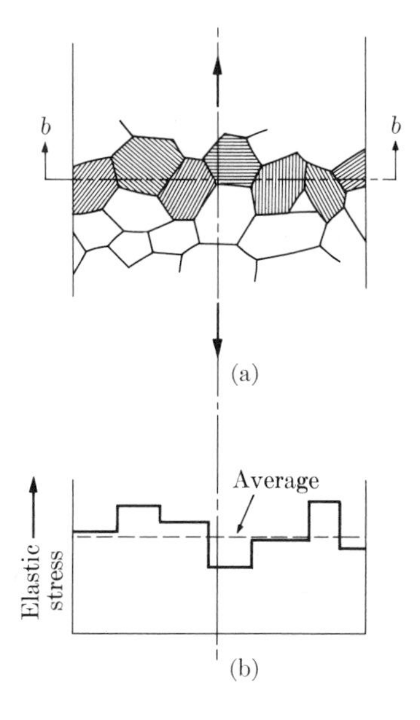

Fig. 6–3.4. Stress heterogeneities (schematic). Elastic stresses vary with grain orientation, because the moduli of elasticity are not isotropic.

Example 6–3.1 A plate of steel has a 100.0-cm × 100.0-cm square scribed on its surface. It is loaded in one direction (parallel to two opposite edges of the square) with a 200-MPa (29,000-psi) stress. a) What are the dimensions of the scribed area? (Poisson's ratio of iron = 0.29.)

Without removing the initial stress, a second tension stress of 410 MPa (60,000 psi) is applied at right angles to the first, i.e., parallel to the other edges of the original square. b) What are the new dimensions of the scribed area?

Solution: Since no preferred orientation is indicated, we will assume the random grain value of Young's modulus (Table 6–3.1).

a) With Eq. (1–2.1): $\epsilon_y = 200 \text{ MPa}/205{,}000 \text{ MPa} = 0.000975;$

With Eq. (6–3.1): $\epsilon_x = -0.29(0.000975) = -0.00028;$

$100.0 \text{ cm } (1 + 0.000975) \times 100.0 \text{ cm } (1 - 0.00028) = 100.10 \text{ cm} \times 99.97 \text{ cm}.$

b) $\epsilon_x = -0.00028 + 410 \text{ MPa}/205{,}000 \text{ MPa} = 0.00172;$

$\epsilon_y = 0.000975 - 0.29(410/205{,}000) = 0.00040;$

$100.0 \text{ cm } (1 + 0.0004) \times 100.0 \text{ cm } (1 + 0.00172) = 100.04 \text{ cm} \times 100.17 \text{ cm}.$

Comments. We can write a general equation for elastic deformation in three dimensions from Eqs. (1–2.1) and (6–3.1):

$$\epsilon_x = \frac{\sigma_x}{E} - \frac{\nu\sigma_y}{E} - \frac{\nu\sigma_z}{E}. \tag{6–3.7}$$ ◀

Example 6–3.2 What is the percentage volume change in iron if it is hydrostatically compressed with 200,000 psi (1400 MPa)? (Poisson's ratio = 0.29.)

Solution: From Eq. (6–3.6),

$$K = (30{,}000{,}000 \text{ psi})/(3(1 - 0.58))$$
$$= 23{,}800{,}000 \text{ psi}.$$
$$\Delta V/V = -200{,}000 \text{ psi}/23{,}800{,}000 \text{ psi} = -0.85 \text{ v/o}.$$

Alternative solution: Since $\sigma_x = \sigma_y = \sigma_z$, Eq. (6–3.7) becomes

$$\epsilon_x = (1 - 2\nu)(-1400 \text{ MPa})/(205{,}000 \text{ MPa})$$
$$= -0.0028 = \epsilon_y = \epsilon_z;$$
$$1 + \Delta V/V = (1 + \epsilon)^3,$$
$$\Delta V/V = -0.85 \text{ v/o}.$$

Comment. Equation (6–3.6) is derived by using the special case of Eq. (6–3.7) where $\sigma_x = \sigma_y = \sigma_z$. ◀

6–4 PLASTIC DEFORMATION OF SINGLE CRYSTALS

Cubic metals and their nonordered alloys deform predominantly by *plastic shear*, or *slip*. This is also one of the methods of deformation in hexagonal metals. Shear

Table 6–4.1
SLIP SYSTEMS IN METALS

Structure	Examples	Slip plane	Slip direction	Number of combinations
bcc	α-Fe, Mo, Na, W	{101}	⟨$\bar{1}$11⟩	12
bcc	α-Fe, Mo, Na, W	{211}	⟨$\bar{1}$11⟩	12
fcc	Ag, Al, Cu, γ-Fe, Ni, Pb	{111}	⟨$\bar{1}$10⟩	12
hcp	Cd, Mg, α-Ti, Zn	{0001}*	⟨11$\bar{2}$0⟩	3
hcp	α-Ti	{10$\bar{1}$0}*	⟨11$\bar{2}$0⟩	3

deformation occurs even when compression or tension stresses are applied, because these stresses may be resolved into shear stresses.

Slip occurs more readily along certain crystal directions and planes than along others. This is illustrated in Fig. 6–4.1, where a single crystal of an hcp metal was deformed plastically. The shear stress required to produce slip on a crystal plane is called the *critical shear stress* τ_c.

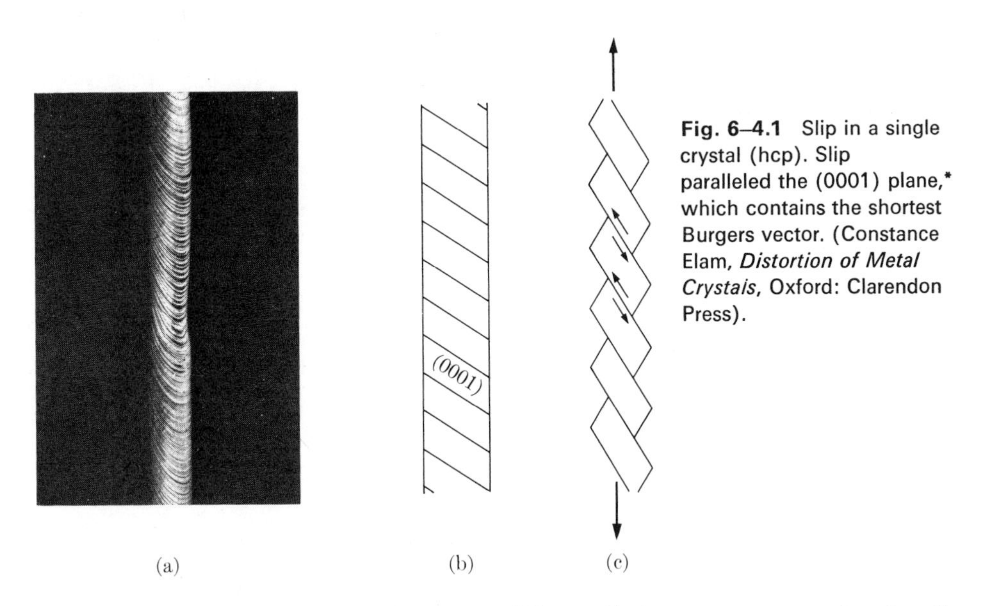

Fig. 6–4.1 Slip in a single crystal (hcp). Slip paralleled the (0001) plane,* which contains the shortest Burgers vector. (Constance Elam, *Distortion of Metal Crystals*, Oxford: Clarendon Press).

When these and other metals are carefully studied, we can summarize the *slip systems* in several familiar metals in Table 6–4.1. A slip system includes the *slip plane* {*hkl*}*, and a *slip direction* ⟨*hkl*⟩. There are a number of slip combinations

* See Section 3–8 for {*hkil*} indices of hexagonal crystals.

within a slip system, because of the multiple planes in a form and multiple directions in a family (Sections 3–8 and 3–7). Two facts stand out in Table 6–4.1.

1. The slip direction in each metal crystal is the direction with the highest linear density of equivalent points, or the shortest distance between equivalent points (called the *Burger's vector*, **b**).
2. The slip planes are planes which have wide interplanar spacings (Section 3–9).

Resolved shear stresses. The force required to produce slip not only is a function of the critical shear stress, but also depends on (1) the angle between the slip plane and the direction of applied force F, and (2) the angle between the slip direction and the direction of force. Consider Fig. 6–4.2, where A is the cross-sectional area perpendicular to the direction of force F, so that F/A is the axial stress. Accordingly, the *resolved shear stress* τ in the slip direction is

$$\tau = \frac{F \cos \lambda}{A/\cos \phi} = \frac{F}{A} \cos \lambda \cos \phi. \qquad (6\text{–}4.1)$$

In this equation, known as *Schmid's law*, ϕ is the angle between the direction of the force and the normal to the slip plane, and λ is the angle between the direction of force and the slip direction. Slip occurs with the minimum axial force when both λ and ϕ are 45°. Under these conditions τ is equal to one-half the axial stress F/A. The resolved shear stress is less in relation to the axial stress for any other crystal orientation, dropping to zero as either λ or ϕ approaches 90°.

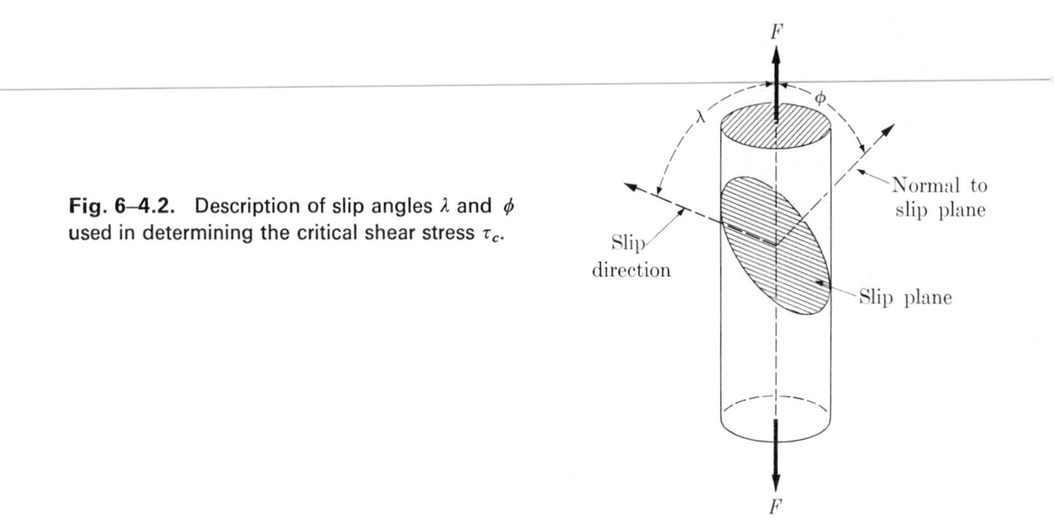

Fig. 6–4.2. Description of slip angles λ and ϕ used in determining the critical shear stress τ_c.

Mechanism of slip. Figure 6–4.3 shows a simplified mechanism of slip. If we attempted to calculate the strength of metals on this basis, the result would indicate that the strength of metals should be about $E/20$ psi, where E is the modulus of elasticity.

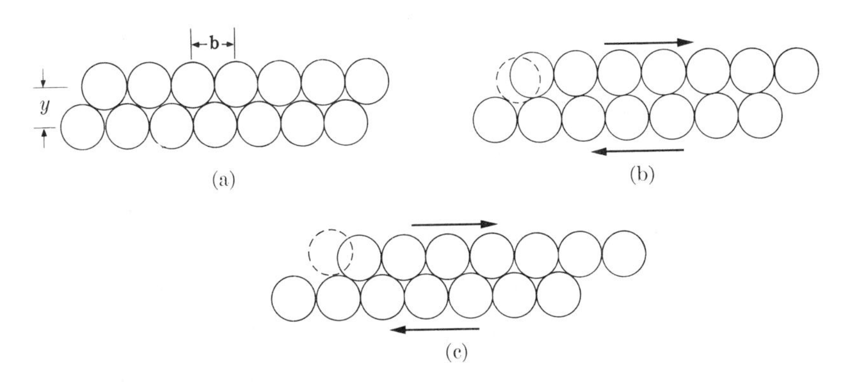

Fig. 6–4.3. An assumed mechanism of slip (simplified). Metals actually deform with less shear stress than this mechanism would require.

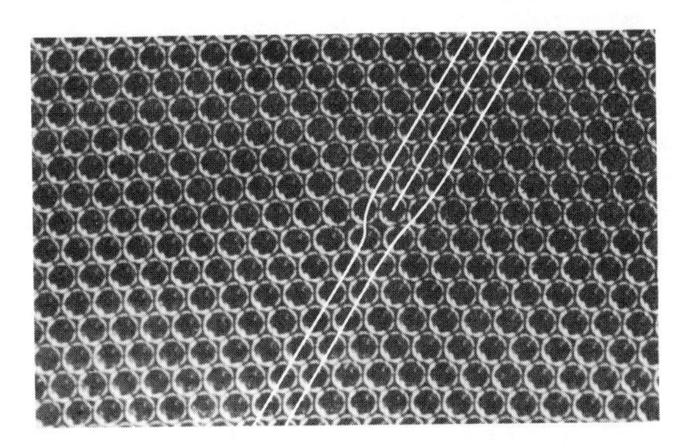

Fig. 6–4.4. Edge dislocation. "Bubble-raft" model of an imperfection in a crystal structure. Note the extra row of atoms. (Bragg and Nye, *Proc. Roy. Soc. (London)* A190, 1947, p. 474.)

Since metals are not that strong, it is apparent that another slip mechanism is operative. All experimental evidence supports a mechanism involving dislocation movements. If we use Fig. 6–4.4 as a model of a dislocation and place a shear stress along the horizontal direction, the dislocation can be moved (Fig. 6–4.5) with a shearing displacement within the crystal. (See also Fig. 4–4.5.) The shear stress required for this type of deformation is only a fraction of the previously cited value of $E/20$. In this respect it matches the shear strengths encountered in laboratory testing.

Since the mechanism of slip involves dislocation movements, the direction in which the critical shear stress is least is the direction with the shortest Burgers vector, i.e., the shortest displacement distance **b**, and the greatest atomic density (Fig. 4–4.3). In that direction, the energy required to move a dislocation is the least, because the energy E is proportional to the product of the shear modulus G and the square of the Burgers vector **b**:

$$E \propto G\mathbf{b}^2. \qquad (6\text{–}4.2)$$

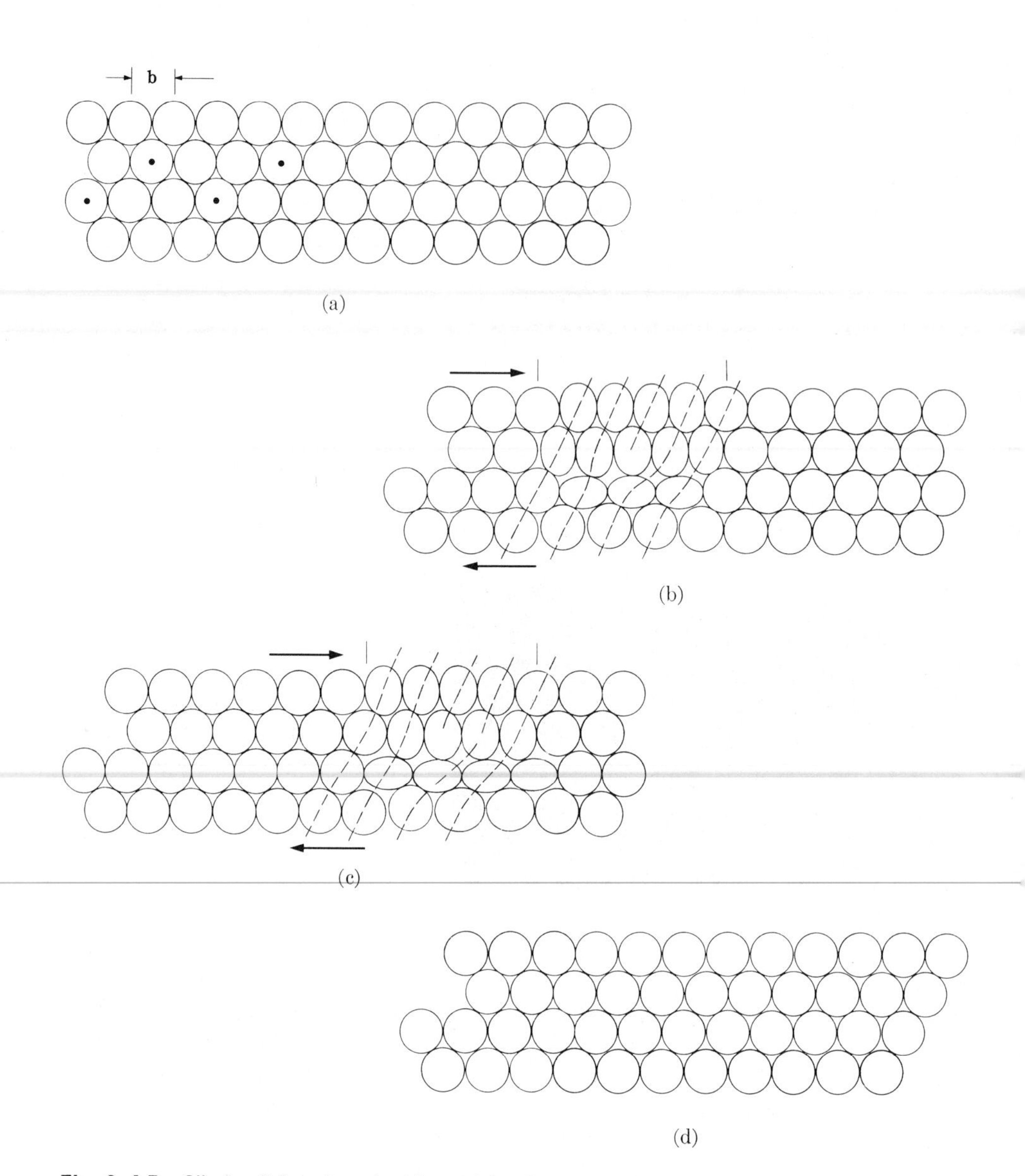

Fig. 6–4.5. Slip by dislocation. In this model only a few atoms at a time are moved from their low-energy positions. Less stress is therefore required to produce slip. Compare with the model of Fig. 6–4.3.

Dislocation movements in solid solutions. The energy associated with an edge dislocation (Fig. 4–4.3) is the same, whether the dislocation is at point (b) or point (c) of Fig. 6–4.5. Therefore, no net energy is required for the movement between the two points.* Such is not the case when solute atoms are present. As shown in Fig. 6–4.6, when an impurity atom is present the energy associated with a dislocation is less than it is in a pure metal. Thus, when a dislocation encounters foreign atoms, its movement is restrained because energy must be supplied to release it for further slip. As a result, solid-solution metals always have higher strengths than do pure metals (Figs. 6–1.2 and 6–1.3). We call this *solution hardening.*

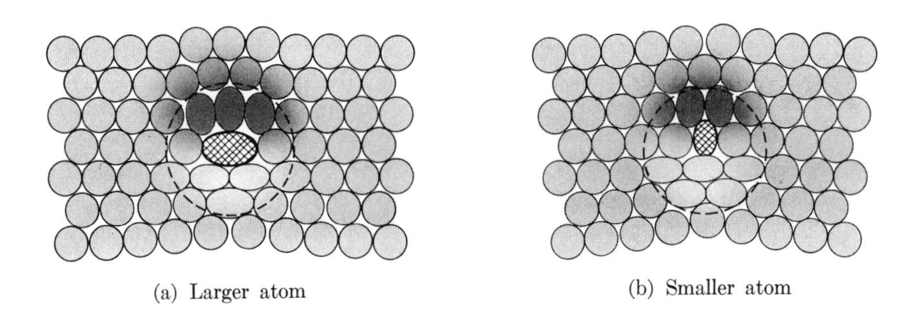

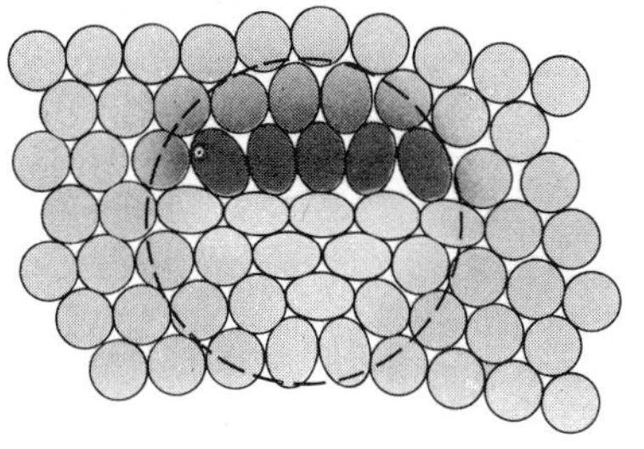

Fig. 6–4.6. Solid solution and dislocations. An odd-sized atom decreases the stress around a dislocation. As a result the dislocation is more stable and requires more stress to be moved. (Cf. Fig. 6–4.4.)

•**Hard intermetallic phases.** Phases such as Al_5Mg_3, Cu_2Al, $CuAl_2$, and β'-brass (CuZn) receive mention elsewhere in the text. Usually they are not stoichiometric, but have a small range of compositions. Even so they are commonly called intermetallic compounds because they are at least partially ordered (Fig. 4–2.2). Example 4–3.2 showed that β'-brass has a preference for copper atoms to be surrounded by zinc atoms, and zinc atoms to be surrounded by copper atoms. Additional temperature destroys this ordering by providing extra energy to the atoms so they may move out of their preferred low-energy arrangements.

* This statement does not apply if (1) the movement includes an increase in the length of the dislocation loop, or (2) there is a pile-up of dislocations (Fig. 6–5.2).

Deformation will also destroy the order in these phases by displacing adjacent atoms (see Fig. 8–8.1); however, extra energy must be supplied here too. This makes these intermetallic phases harder (and less ductile) than pure metals, or alloys with a random solid solution. As an example, an alloy with equal numbers of copper and zinc atoms is soft and ductile above ~460°C, where they form a random bcc solid solution (β-brass). Below that temperature they become ordered as indicated in Example 4–3.2 (β'-brass). The ordered alloy β' cannot be rolled into wire because it is too hard and brittle; the high-temperature β structure can be deformed.*

These hard phases cannot be used readily by themselves; however, they have strengthening effects when they are present as very fine particles in alloys which are otherwise soft (Chapter 11).

Example 6–4.1 The critical shear stress τ_c in the $\langle\bar{1}10\rangle\{111\}$ slip system of *pure* copper was found to be 1 MPa (145 psi). a) What stress must be applied in the [001] direction to produce slip in the [101] direction on the $(\bar{1}11)$ plane? b) In the [110] direction on the $(\bar{1}11)$ plane?

Solution: From Fig. 3–8.3(c),

a)

$$\cos\phi = \frac{\text{Edge of unit cell}}{\text{Long diagonal of unit cell}} = \frac{a}{a\sqrt{3}} = 0.577;$$

$$\cos\lambda = \frac{\text{Edge of unit cell}}{\text{Short diagonal of unit cell}} = \frac{a}{a\sqrt{2}} = 0.707;$$

$$\sigma = \frac{F}{A} = \frac{1\text{ MPa}}{(0.577)(0.707)} = 2.45\text{ MPa } (= 355\text{ psi}).$$

b) Since $\lambda = 90°$ by inspection, $\cos\lambda = 0$, and $\sigma = \infty$. (Slip cannot occur in this direction when the stress is applied in the [001] direction.)

Comments. The dot product is useful here, but can only be used for *cubic* crystals.

The reader is asked to verify that the 12 combinations (Table 6–4.1) in the $\langle\bar{1}10\rangle\{111\}$ slip system for fcc include the following and no others.

$[110](\bar{1}11)$ $[110](1\bar{1}1)$ $[\bar{1}10](11\bar{1})$ $[\bar{1}10](111)$
$[101](\bar{1}11)$ $[10\bar{1}](1\bar{1}1)$ $[101](11\bar{1})$ $[10\bar{1}](111)$
$[0\bar{1}1](\bar{1}11)$ $[011](1\bar{1}1)$ $[011](11\bar{1})$ $[0\bar{1}1](111)$. ◀

Example 6–4.2 Both the [011] and the [112] lie in the $(11\bar{1})$ plane of fcc aluminum. Compare the energy associated with dislocations for $[011](11\bar{1})$ and $[112](11\bar{1})$ slip.

---

* The ordered alloy (β') is simple cubic with the shortest repetition distance between equivalent sites being a (~3 Å) in the six $\langle 100\rangle$ directions. The high temperature form (β) has a repeating distance of $2R$ (~2.5 Å) in the eight $\langle 111\rangle$ directions. Not only is $\mathbf{b}^2$ of Eq. (6–4.2) smaller in the ductile β than in β', but also there are more directions of possible slip during deformation of β.

Solution: From Eq. (6–4.2), the energy is proportional to $\mathbf{b}^2$, where $\mathbf{b}$ is the displacement distance to the next equivalent site.

$$\mathbf{b}_{011} = a/\sqrt{2}$$
$$\mathbf{b}_{112} = a\sqrt{\tfrac{3}{2}}.$$
$$(\mathbf{b}_{011}/\mathbf{b}_{112})^2 = (a^2/2)/(3a^2/2)$$
$$E_{011} = 0.33E_{112}.$$

Comment. The displacement in the [011] direction is shorter and therefore requires less energy. ◀

6–5 DEFORMATION OF POLYCRYSTALLINE METALS (LOW TEMPERATURES)

Data are readily available which show that fine-grained metals are stronger and less ductile than coarse-grained metals. Since we attribute this to the interference of grain boundaries with slip, we will plot these properties as a function of the grain-boundary area (Fig. 6–5.1). The grain boundaries (Section 4–4) interfere with slip because they are the terminus of the crystal planes on which dislocations move. Thus there is a pile-up of dislocations which will have the same effect as a traffic jam on an arterial highway (Fig. 6–5.2). It takes more force (higher stresses) to continue plastic deformation.*

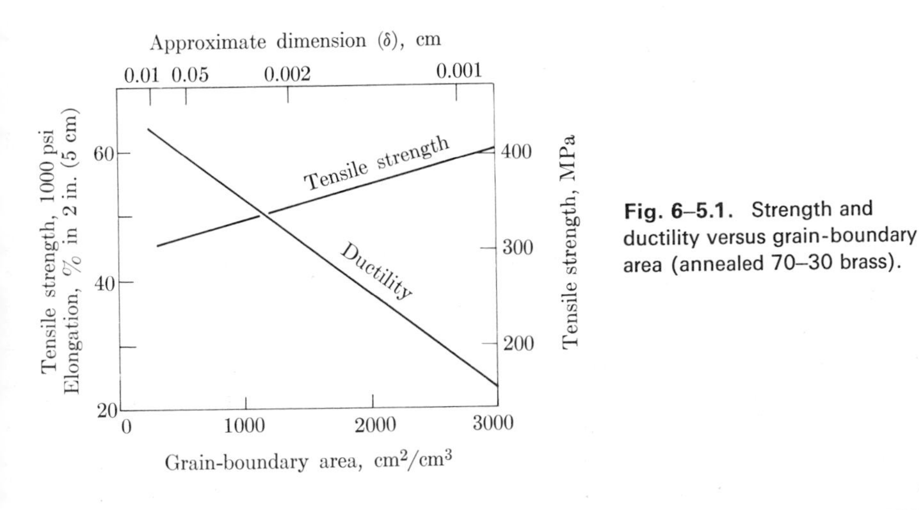

Fig. 6–5.1. Strength and ductility versus grain-boundary area (annealed 70–30 brass).

---

* We will see in Section 6–8 that the grain boundaries can serve as a sink (escape) for dislocations at temperatures which are high enough for diffusion to become significant. Thus the boundary reverses its role from one of a metal strengthener to a metal weakener. This fact is important in gas turbines and related high-temperature applications.

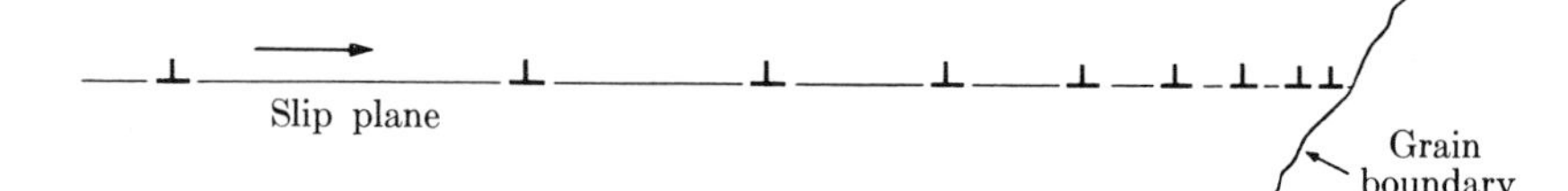

Fig. 6–5.2. Dislocation pile-up. A boundary or surface hinders continued dislocation movements. $\perp$ = edge dislocation.

Polycrystalline metals deform differently from single-metal crystals in another respect. The single crystal of Fig. 6–4.1 was not constrained by adjacent crystals. In contrast, observe Fig. 6–5.3 where it is evident that the large grain of copper in the center did not yield independently of its neighboring grains. The metallurgist can show that at least five combinations of slip (Table 6–4.1) must operate simultaneously within a grain if that grain is to deform in concert with its neighbors and not introduce cracks or gaps. Since not all slip planes are favorably oriented (Example 6–4.1(b)), we can appreciate why bcc and fcc metals, with a larger number of slip combinations (Table 6–4.1), are the more ductile metals; and hcp metals are less ductile.

Fig. 6–5.3. Plastically deformed, polycrystalline copper (X25). The traces of the slip planes are revealed at the polished surface of the metal. (National Bureau of Standards. Reproduced by permission from B. Rogers, *The Nature of Metals*, 2nd ed., American Society for Metals, 1964).

6–6 PROPERTIES OF PLASTICALLY DEFORMED METALS

Plastic deformation changes the internal structure of a metal; therefore, it is to be expected that deformation also changes the *properties* of a metal. Evidence of such property changes may be obtained through resistivity measurements. The distorted structure reduces the mean free path of electron movements (Section 5–2), and therefore increases the resistivity (Fig. 6–6.1).

In the figure just cited, as well as in other cases, it is convenient to refer to the amount of *cold work* as an index of plastic deformation. Cold work is the amount of plastic deformation during processing, expressed by the reduction in cross-sectional area,

$$\mathrm{CW} = \left[\frac{A_o - A_f}{A_o}\right] 100, \tag{6–6.1}$$

where A_o and A_f are the original and final areas respectively.

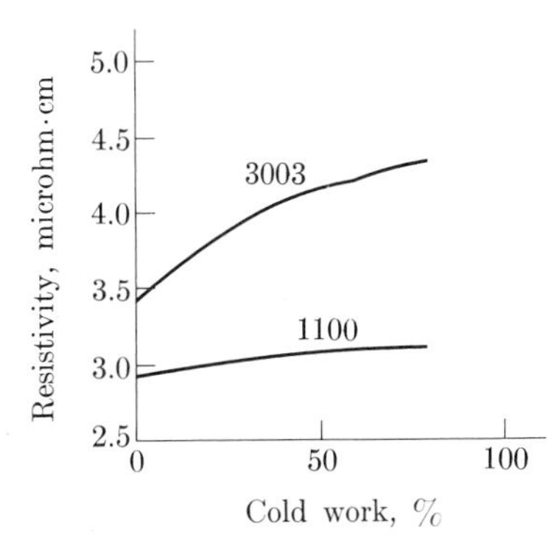

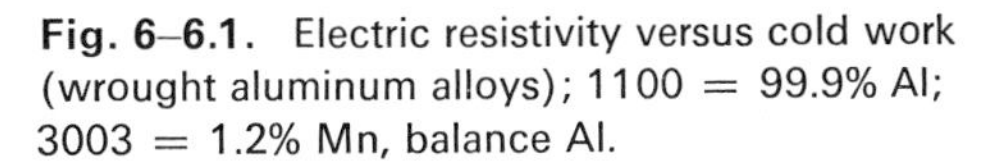
Fig. 6–6.1. Electric resistivity versus cold work (wrought aluminum alloys); 1100 = 99.9% Al; 3003 = 1.2% Mn, balance Al.

Strain hardening. The traces of slip planes of cold-worked copper in Fig. 6–5.3 show that deformation has occurred. Dislocation movements along the slip planes and the distortion of the planes arising from restraint by adjacent grains have modified the regular crystalline pattern that was initially present. Therefore, additional slip occurs less readily, and the hardness of the metal is increased (Figs. 6–6.2 and 6–6.3).

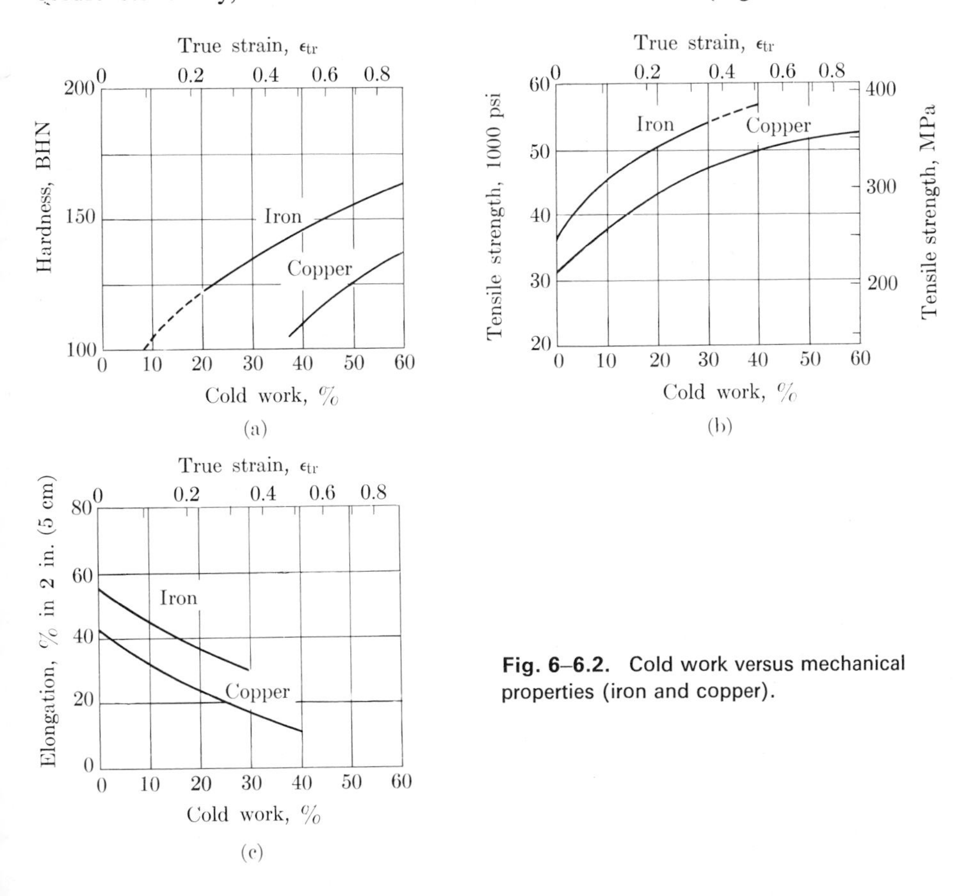

Fig. 6–6.2. Cold work versus mechanical properties (iron and copper).

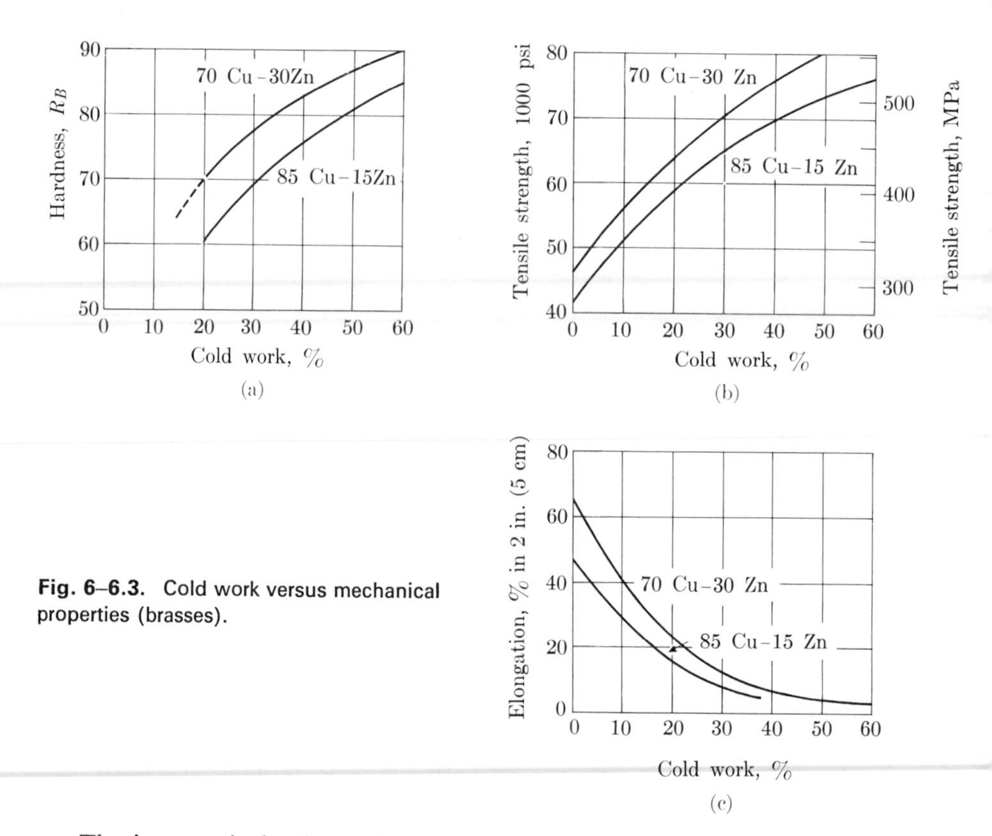

Fig. 6–6.3. Cold work versus mechanical properties (brasses).

The increase in hardness that results from plastic deformation is called *strain hardening*. Laboratory tests show that an increase in both tensile strength and yield strength accompanies this increase in hardness. On the other hand, strain hardening reduces ductility, because part of the "elongation" takes place during cold work,

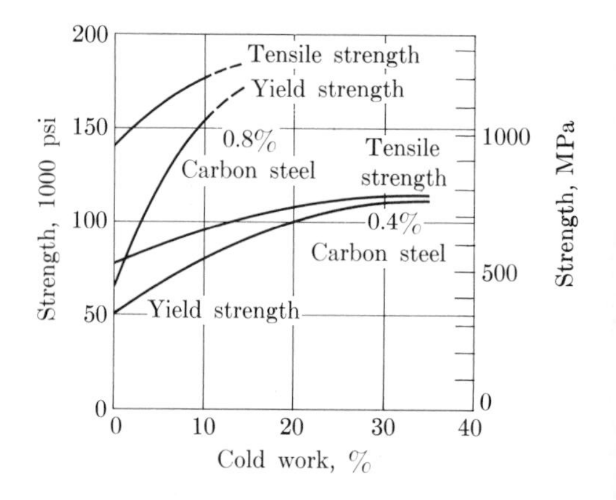

Fig. 6–6.4. Cold work versus strength of plain-carbon steels.

before the gage marks (Fig. 1–2.2) are placed on the test bar. Thus less elongation is observed during testing. The process of strain hardening increases yield strength more than tensile strength (Fig. 6–6.4), and the two approach the true breaking strength (Fig. 1–2.1(d)) as the amount of cold work is increased.

Example 6–6.1 A cold-worked copper or brass may be used in an application which has the specifications of TS ≥ 50,000 psi (≥345 MPa) and an elongation of 20% in 2 in. (5 cm).

Solution: The cold-work data provide the following specification windows:

	Copper	Brass (85–15)	Brass (70–30)
Figure	6–6.2	6–6.3	6–6.3
Tensile strength	≥40% CW	≥ 9% CW	≥ 4% CW
Elongation	≤24% CW	≤16% CW	≤23% CW
Window	—	9–16% CW	4–23% CW

Comments. The designer has a choice of brasses. Other factors equal, he would choose the higher-zinc brass (70Cu–30Zn) because zinc is cheaper than copper. Furthermore, the specification window of 4–23% CW gives him more flexibility in processing the metal. ◀

6–7 RECRYSTALLIZATION

Crystals which have been plastically deformed, like those in Fig. 6–5.3, have more energy than unstrained crystals because they are loaded with dislocations and other imperfections. Given a chance, the atoms will move to form a more perfect, unstrained array. Such an opportunity arises when the crystals are subjected to high temperatures, through the process called *annealing*. The greater thermal vibrations of the lattice at high temperatures permit a reordering of the atoms into less distorted grains. Figure 6–7.1 shows the progress of this *recrystallization*. (See page 192.)

Recrystallization temperatures. Since recrystallization forms softer crystals, hardness values are good indexes of the occurrence of recrystallization. Figure 6–7.2 shows the variation of hardness with increased temperatures in strain-hardened 65Cu–35Zn brass. The temperature of marked softening is called the *recrystallization temperature*. As indicated in this figure, a highly strain-hardened metal (60% CW) is crystallographically more unstable than a metal with less cold work (20% CW), because the metal with more cold work will soften at significantly lower temperatures (275°C vs. 330°C). Recrystallization temperature is also affected by length of time of heating. Since longer heating times give the atoms more opportunity to realign themselves, recrystallization occurs at lower temperatures.

Recrystallization requires relocation of the atoms in a material by diffusion; therefore, the temperature required for recrystallization depends on the forces holding the atoms together. This conclusion is consistent with the fact that the thermal energy required for melting is related to the forces holding the atoms together.

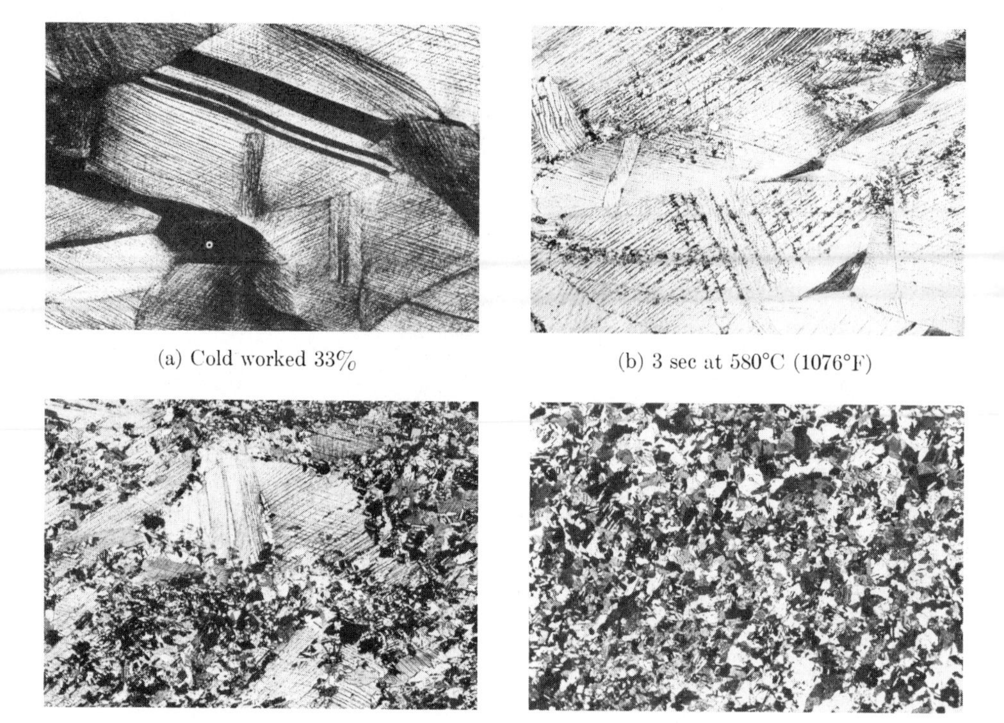

Fig. 6–7.1. Recrystallization of strain-hardened brass (X40). (Courtesy of J. E. Burke, General Electric Co.)

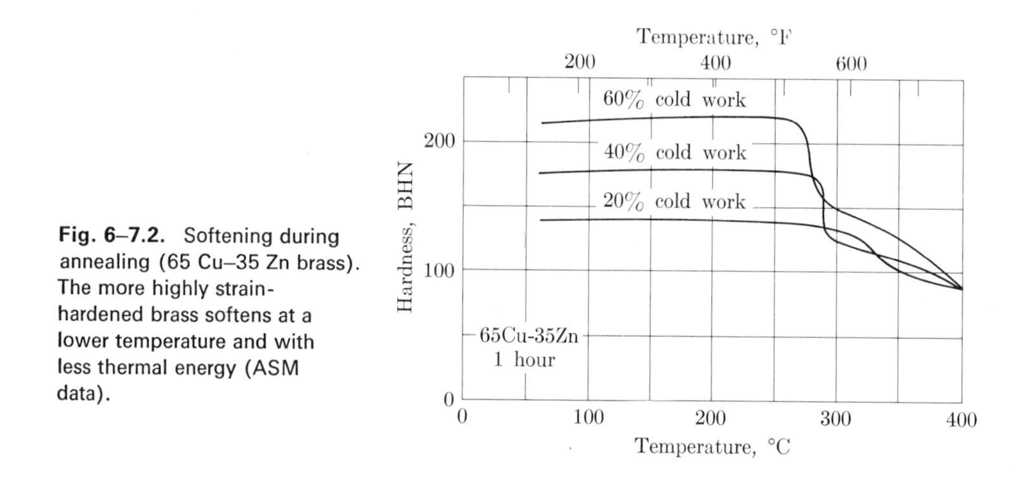

Fig. 6–7.2. Softening during annealing (65 Cu–35 Zn brass). The more highly strain-hardened brass softens at a lower temperature and with less thermal energy (ASM data).

the recrystallization and melting temperatures. Figure 6–7.3 compares these temperatures for a number of common metals. Although there are exceptions, the recrystallization temperature is between one-third and one-half of the (absolute) melting temperature.

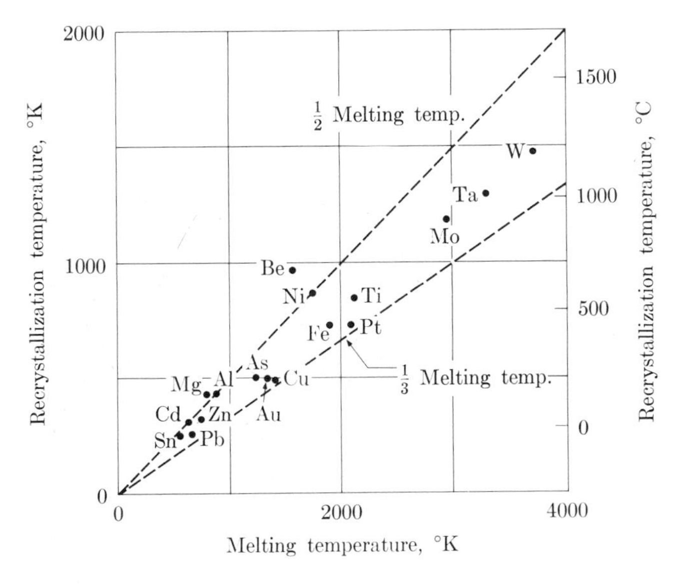

Fig. 6–7.3. Recrystallization temperature versus melting temperature. The average recrystallization temperature is roughly one-half the absolute melting temperature.

Hot-working versus cold-working of metals. In production operations, the distinction between *hot-working* and *cold-working* does not rest on temperature alone, but on the relationship of the processing temperature to the recrystallization temperature. Hot-working is performed above the recrystallization temperature; cold-working is performed below it. Thus the temperature for cold-working copper may be higher than that for hot-working lead.

The choice of the recrystallization temperature as the point for distinguishing between hot- and cold-working is quite logical from the production point of view. Below the recrystallization temperature, the metal becomes harder and less ductile with additional mechanical working. More power is required for deformation and there is a greater chance for cracking during the process. Above the recrystallization temperature, the metal will anneal itself during, or immediately after, the mechanical working. It remains soft and relatively ductile.

Engineering significance of cold-working and annealing. Strain hardening by cold work is of prime importance to the design engineer. It permits him to use smaller parts with greater strength. Of course, the product must not be used at temperatures which will anneal the metals.

Cold work limits the amount of plastic deformation which a metal can subsequently undergo during a shaping operation. The hardened, less ductile metal requires more power for working and is subject to cracking. Therefore, *cold-work-anneal cycles* are used to assist production (Example 6–7.1).

The loss of ductility during cold-working has a useful side-effect in machining. With less ductility, the chips break more readily (Fig. 6–7.4) and facilitate the cutting operation.

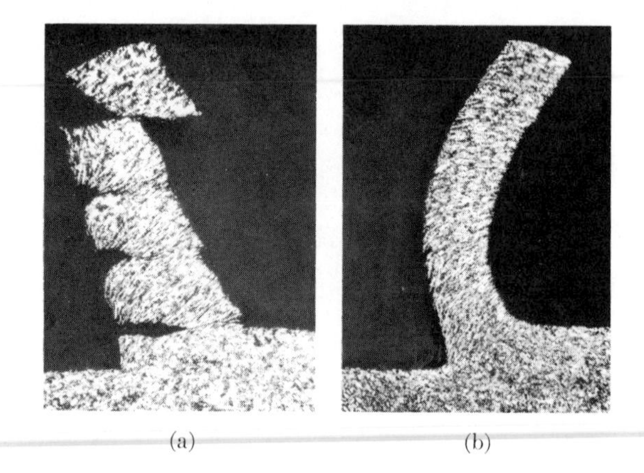

Fig. 6–7.4. The cutting of metal turnings by a machine tool. The strain-hardened metal in (a) formed the more desirable chips, while the annealed metal in (b) formed continuous turnings. (Hans Ernst, Cincinnati-Milacron.)

Example 6–7.1 A 70–30 brass rod is required to have a diameter of 0.197 in. (0.5 cm), a tensile strength of more than 60,000 psi (420 MPa), and a 2-in. elongation of more than 18%. The rod is to be drawn from a larger 0.355-in. (0.9-cm) rod. Specify the final processing steps for making the 0.197-in. rod.

Solution: From Fig. 6–6.3,

$$\text{CW} > 15\% \text{ for TS}, \qquad \text{CW} < 25\% \text{ for El.}$$

Use 20% cold work as *last* drawing step. By Eq. (6–6.1),

$$0.20 = \frac{d^2\pi/4 - (0.197)^2\pi/4}{d^2\pi/4},$$

$$d = 0.22 \text{ in.} \ (= 0.56 \text{ cm}).$$

Comments. Hot-work from 0.355 in. to 0.22 in. (or cold-work and anneal in one or more cycles). The rod should be annealed at 0.22 in. diameter. Cold-draw 20% to 0.197 in. diameter. ◀

•6–8 DEFORMATION OF POLYCRYSTALLINE METALS (HIGH TEMPERATURES)

In Section 6–5 we observed that fine-grained metals are stronger than coarse-grained metals—at low temperatures. At elevated temperatures diffusion may become significant and we find that *creep* takes place. As implied by the name, this is a slow process of strain. The rates range from a few percent per hour at high stress levels or at high temperatures, down to less than 10^{-4}%/hr* (Fig. 6–8.1). These are small; however, consider their importance when designing a steam power plant or nuclear reactor, which must be at high-temperature service for many years. Likewise, creep becomes important in gas-turbine design and other applications which must be operated without change in dimensions at high stresses and elevated temperature to maximize energy-conversion efficiencies.

Since creep is time-dependent, we may plot strain as a function of temperature. Figure 6–8.2 is typical. When a metal is stressed, it undergoes immediate elastic

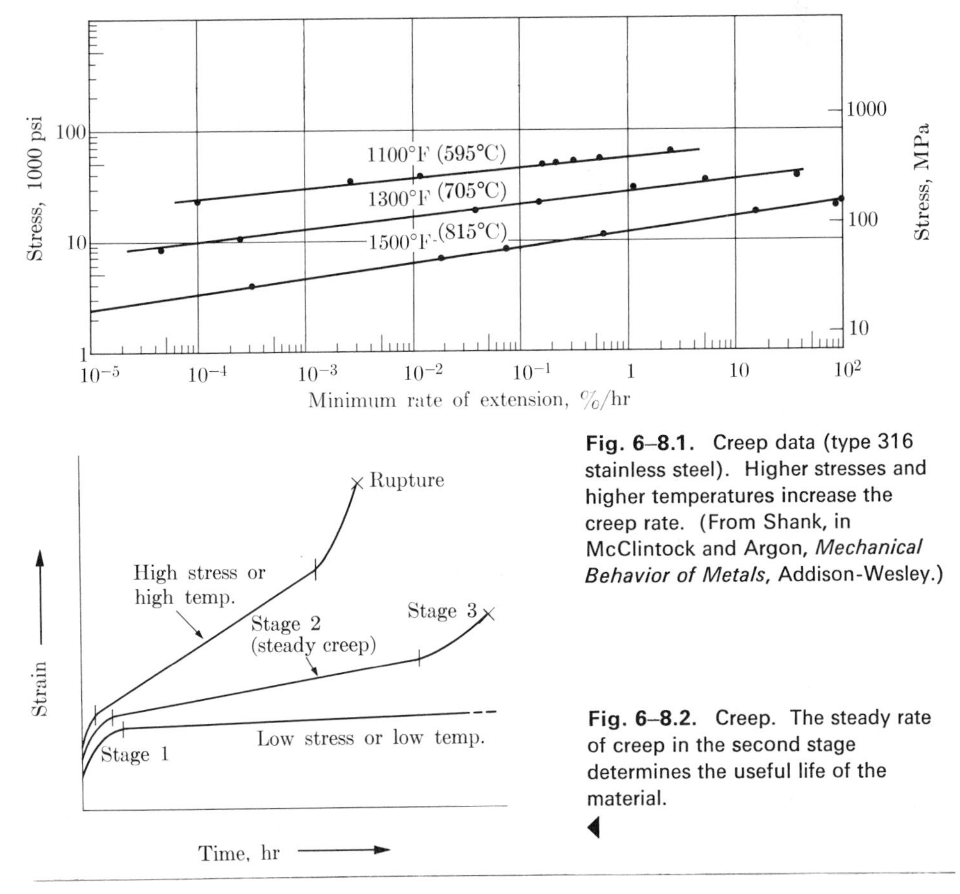

Fig. 6–8.1. Creep data (type 316 stainless steel). Higher stresses and higher temperatures increase the creep rate. (From Shank, in McClintock and Argon, *Mechanical Behavior of Metals*, Addison-Wesley.)

Fig. 6–8.2. Creep. The steady rate of creep in the second stage determines the useful life of the material.

* ~1%/year.

deformation, which is greater when either the stress or the temperature is high. In the first short period of time (Stage 1), it makes additional, relatively rapid, plastic adjustments at points of stress concentrating along grain boundaries and at internal flaws. These initial plastic adjustments give way to a slow, nearly steady rate of strain which we define as the *creep rate* $\dot{\epsilon}$. The second stage of *steady creep* continues over an extended period of time, until sufficient strain has developed so that a necking-down and area-reduction occurs. With this area change at a constant load, the rate of strain accelerates until rupture occurs (Stage 3). If the load could be adjusted to match the reduction in area and thus maintain a constant stress, the creep rate of stage 2 would continue until rupture.

In Fig. 6–8.2 the following relationships are shown schematically.

1. The steady creep rate increases with both increased temperature and stress.
2. The total elongation at rupture also increases with these variables.
3. However, the time before eventual failure by *stress-rupture* is decreased as the temperatures and applied stresses are increased (Fig. 6–8.3).

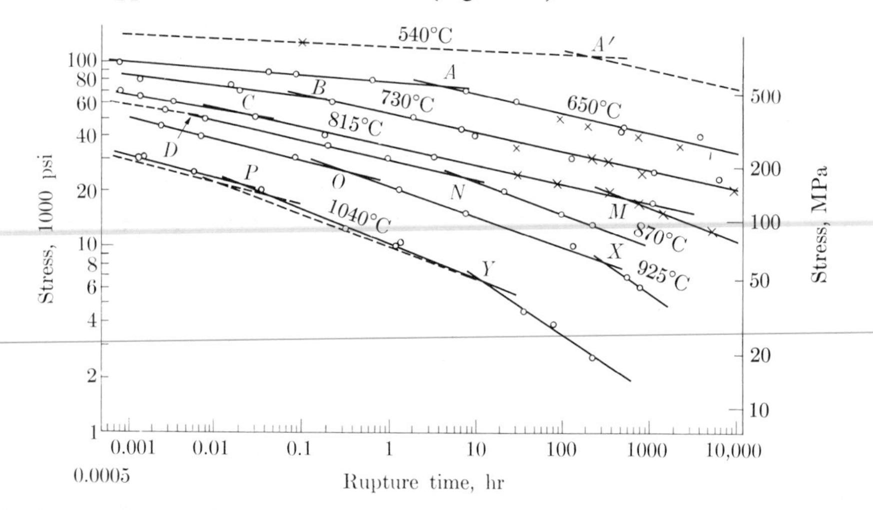

Fig. 6–8.3. Stress-rupture data. The time to rupture is less with higher stresses and higher temperatures. (After N. J. Grant, American Society for Metals.)

Creep mechanisms. The mechanism of creep is related to the movements of dislocations. At low temperatures, strain is restricted because the dislocation movements are stopped by grain boundaries (Fig. 6–5.2), or by impurity atoms. However, at higher temperatures, atomic movements permit the dislocations to *climb*, *jog*, or even be annihilated, as shown in Fig. 6–8.4. As vacancies and atoms move to and from the dislocation, the dislocation can climb out of the initial slip plane, thus permitting the continuation of incremental strain, or creep.

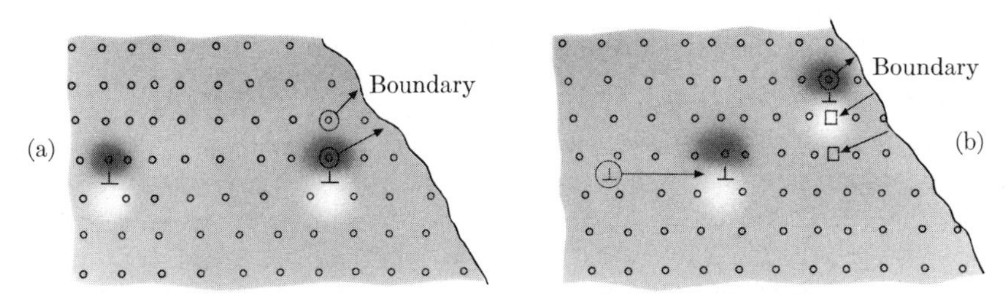

Fig. 6–8.4. Dislocation climb. At higher temperatures, the diffusion of vacancies (□) to edge dislocations (or of atoms away from the dislocations) permits dislocations to climb to another plane, thus relieving dislocation pile-ups and permitting the continuation of incremental strain, or creep, at low stresses.

Although at low temperatures grain boundaries provide interference to dislocation movements, they also provide sources (and sinks) of atoms, and vacancies which permit dislocation climbs during creep. Therefore, we find that, as the temperature is increased, the role of the grain boundary is reversed from one of resisting deformation to one of aiding deformation. The temperature of inversion is called the *equicohesive temperature* (ECT), and of course is important in the design of materials to be used at high temperatures (Fig. 6–8.5). In these two examples the equicohesive temperatures are 250°C and 175°C. The ECT increases with the melting temperatures of the alloy.

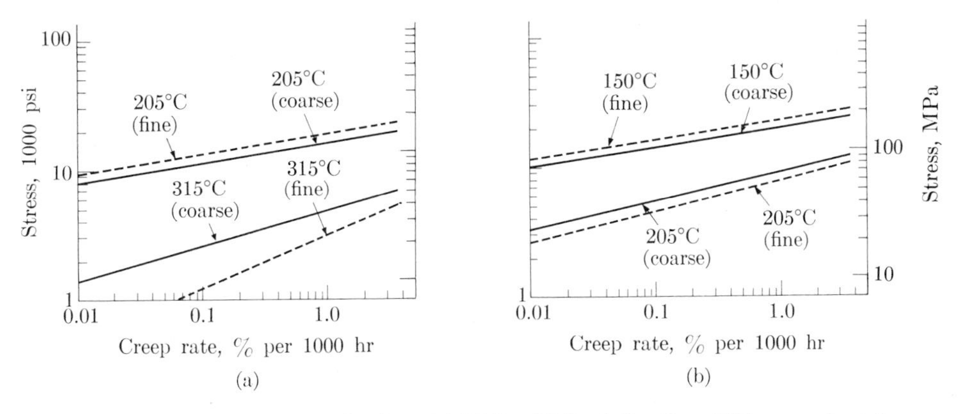

Fig. 6–8.5. Creep rate versus grain size. (a) 77 Cu–22 Zn–1 Sn alloy; ECT, approximately 250°C. (b) 59 Cu–40 Zn–1 Sn alloy; ECT, approximately 175°C. Above the equicohesive temperature, coarse-grained metals are stronger than fine-grained metals. The opposite is true at lower temperatures. (Adapted from ASM data.)

•6–9 FRACTURE

The end of the stress–strain curve (Fig. 1–2.1) is punctuated by fracture. This may or may not have been preceded by some plastic deformation. Thus we refer to *ductile fracture* and *brittle fracture*, and speak of toughness as measured either through (1) the amount of elongation and reduction-in-area or (2) the impact energy. These can be determined quantitatively.

Transition temperatures. Many materials exhibit a brittle–ductile transition and have a characteristic temperature of change (Figs. 6–9.1 and 6–9.2). At low temperatures, a crack can propagate faster than the plastic deformation can occur, so little energy is absorbed. At higher temperatures cracking is preceded by energy-consuming deformation. This discontinuity in energy absorption is particularly characteristic

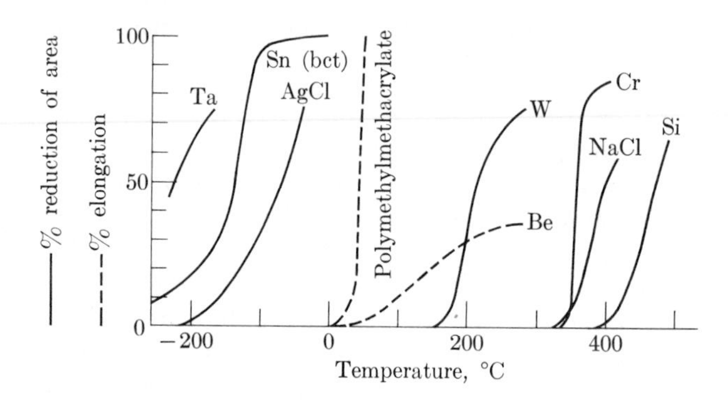

Fig. 6–9.1. Ductility versus temperature (tensile tests—schematic). Except for fcc metals, most materials lose ductility at low temperatures. For a given material the transition temperature is higher for higher strain rates, e.g., impact loading. (After data by A. H. Cottrell, *The Mechanical Properties of Matter*. New York: Wiley, 1964.)

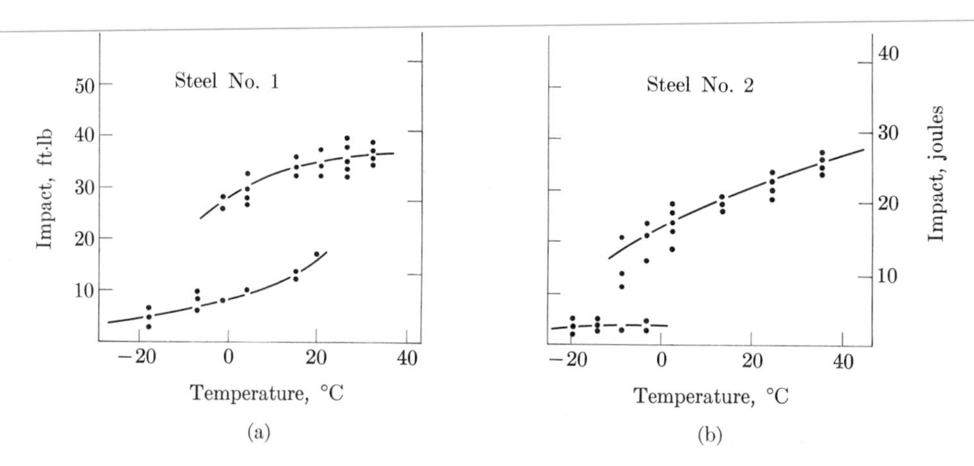

Fig. 6–9.2. Transition temperatures (ship-plate steel). For each steel, there is a marked change in toughness at lower temperatures. The transition temperature is significantly lower for Steel No. 2 than for Steel No. 1. (Adapted from N. A. Kahn and E. A. Imbembo, "Reproducibility of the Single-Blow Charpy Notch-Bar Test," *ASTM Bull.*)

of bcc metals; as a result the designer should be alert to the behavior of steels at low temperatures.*

The transition temperature varies with the rate of loading. Thus a slowly deformed steel may fail ductilely, while it would fail in a brittle manner under impact, because there is no opportunity for plastic deformation to occur. The data of Fig. 6–9.2 are for impact loading. These data are for samples of standardized geometry, since we also find that the energy requirements depend on the three-dimensional distribution of stresses in the neighborhood of the crack. Although the exact transition temperature varies with geometry, the naval architect will prefer Steel No. 2 in Fig. 6–9.2 over Steel No. 1 when he designs a welded ship for use in North Atlantic winter waters. Otherwise a crack, once started, could continue to propagate until it comes out the other side of the ship. There were a number of unfortunate catastrophes which occurred before the designer learned how to make appropriate adjustments, and the metallurgist found that fine-grained steels had lower transition temperatures than coarse-grained steels.

Fatigue. There are many documented examples of eventual failure of rotating shafts on power turbines and on other mechanical equipment, which had initially performed satisfactorily for long periods of time. Figure 6–9.3 shows one such fracture.†

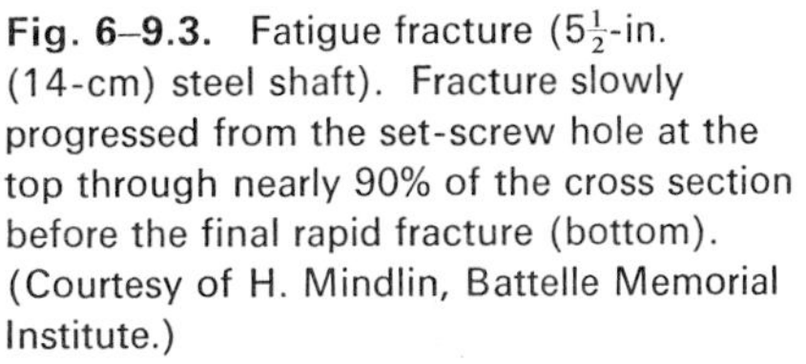

Fig. 6–9.3. Fatigue fracture ($5\frac{1}{2}$-in. (14-cm) steel shaft). Fracture slowly progressed from the set-screw hole at the top through nearly 90% of the cross section before the final rapid fracture (bottom). (Courtesy of H. Mindlin, Battelle Memorial Institute.)

The stresses which a material can tolerate under cyclic loading is much less than it is under static loading. The tensile strength can be used as a guide in design only

* Metals such as copper and aluminum (fcc) do not change abruptly in toughness as a function of temperature.

† The early explanation was that the metal became "tired" and failed from fatigue. We now know that fatigue fracturing is a result of very localized microstructural alterations that produce cracks.

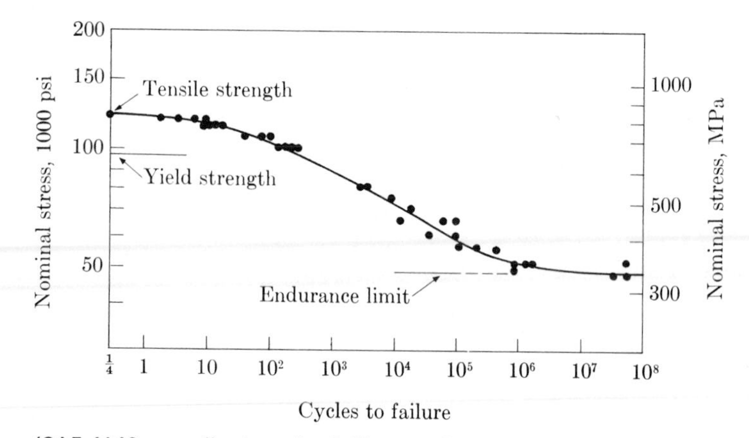

Fig. 6–9.4. S–N curve (SAE 4140 normalized steel). S–N = cyclic *stress* versus *number* of cycles to failure. At the endurance limit, the number of cycles becomes indeterminately large. (Adapted from R. E. Peterson, *ASTM Materials Research and Standards*.)

for structures which are in service under static loading. The number of cycles N which a metal will endure decreases with increased stresses S. Figure 6–9.4 is a typical S-N curve for fatigue fracture. In designing for unlimited cyclic loading, it is necessary to restrict the stresses to values below the *endurance limit* of this curve.

Figure 6–9.5 shows three examples of cyclic loading. The axle of a train has many sinusoidal cycles. Examples of low-cycle stresses are found in the rotor of a generator which is used as a "topping" unit to meet peak demand for electricity.

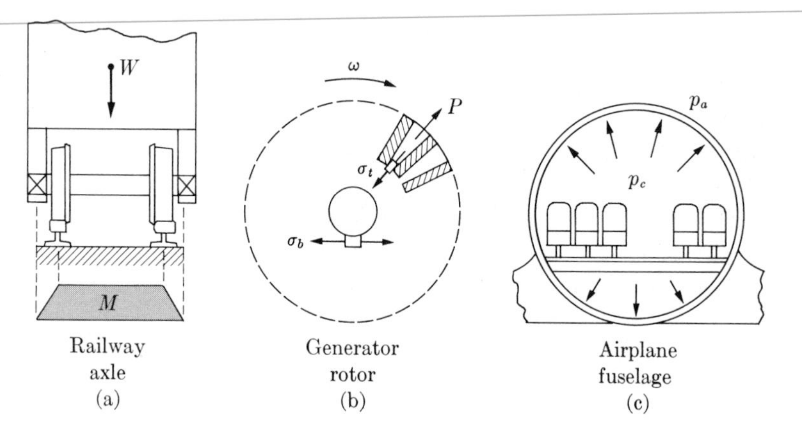

Fig. 6–9.5. Examples of cyclic loading. a) Axle of rail car. b) Rotor of generator during starting and stopping. c) Pressurization and depressurization of plane. The latter two may encounter only a few thousand cycles; however, as indicated by Fig. 6–9.4, the yield strength cannot be used by designers. (Courtesy *ASTM*, R. E. Peterson, "Fatigue in Metals," *Materials Research and Standards*.)

Table 6–9.1
SURFACE FINISH VERSUS ENDURANCE LIMIT
(SAE 4063 steel, quenched and tempered to R_C 44)*

Type of finish	Surface roughness,		Endurance limit,	
	μin.	μm	psi	MPa
Circumferential grind	16–25	0.4–0.6	91,300	630
Machine lapped	12–20	0.3–0.5	104,700	720
Longitudinal grind	8–12	0.2–0.3	112,000	770
Superfinished (polished)	3–6	0.08–0.15	114,000	785
Superfinished (polished)	0.5–2	0.01–0.05	116,750	805

* Adapted from M. F. Garwood, H. H. Zurburg, and M. A. Erickson, "Correlation of Laboratory Tests and Service Performance," *Interpretation of Tests and Correlation with Service.*

It starts and stops up to 1000 times per year, each time introducing a stress cycle at the base of the winding slot and at the arbor. Likewise an airplane fuselage has stresses imposed during pressurization following each take-off; these are removed as it returns to the ground.

Fatigue cracks usually start at the surface where (1) bending or torsion will cause the highest stresses to occur and (2) surface irregularities introduce stress concentrations. As a result, the endurance limit is very sensitive to surface finish (Table 6–9.1).

A close examination of the early stages of crack development shows that microscopic and irreversible slip occurs within individual grains.* There is a gradual reduction in ductility in the regions of these slip planes, causing microscopic cracks to form. This may be after only 10% or 20% of the eventual fatigue life. The cracks progress slowly through the remaining cycles. Eventually they have reduced the cross-sectional area sufficiently so that a final catastrophic failure occurs (Fig. 6–9.3).

Any design factor which concentrates stresses can lead to premature failure. We have already seen, in Table 6–9.1, that surface finish is important. Keyways and other notches (Fig. 6–9.3) are also critical. Finally the generous use of fillets is recommended in mechanical engineering design.

•6–10 RADIATION DAMAGE

Materials selection is the key to successful nuclear-reactor design. First, they must be "fail-safe" in case there are any service problems. Secondly, the radiation environment will alter the internal structure of many materials. Since these alterations generally introduce undesired property changes, we speak of *radiation damage*. We will limit

* This will cause extrusions and intrusions on the external surface of the grains even when highly polished. Then these further concentrate stresses. See Fig. 21–20, Van Vlack, *Materials Science for Engineers*, Addison-Wesley.

our discussion in this chapter to neutron radiation and its effect on metals; however, other types of radiation such as α-particles (He^{2+} ions), β-rays (energetic electrons), and γ-rays are also present. We will see, in Chapter 7, that polymers are affected by radiation too.

Collisions of neutrons with atoms. Since neutrons are not charged, if they are given sufficient energy they can move directly through a material without being preferentially attracted to ions within the material, as are β-rays, protons, or α-particles. Neutrons interact with the atoms in a material only when they happen to "collide" with a nucleus, and such collisions occur only after many atoms have been passed. When a collision does occur in a crystal, a neutron is deflected and the atom (or ion) may be displaced from its position in the crystal to produce a vacancy and interstitialcy (Fig. 6–10.1). Because each collision slows down the neutron, collisions become more frequent until the neutron is finally captured* by the atom.

The displacement of an atom produces defects and distortion in the structure of the solid. The resulting changes in properties are somewhat similar to those arising from cold-work and strain-hardening.

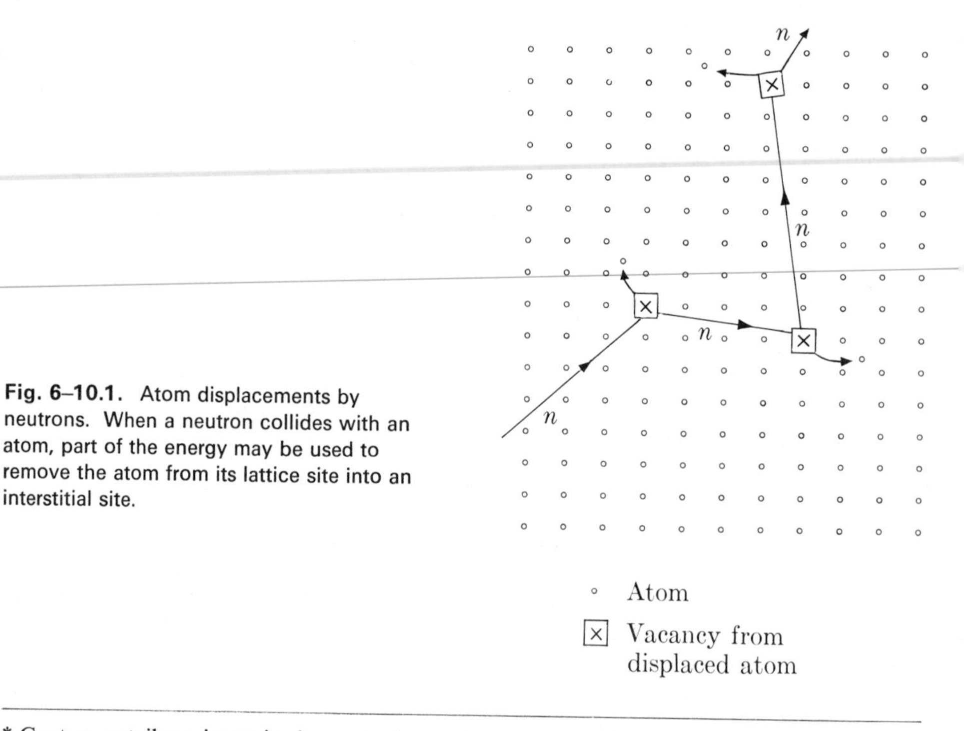

Fig. 6–10.1. Atom displacements by neutrons. When a neutron collides with an atom, part of the energy may be used to remove the atom from its lattice site into an interstitial site.

---

* Capture entails an isotopic change in the capturing atom. For example, when a manganese nucleus with 30 neutrons and 25 protons captures a neutron, it will contain 31 neutrons and 25 protons. This particular isotope of manganese happens to be unstable and will sooner or later (half-life = 2.59 hr) lose an electron (β-ray) from a neutron in the nucleus to form iron, which contains 30 neutrons and 26 protons: $n \rightarrow p^{+} + e^{-}$.

Figure 6–10.2 shows the effect of neutron exposure on the mechanical properties of a carbon–silicon steel. Each of the altered properties is a consequence of a decrease in ease of slip in the distorted lattice structure.

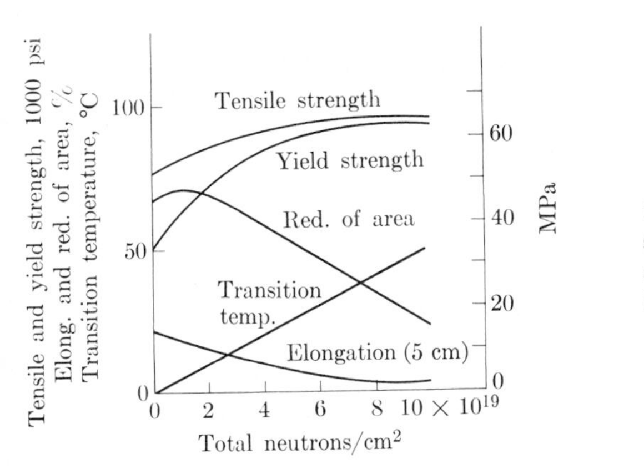

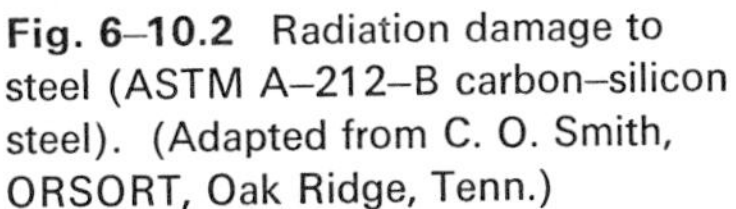
Fig. 6–10.2 Radiation damage to steel (ASTM A–212–B carbon–silicon steel). (Adapted from C. O. Smith, ORSORT, Oak Ridge, Tenn.)

Irradiation as a method of hardening or strengthening metals looks attractive at first sight. However, there are several inherent disadvantages. First, the effect of irradiation is logarithmic with exposure, as indicated in Fig. 6–10.3 for type 347 stainless steel. The exposure required for each succeeding increment of hardness is progressively greater. Secondly, the neutron capture and gamma-ray activation which accompanies irradiation may produce a radiologically "hot" material.

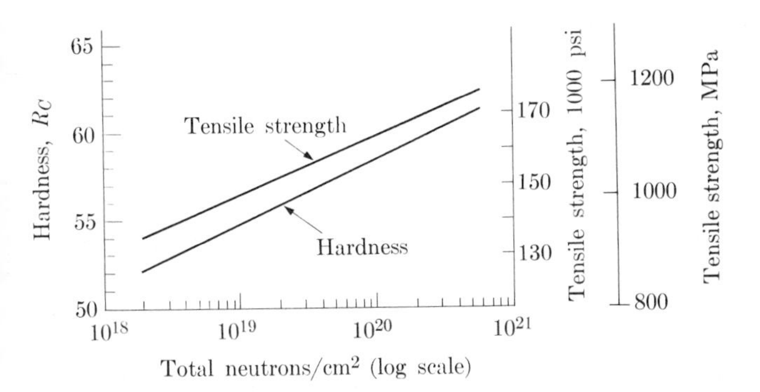

Fig. 6–10.3. Radiation hardening (Type 347 stainless steel). Neutrons dislodge atoms and therefore restrict slip in metals. Since the neutron flux is plotted on a logarithmic scale, each cycle requires appreciably longer exposures. (Adapted from C. O. Smith, ORSORT, Oak Ridge, Tenn.)

Thermal and electrical resistivities increase with neutron irradiation. Figure 6–10.4 illustrates the effect on thermal resistivity. The increase of both resistivities could be predicted from knowledge of the altered electron mobility in distorted solids (Section 5–2).

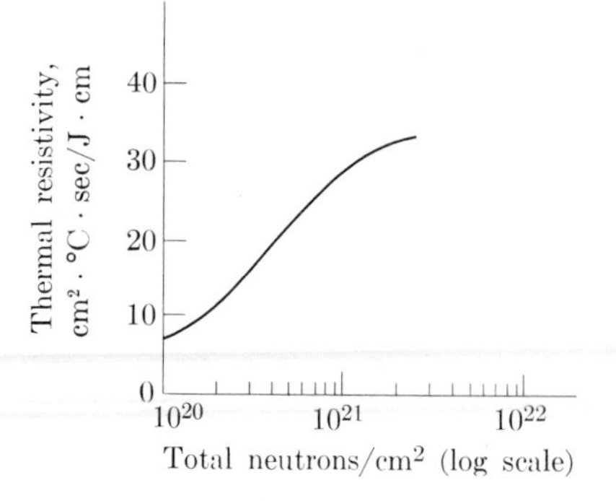

Fig. 6–10.4. Resistivity versus neutron radiation (graphite parallel to the direction of preferred crystal orientation). (Adapted from J. F. Fletcher and W. A. Snyder, "Use of Graphite in the Atomic Energy Program," *Bulletin Amer. Cer. Soc.*)

Recovery from radiation damage. Radiation damage may be erased by appropriate annealing at elevated temperatures. The mechanism of damage removal is somewhat analogous to the mechanism of recrystallization (Section 6–7). However, the required temperature is usually lower, apparently because the atoms were displaced individually into positions of higher energy than is the average position along a dislocation.

REVIEW AND STUDY

Metals have high electrical conductivity and thermal conductivity because of their delocalized electrons. They also have very useful strengths, commonly with a high degree of ductility. This permits them to be shaped into components of industrial and consumer products. Thus in this chapter we have focused on the mechanical behavior of metals as a function of alloy content, deformation, and elevated temperatures. Our attention was limited to metals with only one phase. Multiphase materials will be considered in later chapters.

SUMMARY

Single-phase alloys are stronger than either metal component alone. We call this solution hardening. It is one of two prime means we have for strengthening copper alloys, and is also used to increase the strength of high-temperature alloys. Solution hardening occurs because impurity atoms inferfere with dislocation movements along slip planes.

The variables of single-phase microstructures include (1) grain size, (2) grain shape, and (3) grain orientation. Grain growth occurs at all temperatures, but is insignificantly slow until the temperature exceeds one-half of the melting temperature. At low temperatures, grain boundaries interfere with plastic deformation; therefore, fine-grained steels are somewhat stronger than coarse-grained steels. The situation is reversed at high temperatures, when atoms can move more readily within the solid metal.

The shear modulus is the ratio of shear stress to elastic shear strain. The bulk modulus relates hydrostatic pressure to elastic volume change. Both of these moduli can be calculated from Young's modulus (Section 1–2) through Poisson's ratio. Elastic moduli are anisotropic.

Metals deform by shear. Thus it is the shear component of axial stresses which is critical for plastic deformation.

Strain hardening is a consequence of prior plastic deformation. Any cold-forming process therefore alters the mechanical properties. Along with solution hardening, this is the second major way we have to modify the properties of metals. Unfortunately there is generally a concurrent decrease in ductility. At higher temperatures, atoms can rearrange themselves and eliminate the effects of cold work, thereby softening the metal and making it more ductile. We call this annealing and recrystallization because new crystals form. Deformation processes above the recrystallization temperature are called hot-working processes. At these elevated temperatures, the metal is annealed as fast as it is deformed; thus it does not show strain hardening.

Service conditions may produce creep at high temperatures, or fatigue with cyclic loading. Fracturing is brittle in the absence of plastic deformation. It is ductile when it is accompanied by energy-consuming plastic deformation. These various factors must be taken into account in specifying metals for products and structures.

TERMS FOR REVIEW

Alloy, single-phase
Annealing
Cold work
•Creep
•Creep rate
Deformation
  elastic
  plastic
•Endurance limit
•Equicohesive temperature, ECT
•Fatigue
•Fracture
  brittle
  ductile
Grain growth
Hot work
•Intermetallic phase
Microstructures
Modulus
  bulk, K
  shear, G
  Young's, E
Poisson's ratio
•Radiation damage
Recrystallization
Recrystallization temperature
Schmid's law
Shear strain, γ
Shear stress, τ
  critical, τ_c
  resolved
Slip direction
Slip plane
Slip system
Solution hardening
Strain hardening
•Transition temperature
Wiedemann–Franz ratio

FOR DISCUSSION

A_6 Why are alloys of copper and nickel stronger than either metal alone?

B_6 Why do alloys of copper and nickel have lower conductivity than either metal alone?

C_6 Why does the Wiedemann–Franz ratio not apply to nonmetallic materials?

D_6 How does the fact that an atom has lower energy when it coordinates with the grain on the concave surface lead to boundary movement?

E_6 Distinguish between grain shape and grain orientation in single-phase materials.

F_6 Derive Eq. (6–3.6) from Eq. (6–3.7).

G_6 Distinguish between the three moduli of elasticity.

H_6 The engineer commonly defines the stress for yield strength after an 0.2% offset strain has occurred (Fig. 1–2.1(c)). Based on Fig. 6–3.4, why is this necessary?

I_6 Alloys of zinc and of magnesium are more widely used as casting alloys than as wrought (plastically deformed) alloys. Suggest a valid reason.

J_6 Identify the 12 combinations of slip for the bcc $\langle\bar{1}11\rangle\{101\}$.

K_6 At what combination of angles λ and ϕ (Eq. 6–4.1) is the resolved shear stress the greatest?

L_6 Differentiate between resolved shear stress and critical shear stress.

M_6 An impurity atom stops a dislocation movement. Explain.

•N_6 Consider that Fig. 6–4.5(a) is an intermetallic compound of AB_2, with every third atom in each row being an A atom. (Continue the pattern started by the four dots at the left, and note that each A atom (dot) is surrounded only by B atoms (without dots).) Explain why the critical shear stress is higher in this ordered compound, where the surrounding of each A atom with B atoms gives lower energy. (Dot the atoms in Fig. 6–4.5(c) to show the dislocation movement.)

O_6 Explain why a single-crystal metal is softer and more plastic than a polycrystalline metal.

P_6 A rod is cold-drawn through a die to twice its original length. Why is its elongation decreased? How much is it cold worked? Will its reduction in area be increased or decreased? (This question was asked in order to review the difference between *lengthening* and *elongation* as a property, and between decrease in area arising from *cold work* and *reduction-in-area* as a property.)

Q_6 Use the data of Figs. 6–1.2(b) and 6–6.3. Sketch a plausible curve of yield strength-versus-cold work for a 70–30 brass.

R_6 How will longer times (>1 hour) affect the curves of Fig. 6–7.2? How will the curves differ for an 85–15 brass?

S_6 Distinguish between hot-working and cold-working of tin; of tungsten.

- $\bullet T_6$ Explain the basis for the various stages of the creep shown in Fig. 6–8.2.
- $\bullet U_6$ Other things being equal, which will have the lowest creep rate: (a) steel in service with a high tensile stress and low temperature, (b) steel in service with a low tensile stress and high temperature, (c) steel in service with a high tensile stress and high temperature, (d) steel in service with a low tensile stress and low temperature? Why?
- $\bullet V_6$ Why are coarse-grained materials stronger than fine-grained materials at high temperatures?
- $\bullet W_6$ Predict how the equicohesive temperature relates to the recrystallization temperature. Explain your conclusion.
- $\bullet X_6$ Give examples where ductile fracture provides a "fail-safe" design feature.
- $\bullet Y_6$ Transition temperatures were not given the attention in the design of riveted ships which are given in the design of welded ships. Why?
- $\bullet Z_6$ (a) Look into the engine of an automobile. Name as many components as you can that would have to have endurance-limit specifications. (b) Make a two-page report on the failure of the British Comets. [T. Bishop, "Fatigue and the Comet Disasters," *Metal Progress*, **67**, pp. 77–85 (May 1955) or R. E. Peterson, "Fatigue of Metals," *Materials Research and Standards*, pp. 122–139 (February 1963).]

STUDY PROBLEMS

6–1.1 Based on Fig. 6–1.2, what is the electrical resistivity of annealed 70–30 brass?

Answer: 6 microohm · cm

6–1.2 A copper wire has a resistance of 0.5 ohm per 100 ft. Consideration is being given to the use of a 75–25 brass wire instead of copper. What would be the resistance, if the brass wire were the same size?

6–1.3 A brass alloy is to be used in an application which will have tensile strength of more than 40,000 psi (275 MPa) and an electrical resistivity of less than 5×10^{-6} ohm · cm (resistivity of Cu $= 1.7 \times 10^{-6}$ ohm · cm). What percent zinc should the brass have?

Answer: 14 to 23% zinc

6–1.4 A motorboat requires a seat brace. Iron is excluded because it rusts. Select the most appropriate alloy from Figs. 6–1.2 and 6–1.3. The requirements include: a tensile strength of at least 310 MPa (45,000 psi); a ductility of 45% elongation (in 5 cm); and low cost. [*Note:* Zinc is cheaper than copper.]

6–1.5 A brass wire must carry a load of 10 lb (45 N) without yielding, and have a resistance of less than 0.01 ohm per foot (0.00033 ohm/cm). a) What is the smallest wire that can be used if it is made of 60–40 brass? b) 80–20 brass? c) 100% Cu?

Answer: a) 0.07 in. (1.8 mm) dia b) 0.055 in. (1.4 mm) dia c) 0.05 in. (1.2 mm) dia

6–1.6 A certain application requires a piece of metal having a yield strength greater than 100 MPa and a thermal conductivity greater than 0.4 $J \cdot cm/cm^2 \cdot s \cdot °C$. Specify either an annealed brass or an annealed Cu–Ni alloy that could be used.

•**6–2.1** A metal has a grain-growth exponent of $\frac{1}{4}$. How much longer will it take for the average grain dimensions to grow to 1 mm than to 0.25 mm?

Answer: 256 times as long

6–3.1 A test bar 0.5051 in. (1.283 cm) in dia with 2-in. (5 cm) gage length is loaded elastically with 35,000 lb (156 kN) and is elongated 0.014 in. (0.0356 cm). Its diameter is 0.5040 in. (1.280 cm) under load. a) What is the bulk modulus of the bar? b) The shear modulus?

Answer: a) 22,000,000 psi (152,000 MPa) b) 9,500,000 psi (65,000 MPa)

6–3.2 Assume that copper has a modulus of elasticity of 110,000 MPa (16,000,000 psi) and a Poisson's ratio of 0.3, and is under a tensile stress of 83 MPa (12,000 psi). What would be the dimensions of the unit cell? (Stress is parallel to the axes.)

6–3.3 If copper has an axial stress of 97 MPa (14,000 psi), what will be the highest local stress within a polycrystalline copper bar?

Answer: 170 MPa (24,500 psi)

6–3.4 When iron is compressed hydrostatically with 30,000 psi (205 MPa), its volume is changed by 0.13%. How much will its volume change when it is stressed axially with 60,000 psi (415 MPa)?

6–4.1 An aluminum crystal slips on the (111) plane and in the $[1\bar{1}0]$ direction with a 3.5-MPa (500-psi) stress applied in the $[1\bar{1}1]$ direction. What is the critical resolved shear stress?

Answer: 136 psi (0.95 MPa)

6–4.2 a) What is the normal stress perpendicular to the (110) plane in the above problem? b) Perpendicular to the (001) plane?

6–6.1 A copper wire 0.10 in. (0.25 cm) in diameter was annealed before cold-drawing it through a die 0.08 in. (0.20 cm) in diameter. What tensile strength does the wire have after cold-drawing?

Answer: 48,000 psi (330 MPa)

6–6.2 A pure iron sheet 0.25 cm (0.10 in.) thick is annealed before cold-rolling it to 0.20 cm (0.08 in.), with negligible change in width. a) What would be the ductility of the iron after cold-rolling? b) Estimate the approximate temperature of recrystallization for this iron. c) Give two reasons why the recrystallization temperature of any metal is not fixed.

6–6.3 Copper is to be used in a form with at least 45,000 psi (310 MPa) tensile strength and at least 18% elongation (2 in., or 5 cm). How much cold work should the copper receive?

Answer: 25% cold work

6–6.4 Iron is to have a BHN of at least 125 and an elongation of at least 32% (2 in., or 5 cm). How much cold work should the iron receive?

6–7.1 A 70–30 brass (Fig. 6–6.3) wire with a tensile strength of more than 60,000 psi (415 MPa), a hardness of less than 75 R_B, and an elongation (2 in. or 5 cm) of more than 25% is to be made by cold-drawing. The diameter as received is 0.1 in. (0.25 cm). The diameter of the final product is to be 0.04 in. (0.10 cm). Prescribe a procedure for obtaining these specifications.

Answer: 14 to 19% cold work; therefore it should be annealed when the diameter is 0.044 in. (0.11 cm) before the final 17% cold work.

6–7.2 A round rod of 85 Cu-15 Zn alloy 0.5 cm in diameter is to be cold-reduced to a bar 0.125 cm in dia. Suggest a procedure to be followed if a final tensile strength of 415 MPa (or greater) is to be achieved along with a final ductility of at least 10% elongation (in 5-cm gage length).

6–7.3 A rolled 66 Cu–34 Zn brass plate 0.500 in. (1.27 cm) thick has a ductility of 2% elongation (in 2-in. (5 cm) gage length) when it is received from the supplier. This plate is to be rolled to a sheet with a final thickness of 0.125 in. (0.32 cm). In this final form it is to have a tensile strength of at least 70,000 psi (483 MPa) and a ductility of at least 7% elongation (in 2-in. gage length). Assuming that the rolling process which reduces the plate to a sheet does not change the width, specify *all steps* (including temperature, times, thickness, etc.) which are required.

6–7.4 A round bar of brass (85% Cu, 15% Zn) 0.20 in. (0.5 cm) in diameter is to be cold-drawn to wire that is 0.10 in. (0.25 cm) in diameter. Specify a procedure for the drawing process such that the wire will have a hardness less than 72 R_B, a tensile strength greater than 60,000 psi (415 MPa), and a ductility of greater than 10% elongation in the standard gage length.

6–7.5 A copper wire must have a diameter of 0.7 mm and a tensile strength of 50,000 psi (345 MPa). It is to be processed from a 10-mm copper rod. What should be the diameter for annealing prior to the final cold-draw?

•6–8.1 The following data were obtained in a creep-rupture test of Inconel "X" at 815°C (1500°F): (a) 1% strain after 10 hr, (b) 2% strain after 200 hr, (c) 4% strain after 2000 hr, (d) 6% strain after 4000 hr, (e) "neck-down" started at 5000 hr and the rupture occurred at 5500 hr. What was the creep rate?

Answer: 0.001% hr.

•6–8.2 Design considerations would permit a pressure tube to have 3 l/o strain during a year of service. What maximum creep rate is tolerable when reported in the normal %/hr figures?

•6–9.1 Sometimes it is assumed that if failure does not occur in a fatigue test in 10^8 cycles, the stress is below the endurance limit. A test machine is connected directly to a 1740-rpm electric motor. How long will it take to log that number of cycles?

Answer: 40 days

•6–9.2 Examine a crankshaft of a car. Point out design features which alter the resistance to fatigue.

CHAPTER SEVEN

MOLECULAR PHASES

PREVIEW

Plastics have rapidly come into technical importance. While they have the major advantage of being lightweight, their greatest value is their ability to be processed into specific geometric shapes with desired properties, using the minimum of labor. They constitute one of the fastest-growing areas of materials development and applications. Since plastics may be processed easily, it is common to find them used by designers without full knowledge of their characteristics and limitations. The purpose of this chapter is to outline the structural nature of these nonmetallic materials and to relate these structures to their properties and service behavior.

The technical name for plastics is *polymers* because they contain large molecules made up of many repeating units, or *mers*. They are also called *macromolecules*. We will examine these mers and how they combine into polymeric materials. We will look first at the ideal molecular arrangements. However, as with crystals, we will also look at irregularities, because they have a pronounced influence on the behavior of the polymer.

Crystallization does not occur readily because we have large molecules rather than individual atoms. In addition, the molecules have relatively weak *inter*molecular bonds. Furthermore, polymeric materials are insulators, because they have no free electrons. As a result, the properties and service behavior of plastics differ from those of metals in many respects.

7–1 GIANT MOLECULES

Our next principal category of materials following metals is molecular solids. These are nonmetallic compounds (Section 2–7) that commonly originate from organic raw materials. Organic substances have served as engineering materials from the time of the first engineer. Wood has long been a common construction material, and such natural organic substances as leather for gaskets, felt for packings, cork for insulation, fibers for binding, oils for lubrication, and resins for protective coatings are extensively used by engineers.

Early in the history of the use of organic materials, attempts were made to improve their engineering properties. For example, the properties of wood are highly directional; the strength parallel to the grain is several times greater than in the perpendicular direction. The development of plywood has helped to overcome this difficulty, and still better physical properties are obtained when the pores of the wood are impregnated with a thermosetting resin.

The ingenuity of technologists in working with organic materials has not been limited to improving natural organic materials; many synthetic substances have been developed as well. For example, the field of *plastics** has given the engineer an infinitely greater variety of materials for his applications. Great strides have been and continue to be made in the utilization of such materials.

Whether the engineer is working with natural organic materials or with artificial ones, he is concerned primarily with the nature and characteristics of *large molecules.* In natural materials, large molecules are "built in" by nature; in artificial materials, they are built by *deliberately joining small molecules.*

Molecular weights. The reader who has had chemistry knows that the melting temperature of a paraffin is related to the size of the molecule (Fig. 7–1.1). In general,

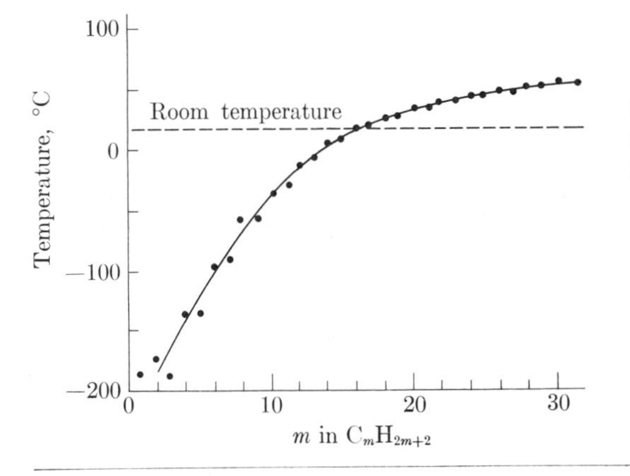

Fig. 7–1.1. Melting temperatures versus molecule size (paraffin series of hydrocarbons).

---

* Strictly speaking, *plastic* is an adjective defining a permanently deformable material (Section 1–2), but by common usage "plastics" denotes organic materials which have been shaped by plastic deformation. This is the usage we shall follow here.

those plastics made up of large molecules are stronger and more resistant to thermal and mechanical stresses than are those composed of small molecules (Fig. 7–1.2). Examples of this type of relationship between molecule size and properties exist for such artificial organic plastics as polyvinyl chloride, nylon, saran, etc., and such natural substances as cellulose, rubber, waxes, and shellac.

The size of a molecule is determined by dividing the molecular weight by the mer weight. This number is called the *degree of polymerization*, DP:

$$\mathrm{DP} = \frac{\text{Molecular weight}}{\text{Mer weight}}. \tag{7–1.1}$$

The units for this expression are

$$\frac{\text{amu/molecule}}{\text{amu/mer}} = \frac{\text{mers}}{\text{molecule}}.$$

For example, a polyvinyl chloride molecule (Fig. 7–1.3(b)) containing 1000 carbons, 1500 hydrogens, and 500 chlorines contains 500 mers, each with 2, 3, and 1 of the above atoms; for this molecule DP = 500. In commercial plastics the degree of polymerization normally falls in the range of 75 to 750 mers per molecule.

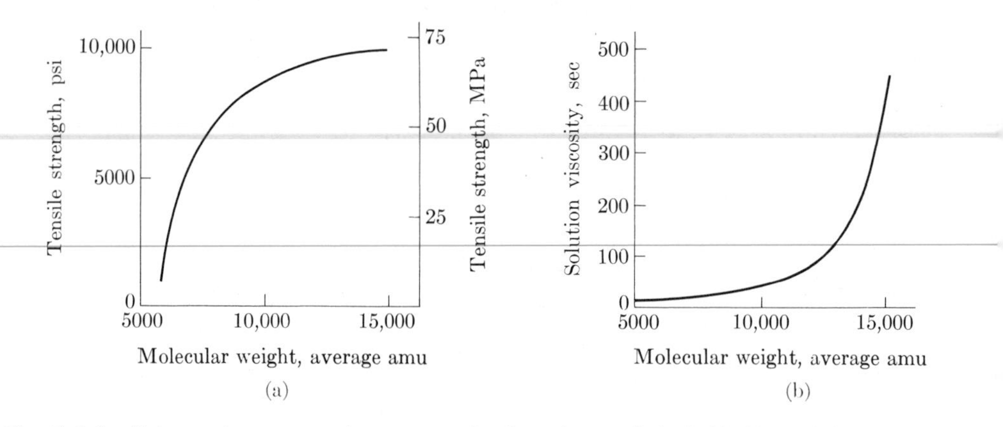

Fig. 7–1.2. Polymer size versus polymer properties (copolymer of vinyl chloride and vinyl acetate). (a) Tensile strength. (b) Viscosity. (Adapted from G. O. Crume and S. D. Douglas, *Ind. Eng. Chem.*)

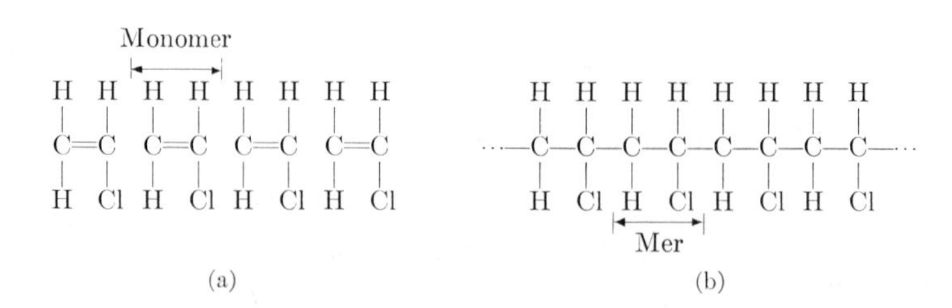

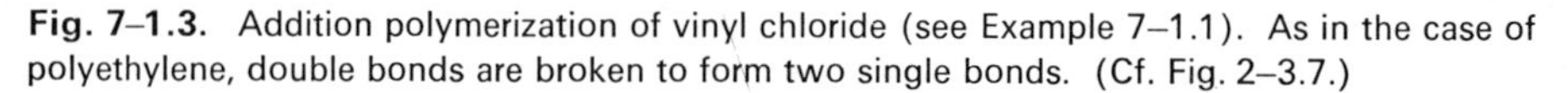

Fig. 7–1.3. Addition polymerization of vinyl chloride (see Example 7–1.1). As in the case of polyethylene, double bonds are broken to form two single bonds. (Cf. Fig. 2–3.7.)

The molecule described above has a molecular weight of more than 31,200. Such a value differs by a major order of magnitude from that of other molecules (Fig. 2–3.1). However, as large as the polymer molecule appears to be by weight, it is still smaller than the resolving power of an optical microscope, and only under certain circumstances can it be resolved even by an electron microscope. Consequently, molecular-weight determinations are usually made indirectly by such physical means as measurement of viscosity, osmotic pressure, or light-scattering, all of which are affected by the number, size, or shape of molecules in a suspension or in a solution (Fig. 7–1.2(b)).

Molecular-size distribution. When a material like polyethylene (Fig. 7–1.4) or polyvinyl chloride is formed from small molecules, not all the resulting large molecules are identical in size. As might be expected, some grow larger than others. As a result, plastics contain a range of molecular sizes, somewhat analogous to the mixture of propane, hexane, octane, and other paraffin hydrocarbons in crude oils. Hence it is necessary to calculate the *average* degree of polymerization if a single index is desired.

One procedure for determining average molecular weights (Fig. 7–1.5(a)) utilizes the weight fraction of the polymer that is in each of several size intervals. The "*weight-*

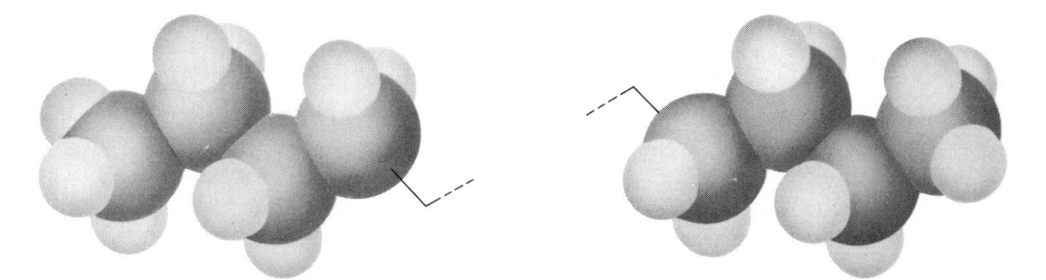

Fig. 7–1.4. Linear polymer (polyethylene). The molecular length varies from molecule to molecule. Typically they contain hundreds of carbon atoms.

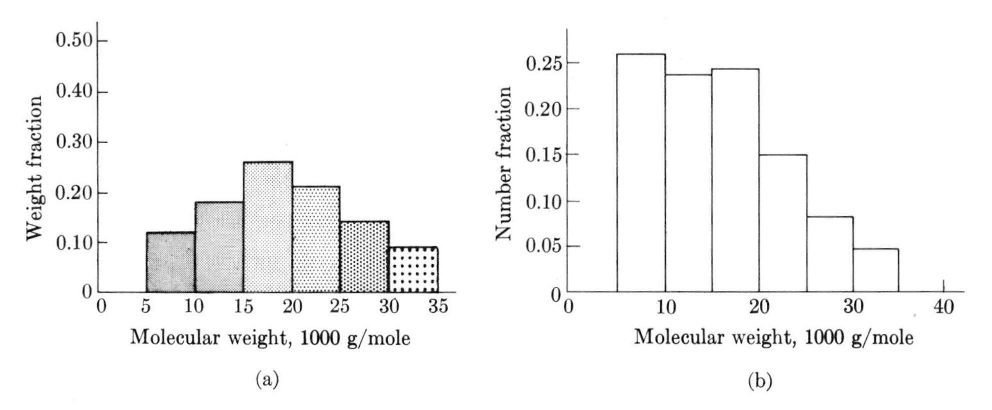

Fig. 7–1.5. Polymer size distribution (see Example 7–1.1).

average" molecular weight $\overline{M}_w$ is calculated as follows:

$$\overline{M}_w = \sum [(W_i)(MW)_i], \qquad (7\text{–}1.2)$$

where W_i is the weight fraction in each size interval and $(MW)_i$ is the middle of each size interval. The "weight-average" molecular weight is particularly significant in the analysis of properties such as viscosity, where the *weight* of the individual molecules is important.

Properties such as strength (Fig. 7–1.2(a)) are more sensitive to the *numbers* of large molecules than to the actual weight. Hence, a "*number-average*" *molecular weight* $\overline{M}_n$ has significance:

$$\overline{M}_n = \sum [(X_i)(MW)_i]. \qquad (7\text{–}1.3)$$

The value X_i is the numerical fraction of molecules in each size interval (Fig. 7–1.5(b)). (See Example 7–1.1.)

Molecular length. We could make a calculation from Table 2–3.1 that the average length of a polyvinyl chloride molecule with DP of 307 mers (Example 7–1.1) is 946 Å, because each C—C bond is 1.54 Å and there are twice as many of these bonds as there are mers. Immediately, however, we realize that this calculation needs a correction because the sketch of Fig. 7–1.3 is overly simplified.* The C—C—C bond angles are not 180°, but 109.5° (Fig. 7–1.4). Thus a "sawtooth" length would be (946 Å) sin 54.7°, or 772 Å. Even more important, however, is the fact that the single bonds of the carbon chain are free to rotate (Fig. 7–1.6). With only three bonds, the end-to-end length can vary from less than 3 Å, *a–d′* to practically 4 Å, *a–d*. The length varies randomly between these limits as the thermal agitation rotates the bond angles. With the above molecule of 307 mers, the end-to-end length could be as long as 772 Å

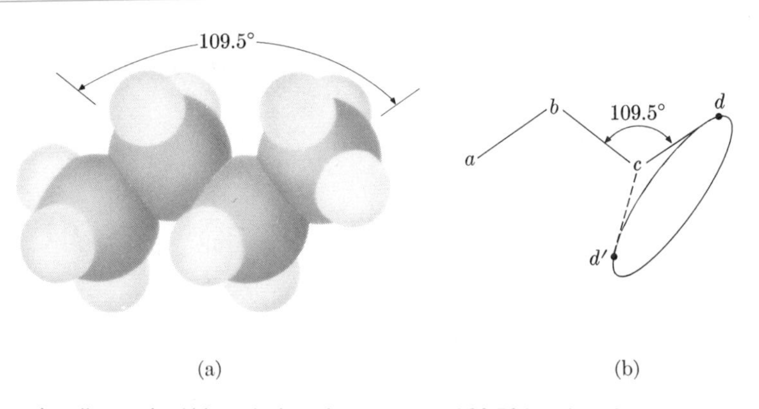

Fig. 7–1.6. Bond rotation (butane). Although there is a constant 109.5° bond angle, the end-to-end distance can vary from a–d to a–d′. Large molecules will have much more variation.

* We will still make use of this presentation, however, since it meets many of our needs.

if it were like Fig. 7–1.4, or could be as short as a couple of angstroms if the two ends of the molecule happen to be adjacent to each other. Actually, thermal agitation keeps the molecule continuously *kinked* and changing (Fig. 7–1.7). The statistically most probable, or *root-mean-square*, length $\bar{L}$ is:

$$\bar{L} = l\sqrt{m}. \qquad (7\text{–}1.4)^*$$

In this equation, l is the bond length, that is, 1.54 Å for C—C (Table 2–3.1), and m is the number of bonds. For polyvinyl chloride (Fig. 7–1.3), there are two bonds per mer; therefore $m = 2\text{DP}$, and according to Eq. 7–1.4, $\bar{L}$ becomes 38 Å for a molecule with 307 mers.

The kinked *conformation* of Fig. 7–1.7 becomes very important to us because it is the basis for the stretching and contraction of rubbers (Section 7–5). Also, it explains why *stretched* rubber has the abnormality of a negative coefficient of thermal expansion.

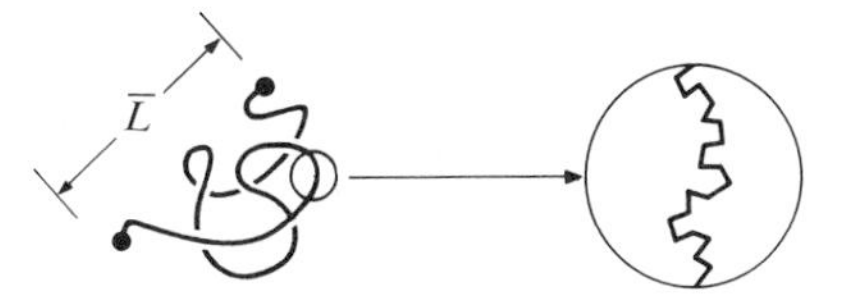

Fig. 7–1.7. Kinked conformation. Since each C—C bond can rotate (Fig. 7–1.6), a long molecule is normally kinked and has a relatively short mean length, $\bar{L}$ (Eq. (7–1.4).)

Lineal and network polymers. The polyethylene (PE) of Fig. 7–1.4 and the polyvinyl chloride (PVC) of Fig. 7–1.3 are lineal. Their growth to giant molecules is one-dimensional. Each mer is *bifunctional*; i.e., each mer has two connection points. We shall examine these more closely in Section 7–2.

Giant molecules can also be *trifunctional* (or even tetrafunctional) with three or more connection points for each mer. These polymers form a three-dimensional network, and have markedly different properties from lineal polymers. We shall examine these polymers in Section 7–4.

Example 7–1.1 It has been determined that polyvinyl chloride (PVC) has the molecular size distribution shown in Figs. 7–1.5(a) and (b)—for weight fraction and number fraction, respectively. (a) What is the "weight-average" molecular weight? (b) What is the "number-average" molecular weight? (c) What is the degree of polymerization (DP) based on $\bar{M}_w$?

---

* The mathematics required to derive this equation is beyond the scope of this book. It utilizes the statistics of a "random-walk" process.

Solution: (a) and (b)

Mol. wt. size interval, amu	$(MW)_i$, mid-value, amu	W_i, Weight fraction (Fig. 7–1.5(a))	$(W_i)(MW_i)$, amu	X_i, Number fraction (Fig. 7–1.5(b))	$(X_i)(MW_i)$, amu
5–10,000	7,500	0.12	900	0.26	1,950
10–15,000	12,500	0.18	2,250	0.23	2,875
15–20,000	17,500	0.26	4,550	0.24	4,200
20–25,000	22,500	0.21	4,725	0.15	3,375
25–30,000	27,500	0.14	3,850	0.08	2,200
30–35,000	32,500	0.09	2,925	0.04	1,300
			Σ = 19,200 (a)		Σ = 15,900 (b)

c) Mer weight of PVC (Fig. 7–1.3):

$$(C_2H_3Cl) = 24 + 3 + 35.5 = 62.5 \text{ amu/mer},$$

$$DP = \frac{(19{,}200 \text{ amu/molecule})}{(62.5 \text{ amu/mer})} = 307 \text{ mers/molecule}.$$

Comments. Whenever there is a distribution of sizes, the "number-average" molecular weight is always less than the "weight-average" value, because of the large number of smaller molecules per gram in the smaller weight intervals. The two averages diverge more as the range of the size distribution increases. ◀

Example 7–1.2 A polyethylene $(C_2H_4)_n$ molecule (see Fig. 7–1.4), with a molecular weight of 22,400 amu, is dissolved in a liquid solvent. a) What is the longest possible end-to-end distance of the polyethylene (without altering the 109.5° C—C—C bond angle)? b) The shortest? c) The most probable?

Solution:

$$DP = (22{,}400 \text{ amu/molecule})/(28 \text{ amu/mer}) = 800 \text{ mers/molecule}.$$

Therefore,

$$\text{bonds} = 1600.$$

From Table 2–3.1,

$$l = 1.54 \text{ Å}.$$

a) "Sawtooth length" $= (1600)(1.54 \text{ Å})(\sin 109.5°/2) = 2000 \text{ Å};$

b) Shortest distance $= \sim 3$ Å with ends in contact.

c) Eq. (7–1.4): $L = 1.54 \text{ Å}\sqrt{1600} = 62 \text{ Å}.$

Comments. Since the molecule is under continuous thermal agitation, the probability of its attaining a length of 2000 Å would be extremely remote. A force would have to be used to stretch it out to that length. Furthermore the required force would become greater at higher temperatures because the kinking becomes more persistent with increased thermal agitation. ◀

7–2 LINEAR POLYMERS

Large one-dimensional molecules make up the basis of many of our common polymers. Examples include the *vinyls*, which have the structure

$$\left[\begin{array}{cc} H & H \\ | & | \\ -C- & C- \\ | & | \\ H & \mathbf{R} \end{array}\right]_n \tag{7–2.1}$$

as a mer. (See Table 7–2.1 where **R** is one of a variety of radicals.) Many readers have also heard the terms *polyesters*, *polyurethanes*, and *polyamides*, which have linear molecules. Finally, *rubbers* have linear (and very highly kinked) molecules. Three of the simpler rubber mers are listed in Table 7–2.2. There are two main categories of polymerization reactions which we shall consider, *addition* and *condensation*. We will observe that the mers along linear molecular chains are not perfectly regular, but have various *configurations*.

Addition polymerization. A molecule with a single mer (*monomer*) such as one of the vinyls,

$$\begin{array}{cc} H & H \\ | & | \\ C= & C, \\ | & | \\ H & \mathbf{R} \end{array} \tag{7–2.2}$$

is bifunctional because it can open up to give both forward and rear coupling. For polyvinyl alcohol, where —**R** of Eq. (7–2.2) is —OH, this reaction is

$$n\left[\begin{array}{cc} H & H \\ | & | \\ C= & C \\ | & | \\ H & OH \end{array}\right] \rightarrow \left[\begin{array}{cc} H & H \\ | & | \\ -C- & C- \\ | & | \\ H & OH \end{array}\right]_n . \tag{7–2.3}$$

By the simplest analogy, the growth of the polymer is comparable to coupling railroad cars. The process, however, is more complex, because simply placing monomers close together does not automatically produce an addition polymerization reaction. The reaction must be *initiated*, followed by *propagation*; finally, it is *terminated*.

An initiator is commonly a free radical such as a split H_2O_2 molecule:

$$\text{H–O—O–H} \rightarrow 2\ \text{HO}\bullet, \tag{7–2.4}$$

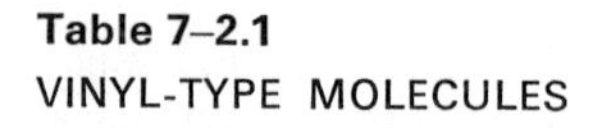

Table 7–2.1
VINYL-TYPE MOLECULES

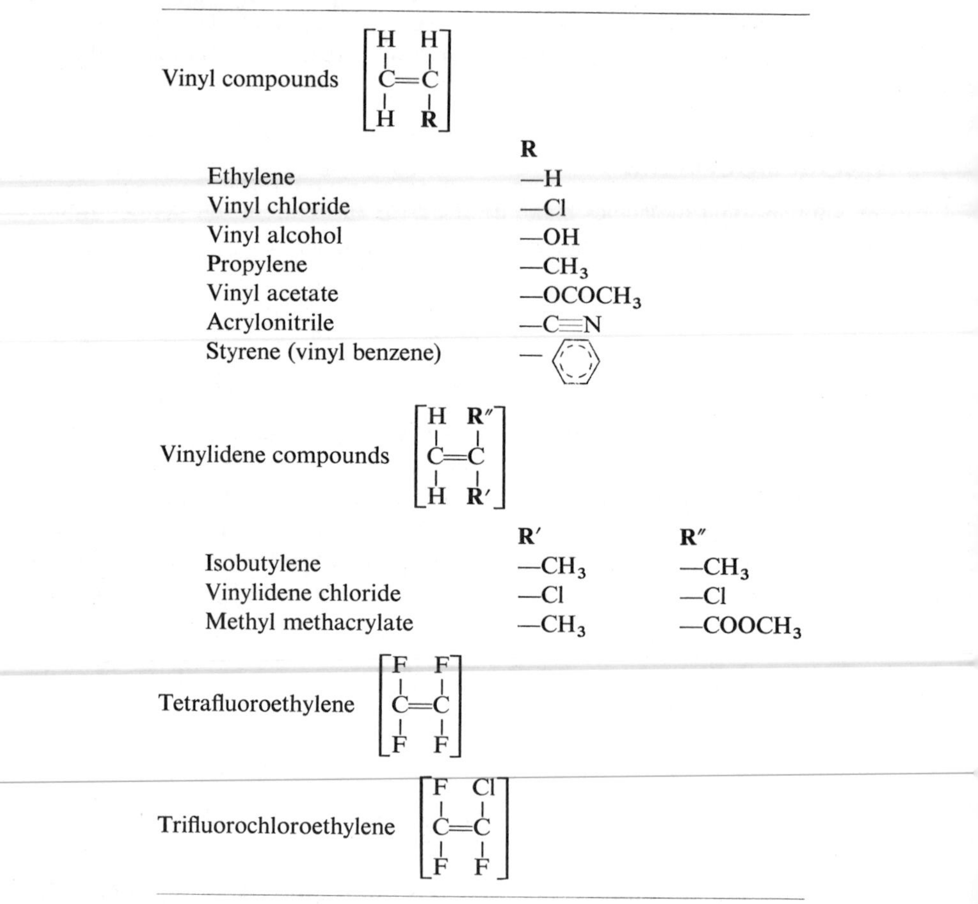

Vinyl compounds

	R
Ethylene	—H
Vinyl chloride	—Cl
Vinyl alcohol	—OH
Propylene	$—CH_3$
Vinyl acetate	$—OCOCH_3$
Acrylonitrile	—C≡N
Styrene (vinyl benzene)	— (benzene ring)

Vinylidene compounds

	R′	R″
Isobutylene	$—CH_3$	$—CH_3$
Vinylidene chloride	—Cl	—Cl
Methyl methacrylate	$—CH_3$	$—COOCH_3$

Tetrafluoroethylene

Trifluorochloroethylene

Table 7–2.2
BUTADIENE-TYPE MOLECULES

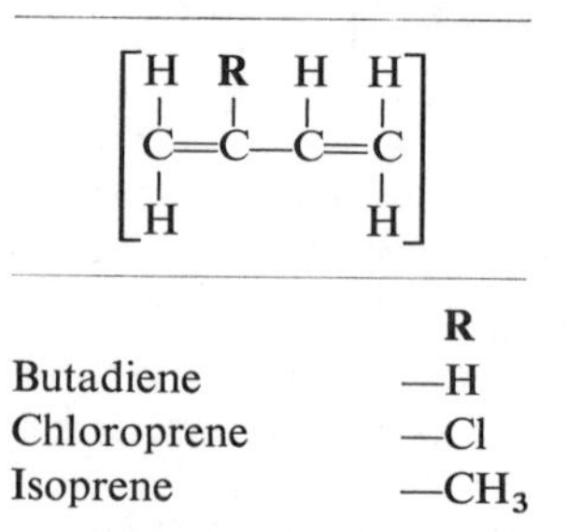

	R
Butadiene	—H
Chloroprene	—Cl
Isoprene	$—CH_3$

where the reactive dot is an unfilled orbital of the HO pair.* Therefore it will react readily. For example, when it encounters a vinyl molecule,

$$\mathrm{H{-}O}\bullet + \begin{array}{c}\mathrm{H}\\|\\\mathrm{C}\\|\\\mathrm{H}\end{array}{=}\begin{array}{c}\mathrm{H}\\|\\\mathrm{C}\\|\\\mathrm{R}\end{array} \rightarrow \mathrm{H{-}O{-}}\begin{array}{c}\mathrm{H}\\|\\\mathrm{C}\\|\\\mathrm{H}\end{array}{-}\begin{array}{c}\mathrm{H}\\|\\\mathrm{C}\\|\\\mathrm{R}\end{array}\bullet. \qquad (7\text{–}2.6)$$

We still have a free radical (admittedly longer), so it can react with another;

$$\mathrm{H{-}O{-}}\begin{array}{c}\mathrm{H}\\|\\\mathrm{C}\\|\\\mathrm{H}\end{array}{-}\begin{array}{c}\mathrm{H}\\|\\\mathrm{C}\\|\\\mathrm{R}\end{array}\bullet + \begin{array}{c}\mathrm{H}\\|\\\mathrm{C}\\|\\\mathrm{H}\end{array}{=}\begin{array}{c}\mathrm{H}\\|\\\mathrm{C}\\|\\\mathrm{R}\end{array} \rightarrow \mathrm{H{-}O{-}}\begin{array}{c}\mathrm{H}\\|\\\mathrm{C}\\|\\\mathrm{H}\end{array}{-}\begin{array}{c}\mathrm{H}\\|\\\mathrm{C}\\|\\\mathrm{R}\end{array}{-}\begin{array}{c}\mathrm{H}\\|\\\mathrm{C}\\|\\\mathrm{H}\end{array}{-}\begin{array}{c}\mathrm{H}\\|\\\mathrm{C}\\|\\\mathrm{R}\end{array}\bullet. \qquad (7\text{–}2.7)$$

A chain reaction continues to propagate, to produce molecules with many units, i.e., polymers.

The driving force for Eqs. (7–2.7) and continuing reactions is the change in energy. For example the left side of Eq. (7–2.7) contains a double C═C bond, whereas the right side contains an additional *pair* of single C—C bonds. All other bonds are unchanged. Table 2–3.1 indicates that the C═C bonds have 680 kilojoules per mole of 0.6×10^{24} bonds. Each C—C bond has 370 kJ per mole. By eliminating the double bond of Eq. (7–2.7), we must supply $+1.1 \times 10^{-18}$ joules; by forming *two* single bonds, $-(2)(0.6 \times 10^{-18}$ joules) are released. Thus 0.1×10^{-18} joules of energy are given off $(-)$ in Eq. (7–2.7), or 60 kJ/mole. The lower-energy polymer compounds are more stable than the original monomers.†

The propagation of Eqs. (7–2.6) and 7–2.7) continues until termination because (1) the supply of monomers is exhausted or (2) the ends of the two growing chains encounter and join. A range of molecular sizes is produced, since different molecules terminate their growth at different times (Fig. 7–1.5).

Copolymers. Each of the vinyl polymers considered so far included only one type of mer. Polyethylene has only (C_2H_4) mers; polyvinyl chloride has only (C_2H_3Cl) mers, and polyvinyl alcohol has only (C_2H_3OH) mers. A marked advance in the technology of producing plastics occurred when the scientist learned that addition polymers containing mixtures of two or more different mers frequently have improved physical and mechanical properties.

---

* The HO• radical is not charged. Therefore, it differs from an OH^- ion. Rather, the oxygen has only seven of its possible complement of eight electrons:

$$\mathrm{H}:\overset{..}{\underset{..}{\mathrm{O}}}:\overset{..}{\underset{..}{\mathrm{O}}}:\mathrm{H} \rightarrow \mathrm{H}:\overset{..}{\underset{..}{\mathrm{O}}}\cdot + \cdot\overset{..}{\underset{..}{\mathrm{O}}}:\mathrm{H}. \qquad (7\text{–}2.5)$$

† The chemist is aware that another factor, that of entropy, must also be considered. However, it does not alter this basic picture.

Table 7–2.3

VINYL CHLORIDE-ACETATE COPOLYMERS: CORRELATION BETWEEN COMPOSITION MOLECULAR WEIGHT AND APPLICATIONS*

Item	w/o of vinyl chloride	No. of chloride mers per acetate mer	Range of average mol. wts.	Typical applications
Straight polyvinyl acetate	0	0	4,800–15,000	Limited chiefly to adhesives.
Chloride-acetate copolymers	85–87	8–9	8,500– 9,500	Lacquer for lining food cans; sufficiently soluble in ketone solvents for surface-coating purposes.
	85–87	8–9	9,500–10,500	Plastics of good strength and solvent resistance; molded by injection.
	88–90	10–13	16,000–23,000	Synthetic fibers made by dry spinning; excellent solvent and salt resistance.
	95	26	20,000–22,000	Substitute rubber for electrical-wire coating; must be plasticized; extrusion-molded.
Straight polyvinyl chloride	100	—	—	Limited, if any, commercial applications *per se*.

* Adapted from A. Schmidt and C. A. Marlies, *Principles of High Polymer Theory and Practice.* New York: McGraw-Hill.

It is possible, for example, to have a polymer chain composed of mers of vinyl chloride and vinyl acetate (Fig. 7–2.1). The resulting structure, called a *copolymer*, is comparable to a solid solution in crystals (Section 4–2). A copolymer may have properties quite different from those of either component member. Table 7–2.3 shows the variety of properties and applications of vinyl chloride–vinyl acetate mixtures with different degrees of copolymerization. The range is striking. It means that the engineer may tailor his plastics to a wide variety of requirements.

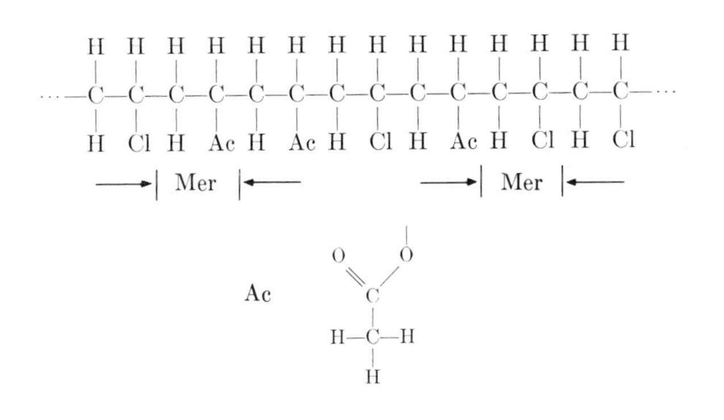

Fig. 7–2.1. Copolymerization of vinyl chloride and vinyl acetate. This is comparable to a solid solution in metallic and ceramic crystals.

The ABS plastics are triple copolymers of **a**crylonitrile, **b**utadiene, and **s**tyrene (Tables 7–2.1 and 7–2.2). Copolymerization has been applied extensively in the field of artificial rubbers. For example, the *buna-S* rubbers are copolymers of butadiene and styrene (Fig. 7–2.2).

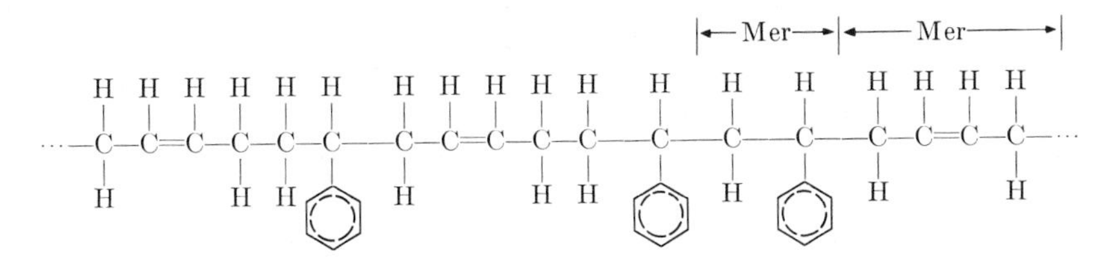

Fig. 7–2.2. Copolymerization of butadiene and styrene. This is the basis for many of our artificial rubbers. (Hydrogens are not shown on the benzene ring.)

Condensation polymerization. In contrast to addition reactions, which are primarily a summation of individual molecules into a polymer, *condensation reactions* form a second, nonpolymerizable molecule as a by-product. Usually the by-product is water or some other simple molecule such as HCl or CH_3OH. A familiar example of a condensation polymer is one of the nylons, which is formed as indicated below.

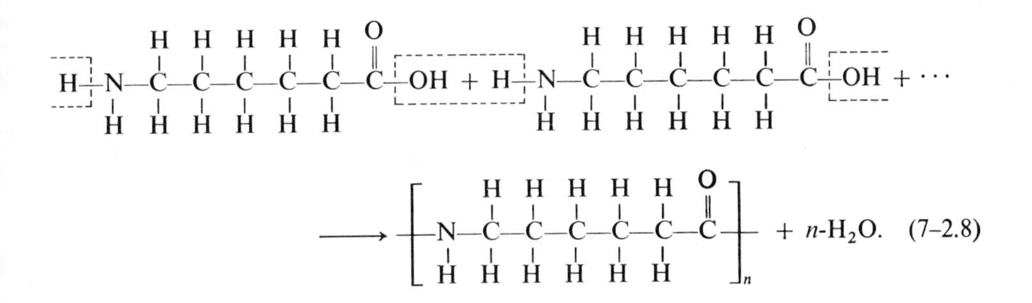

Small H_2O molecules are formed as a by-product. These can be removed, leaving the linear polymer of mers with several —CH_2— units plus a

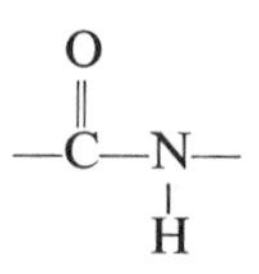

connection. One of the familiar polyesters is a product of two reactants.

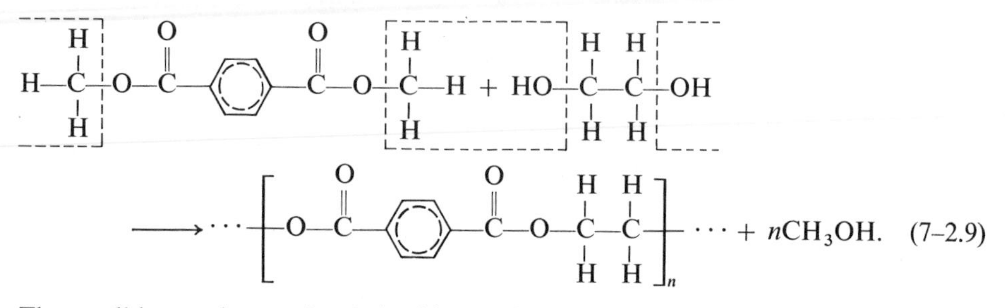

The small by-product molecule in this case is methyl alcohol. The polymer can continue to grow to produce a large lineal molecule.*† The condensation reactions of Eqs. (7–2.8) and (7–2.9) release energy just as does an addition reaction (Example 7–2.1). This is the driving force which activates the process.

There is no absolute termination in the growth of condensation polymers. Both ends of the growing molecule remain functional and can always join to another molecule. This is unlike a growing molecule in the addition reaction of Eq. (7–2.7),

---

* The product of Eq. (7–2.8) is commonly called Nylon-6 because each mer has six carbons. There are other nylons, because other sizes of mer can be used. The product of Eq. (7–2.9) goes under various trade names such as Dacron, Mylar, Terylene, Tergal, or Tervira, depending on the producing company and its physical form such as fiber or film.

† The product of Eq. (7–2.9) is a *polyester* since it contains a

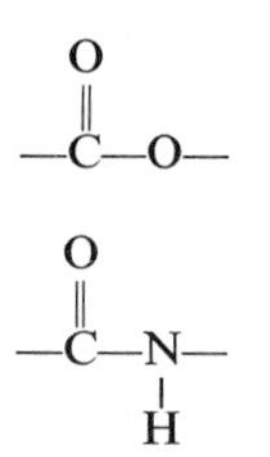

linkage; comparably, the linkage of Eq. (7–2.8) identifies it as one of the *polyamides*. The *polyurethanes* possess

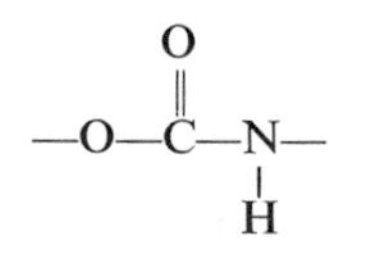

groups along their molecules.

where the chain reactions can proceed in only one direction, thus terminating the reaction if two chains join. This difference affects the molecular size distributions which are eventually achieved.

Example 7–2.1 Nylon-66 is a condensation polymer of

$$\mathrm{HO{-}\overset{\overset{\displaystyle O}{\|}}{C}(CH_2)_4\overset{\overset{\displaystyle O}{\|}}{C}{-}OH} \qquad \text{and} \qquad \mathrm{H_2N(CH_2)_6NH_2}.$$

(a) Sketch the structure of these two molecules. (b) Show how condensation polymerization can occur. (c) What is the amount of energy released per mole of H_2O formed?

Solution:

a) and b)

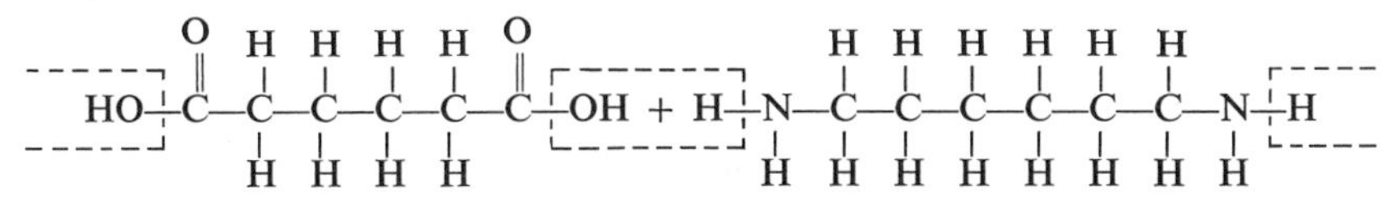

c) A mole (0.6×10^{24}) of water requires elimination of that number of C—O and H—N bonds, and the formation of equal numbers of C—N and H—O bonds.

From Table 2–3.1:

Bonds removed		Bonds formed	
C—O	+360 kJ/mole	C—N	−305 kJ/mole
H—N	+430	H—O	−500
	+790		−805

The net energy difference is −15 kJ/mole, or −3.5 kcal/mole. ◀

7–3 MOLECULAR IRREGULARITIES

A lineal polymer seldom, if ever, has the geometric regularity depicted in Fig. 7–1.4. Unless the polymer chain is constrained, thermal agitation will introduce kinks (Fig. 7–1.7). Also, as we have recently observed, it is possible to produce solid solutions and obtain copolymers. Some irregularities arise because there may be several permutations for joining monomers into a polymeric chain. Others, such as polar groups, are inherent in the chains. Each of these irregularities has an effect on the way adjacent molecules pack together and on the mechanical behavior of the resulting products.

Crystallization is difficult in polymers because the molecules are long and cumbersome, and the intermolecular attractions are weak. We find that any molecular

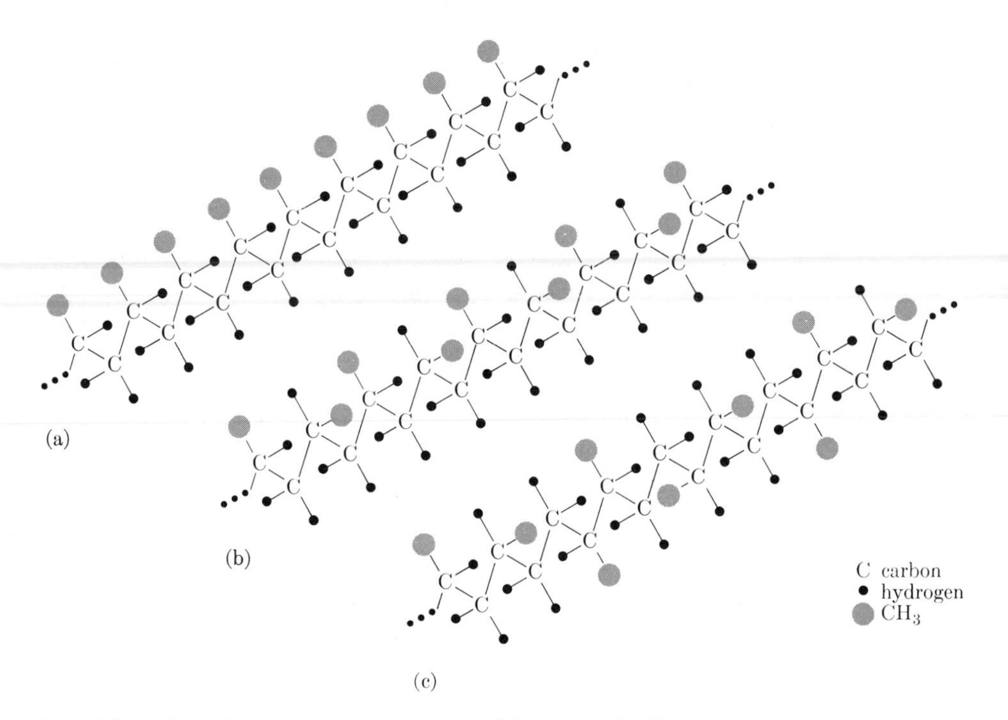

Fig. 7–3.1. Stereoisomers (polypropylene). (a) Isotactic. (b) Syndiotactic. (c) Atactic.

irregularity adds still further interference. Therefore we will want to look closely at polymer crystallization (and lack thereof).

Stereoisomers. The molecular chains of Fig. 7–3.1(a) show a high degree of regularity along the polymer. Not only is there an addition sequence of monomers which form a linear polymer, but there is also identical ordering of the propylene mers so that the radicals are always at the corresponding position within the mer. Such an ordering is called *isotactic*, as contrasted to the *atactic* arrangement in Fig. 7–3.1(c), and the *syndiotactic* arrangement in Fig. 7–3.1(b).

A second example of polymer arrangement is found in rubbers which are made of butadiene-type molecules (Table 7–2.2). Natural rubber has polymerized isoprene with the structure

(7–3.1)

as a mer. In the resulting polymer, the unsaturated positions adjacent to the two center carbons of the molecule in Table 7–2.2 are on the same side of the chain. This positioning is called *cis* (same side), and has important consequences in the chain behavior, because it arcs the mer (Eq. 7–3.1). Another modification, which is called *trans*, has the unsaturated positions on opposite sides of the chain:

$$\begin{array}{cccc} & CH_3 & & H \\ & | & & | \\ H & C & & C \\ | & & & | \\ C & & C & H \\ | & & | & \\ H & & H & \end{array} \qquad (7\text{–}3.2)$$

The two isomers have different chain structures. The cis type, polyisoprene, has a very highly kinked chain, as a result of the arc within the mer. The polymer of the trans isomer, called *gutta percha*, has a bond-angle pattern that is more typical of the previously cited plastics (Fig. 7–1.4). In effect, the unsaturated positions balance each other across the double bond.

Polar groups. Adjacent polyethylene chains have relatively little attraction to one another. In effect, they are very long paraffins. They have the same waxy characteristics. Internally, the molecules slide by one another when a shear stress is applied. Other molecules exhibit greater intermolecular attractions, particularly if they contain polar groups. One such polar group is the

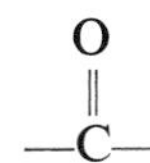

in Eqs. (7–2.8) and (7–2.9). It possesses an electric dipole with the oxygen at the negative end. The two lone-pairs of electrons with the oxygen (cf. Fig. 2–3.4(b)) strongly attract hydrogen atoms in adjacent molecules because those atoms are exposed protons (+) on the end of a covalent bond. The presence of these polar groups significantly affects the properties of a polymer. They must be credited with the difference between nylon and polyethylene, since the balance of the molecule contains the same CH_2 units.

Molecular crystals. In many metallic crystals, lattice points contain only one atom each. In crystalline methane, there is one molecule or five atoms per fcc lattice point (Fig. 3–6.1). Crystalline iodine (Fig. 7–3.2) is simple orthorhombic.* According to that figure, it has four times as many molecules as unit cells.

* In general, elongated molecules do not form cubic crystals, since a directionality is established when they are packed together.

The lattice of Fig. 7–3.2 is not face-centered orthorhombic because the molecules at the face-centered positions have a different orientation than those in the corner positions and therefore are not fully equivalent.

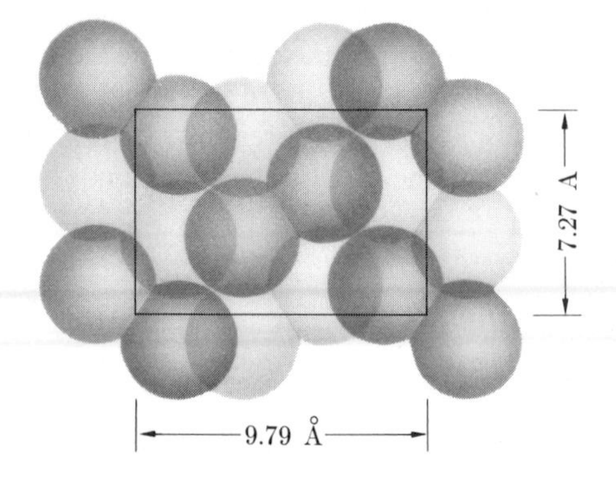

Fig. 7–3.2. Molecular crystal (iodine). The molecule of I_2 acts as a unit in the repetitive crystal structure. This lattice is *simple orthorhombic* because $a \neq b \neq c$, and the face-centered positions are *not* identical to the corner positions. (The molecules are oriented differently.)

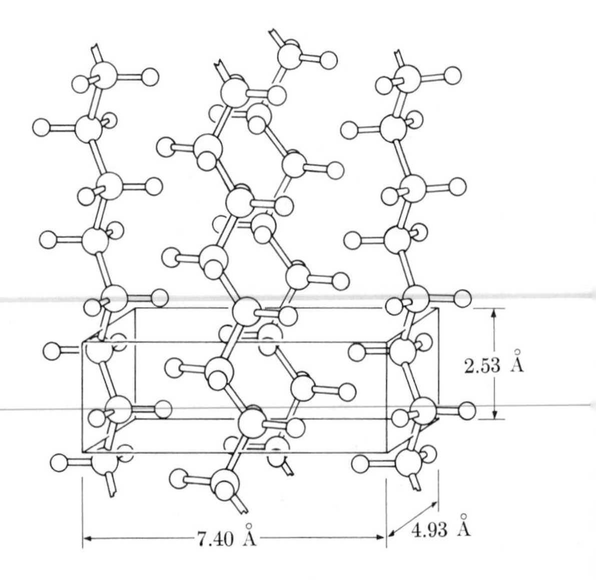

Fig. 7–3.3. Molecular crystal (polyethylene). The chains are aligned longitudinally. The unit cell is orthorhombic with 90° angles. (M. Gordon, *High Polymers*, Iliffe, and Addison-Wesley. After C. W. Bunn, *Chemical Crystallography*, Oxford.)

More complex situations arise when polyethylene crystallizes (Fig. 7–3.3) because the molecule extends beyond the unit cell. However, from this figure we can determine that there are two mers per unit cell of polyethylene. In Example 7–3.1 we will calculate the density of polyethylene and obtain an answer of 1.01 g/cm^3. In laboratory experiments, the measured densities are normally 5 to 10% less than this value (Table 7–3.1). Scientists can show that this discrepancy occurs because the long polyethylene molecules seldom have perfect alignment. There is 5 to 10 v/o of "free space" in the solid. To see how this occurs, note that the chains have a zipperlike coordination in the crystal structure of Fig. 7–3.3. If they are not perfectly aligned, their packing factor decreases.

Table 7–3.1
CHARACTERISTICS OF POLYETHYLENES

	Low-density polyethylene (LDPE)	High-density polyethylene (HDPE)
Density, g/cm^3	0.92	0.96
Crystallinity, v/o	Near zero	~50
Thermal expansion, C°$^{-1}$	180×10^{-6}	120×10^{-6}
Thermal conductivity (J · cm)/(cm^2 · °C · sec)	0.0034	0.0052
Heat resistance for continuous use, °C	55–80	80–120

Crystals are difficult to form in polyethylene* because the molecules extend through hundreds of unit cells. Over these distances it is highly unlikely that adjacent chains will remain perfectly aligned, particularly when they are not of the same length and are subject to thermal vibration and folding. Furthermore, the bonds that hold them together come from weak van der Waals forces of attraction. Diffraction analyses (Section 3–9) will reveal negligible evidence of crystallinity in the low-density polyethylenes ($\rho_{LDPE} \sim 0.9$ g/cm^3). The higher-density polyethylenes ($\rho_{HDPE} \sim 0.96$ g/cm^3) have local regions which are well crystallized, but still have intervening amorphous areas. Neither full density nor complete crystallization are observed.

In view of the above complications, it is not surprising that the majority of polymers may be cooled past their melting temperatures and on down to room temperature without molecules aligning themselves into a nice crystalline pattern. Supercooling is common among polymers, especially if there are structural irregularities along the molecular chain. For example, the atactic chain shown in Fig. 7–3.1(c) crystallizes less readily than the isotactic chain of Fig. 7–3.1(a).

Glass temperature in polymers. Even though the molecules of the polymer are entangled, when there is thermal agitation, there is a continuous rearrangement of the atoms and molecules within a polymer liquid. And because of this, there has to be "free space" in the liquid. As the temperature decreases, the thermal agitation lessens, and there is a decrease in both the "free space" and the vibrational amplitude. The resulting decrease in volume continues below the freezing point into the super-cooled liquid range (Fig. 4–5.3). The liquid structure is retained. As with liquid at a higher temperature, flow can occur; however, naturally flow is more difficult (the viscosity increases) as the temperature drops and the excess space between molecules decreases.

Those polymers which are cooled without crystallizing eventually reach a point at which the thermal agitation is not sufficient to allow for rearrangement of the

* But not as unlikely as in the majority of other polymers.

molecules. Although not crystalline, the polymer becomes markedly more rigid. It also becomes more brittle. When there cannot be continued rearrangement of molecules so that there is less "free space" and better packing, additional decreases in temperature mean only that the vibrations of the molecules are of smaller amplitude. Thus the slope of the curve in Fig. 4–5.3 becomes much less steep. This point of change in the slope is called the *glass point*, or the *glass transition temperature*, because this phenomenon is typical of all glasses. In fact, below this temperature, just as with normal silicate glasses, a noncrystalline polymer *is* a glass (although admittedly an organic one). Conversely, we shall observe in Chapter 8 that a normal glass is an inorganic polymer.

The glass transition temperature or, more simply, the glass temperature, T_g, is as important to polymers as the melting (or freezing) temperature, T_m, is. The glass temperature of polystyrene (Table 7–2.1) is at approximately 100°C; therefore it is glassy and brittle at room temperature. In contrast, a rubber whose T_g is at -73°C is flexible even in the most severe winter temperatures.

Example 7–3.1 From Fig. 7–3.3, calculate the density of polyethylene.

Solution: A $(C_2H_4)_n$ mer is parallel to the two ends of the rectangular cell, for an equivalent of one mer per unit cell. Likewise, a mer is parallel to the two sides. Together, there are a total of two mers per unit cell.

$$\rho = \frac{2(24 + 4 \text{ amu})/(0.602 \times 10^{24} \text{ amu/g})}{(2.53 \times 7.40 \times 4.93)(10^{-24} \text{ cm}^3)} = 1.01 \text{ g/cm}^3.$$

Comments. Densities normally lie in the 0.92 to 0.96 g/cm^3 range, depending on the degree of crystallinity (see Study problem 7–3.1 and Table 7–3.1). ◀

7–4 THREE-DIMENSIONAL POLYMERS

As pointed out in Section 7–1, polyfunctional mers can produce *network* structures. A familiar condensation polymer, which goes by various trade names, is formed from formaldehyde (CH_2O) and phenol (C_6H_5OH). The atom arrangements within these molecules (Fig. 2–3.1) are shown in Fig. 7–4.1(a). At room temperature formaldehyde is a gas; phenol is a low-melting solid. The polymerization which results from the interaction of these two compounds is shown in Fig. 7–4.1(b). The formaldehyde has supplied a CH_2 unit which serves as a bridge between the benzene rings in two phenols. Stripping two hydrogens from the benzene rings and one oxygen from the formaldehyde (to permit the connection) forms water, which can volatilize and leave the system. The reaction of Fig. 7–4.1 can occur at several points around the phenol molecule.* As a result of this polyfunctionality, a molecular network is formed, rather than a simple linear chain (Fig. 7–4.2).

---

* Three is the normal maximum, because there simply is not space to attach more than three CH_2 bridges. The number is limited by *stereohindrance*.

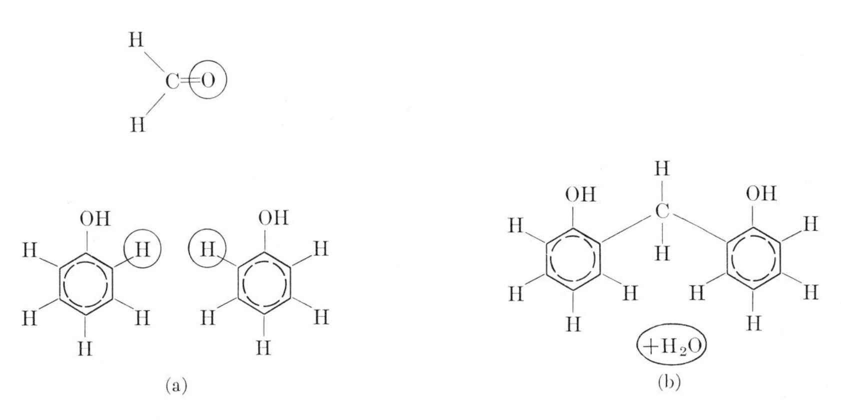

Fig. 7–4.1. Phenol-formaldehyde reaction. The phenols (C_6H_5OH) contribute hydrogen and the formaldehyde (CH_2O) contributes oxygen to produce water as a byproduct. The two rings are joined by a $—CH_2—$ bridge.

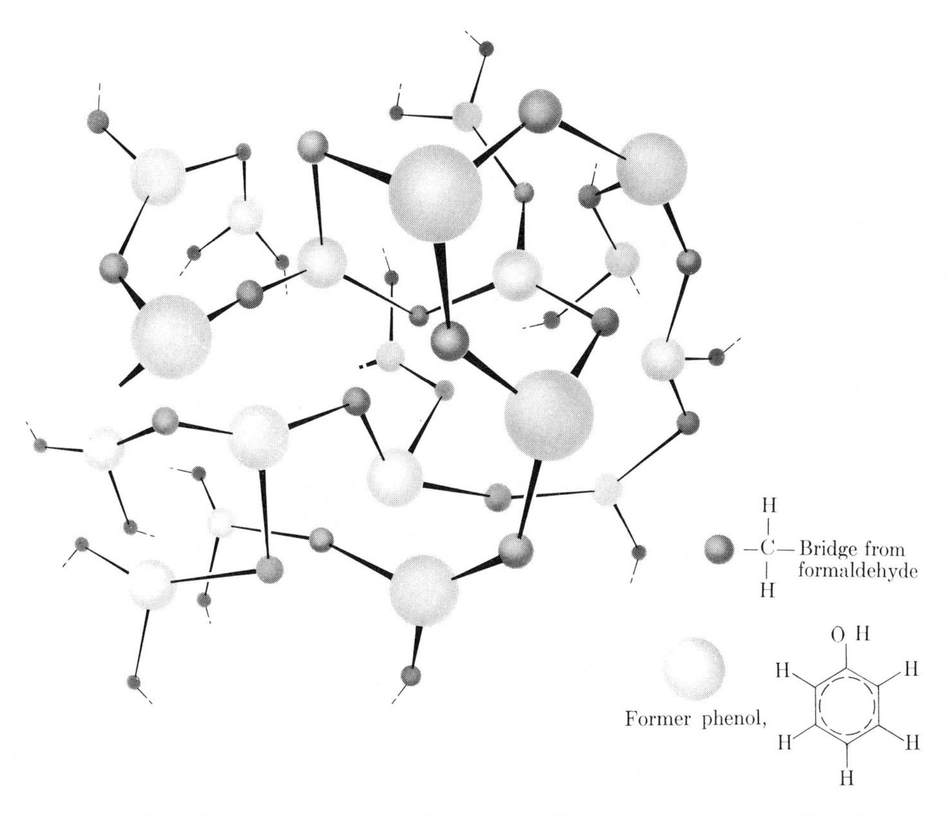

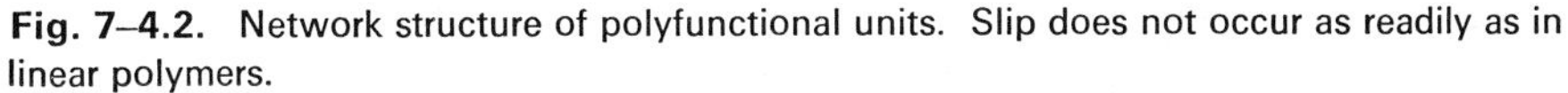

Fig. 7–4.2. Network structure of polyfunctional units. Slip does not occur as readily as in linear polymers.

Thermosets and thermoplasts. The polymer just described is phenol-formaldehyde (PF). It was one of the earlier synthetic plastics. It has the attribute of not softening when heated because its three-dimensional structure keeps it rigid. Of course, this means that it cannot be molded in the same manner as lineal polymers. In order to mold phenol-formaldehyde, it is necessary to start with a mixture which is only partially polymerized. At this stage, the combination is solid but averages less than three CH_2 bridges between adjacent phenols. Thus, it can be deformed under pressure. As it is held at 200–300°C, it completes its polymerization into a *more* rigid three-dimensional structure. It is thermal setting, so the manufacturer calls these *thermosets*. Once set, the product can be removed from the mold without waiting for cooling to occur.

The thermosets are in contrast to *thermoplasts*, which are lineal polymers that can be injected into a mold when warm because they soften at higher temperatures. The thermoplasts must be cooled before they can leave the mold, or they will lose their shape. There is no further polymerization in the molding process. The difference between thermosets and thermoplasts is mainly that the former has *three-dimensional* and the latter *lineal* molecular structure. This difference has considerable technical implications to both processing and engineering applications.

Cross-linking. Some linear molecules, by virtue of their structure, can be tied together into three dimensions. Consider the molecule of Fig. 7–4.3(a) and its polymerized combination in Fig. 7–4.3(b). Intentional additives (divinyl benzene in this case, but we do not have to remember the name) tie together two chains of polystyrene. This causes restrictions with respect to plastic deformation.

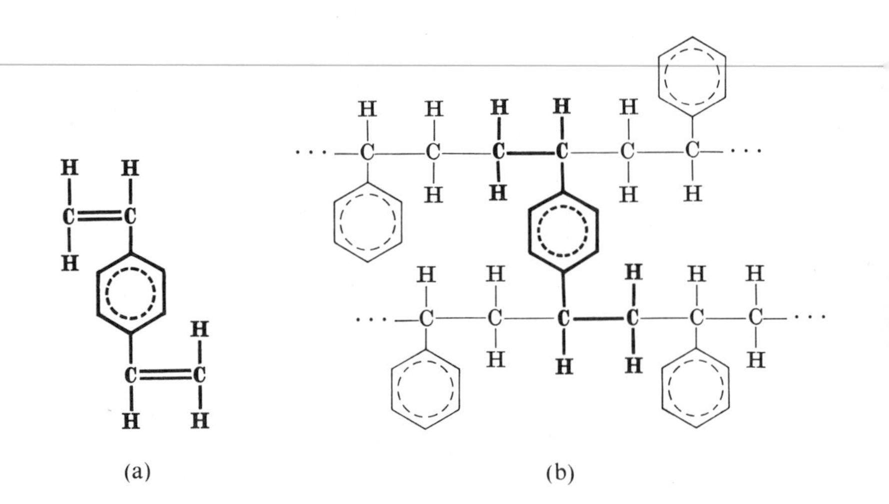

Fig. 7–4.3. Cross-linking of polystyrene. The divinyl benzene (a) becomes part of two adjacent chains because it is tetrafunctional; i.e., it has four reaction points.

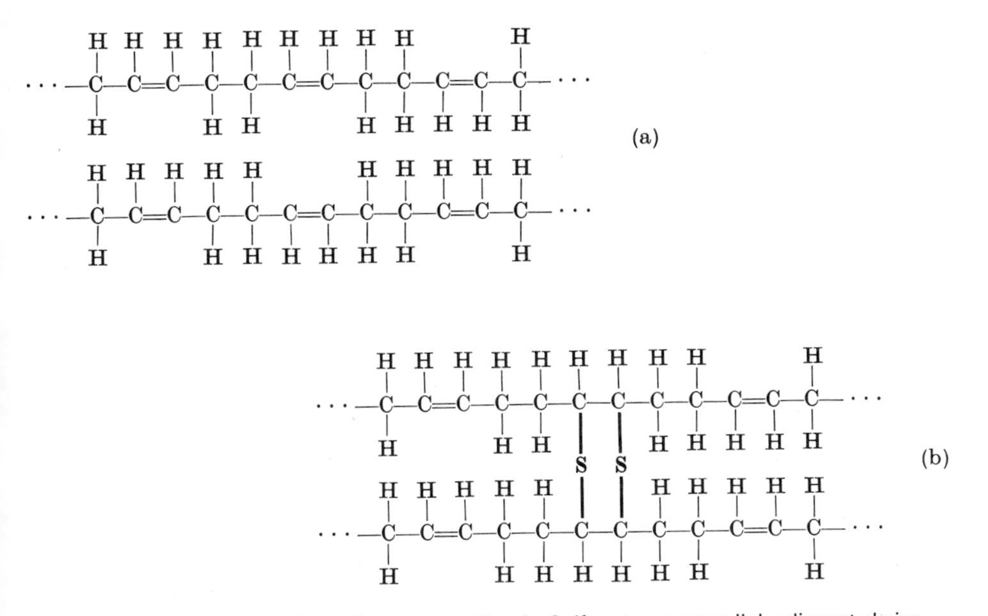

Fig. 7–4.4. Vulcanization (butadiene-type rubbers). Sulfur atoms cross-link adjacent chains. In this simplest version of vulcanization, two sulfurs are required for each pair of mers. (Other cross-linking arrangements are possible.)

The *vulcanization* of rubber is a result of cross-linking by sulfur, as shown schematically in Fig. 7–4.4. The effect is pronounced. Without sulfur, rubber is a soft, even sticky, material which, when it is near room temperature, flows by viscous deformation. It could not be used in automobile tires because the service temperature would make it possible for molecules to slide by their neighbors, particularly at the pressures encountered. However, cross-linking by sulfur at about 5% of the possible sites gives the rubber mechanical stability under the above conditions, but still enables it to retain the flexibility which is obviously required. Hard rubber has a much larger percentage of sulfur and appreciably more cross-links. You can appreciate the effect of the addition of greater amounts of sulfur on the properties of rubber when you examine a hard-rubber product such as a pocket comb.

Branching. Ideally, lineal polymer molecules such as we have studied to date are two-ended chains. There are cases, however, in which a polymer chain branches. We can indicate this schematically as in Fig. 7–4.5. Although a branch is unusual, once formed it is stable because each carbon atom has its complement of four bonds and each hydrogen atom has one bond. The significance of branching lies in the three-dimensional entanglements which can interfere with plastic deformation. Think of a pile of tree branches compared with a bundle of sticks; it is more difficult to move a branch with respect to its neighbors, than to move the individual sticks.

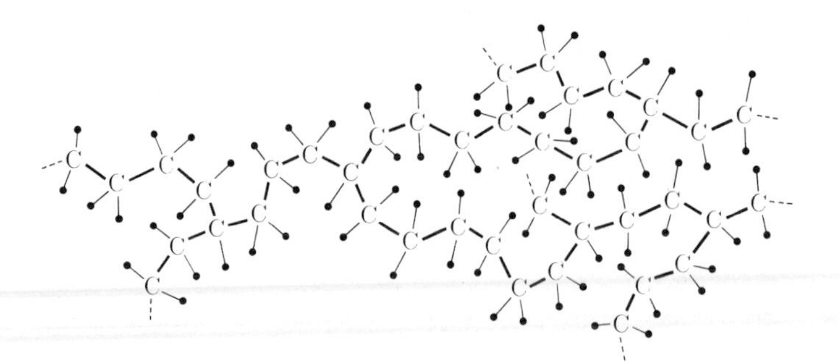

Fig. 7–4.5. Branching (polyethylene). The linear molecule of Fig. 7–1.4 branches. One mechanism for branching is shown in Fig. 7–7.3. (The hydrogen atoms are represented by dots.)

Example 7–4.1 How many grams of sulfur are required per 100 g of final rubber product to completely cross-link butadiene (C_4H_6) rubber with one sulfur per connection?

Solution: With one atom of sulfur (32) required per mer of butadiene:

$$(4)(12) + (6)(1) = 54.$$

$$\text{Fraction sulfur} = \frac{32}{32 + 54} = 0.37 \quad \text{or} \quad 37 \text{ g S/100 g product.}$$ ◀

Example 7–4.2 What fraction of butadiene (C_4H_6) is cross-linked if the product contains 18.5% sulfur? (Assume all the sulfur is utilized in cross-linking, and only one sulfur per connection.)

Solution: Start with the assumption that each mer of butadiene $[= (4)(12) + (6)(1) = 54]$ requires one atom of sulfur (= 32). Therefore,

$$\frac{54}{1 - 0.185} = \frac{x}{0.185},$$

$$x = 12.25 = \text{amount of sulfur present per mer wt of butadiene;}$$

$$\frac{12.25}{32} = \text{fraction of cross-links} = 0.383.$$

Comment. The answer can*not* be obtained from Example 7–4.1 as 18.5/37. ◀

7–5 DEFORMATION OF PLASTICS

Polymers undergo both elastic and plastic deformation, just as metals do. As a class, polymers have lower elastic moduli than metals. (Cf. Parts 1 and 3 of Appendix C.*)

---

* There are, of course, exceptions. Lead and other low-melting metals have Young's moduli of less than 10×10^6 psi (70,000 MPa). Certain polymers with aliphatic rings in their backbones have moduli of higher figures. However, both of these examples are extremes.

Also, as a class, plastics undergo permanent deformation more readily than the average metal. Often we use these characteristics to advantage. The low stress per unit strain is a design feature of rubber. Likewise, the term *plastic* emphasizes the processing advantages of polymers. Our class comparisons are not to rate one higher than the other, but to give us an index for making technical specifications. By knowing what makes a rubber visibly elastic, and what accounts for the viscous flow of polystyrene into a mold, we can better select plastics and know their limitations.

Elastomers. Polymers with low elastic moduli (large strain per unit stress) are called elastomers. The basis of the large strains is the kinked conformation which was discussed in Section 7–1. (See Fig. 7–1.7.) When a stress is applied, the molecules are extended from the mean end-to-end distance of 62 Å calculated in Example 7–1.2. Of course, we never realize the full extension to 2000 Å, even in nonvulcanized rubbers, but several hundred percent of elastic strain are common. The stretched conformation of rubber is not natural; therefore, the molecules quickly rekink by bond rotation after the stress is removed.

In order for this unkinking and rekinking to occur, the elastomer or rubber must be above its *glass temperature*, T_g. This permits the rearrangements between adjacent molecules. Below the T_g, the elastomer is brittle and hard.*

The elastic portion of the stress-strain curve for a rubber is unlike that for metals. The first strain occurs easily with little stress, because the molecules are simply unkinking. As the molecules become straightened and are aligned together, the stress requirements increase for each additional increment of strain (Fig. 7–5.1). Thus the modulus of elasticity increases. The data for rubbers in Appendix C are for the initial stress-strain ratios, since that is the range in which elastomers are usually used.

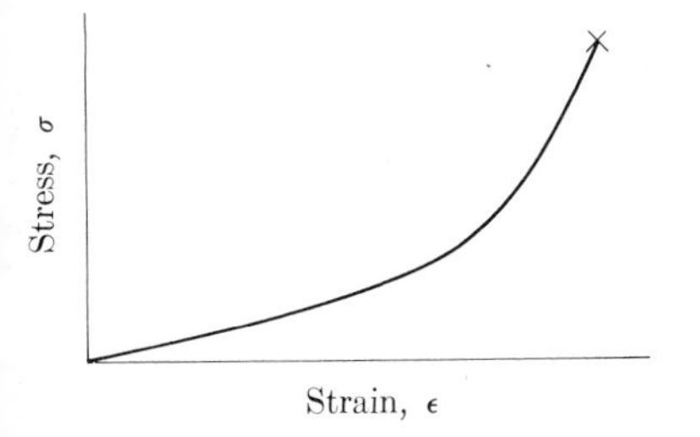

Fig. 7–5.1. Stress-strain curve (rubber). The slope (elastic modulus) of the σ-curve increases as the molecules are unkinked. (Cf. Fig. 1–2.1 for metals.)

The molecules of rubber are stretched by restricting the bond rotation that causes kinking. Now, if a stretched rubber band is heated, the thermal agitation for rotation and the retractive forces increase; the band will *contract* if the external load is not changed. A stretched rubber band has a *negative* coefficient of thermal expansion, and Young's modulus *increases* with temperature (Fig. 7–5.2).

* An interesting demonstration is to place an *unstretched* rubber band into liquid nitrogen (−196°C, or 77°K) and then break it like glass. The glass temperature T_g of common rubbers is about −75°C or 200°K.

Finally, partial *crystallization* occurs more readily in any polymer which has been deformed so that its molecules are aligned. Thus, a rubber which is normally completely amorphous and noncrystalline, coordinates its molecules into an oriented array when it is stressed. Figure 7–5.3 reveals this by x-ray diffraction. The crystallographer can relate the diffraction spots in Fig. 7–5.3(b) to molecular alignment. When the stress is removed and the rubber contracts, the diffraction spots disappear.

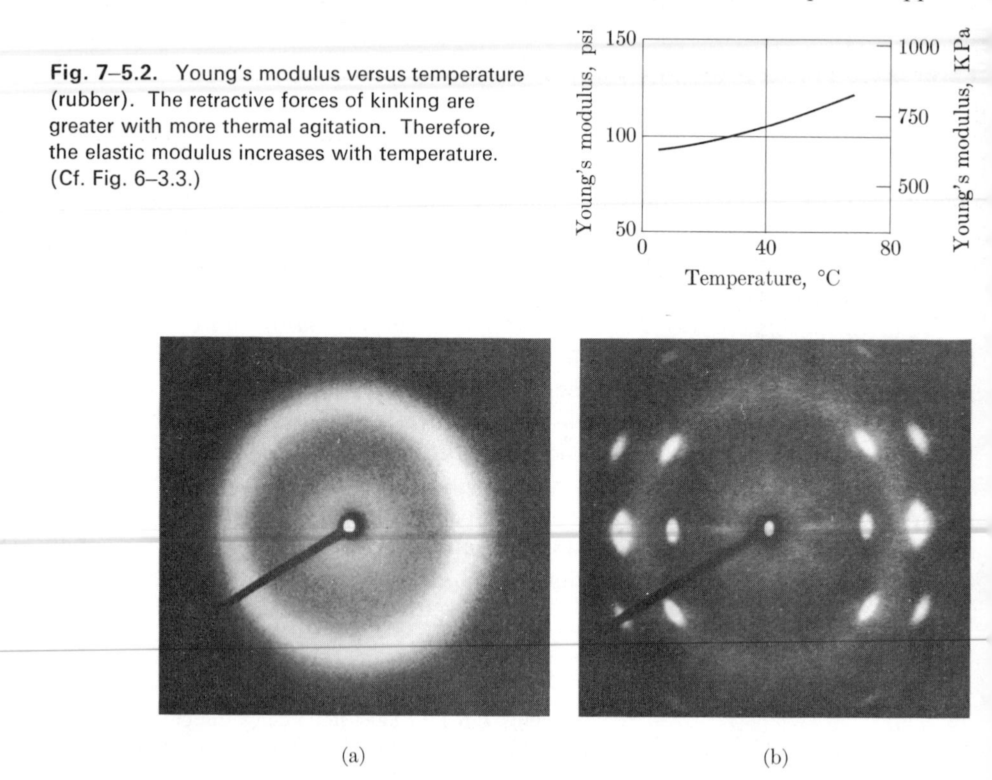

Fig. 7–5.2. Young's modulus versus temperature (rubber). The retractive forces of kinking are greater with more thermal agitation. Therefore, the elastic modulus increases with temperature. (Cf. Fig. 6–3.3.)

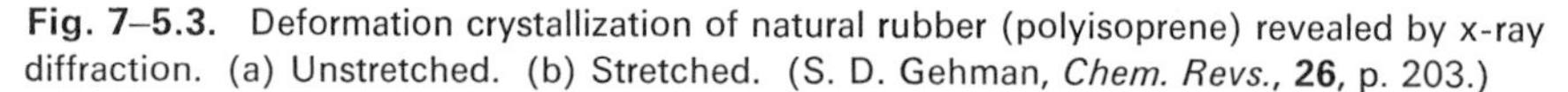

Fig. 7–5.3. Deformation crystallization of natural rubber (polyisoprene) revealed by x-ray diffraction. (a) Unstretched. (b) Stretched. (S. D. Gehman, *Chem. Revs.*, **26**, p. 203.)

Viscous deformation. Shear stresses make a liquid flow. The rate of flow depends inversely upon the viscosity. Those plastics that lie between the glass temperature T_g and the melting temperature T_m as supercooled liquids (Fig. 7–5.4(b)) are therefore subject to viscous flow in addition to elastic deformation. The rate may be slow because the viscosity η is commonly high. However, this does lead to softening when a plastic carries a long-term load. Likewise, it permits processing by deforming the plastic within a mold at high temperatures.

Figure 7–5.4(a) shows the shear stress τ required to deform a polymethyl methacrylate (PMM*) plastic one linear percent (one l/o) as a function of temperature. It

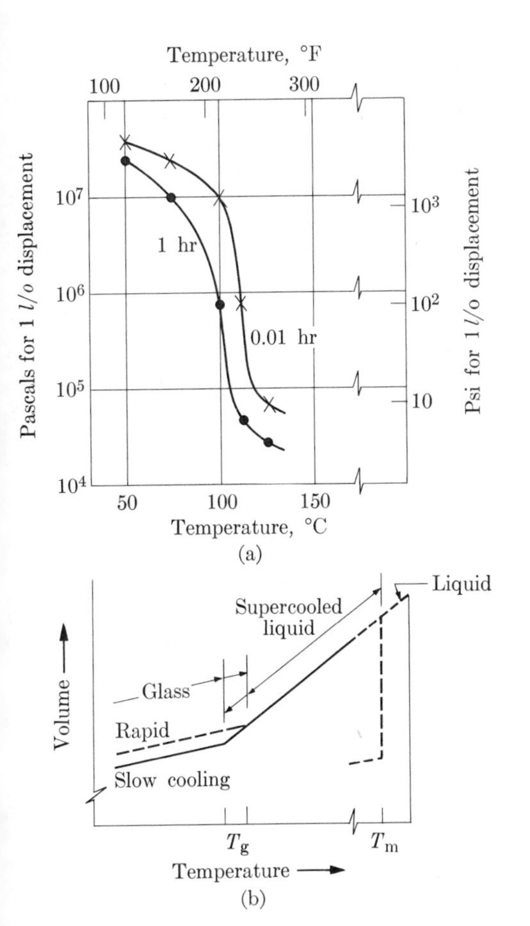

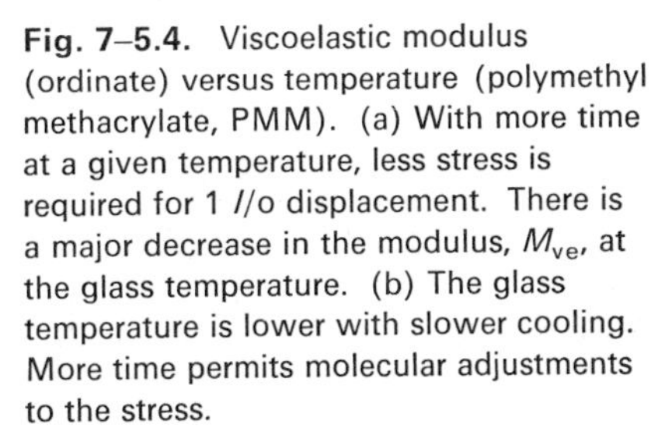
Fig. 7–5.4. Viscoelastic modulus (ordinate) versus temperature (polymethyl methacrylate, PMM). (a) With more time at a given temperature, less stress is required for 1 l/o displacement. There is a major decrease in the modulus, M_{ve}, at the glass temperature. (b) The glass temperature is lower with slower cooling. More time permits molecular adjustments to the stress.

is obvious that a marked softening occurs slightly above 100°C. This temperature corresponds to the glass temperature, T_g, of Fig. 7–5.4(b). Recall from that earlier discussion that, above the glass temperature, the molecules have the freedom to kink and turn by thermal agitation. When they drop below that temperature, there is insufficient thermal agitation to permit rearrangements of molecules into closer packing. Thus this represents a discontinuity in the thermal behavior of the material. Return to Fig. 7–5.4(a) and observe that the stress required for a deformation of one l/o changes by more than two orders of magnitude at the glass temperature. Obviously the glass temperature is a temperature important for polymer behavior.

The two curves of Fig. 7–5.4(a) indicate that less stress is required when the time at stress is increased from 36 sec (0.01 hr) to 1 hr. The two curves also indicate that the glass temperature drops ~10°C as the time is increased from 0.01 to 1.0 hr. This

* Lucite is one trade name. The composition is given in Table 7–2.1, for those who are interested.

change is reflected in Fig. 7–5.4(b) as a change in the glass transition temperature with slower cooling. With the slower cooling rates, the molecules continue to rearrange themselves until somewhat lower temperatures are reached.

We may compare different molecular structures and their effect on deformation. In Fig. 7–5.5, the ordinate shows the *viscoelastic modulus* M_{ve} where

$$M_{ve} = \tau/(\gamma_e + \gamma_f). \tag{7–5.1}$$

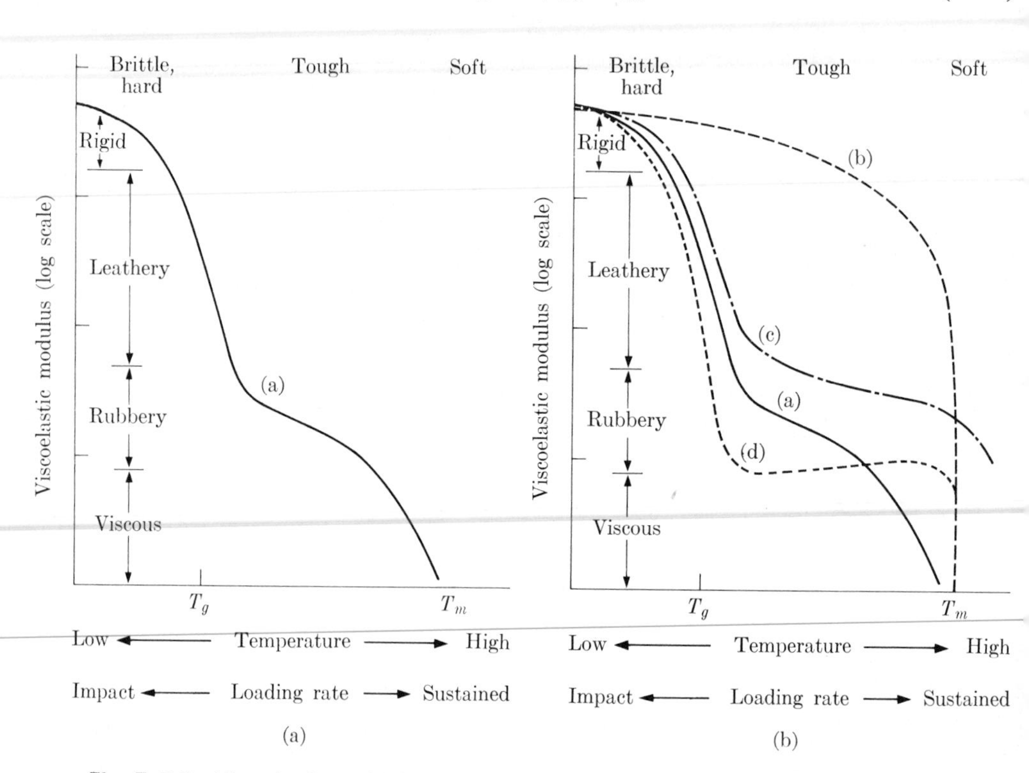

Fig. 7–5.5. Viscoelastic modulus versus structure. (a) Amorphous linear polymer. (b) Crystalline polymer. (c) Cross-linked polymer. (d) Elastomer (rubber).

As in Chapter 6, τ is shear stress and γ is shear deformation, γ_e being elastic deformation and γ_f being displacement by viscous flow. The abscissa has been generalized. The right end includes higher temperatures and/or longer times, both of which introduce more deformation (and therefore lower values for M_{ve}). At the left end of Fig. 7–5.5(a) and below the glass temperature T_g, where only elastic deformation can occur, the material is comparatively *rigid*; a clear plastic triangle used by a draftsman is an example. In the range of the glass temperature, the material is *leathery*; it can be deformed and even folded, but does not spring back quickly to its original shape. In the *rubbery plateau*, the plastic deforms readily but quickly regains its previous shape if the stress is removed. A rubber ball and a polyethylene "squeeze" bottle serve as excellent examples because they are soft and quickly elastic. At still higher temperatures, or under sustained loads, the plastic deforms extensively by *viscous flow*.

The second part of Fig. 7–5.5 compares the deformation behavior for the different structural variants cited earlier with the amorphous polymer just described. A highly *crystalline* polymeric material (b) does not have a glass temperature. Therefore it softens more gradually as the temperature increases until the melting temperature is approached, at which point fluid flow becomes significant. The higher-density polyethylenes (Table 7–3.1) lie between curves (a) and (b) of Fig. 7–5.5(b) because they possess approximately 50% crystallinity.

The behavior of *cross-linked* polymers is represented by curve (c) of Fig. 7–5.5. A vulcanized rubber, for example, is harder than a nonvulcanized one. Curve (c) is raised more and more as a larger fraction of the possible cross-links are connected. Note that the effects of cross-linking carry beyond the melting point into the true liquid. In this respect, a network polymer like phenol-formaldehyde (Fig. 7–4.2) may be considered as an extreme example of cross-linking, which gains its thermoset characteristics by the fact that the three-dimensional amorphous structure carries well beyond an imaginable melting temperature.

Once the glass temperature is exceeded, *elastomeric* molecules can be rotated and unkinked to produce considerable strain. If the stress is removed, the molecules quickly snap back to their kinked conformations (Fig. 7–1.7). This rekinking tendency increases with the greater thermal agitation at higher temperatures. Therefore the behavior curve increases slightly to the right across the rubbery plateau (Fig. 7–5.5(d). Of course, the elastomer finally reaches the temperature at which it becomes a true liquid, and then flow proceeds rapidly.

•**Stress relaxation.** In the discussion of Fig. 7–5.5, we assumed constant stress and increasing strain. In some situations, strain is constant. Since viscous flow proceeds, the stress is reduced. The reader has undoubtedly observed such a phenomenon if he has removed a stretched rubber band from a book or bundle of papers after a period of time. The rubber band does not return to its original length. For this reason, it was not holding the papers as tightly. Some of the stress had disappeared.

The stress decreases in a viscoelastic material that is under constant strain because the molecules can gradually flow by one another. The rate of stress decrease $(-d\sigma/dt)$ is proportional to the amount of stress:

$$(-d\sigma/dt)\lambda = \sigma \tag{7–5.2}$$

Rearranging and integrating,

$$\begin{aligned} d\sigma/\sigma &= -dt/\lambda; \\ \ln \sigma/\sigma_0 &= -t/\lambda, \end{aligned} \tag{7–5.3(a)}$$

or

$$\sigma = \sigma_0 e^{-t/\lambda}. \tag{7–5.3(b)}$$

The stress ratio, σ/σ_0 is between the stress at time t and the original stress σ_0 at t_0. The proportionality constant λ of Eq. (7–5.2) must have the units of time, and is called the *relaxation time*. When $t = \lambda$, $\sigma/\sigma_0 = 1/e$.

Since stress relaxation is a result of molecular movements, we find that temperature affects stress relaxation in much the same manner as it affects diffusion:

$$1/\lambda \propto e^{-Q/RT}, \tag{7–5.4}$$

or

$$\log 1/\lambda = \log 1/\lambda_0 - Q/4.575T. \tag{7–5.5}$$

In these equations, Q, R, and T have the same meanings and units as in Eq. (4–7.4); $1/\lambda$, like D, contains $\sec^{-1}$ in its units.

Example 7–5.1 Detect evidence of orientation of rubber molecules by detecting a temperature change.

Answer: This is a simple experiment using a heavy but easily deformable rubber band. Your lip can serve as a sensitive detector of temperature changes. Place the band in contact with your lower lip. Stretch it rapidly, then quickly (without snapping) return it to its original length, repeating this cycle several times. A little care will permit one to detect a temperature increase on stretching and a temperature decrease on release. These temperature changes occur because a heat of fusion is released from the rubber band to your lip during orientation. Energy is absorbed (as entropy) during deorientation when the stress is removed. ◀

•**Example 7–5.2** A stress of 1200 psi (8.3 MPa) is required to stretch a 10 cm rubber band to 14 cm. After 42 days at 20°C in the same stretched position, the band exerts a stress of only 600 psi (4.13 MPa). (a) What is the relaxation time? (b) What stress would be exerted by the band in the same stretched position after 90 days?

Solution:

a) From Eq. (7–5.3(a)),

$$\ln \frac{600}{1200} = -\frac{42}{\lambda},$$

$$\lambda = 61 \text{ days};$$

b) Eq. (7–5.3(b)),

$$\sigma_{90} = 1200e^{-90/61}$$

$$= 270 \text{ psi } (1.9 \text{ MPa}).$$

Alternative answer for (b), with 48 *additional* days:

$$\sigma_{48} = 600e^{-48/61} = 270 \text{ psi.}$$ ◀

•**Example 7–5.3** The relaxation time for 25°C is 50 days for the rubber band in Example 7–5.2. What will be the stress ratio, σ/σ_0, after 38 days at 30°C?

Solution: From Eq. (7–5.5), and since $\lambda = 61$ days at 20°C,

$$\log 1/\lambda_0 = \log 1/61 + Q/4.575(293°\text{K})$$
$$= \log 1/50 + Q/4.575(298°\text{K}).$$

Solving simultaneously,

$$Q = 6900 \text{ cal/mol},$$
$$\log 1/\lambda_0 = 3.36.$$

At 30°C,

$$\log 1/\lambda = 3.36 - 6900/(4.575)(303),$$
$$\lambda = 41 \text{ days},$$
$$\sigma_{38} = \sigma_0 e^{-38/41},$$
$$\sigma_{38}/\sigma_0 = 0.4.$$

Comments. The relaxation time shortens at higher temperatures.

A rubber is more subject to oxidation when it is under stress. When this occurs, the structure is modified, and we observe other changes such as hardening and/or cracking. ◀

7–6 ELECTRICAL BEHAVIOR OF POLYMERS

Plastics are widely used for electrical insulation. Plastics have obvious advantages. They may be either rigid or flexible; they may be made as a thin film; and they may even be applied as a fluid and polymerized in place, e.g., around a wire during processing, or as a potting compound after the device has been assembled. Of primary importance is the fact that the predominantly covalent bonds of all polymers generally limit electrical conduction.

• **Conduction.** Although polymers are inherently insulators, their compositions can be adjusted to permit some conductivity. Conductivity is achieved in specialized rubbers through the addition of finely powdered graphite, which provides a path for electron movements. Thus this conductivity does not arise from the polymer *per se*, but results from the inclusion of a second, conducting phase.

Nature produces electrical signals, presumably by conductivity, through nerve cells which, of course, are molecular in structure. One may speak of delocalized electrons which move through the molecular paths in nerve cells. However, as yet, the scientist has not analyzed this mechanism sufficiently so that it may be synthesized. This topic remains as another of the challenging areas of materials science and engineering.

Dielectric constant. Plastics contain electrical charges in the form of atomic nuclei, electrons and polar groups. These respond to electric fields. Although they cannot leave their parent molecules, electrons will shift their centers of motion a distance, d, into the positive direction. The protons of hydrogen atoms shift their centers of vibration toward the negative electrode. Polar groups (Section 7–3) and polar

Table 7–6.1
RELATIVE DIELECTRIC CONSTANTS OF PLASTICS (20°C, UNLESS OTHERWISE STATED)

	At 60 Hz*	At 10^6 Hz
Nylon 6/6	4.0	3.5
Polyethylene, PE	2.3	2.3
Polytetrafluorethylene, PTFE	2.1	2.1
Polystyrene, PS	2.5	2.5
Polyvinyl chloride		
plasticized ($T_g \approx 0°C$)†	7.0	3.4
rigid ($T_g = 85°C$)	3.4	3.4
Rubber (12 w/o S, $T_g \approx 0°C$)		
−25°C	2.6	2.6
+25°C	4.0	2.7
+50°C	3.8	3.2

* Hz = cycles per sec.
† T_g = glass temperature.

molecules (Section 2–4) align with the electric field. As a result, plastics have *relative dielectric constants* κ that are greater than unity (Table 7–6.1).

The adjustments of the charges Q which have just been described produce *polarization* $\mathscr{P}$ within the material (Section 1–4) which is the sum of all the dipole moments, Qd per unit volume V:

$$\mathscr{P} = \sum Qd/V. \tag{7–6.1}$$

The amount of polarization depends upon the structure of the material and affects the *charge density* $\mathscr{D}$ which a capacitor can carry. Thus, in Fig. 1–4.1, the capacitor builds up a charge $\mathscr{D}_0$ that is proportional to the electric field $\mathscr{E}$ if only a vacuum is present (Eq. 1–4.4). With a material present, the charge is increased to $\mathscr{D}_m$.

$$\mathscr{P} = \mathscr{D}_m - \mathscr{D}_0, \tag{7–6.2}$$

or, from Eqs. (1–4.4) and (1–4.5),

$$\mathscr{P} = (\kappa - 1)\varepsilon_0\mathscr{E}, \tag{7–6.3}$$

where κ is the relative *dielectric constant* and ε_0 is 8.85×10^{-12} C/V · m, the proportionality constant called the *permittivity* of free space.

At high frequencies, the polarization $\mathscr{P}$ arises solely from the displacement of electrons, because only the electrons can respond to an electric field reversing at 10^{15} hertz; with a material such as nylon (Eq. 7–2.8), the protons on the side hydrogen atoms can be induced to respond when the frequency drops to about 10^{10} hertz. The chief polarization within nylon occurs if the molecular dipole involving

$$\overset{\displaystyle O}{\underset{}{\overset{\|}{—C—}}}$$

can respond to the electric field. Not only does the oxygen carry a number of charges in the form of lone-pair electrons, but also the *dipole length d* of this polar group is significant. However, this molecular dipole responds relatively slowly to the electric field, and then only if the temperature is above the glass transition temperature T_g. Figure 7–6.1 shows schematically the effect of polarization $\mathscr{P}$ on the dielectric constant κ, as a function of frequency.

Temperature changes have two effects on the polarization and dielectric constant. As the temperature is raised, the increased thermal agitation destroys the alignment of the dipoles with the electric field and decreases the dielectric constant (Fig. 7–6.2).

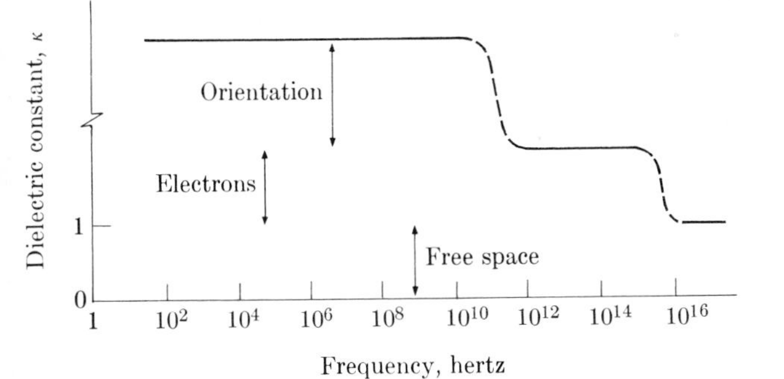

Fig. 7–6.1. Dielectric constant versus frequency (schematic). Electrons respond to the alternating electric field below $\sim 10^{15}$ Hz. Molecular dipoles can respond below about 10^{10} Hz.

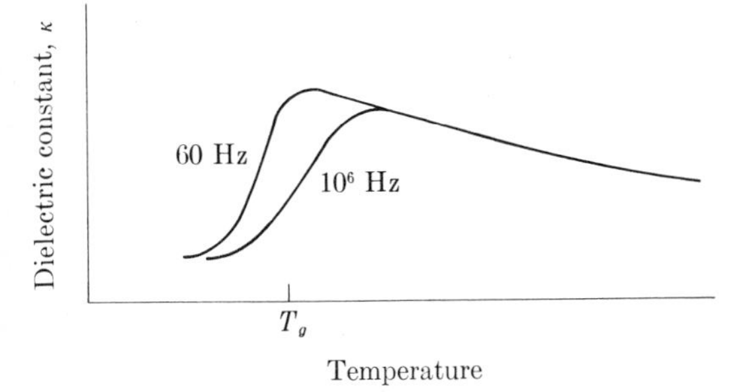

Fig. 7–6.2. Dielectric constant versus temperature. Below the glass temperature, T_g, the molecular dipoles cannot respond to the alternating electric fields. (The glass temperature is lower with low frequencies—Fig. 7–5.4(b).) Above T_g, thermal agitation destroys the polarization and therefore reduces the dielectric constant κ.

A more pronounced effect is observed below T_g, the glass temperature. First, the molecular dipoles and then the induced dipoles fail to respond, because the plastic simply becomes too rigid. This is particularly significant at higher frequencies (Fig. 7–6.3).

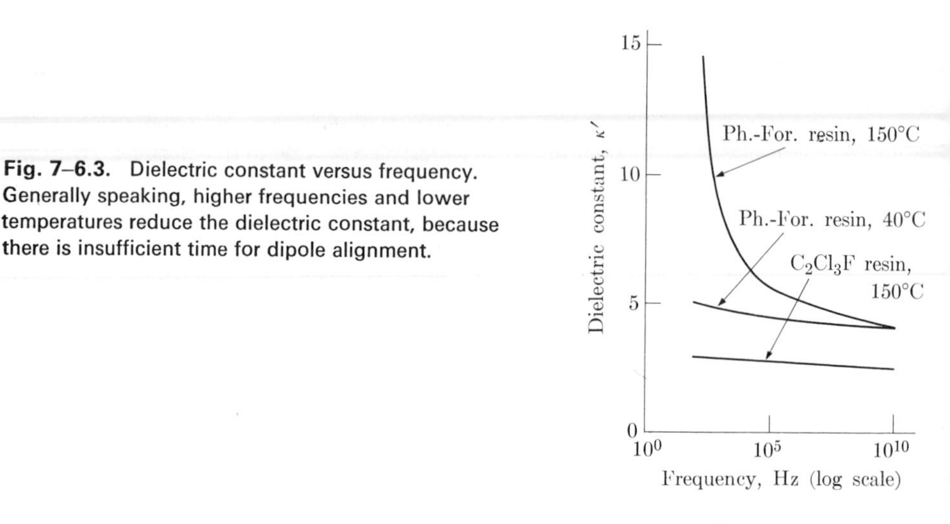

Fig. 7–6.3. Dielectric constant versus frequency. Generally speaking, higher frequencies and lower temperatures reduce the dielectric constant, because there is insufficient time for dipole alignment.

Example 7–6.1 The capacitance of a parallel-plate capacitor can be calculated from Eq. (1–4.6). The dielectric constants for polyvinyl chloride (PVC) and polytetrafluoroethylene (PTFE) are as follows:

Frequency, hertz	PVC	PTFE
10^2	6.5	2.1
10^3	5.6	2.1
10^4	4.7	2.1
10^5	3.9	2.1
10^6	3.3	2.1
10^7	2.9	2.1
10^8	2.8	2.1
10^9	2.6	2.1
10^{10}	2.6	2.1

a) Plot the capacitance-versus-frequency curves for three capacitors with 3.1 cm × 102 cm effective area separated by 0.025 mm of (1) vacuum, (2) PVC, and (3) PTFE.

b) Account for the decrease in the dielectric constant of PVC with increased frequency, and the constancy in the dielectric constant of PTFE.

Solution: (a) Sample calculations at 10^2 cycles per second:

$$C_{\text{vac}} = (1)(8.85 \times 10^{-12}\ \text{C/v}\cdot\text{m})(0.031\ \text{m})(1.02\ \text{m})/(25 \times 10^{-6}\ \text{m}) = 0.0112\ \mu\text{f};$$

$$C_{\text{pvc}} = (6.5)(0.0112\ \mu\text{f}) = 0.073\ \mu\text{f}.$$

See Fig. 7–6.4(a) for the remainder of the results.

b) The dielectric constant of PVC is high at low frequencies because the molecular polarization of PVC is in an asymmetric mer (Fig. 7–6.4(b)). At high frequencies the resulting large dipole cannot maintain alignment with the alternating field. On the other hand, PTFE has a symmetric mer, and therefore its polarization is only electronic and atomic. Although the dipoles in PTFE are weaker, they can be oscillated at the indicated frequencies. ◀

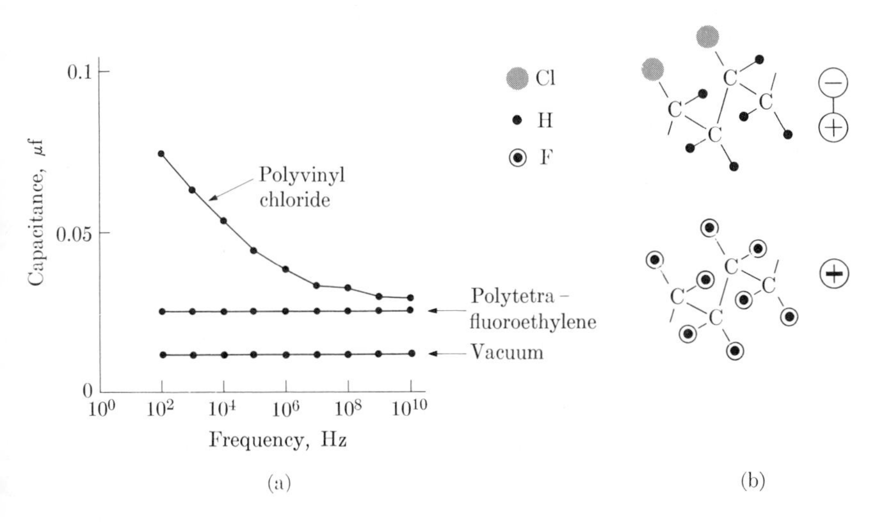

Fig. 7–6.4. Capacitance versus frequency. (a) See Example 7–6.1. (b) Symmetry of polyvinyl chloride and polytetrafluorethylene mers.

7–7 STABILITY OF POLYMERS

Plastics soften by viscoelastic deformation (Section 7–5). This softening does not break the primary covalent bonds within the molecule. Under more severe conditions, however, these bonds may be ruptured. Of course, any resulting change in structure affects the properties. Excessive heat can degrade the polymer by breaking bonds. Oxidizing environments can bring about chemical changes. Finally, radiation can induce scission, and in certain cases branching or cross-linking.

Degradation. The most obvious degradation of plastics is the charring of a polymer. If the side radicals and the hydrogen atoms of a vinyl polymer (Table 7–2.1) are literally torn loose by thermal agitation, only the backbone of carbon atoms remains. We see this occur in the starch of our morning toast. It is generally to be avoided in a plastic.* Carbonization is accelerated in the presence of air, because oxygen reacts with the hydrogen atoms along the side of the polymer chain.

Oxidation. The degradation just described is accentuated by oxygen. In addition, oxygen can have other effects. For example, many rubbers are vulcanized with only 5–20 percent of the possible cross-links anchored by sulfur. This permits the rubber to remain soft and "elastic." Over a period of time, the rubber may undergo further cross-linking by oxygen. The result is identical to Fig. 7–4.4, except that oxygen is the connecting link rather than sulfur. Naturally the rubber becomes harder and less deformable.

Several factors accelerate the oxidation reaction just described.

1. Oxygen in the form of ozone, O_3, is much more reactive than normal O_2.

2. The radiation of ultraviolet light can provide the energy to break existing bonds so that the oxidation reaction can proceed.

3. Likewise, the existing bonds are broken more readily when the molecules are stressed.

Because of these features, the tiremaker commonly adds carbon black or similar light absorbers to decrease the oxidation rate of his product. Applying the same principle, accelerated testing procedures for product stability commonly expose the polymer to ozone and/or ultraviolet light.

Scission. Radiation by ultraviolet light as just described, and in the form of neutrons, can readily break a C—C bond of a vinyl-type polymer (Fig. 7–7.1). This process, which produces smaller molecules, is called scission. This, of course, affects properties such as strength, viscosity, etc. One such example is given in Fig. 7–7.2 where the intrinsic viscosity of a polymer melt decreases with a mild exposure of 8×10^{19} neutrons/cm^2.

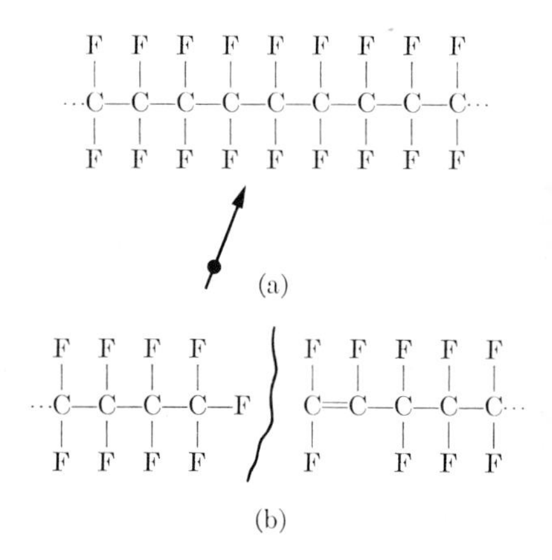

Fig. 7–7.1. Degradation by irradiation (polytetrafluorethylene). Most polymers react in this manner rather than as shown in Fig. 7–7.3. As a result, most polymers lose strength through radiation damage.

* Under controlled conditions, a graphite fiber can be formed from a polymer fiber. Such fibers hold considerable promise as a high-temperature reinforcement for composites.

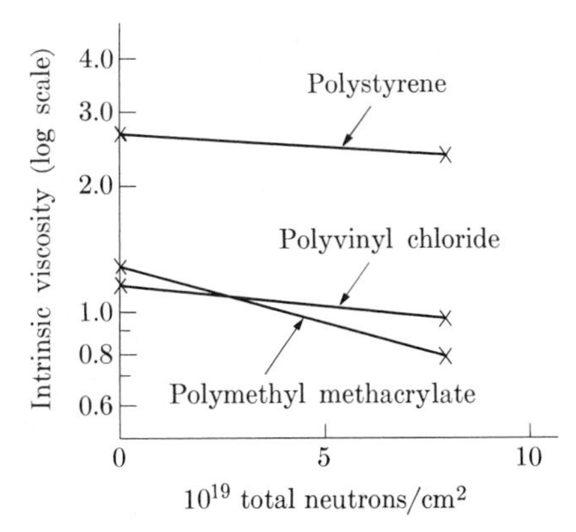

Fig. 7–7.2. Degradation by neutron exposure. The intrinsic viscosity is lowered because the polymers are ruptured into smaller molecules. (Adapted from L. A. Wall and M. Magot, "Effects of Atomic Radiation on Polymers," *Modern Plastics*.)

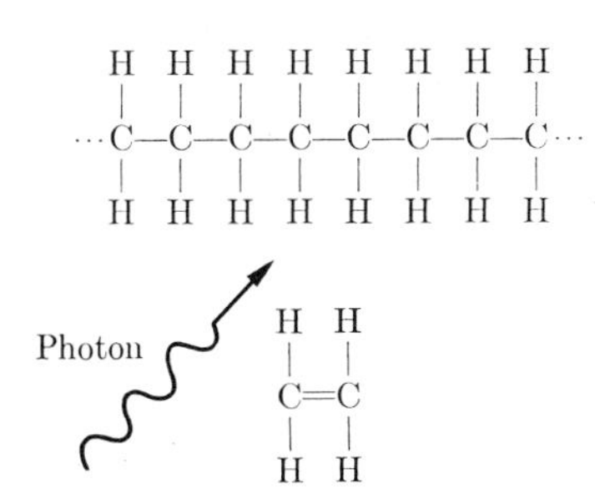

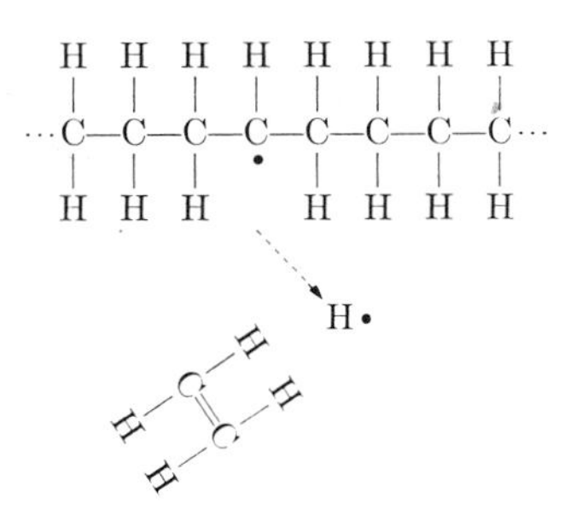

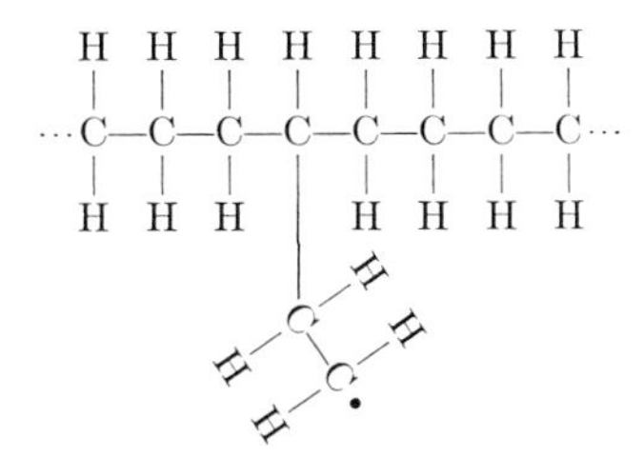

Fig. 7–7.3. Branching by irradiation. A photon can supply the activation energy necessary to cause branching. A neutron can produce the same effect.

In certain cases (e.g., polyethylene), radiation can induce branching. If this occurs before the polymerization is complete, it can produce a structure represented by Fig. 7–7.3. The consequent effect on properties can be significant, as shown by the two pairs of squeeze bottles in Fig. 7–7.4. Whether radiation produces the desired effects of Fig. 7–7.4 or the scission of Fig. 7–7.1 is not easily predictable.

Fig. 7–7.4. Irradiation of polyethylene. All four squeeze bottles have been heated to 120°C(250°F) for 20 minutes. The two rear bottles had been previously exposed to gamma irradiation to induce branching (Fig. 7–7.3); the slumped bottles had not. Excessive radiation, however, may reverse the effect by introducing scission (Fig. 7–7.1.) More recent technology favors high-density polyethylene (HDPE) as a means of achieving boiling-water stability for sterilization (Table 7–3.1). (General Electric Co.)

Example 7–7.1 Assume that all the energy required to remove the hydrogen atoms of Fig. 7–7.3 comes from a photon (and that none of the energy is thermal). (a) What is the maximum wavelength which can be used? (b) How many eV are involved?

Solution: a) From Table 2–3.1,

$$\begin{aligned} \text{C—H} &= 435{,}000 \text{ J}/(0.6 \times 10^{24} \text{ bonds}) \\ &= 0.725 \times 10^{-18} \text{ J/bond.} \end{aligned}$$

From Eq. (2–2.1), $E = h\nu = hc/\lambda$,

$$\begin{aligned} \lambda &= \frac{(6.62 \times 10^{-34} \text{ J} \cdot \text{sec})(3 \times 10^{10} \text{ cm/sec})}{0.725 \times 10^{-18} \text{ J}} \\ &= 2.74 \times 10^{-5} \text{ cm} = 2740 \text{ Å}. \end{aligned}$$

b)

$$\begin{aligned} \text{eV} &= (0.725 \times 10^{-18} \text{ J/bond})(6.24 \times 10^{18} \text{ eV/J}) \\ &= 4.5 \text{ eV/bond.} \end{aligned}$$

Comments. This is in the ultraviolet region. Shorter wavelengths will supply more energetic photons, which, of course, can also break the bonds. ◀

REVIEW AND STUDY

Plastics are molecular solids. Chemical bonding, which was considered in Section 2–3, therefore provides the basis of their structure. The response of molecules to thermal, mechanical, and electrical conditions provides the basis for the properties and behaviors of plastics.

SUMMARY: MOLECULAR STRUCTURES

The molecules in plastics are very large. They contain many units (polymers); each unit, or mer, contains the atomic features present in the total molecular structure. These mers are repeated throughout the polymeric molecule. Since not all molecules are identical in size, we must calculate average molecular weights to characterize the plastic. We may use either a weight-based average or a number-based average.

The vinyls,

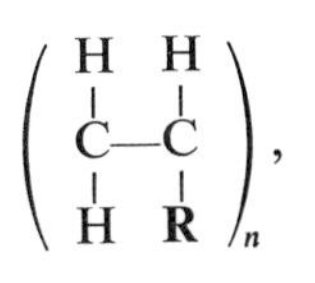

are a major category of addition-type polymers which join mers without any by-product; the **R** may be any of several simple side radicals. The polyesters are a common category of condensation-type polymer. They react to give a small by-product molecule such as water, and connecting links of

$$-\overset{\displaystyle O}{\overset{\|}{C}}-O-$$

in the polymer. Both of the above categories produce lineal polymers if the reacting molecules are bifunctional. Polyfunctional monomers (i.e., small molecules which can connect to three or more other molecules) produce network polymers.

A copolymer includes more than one type of mer within the molecule. It gives a molecular solid solution. Structural variations include stereoisomers (isotactic, atactic), and *cis* and *trans* modifications in rubbers. These arrangements, plus the natural tendency for long molecules to kink by thermal agitation, suppress crystallization.

SUMMARY: MOLECULAR BEHAVIOR

Molecules, particularly giant molecules (macro-molecules), behave differently from metallic atoms. Most basic in this respect is the fact that many macromolecular solids have the structure of supercooled liquids and therefore are amorphous. At best, they form highly imperfect crystals. A supercooled liquid, however, has a transition temperature called a glass temperature, T_g. Below T_g, molecular rearrangements are precluded; above T_g the molecules can respond to mechanical stresses and forces of electric fields. Below T_g the plastic is rigid and brittle; it has a low dielectric constant. Above T_g the plastic is flexible; and when provided opportunity with extended times or a higher temperature, the plastic deforms readily. It is this latter feature which facilitates processing; it also places limits on service environments. Above T_g, the dielectric constant increases because the polar groups are more free to respond to the electric fields.

Lineal polymers are thermoplastic and deform as viscous liquids at high temperatures. Network polymers are less temperature-sensitive and will in fact become stiffer (take on a "set") at elevated temperatures if polymerization, cross-linking, or branching can continue.

Polymers can be degraded by excessive thermal agitation, by oxidation, and by exposure to radiation.

TERMS FOR REVIEW

Atactic
Bifunctional
Branching
Butadiene-type compounds
Cis-
Conformation
  kinked
  stretched
Copolymer
Cross-linking
Crystallization of polymers
Degradation
Degree of polymerization, DP
Dipole moment, Qd
Elastomer
Glass temperature, T_g
Initiation
Isotactic
Mer
Molecules
  lineal
  network
Molecular crystals
Molecular length
Molecular weight
  number-average, $\overline{MW}_n$
  weight-average, $\overline{MW}_w$
Monomer
Plastics
Polar groups
Polarization, $\mathscr{P}$
Polyester
Polymerization
  addition
  condensation
Propagation
Relative dielectric constant, κ
• Relaxation time, λ
Root-mean-square length, $\bar{L}$
Scission
Stereoisomers
• Stress relaxation
Syndiotactic
Termination
Thermoplasts

TERMS FOR REVIEW—CONTINUED

Thermosets
Trans-
Trifunctional
Vinyl-type compounds
Viscoelastic modulus
Vulcanization

FOR DISCUSSION

A_7 Explain why the "number-average" molecular weight is always smaller than the "weight-average" molecular weight.

B_7 Obtain 10 toothpicks. Lay them on your desk end-to-end, but with a 120° angle at each contact point. Flip a coin before adding each toothpick: heads, to the right; tails, to the left. a) What is the final end-to-end distance? b) How does this differ from the results of Eq. (7–1.4)? (This may also be performed with triaxial graph paper.)

C_7 Why are polyfunctional mers necessary for network molecules?

D_7 Sketch the structure of three vinyl monomers. Repeat for the mers of the same three vinyl polymers.

E_7 Show how bonds are altered for the addition polymerization of propylene (CH_2CHCH_3).

F_7 Show how bonds are altered for the addition polymerization of butadiene [$CH_2(CH)_2CH_2$].

G_7 Hydrogen peroxide (H_2O_2) promotes polymerization. "Twice as much is better yet." Discuss.

H_7 Heat removal requires attention by the engineers who design polymerization processes. Why?

I_7 Compare the amounts of heat evolved during the polymerization of polyethylene, of polystyrene, and of a 50–50 copolymer of the two.

J_7 Rewrite Eq. (7–2.8) so that NH_3 is a by-product, rather than H_2O. Suggest why H_2O is the normal by-product, and not NH_3.

K_7 Suggest the compositions of some rubbers other than those shown in Table 7–2.2.

L_7 By a sketch show why the $-\overset{\displaystyle Cl}{\underset{\displaystyle H}{\overset{|}{\underset{|}{C}}}}-$ portion of polyvinyl chloride is a polar group.

M_7 Why is it that metals are seldom glassy but amorphous polymers are common?

N_7 The molecule of polyethylene is "sawtoothed" but otherwise straight in its crystal (Fig. 7–3.3). When it is dissolved in cyclohexane (a liquid solvent), it has the kinked conformation of Fig. 7–1.7. Both forms are stable. Why?

O_7 Explain why the thermal-expansion coefficient (slope of the curve in Fig. 7–5.4(b) is greater above T_g than below.

P_7 The exact value of T_g depends on the cooling rate. Why?

Q_7 Why does polyvinyl chloride seldom crystallize?

R_7 A copolymer of two vinyls may be *random*, or they may be in *blocks* with the mer groups clustered along the chain (e.g., —BBAAAAAABBBAAAAA-BBBBBBBA—). The latter crystallizes more readily than the random copolymer (—BAABABBBABAABAAABABBABAB—). Explain.

S_7 The glass temperature of a copolymer is lower than the average glass temperature of the two component polymers. Explain.

T_7 What is the by-product when phenol and formaldehyde are polymerized? When urea and formaldehyde are polymerized (Fig. 2–3.1)?

U_7 Will thermoplasts or thermosets be more amenable to recycling? Why?

V_7 It is suggested that butadiene (Table 7–2.2) be used to cross-link polyethylene. Discuss pros and cons.

W_7 Explain why rubber is brittle at liquid-nitrogen temperatures (77°K).

X_7 Play for a while with your younger brother's "silly putty." Explain its behavior.

Y_7 Explain the curves of Fig. 7–5.5(b) to someone who is not taking this course.

Z_7 Discuss the relationships between the viscoelastic modulus, glass temperature, and dielectric constant.

STUDY PROBLEMS

7–1.1 The melting temperature, T_m in °K of the paraffins (C_xH_{2x+2}) of Fig. 7–1.1 is sometimes given by the empirical equation

$$T_m = [0.0024 + 0.017/x]^{-1}. \qquad (7\text{–SP.1})$$

What melting point would polyethylene with (a) a DP of 10 have? (b) a DP of 100? (c) a DP of 1000? [*Note:* Mer = C_2H_4.]

Answer: a) 35°C; b) 130°C; c) 142°C

7–1.2 The following data were obtained in a determination of the average molecular weight of a polymer.

Molecular weight, Interval midvalue	Weight
30,000 amu	3.0 g
20,000	5.0
10,000	2.5

Compute the "weight-average" molecular weight.

7–1.3 a) What is the number-fraction of molecules in each of the three size categories of Problem 7–1.2? b) What is the "number-average" molecular weight?

Answer: a) 0.167, 0.417, 0.417; b) 17,500 amu

7–1.4 Two grams of sugar ($C_6H_{12}O_6$) are dissolved in 14 grams of water. (a) What is $\overline{MW}_w$? b) $\overline{MW}_n$?

7–1.5 Refer to discussion item B_7(a). What are the odds that the final end-to-end length will be $10l \sin 60°$? [*Hint:* Any 120° combination of segments one and two is usable.]

Answer: 1/256

7–1.6 Determine the degree of polymerization and the mean-square-length of the average molecule in Problem 7–1.3, if the polymer is polyvinyl alcohol.

7–2.1 What is the net energy change when one mole (0.6×10^{24}) of (a) butadiene molecules polymerize to polybutadiene? (b) Chloroprene to polychloroprene?

Answer: a) and b) −60 kJ (14 kcal)

7–2.2 Two-tenths of one percent by weight of H_2O_2 was added to ethylene prior to polymerization. What would the average DP be if all the H_2O_2 were used as terminals for the molecules? (Mer = C_2H_4.)

7–2.3 H_2O_2 is added to 280 kg (616 lb) of ethylene prior to polymerization. The average degree of polymerization obtained is 1000. Assuming all the H_2O_2 was used to form terminal groups for polymer molecules, how many grams of H_2O_2 were added? (Mer = C_2H_4.)

Answer: 340 g (0.75 lb)

7–2.4 Which molecule has greater mass, a polystyrene (PS) molecule with DP = 80, or a polyvinyl chloride (PVC) molecule with 130 mers?

7–2.5 What is the ratio of ethylene (C_2H_4) mers to vinyl chloride (C_2H_3Cl) mers in a copolymer of the two in which there is 30 w/o chlorine?

Answer: 2 to 1

7–2.6 The mer ratio of styrene/butadiene is $\frac{1}{3}$. What is the w/o of carbon?

7–3.1 A polyethylene with no evidence of crystallinity has a density of 0.90 g/cm^3. Commercial grades of low-density polyethylene (LDPE) have 0.92 g/cm^3, while HDPE has a density of 0.96 g/cm^3. Estimate the fraction of crystallinity in each case.

Answer: $C_{LDPE} = 0.18$; $C_{HDPE} = 0.55$

7–3.2 From Example 7–3.1 and Problem 7–3.1, calculate the amount of "free space" in LDPE; in HDPE.

7–3.3 Sketch the three orthogonal views of the atoms in the polyethylene unit cell (Fig. 7–3.3).

7–4.1 What percent sulfur would be present if it were used as a cross-link at every possible point (a) in polyisoprene? (b) in polychloroprene?

Answer: a) 32% S; b) 26.5% S

7–4.2 A rubber contains 91% polymerized chloroprene and 9% sulfur. What fraction of the possible cross-links are joined in vulcanization? (Assume that all the sulfur is used for cross-links of the type shown in Fig. 7–4.4.)

7–4.3 A rubber contains 54% butadiene, 34% isoprene, 9% sulfur, and 3% carbon black. What fraction of the possible cross-links are joined by vulcanization, assuming that all the sulfur is used in cross-linking?

Answer: 0.188.

7–4.4 What is the percent weight loss if phenol polymerizes trifunctionally with formaldehyde and all of the byproduct water evaporates?

7–5.1 Lucite (PMM of Fig. 7–5.4) is loaded at 125°C for 1 hr. a) How much would the load have to be increased to give the same strain in 36 sec? b) Repeat for 100°C.

Answer: a) $\sigma_{0.01}/\sigma_1 = 2.5$; b) $14\sigma_1$

•**7–5.2** An initial stress of 1500 psi (10.4 MPa) is required to strain a piece of rubber 50%. After the strain has been maintained constant for 40 days, the stress required is only 750 psi (5.2 MPa). What would be the stress required to maintain the strain after 80 days?

•**7–5.3** The relaxation time for a plastic is known to be 45 days and the modulus of elasticity is 70 MPa (both at 100°C). The plastic is compressed 5% and held at 100°C. What is the stress a) initially? b) after 1 day? c) after 1 month? d) after 1 year?

Answer: a) 3.5 MPa; b) 3.4 MPa; c) 1.8 MPa; d) 1000 Pa

•**7–5.4** The relaxation time for a nylon thread is reduced from 4000 min to 3000 min if the temperature is increased from 20°C to 30°C. a) Determine the activation energy for relaxation. b) At what temperature is the relaxation time 2000 minutes?

7–6.1 a) What is the polarization in the polyvinyl chloride of Example 7–6.1 at 100 Hz when 100 volts is applied to the capacitor? b) What will the electron density be on the capacitor plate?

Answer: a) 0.0002 C/m^2; b) 1.4×10^{15} electrons/m^2

7–6.2 A plate capacitor must have a capacitance of 0.25 μf. What should its area be if the 0.0005 in. (0.013 mm) mylar film which is used as a spacer has a dielectric constant of 3.0?

7–7.1 The average energy of a C—Cl bond is 81 kcal/mole (340 kJ/mole) according to Table 2–3.1. Will visible light [4000 Å (violet) to 7000 Å (red)] have enough energy to break the average C—Cl bond?

Answer: 3500 Å required

7–7.2 Determine the (a) mean, (b) median, and (c) standard deviation of the following set of strength data for a polystyrene:

11.8 MPa	12.8	12.2	11.7	12.1	12.0	11.5
12.3	12.7	12.2	11.9	11.6	12.4	
12.1	12.9	12.3	11.7	13.0	12.3	
12.2	13.1	11.8	12.7	12.7	12.1	
12.5	12.4	11.4	11.9	11.5	12.1	
12.9	12.5	12.2	12.5	11.0	12.3	
12.0	12.2	11.9	12.4	12.2	12.1	

7–7.3 What are the (a) mean and (b) median of the following density data for a polyethylene? 0.926, 0.930, 0.923, 0.924, 0.921, 0.926, 0.927, 0.929, 0.922, 0.928 g/cm^3

Answer: a) 0.9256 g/cm^3; b) 0.926 g/cm^3

•**7–7.4** What is the standard deviation of the data in Study Problem 7–7.3?

7–7.5 Raw polyisoprene (i.e., nonvulcanized natural rubber) gains 2.3 w/o by being cross-linked by oxygen of the air. What fraction of the possible cross-links are established?

Answer: 0.10

CHAPTER EIGHT

CERAMIC MATERIALS

PREVIEW

Ceramic materials contain phases which are *compounds of metallic and nonmetallic elements.* We can cite many of these compounds ranging from Al_2O_3 to inorganic glasses, to clay products, and on to sophisticated piezoelectrics such as $Pb(Zr, Ti)O_3$. In general, we will find that ceramic compounds are more stable with respect to thermal and chemical environments than their components, for example, Al_2O_3 as a compound versus aluminum and oxygen separately. Since compounds inherently involve more complex atomic coordinations than their corresponding components, we will find that there is more resistance to slip, so that ceramics are generally harder, often stronger, and always less ductile than their metallic or polymeric counterparts. The dielectric, semiconductive, and magnetic characteristics of selected ceramics are especially valuable to the scientists and engineers who design or utilize devices for electronic circuits.

8–1 CERAMIC PHASES

The term *ceramic* is most familiar as an adjective describing artware. For the engineer, however, ceramics include a wide variety of substances such as glass, brick, stone, concrete, abrasives, porcelain enamels, dielectric insulators (Fig. 8–1.1), nonmetallic magnetic materials, high-temperature refractories, and many others. The characteristic feature all these materials have in common is that they are compounds of metals and nonmetals. The compound MgO is representative of a simple ceramic material; it is used extensively as a refractory because it can withstand exceedingly high temperatures (1500–2500°C, or 3000 to 4500°F) without dissociation or melting. Clay is also a common ceramic material but is more complex than MgO. The simplest clay is $Al_2Si_2O_5(OH)_4$. It forms a crystalline structure of the four different units: Al, Si, O, and the (OH) radical. Although ceramic materials are not as simple as metals, they may be classified and understood in terms of their internal structures.

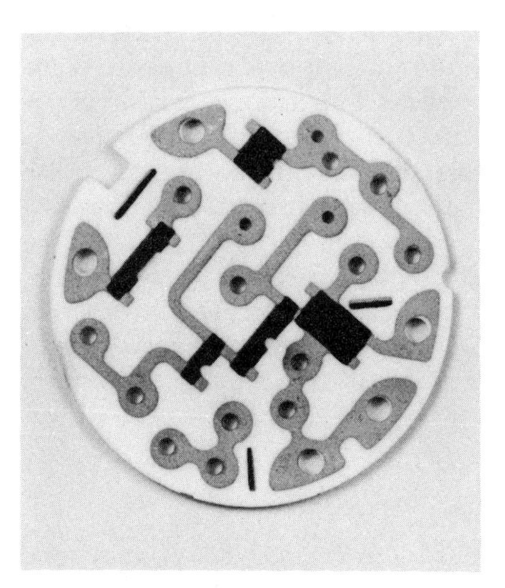

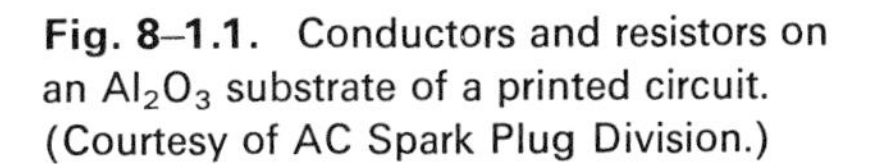

Fig. 8–1.1. Conductors and resistors on an Al_2O_3 substrate of a printed circuit. (Courtesy of AC Spark Plug Division.)

Comparison of ceramic and nonceramic phases. Most ceramic minerals (phases), like metals, have crystal structures. Unlike metals, however, these structures do not contain large numbers of free electrons. Either the electrons are shared covalently with adjacent atoms, or they are transferred from one atom to another to produce an ionic bond, in which case the atoms become ionized and carry a charge.

Ionic bonds give ceramic materials relatively high stability. As a class, they have a much higher melting temperature, on the average, than do metals or organic materials. Generally speaking, they are also harder and more resistant to chemical alteration. Like the organic materials, solid ceramic minerals are usually insulators. At elevated temperatures, with more thermal energy, they do conduct electricity, but poorly as compared with metals. Due to the absence of free electrons, most ceramic materials are transparent, at least in thin sections, and are poor thermal conductors.

Crystalline characteristics may be observed in many ceramic materials. Mica, for example, has cleavage planes which permit easy splitting. In some of the simpler crystals, such as MgO, plastic slip, similar to metallic slip, may occur. Crystal outlines can form during growth, as exemplified by the cubic outline of ordinary table salt. In asbestos, the crystals have a marked tendency toward linearity; in micas and clays, the crystals form two-dimensional sheet structures. The stronger, more stable ceramic materials commonly possess three-dimensional framework structures with equally strong bonding in all three directions.

As compared with those of metals, the crystal structures of ceramic materials are relatively complex. This complexity and the greater strength of the bonds holding the atoms together make ceramic reactions sluggish. For example, at normal cooling rates glass does not have time to rearrange itself into a complicated crystalline structure, and therefore at room temperature it remains as a supercooled liquid for a long time.

The structures and properties of compounds such as the refractory carbides and nitrides fall somewhere between those of ceramic and metallic materials. These include such compounds as TiC, SiC, BN, and ZrN, which contain semimetallic elements and whose structures comprise a combination of metallic and covalent bonds. Ferromagnetic spinels are another example. Because they lack free electrons, they are not good conductors of electricity; however, the atoms can be oriented within the crystal structure so as to possess the magnetic properties normally associated with iron and related metals.

The structure and properties of silicones fall somewhere between those of ceramic and organic materials. They are often called inorganic polymers. Finally, there are close structural relationships between amorphous polymers and commercial glasses, a fact we have already encountered through the examination of the glass temperature T_g in Chapter 7.

8–2 CERAMIC CRYSTALS (AX)

The simplest ceramic compounds for our initial consideration will possess equal numbers of metal and nonmetal atoms. These may be ionic like MgO, where two electrons have been transferred from the metallic to the nonmetallic atoms and have produced cations (Mg^{2+}) and anions (O^{2-}). AX compounds may also be covalent, with a large degree of sharing of the valence electrons. Zinc sulfide (ZnS) provides an example of this type of structure.

The characteristic feature of AX compounds is that the A atoms have only X atoms as immediate neighbors, and the X atoms have only A atoms as *first-neighbors.* Thus the A and X atoms (or ions) are highly *ordered.* There are three principal ways in which AX compounds can form so that the two types of atoms are in equal numbers and possess the ordered coordination just described. The prototypes are (1) CsCl with CN = 8; (2) NaCl with CN = 6; and (3) ZnS with CN = 4. These are shown in Fig. 8–2.1.

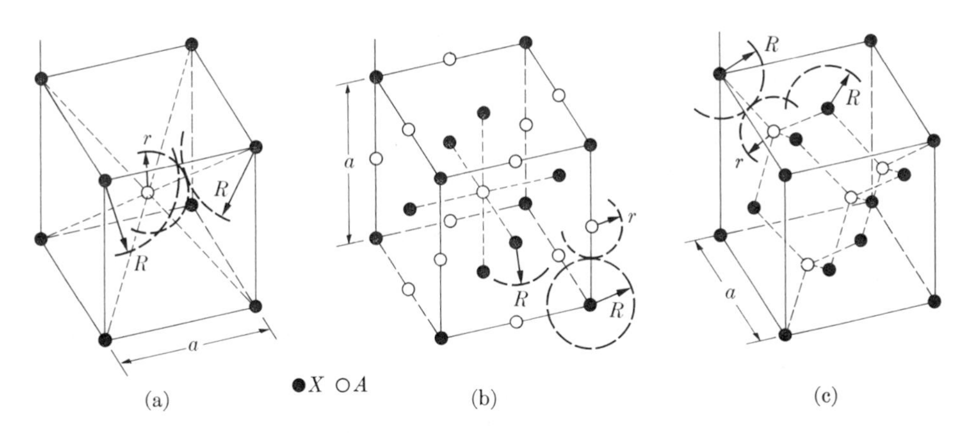

Fig. 8–2.1. AX structures. (a) CsCl-type. The lattice is simple cubic with CN = 8. All of the 8-fold sites are occupied. (Cf. Fig. 3–6.2.) (b) NaCl-type. The lattice is fcc with CN = 6. All of the 6-fold sites are occupied. (Cf. Fig. 3–1.1.) (c) ZnS-type. The lattice is fcc with CN = 4. One-half of the 4-fold sites are occupied. (Cf. Fig. 5–4.3.)

AX structures (CsCl-type). Each A atom of Fig. 8–2.1(a) has eight X neighbors, *and* each X atom has eight A neighbors. Those ceramic compounds which have this structure are strongly ionic. As indicated in Table 2–6.2, we do not find this structure unless $r_A \geq 0.73 R_X$. These are not common compounds because cations, stripped of electrons, tend to be considerably smaller than anions with excess electrons. It was previously noted that CsCl (and β'-brass) have simple cubic structures (Section 3–6 and Fig. 3–6.2). Although there are two ions per unit cell, A^+ and X^-, there is only one equivalent site per unit cell.

AX structures (NaCl-type). This is the most widespread AX structure; it is possessed by several hundred different compounds. Although NaCl is the prototype, we will direct our attention more specifically to MgO, since it has wider technical use. Each A atom or ion of Fig. 8–2.1(b) has six X neighbors, *and* each X ion has six A neighbors. The r_A/R_X ratio must be at least 0.41.

This structure has an fcc lattice, which can best be envisioned by looking at only the X^- ions of Fig. 8–2.1(b). We may view the NaCl-structure as an fcc lattice of anions, with cations in all of the six-fold *interstitial sites.*

Note that the X^- ions do not "touch"* as do fcc metal atoms in Fig. 3–3.5. Thus we cannot use Eq. (3–3.3) to calculate the lattice constant a. By simple inspection, however,

$$a = 2(r + R). \tag{8–2.1}$$

* This is to be expected since there is a mutual repulsion between any two negative ions. In fact, if the r/R ratio is small enough to permit contact, we reach the limit of the stability for that coordination (see Table 2–6.2).

AX structures (ZnS-type). Covalent bonding is more common and ionic bonding less common in this structure (Fig. 8–2.1(c)) than in the previous two structures. In fact, we first considered this structure in Fig. 5–4.3(b) when we were studying semiconducting compounds. Each A atom of Fig. 8–2.1(c) has four X neighbors, *and* each X atom has four A neighbors. As in the other structures of Fig. 8–2.1, like atoms do not come in contact.

Referring to Section 3–6 and Fig. 3–6.3, we will see that this lattice is fcc. We may view the ZnS-structure as an fcc lattice of X atoms with A atoms in four-fold *interstitial sites* (or as an fcc lattice of A atoms with X atoms in four-fold sites). There are, however, eight four-fold sites per unit cell. Therefore, only half of the four-fold sites are filled to maintain the A/X balance of 1/1.*

The lattice constant may be determined for a ZnS-type crystal by noting that 0, 0, 1 and $\frac{1}{4}, \frac{1}{4}, \frac{3}{4}$ of Fig. 8–2.1(c) are separated by a distance $(r + R)$:

$$(r + R)^2 = (a/4)^2 + (a/4)^2 + (-a/4)^2$$

or

$$a = \frac{4(r + R)}{\sqrt{3}}. \tag{8–2.2}$$

Table 8–2.1
PACKING OF AX COMPOUNDS

Structure	X atoms, lattice	A atoms: Site location	A atoms: Fraction filled	Other examples
CsCl	sc	8-fold	all	
NaCl	fcc	6-fold	all	MgO, MnS, LiF
ZnS*	fcc	4-fold	1/2	β-SiC, CdS, AlP
NiAs	hcp	6-fold	all	FeS, MnTe
ZnO	hcp	4-fold	1/2	BeO, ZnS*, AlN

* ZnS, like many ceramic compounds, has more than one polymorph. The cubic polymorph is called *sphalerite* (or *blende*), after the most common zinc ore mineral; the hexagonal polymorph has the name *wurtzite*.

* The four-fold interstitial sites are at:

Set (a) $\frac{3}{4},\frac{3}{4},\frac{3}{4}$ $\frac{1}{4},\frac{1}{4},\frac{3}{4}$ $\frac{1}{4},\frac{3}{4},\frac{1}{4}$ $\frac{3}{4},\frac{1}{4},\frac{1}{4}$.
Set (b) $\frac{1}{4},\frac{1}{4},\frac{1}{4}$ $\frac{3}{4},\frac{3}{4},\frac{1}{4}$ $\frac{3}{4},\frac{1}{4},\frac{3}{4}$ $\frac{1}{4},\frac{3}{4},\frac{3}{4}$.

The sites to be filled are not random, but must be either set (a) *or* set (b), because that provides an fcc pattern for *both* A and X atoms.

•**AX structures (noncubic).** Not all AX compounds are cubic. Two easily visualized examples are those which have hcp arrays of X atoms. In the NiAs-type structure, the A atoms are in all of the six-fold sites. In the ZnO-type structure, the A atoms are in half of the four-fold sites (Table 8–2.1). One may recall from Chapter 3 that fcc and hcp *metals* have equal packing factors (0.74) and the same coordination number (12). The only difference arises in the sequence of stacking of planes, above one another. By like token, the structures of NaCl and NiAs are related; and the structures of ZnS and ZnO are related. However, in each case the symmetry of the hexagonal version is lower than of the cubic version. Thus, we find FeS (with the NiAs-type structure) to be very hard and brittle; in contrast, MnS (with the NaCl-type structure) is among the more ductile AX compounds.

Example 8–2.1 Calculate the density of FeO which has an NaCl-type structure. (Assume stoichiometry.)

Solution: From Table 2–6.1, or Appendix B,

$$r_{Fe^{2+}} = 0.74\ \text{Å} \qquad \text{and} \qquad R_{O^{2-}} = 1.40\ \text{Å}$$

when CN = 6. There are 4 Fe^{2+} and 4 O^{2-} per unit cell. (See Fig. 8–2.1.) Then

$$\begin{aligned} V = a^3 &= [2(0.74 + 1.40) \times 10^{-8}\ \text{cm}]^3 \\ &= 78.4 \times 10^{-24}\ \text{cm}^3; \\ m &= 4(55.8 + 16.0\ \text{amu})/(0.6 \times 10^{24}\ \text{amu/g}) \\ &= 479 \times 10^{-24}\ \text{g}; \\ \rho = m/V &= 479/78.4 = 6.1\ \text{g/cm}^3. \end{aligned}$$

Comments. The measured density is about 5.7 g/cm^3 because the structure contains cation vacancies (Fig. 4–3.2 and Example 4–3.1). ◀

Example 8–2.2 MnS has three polymorphs, (a) the NaCl-type structure (Fig. 8–2.1(b)), (b) the ZnS-type structure (Fig. 8–2.1(c)), and (c) the ZnO-type structure. What percent volume change occurs when the second type (ZnS) changes to the first type (NaCl)? (See Appendix B for radii when CN = 6, and the attached footnote when CN ≠ 6.)

Solution: Each unit cell (Fig. 8–2.1(a) and (b)) has 4 Mn^{2+} ions and 4 S^{2-} ions. Therefore, we can consider one unit cell of each.

a) $V = a^3 = [2(0.80 + 1.84)\text{Å}]^3 = 147\ \text{Å}^3;$

b) $V = a^3 = \left[4(0.73 + 1.67\ \text{Å})/\sqrt{3}\right]^3 = 170\ \text{Å}^3;$

$(\Delta V/V)_{b\to a} = -14$ v/o.

Comment. Only the NaCl-type is stable; however, the other two types can be formed with appropriate starting materials. ◀

8–3 CERAMIC CRYSTALS (A_mX_p)

Not all binary compounds have equal numbers of A and X atoms (or ions). We shall consider only two cases out of many to illustrate the point, the CaF_2 structure and the Al_2O_3 structure. Fluorite (CaF_2) is the basic structure for UO_2, which is used in nuclear fuel elements, and provides the pattern for one of the polymorphs of ZrO_2, which is a useful high-temperature oxide. Corundum, Al_2O_3 is one of the most widely used ceramics for technical purposes. We have already encountered Al_2O_3 in a sparkplug (Fig. 2–7.3). Other uses range from emery grinding wheels to acid pumps, to substrates for printed circuits, and on to high-temperature materials for catalyst supports in exhaust systems.

The CaF_2 structure may be viewed in one of two alternative manners:

1. It is a CsCl-type structure (Fig. 8–2.1(a)), but with only *half* of the eight-fold sites filled by A atoms. That provides the required Ca-to-F ratio of one-to-two.
2. The Ca^{2+} ions have the fcc structure of the metal atoms in ZnS (Fig. 8–2.1(c)) with F^- ions in *all* eight of the four-fold sites. This also gives the required one-to-two ratio, since there are eight four-fold sites per unit cell of four lattice points. This is the alternative sketched in Fig. 8–3.1.

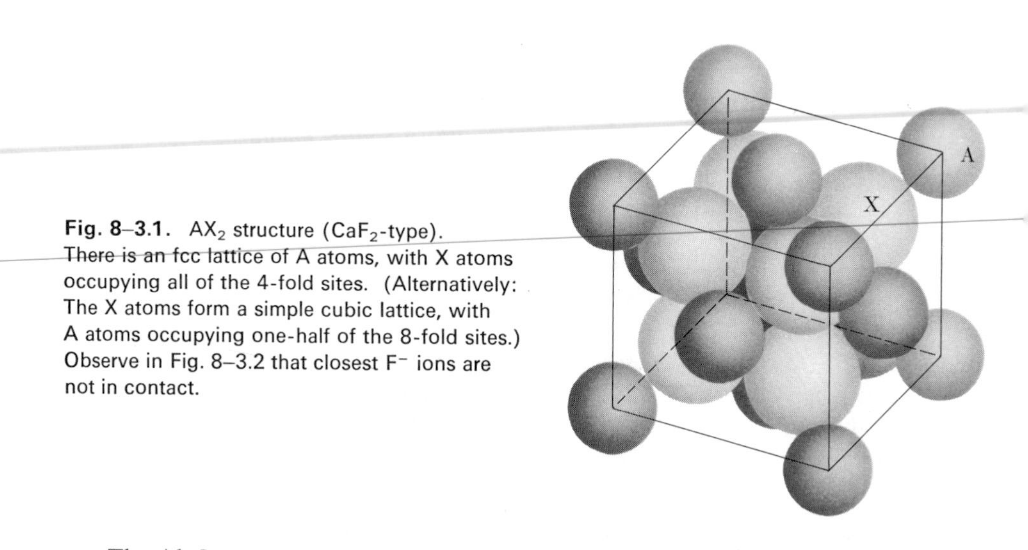

Fig. 8–3.1. AX_2 structure (CaF_2-type). There is an fcc lattice of A atoms, with X atoms occupying all of the 4-fold sites. (Alternatively: The X atoms form a simple cubic lattice, with A atoms occupying one-half of the 8-fold sites.) Observe in Fig. 8–3.2 that closest F^- ions are not in contact.

The Al_2O_3 structure may be viewed as an hcp array of O^{2-} ions with Al^{3+} ions in two-thirds of the six-fold sites.* Since there are as many six-fold sites as hcp positions, the result is the necessary 2-to-3 cation/anion ratio. With an interatomic distance of only 1.91 Å separating the 3^+ and 2^- charges of the aluminum ion and

* This oversimplifies the structure, since it does not take into account a distortion which develops as a consequence of the vacant six-fold sites. However, the above description will meet our needs.

its six oxygen neighbors, it is not surprising that the bonding energies are high.* This is reflected in the melting temperature (>2000°C), hardness (Mohs = 9), and the resistance of Al_2O_3 to a large number of chemicals. Furthermore, its combination of low electrical conductivity and a relatively high thermal conductivity (Appendix C) lets Al_2O_3 be used in many electrical applications.

Example 8–3.1 The lattice constant of CaF_2 is 5.47 Å. (a) Sketch the arrangement of ions on the (110) plane of CaF_2. (b) What is the sum of the two radii ($r_{Ca^{2+}} + R_{F^-}$)? •(c) What is the linear density of equivalent sites in the $[1\bar{1}2]$ direction?

Solution:

a) Figure 8–3.2 is a (110) plane sketched from Fig. 8–3.1.

b) $$r + R = \sqrt{(a/4)^2 + (-a/4)^2 + (a/4)^2} = (a\sqrt{3})/4$$
$$= (5.47\text{ Å})\sqrt{3}/4 = 2.37\text{ Å}.$$

•c) Since CaF_2 is fcc, the repetition distance will be from 0, 0, 0 to $\frac{1}{2}, -\frac{1}{2}, 1$, or $\sqrt{a^2 + (-a/2)^2 + (a/2)^2}$. Therefore:

$$\mathbf{b} = a\sqrt{1.5} = 5.47\text{ Å}(1.225) = 6.7\text{ Å},$$
$$\mathbf{b}^{-1} = 15 \times 10^6/\text{cm}.$$

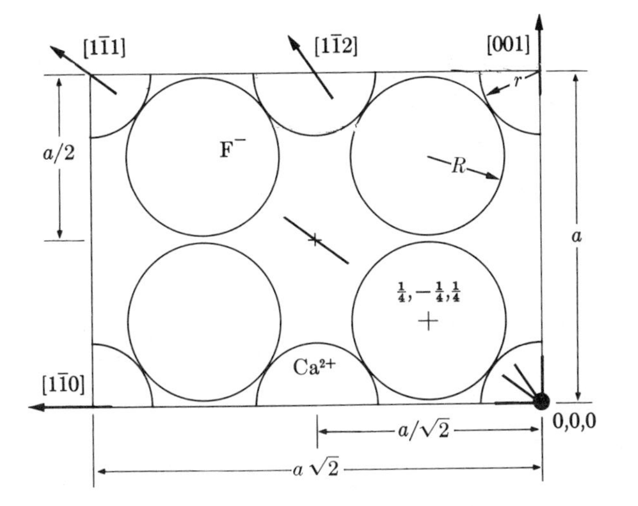

Fig. 8–3.2. CaF_2 structure. [(110) plane.] (See Example 8–3.1.)

Comments. The F^- ions are separated by about 0.3 Å (Fig. 8–3.2).

The data from Appendix B would indicate a = 5.15 Å; however, this does not take into account the mutual repulsion of the 8 F^- ions around the vacant eight-fold sites. This repulsion expands the structure.

---

* Recall from Eq. (2–5.1) that the coulombic attractive forces are proportional to $-Z_1Z_2/a^2$.

This is the same structure as UO_2. The unoccupied eight-fold sites provide space for fission products to reside within the crystal. Discuss the implications of this possibility. ◀

8–4 MULTIPLE COMPOUNDS

$A_mB_nX_p$-type structures. Although the presence of three types of atoms lends additional complexity, several $A_mB_nX_p$ compounds are of sufficient interest for our attention. First among these is $BaTiO_3$, the prototype for the ceramic materials used in applications such as cartridges for record players. Above 120°C, $BaTiO_3$ has a cubic unit cell with Ba^{2+} ions at the corners, O^{2-} ions at the center of the faces, and a Ti^{4+} ion at the center of the cell (Fig. 8–4.1).

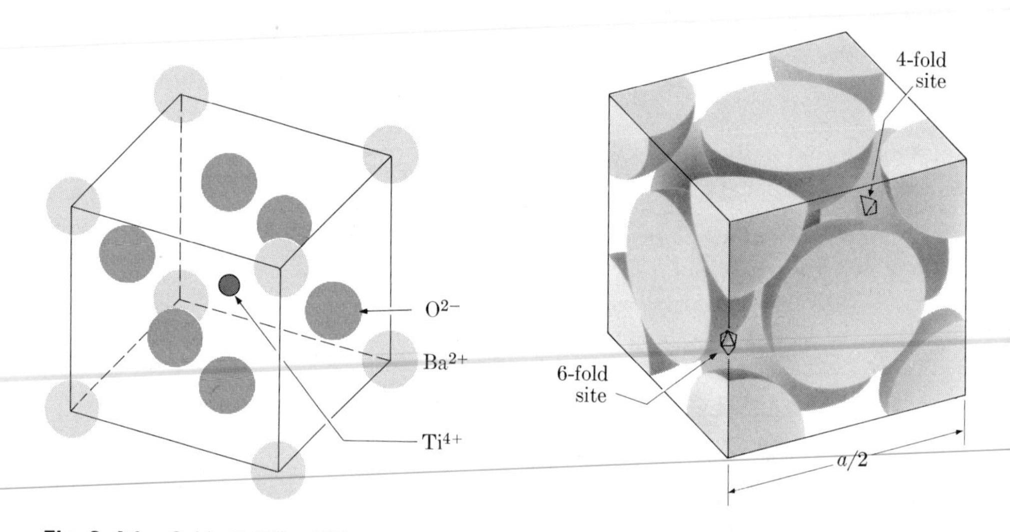

Fig. 8–4.1. Cubic $BaTiO_3$. This structure is stable above 120°C, where it has a Ti^{4+} ion in the center of the cube, Ba^{2+} ions at the corners, and O^{2-} ions at the center of each face.

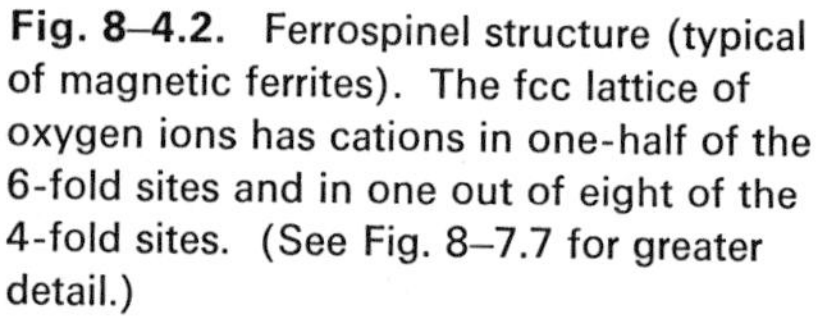

Fig. 8–4.2. Ferrospinel structure (typical of magnetic ferrites). The fcc lattice of oxygen ions has cations in one-half of the 6-fold sites and in one out of eight of the 4-fold sites. (See Fig. 8–7.7 for greater detail.)

Nonmetallic magnets may also be $A_mB_nX_p$ compounds, the most common being a ferrospinel (often called a ferrite) with the composition of MFe_2O_4, where the M are divalent cations with radii of 0.75 ± 0.1 Å. These *spinel* structures have a close-packed (fcc) structure of O^{2-} ions with cations selectively positioned in the 6-fold and 4-fold *interstitial sites* (Fig. 8–4.2). The ferromagnetic characteristics of these structures are influenced by the cation locations.

Solid solutions. Brief attention was given in Section 4–3 to solid solutions in ionic compounds. Two chief requirements were cited as necessary for solid solutions to occur: (1) compatibility in size and (2) balance in charge. These limitations are not as rigid as might be surmised, because compensation may be made in charge. For example, Li^+ ions may replace Mg^{2+} ions in MgO *if* F^- is simultaneously present to replace O^{2-} ions. Conversely, MgO may be dissolved in LiF. We may also find Mg^{2+} dissolved in LiF without comparable O^{2-} ions; however, in this case cation vacancies must be included. As a result, $2Li^+$ are replaced by $(Mg^{2+} + \square)$.

Ceramists depend heavily on solid solutions in the previously mentioned magnetic spinels, because optimum magnetic characteristics exist when part of the divalent ions are zinc ($r = 0.74$ Å) and the balance of the divalent ions are ferromagnetic; for example, Ni^{2+} ($r = 0.69$ Å). In this case it is a simple matter of direct substitution; however, for certain applications it is desirable to replace $2M^{2+}$ with an Li^+Fe^{3+} pair, or to replace $2Fe^{3+}$ with an $Mg^{2+}Ti^{4+}$ pair.

Example 8–4.1 A ferrospinel has a lattice of 32 oxygen ions, 16 ferric ions, and 8 divalent ions. (The unit cell contains 8 times as many oxygen ions as MgO does, so that the repeating pattern can be developed.) If the divalent ions are Zn^{2+} and Ni^{2+} in a 3:5 ratio, what weight fraction of ZnO, NiO, and Fe_2O_3 must be mixed for processing?

Solution: Basis: 8 mol wt of Fe_2O_3.

$$5\,NiO + 3\,ZnO + 8\,Fe_2O_3 \longrightarrow (Zn_3, Ni_5)Fe_{16}O_{32}.$$

			Wt fraction
5 NiO	= 5(58.71 + 16.00)	= 373.5	= 0.197
3 ZnO	= 3(65.37 + 16.00)	= 244.1	= 0.129
8 Fe_2O_3	= 8[2(55.85) + 3(16.00)]	= 1277.6	= 0.674
		1895.2	1.000 ◀

•8–5 SILICATE STRUCTURES

Many ceramic materials contain *silicates*, partly because silicates are plentiful and cheap and partly because they have certain distinct properties which are useful in engineering applications. Probably the most widely known silicate is portland cement, which has the very definite advantage of being able to form a hydraulic bond for rock aggregates. Many other construction materials, such as brick, tile, glass, and vitreous enamel, are also made of silicates. Other engineering applications of silicates include electrical insulators, chemical ware, and reinforcing glass fibers.

Silicate tetrahedral units. The primary structural unit of silicates is the "SiO_4" *tetrahedron* (Fig. 8–5.1), in which one silicon atom fits interstitially among four oxygen atoms. The forces holding these tetrahedra together involve both ionic and covalent bonds; consequently the tetrahedra are tightly bonded. However, with either the ionic or covalent bonding mechanism, each oxygen has only seven electrons rather than the eight available to its outer shell.

Two methods are available to overcome this deficiency of electrons for the oxygen ions: 1) An electron may be obtained from other metal atoms. In this case SiO_4^{4-} ions

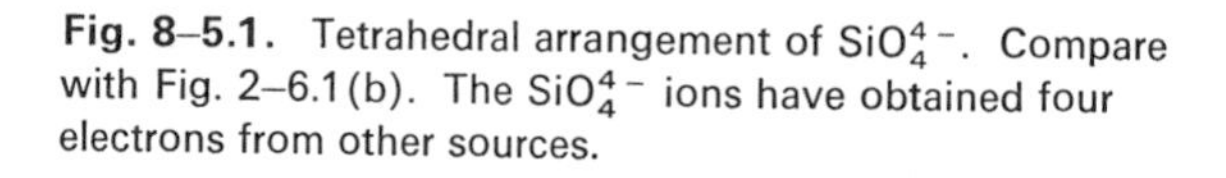

Fig. 8–5.1. Tetrahedral arrangement of SiO_4^{4-}. Compare with Fig. 2–6.1(b). The SiO_4^{4-} ions have obtained four electrons from other sources.

and $Metal^+$ ions are developed. 2) Each oxygen may share an electron pair with a second silicon. In this case multiple 4-fold coordination groups are formed.

SiO_4^{4-} silicate structures. The simplest example of minerals containing SiO_4^{4-} ions is forsterite (Mg_2SiO_4), a mineral frequently used as a high-temperature refractory, since its melting point is 1890°C (3430°F). In Mg_2SiO_4, the SiO_4^{4-} ion has received four electrons from four adjacent magnesium atoms. Each magnesium atom provides a second electron to other SiO_4 units. Consequently, a strong structure is formed in which Mg^{2+} ions serve as connecting links between SiO_4^{4-} ions. Since the SiO_4^{4-} tetrahedral groups are separated from one another, they have received the name "island" structures. When forsterite eventually melts, the melt contains Mg^{2+} and SiO_4^{4-} ions which possess some mobility and therefore ionic conductivity. It should be noted that Mg_2SiO_4 is not a molecule in the structural sense (Section 2–3), because all units are held by primary ionic and by covalent bonds. Secondary, van der Waals forces, while present, are not apparent. Also, the resulting structure has a close-packed pattern of oxygen ions with silicon atoms in part of the 4-fold interstitial sites and magnesium ions in part of the 6-fold sites.

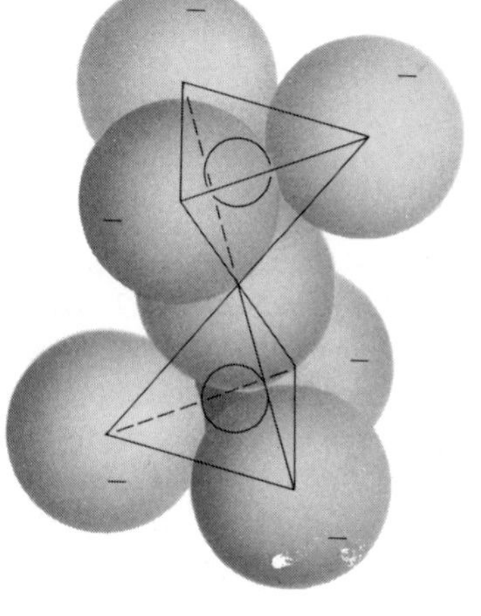

Fig. 8–5.2. Double tetrahedral silicates, $Si_2O_7^{6-}$. The center oxygen receives an electron from each adjacent silicon.

Double tetrahedral units. The second of the two methods available to overcome the deficiency of electrons produces a double tetrahedral unit. One of the oxygens is a member of two units (Fig. 8–5.2). The consequent composition of the double unit is Si_2O_7, which produces an $Si_2O_7^{6-}$ ion when electrons are obtained from adjacent metal atoms.

As in crystal bonding in NaCl (Fig. 3–1.1), both the SiO_4^{4-} and the $Si_2O_7^{6-}$ units may be held rigidly in a solid by mutual attraction with positive metal ions.

Chain silicates (and silicones). It immediately becomes apparent that if one of the oxygens can be shared by two adjacent tetrahedra, similar sharing of oxygens on the other corners of the tetrahedra is possible. Figures 8–5.3 and 8–5.4 show examples in which silica tetrahedra have been built into single and double chains. These *chain*

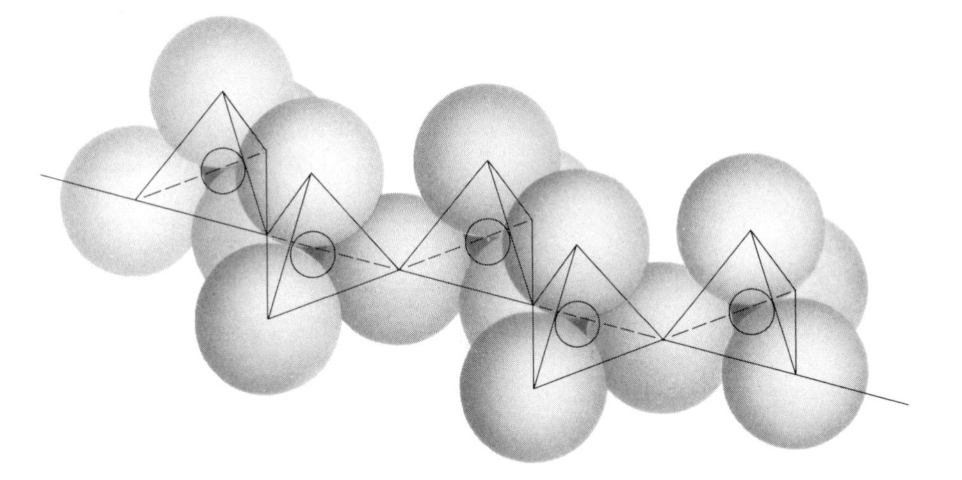

Fig. 8–5.3. Single chain of silica tetrahedra. The sides and top of the chain must receive electrons from adjacent positive ions. Otherwise this chain may be compared with a polymer chain. (Cf. Fig. 7–1.4.)

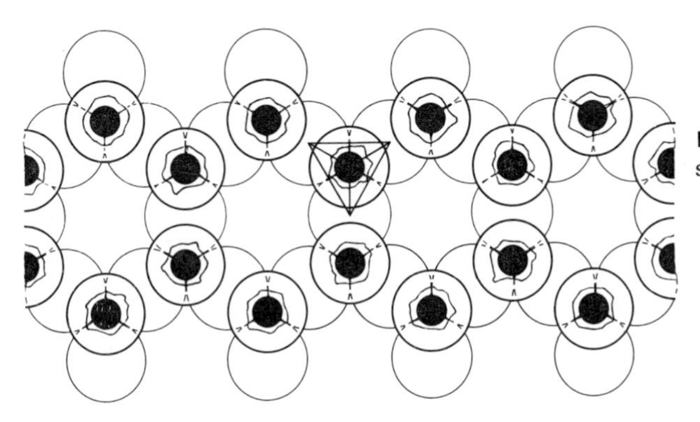

Fig. 8–5.4. Double chain of silica tetrahedra.

structures, theoretically, can be almost infinite in length and may be compared quite directly with polymerization in organic materials (Section 7–2) except for one major difference. With organic materials the adjacent chains are usually held together by weak van der Waals forces, but in ceramics the chains are commonly held together by ionic bonds, as shown schematically in Fig. 8–5.5. Since the ionic bonds between the chains are not quite as strong as the partially covalent Si-O bonds within the chain, fracture, or *cleavage*, occurs parallel to the chain. The rock minerals pyroxene and amphibole show examples of this parting. Also, the fibrous characteristics of *asbestos* are associated with the lower strength between the silicate chains rather than within them.

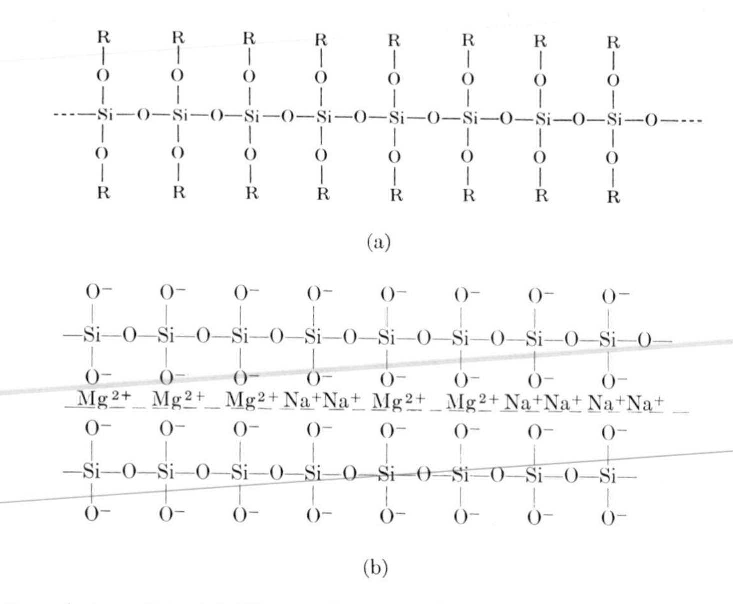

Fig. 8–5.5. Chain structures (schematic). (a) Silicone. One type of silicone has a backbone of [—Si—O—] mers. The side radicals may be of several types. (b) Silicates. The chains of silicates (e.g., asbestos) are held together by coulombic forces between positive ions and polyatomic anions.

Sheet silicates. The extension of silicate tetrahedral units into a plane rather than just along a line makes possible the structures of a number of such ceramic minerals as clays, micas, and talc. The sheet structure of silica tetrahedral units is shown in two representations in Fig. 8–5.6. On the lower side of the sheet, every oxygen is completely satisfied with a full complement of eight available electrons because these oxygens share pairs of electrons with adjacent silicons. Consequently, only secondary bonds are available to hold each sheet to adjacent sheets. The cleavage of mica, the plasticity of clay, and the lubricating characteristic of talc are all consequences of this structural arrangement.

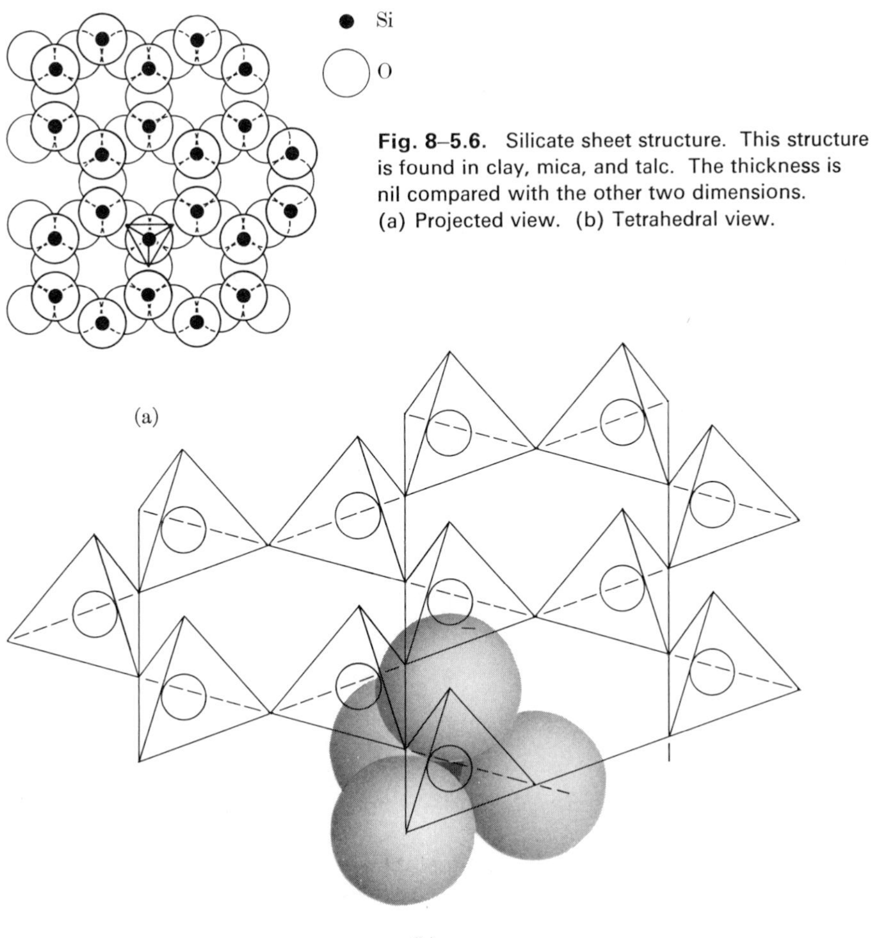

Fig. 8–5.6. Silicate sheet structure. This structure is found in clay, mica, and talc. The thickness is nil compared with the other two dimensions. (a) Projected view. (b) Tetrahedral view.

Network silicates. With pure SiO_2 there are no metal ions, and every oxygen atom is a bridging atom between two silicon atoms (and every silicon atom is among four oxygen atoms, as shown in Fig. 8–5.7). This gives a network-like structure. Silica

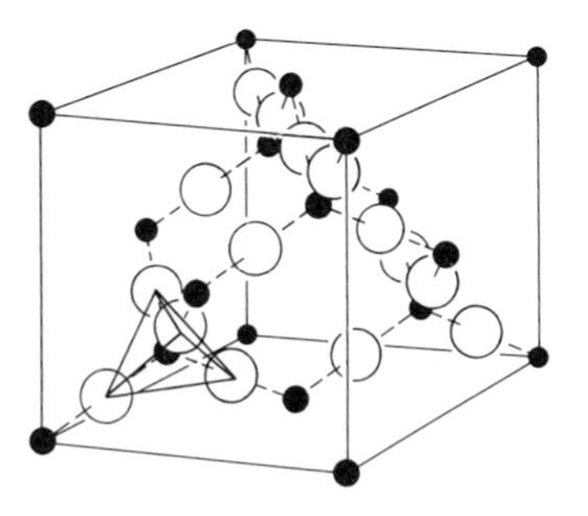

Fig. 8–5.7. Network structure (SiO_2). Each silicon atom is surrounded by four oxygen atoms, and each oxygen atom is part of two SiO_4 tetrahedra. (Cf. Figs. 8–2.1(c) and 5–4.3.) This polymorph of silica is called cristobalite.

(SiO_2) can have several different crystal structures, just as carbon can be in the form of graphite or diamond. The structure shown in Fig. 8–5.7 is a high-temperature form. A more common structure of silica is quartz, the predominant material found in the sands of many beaches. Just as the form of SiO_2 in Fig. 8–5.7 does, quartz contains SiO_4 tetrahedra, but with a more complex lattice.

Another natural silicate is feldspar. The pink phase of granite is $KAlSi_3O_8$, which may be visualized as a network silicate with one of four silicons replaced by an Al^{3+} ion. The latter, however, has only three charges as compared to four for silicon. Thus K^+ is present to balance out the charges. Fortunately most network structures are quite open, so that there is space for extra ions to be present (cf. Fig. 8–5.7). In fact, the K^+ ion may be viewed as an interstitial ion.

Example 8–5.1 If the close-packed arrangement of oxygens in forsterite (Mg_2SiO_4) has the same number of four-fold and six-fold sites as does Fig. 8–4.2, what fraction of these sites is occupied? The Mg^{2+} ions enter six-fold sites; the Si atoms, four-fold sites.

Solution: Basis: 100 oxygen atoms = 100 6-fold sites and 200 4-fold sites.

In Mg_2SiO_4: 100 oxygen atoms = 50 Mg^{2+} ions = 0.50 of 6-fold sites;

and 100 oxygen atoms = 25 Si atoms = 0.125 of 4-fold sites.

Comment. When these interstitial sites are filled as indicated, the unit cell becomes orthorhombic with $a = 10.25$ Å, $b = 6.00$ Å, and $c = 4.75$ Å. ◀

Example 8–5.2 Quartz (SiO_2) has a density of 2.65 g/cm³. (a) How many silicon atoms (and oxygen atoms) are there per cm³? (b) What is the packing factor, given that the radii of silicon and oxygen are 0.38 Å and 1.14 Å, respectively?

Solution:

$$\text{a) } SiO_2/\text{cm}^3 = \frac{2.65\ \text{g/cm}^3}{(28.1 + 32.0)\ \text{g}/(0.6 \times 10^{24}\ SiO_2)} = 2.645 \times 10^{22}\ SiO_2/\text{cm}^3$$

$$= 2.645 \times 10^{22}\ \text{Si/cm}^3 = 5.29 \times 10^{22}\ \text{O/cm}^3.$$

$$\text{b) } V_{Si}/\text{cm}^3 = (4\pi/3)(2.645 \times 10^{22}/\text{cm}^3)(0.38 \times 10^{-8}\ \text{cm})^3 = 0.006$$

$$V_O/\text{cm}^3 = (4\pi/3)(5.29 \times 10^{22}/\text{cm}^3)(1.14 \times 10^{-8}\ \text{cm})^3 = 0.328$$

$$\text{Packing factor} = 0.33$$

Comments. Although there is considerable open space within this structure, most single atoms (except for helium) must diffuse through SiO_2 as ions. Thus their charges prohibit measurable movements. ◀

8–6 GLASSES

The main glasses used commercially are silicates. They have the SiO_4 tetrahedra described in the preceding section, plus some modifying ions. They are *amorphous,*

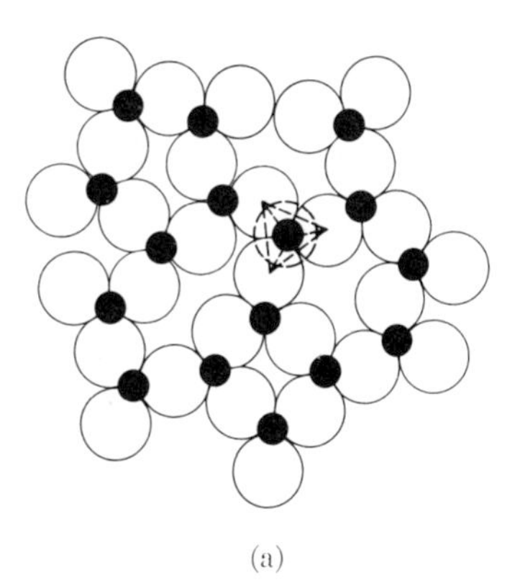

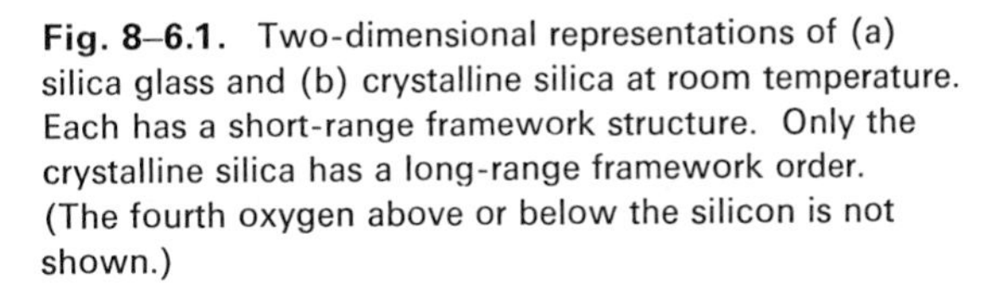
Fig. 8–6.1. Two-dimensional representations of (a) silica glass and (b) crystalline silica at room temperature. Each has a short-range framework structure. Only the crystalline silica has a long-range framework order. (The fourth oxygen above or below the silicon is not shown.)

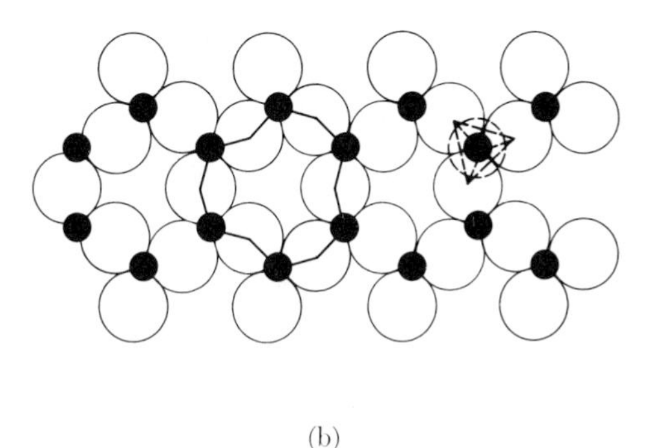

i.e., noncrystalline. Figure 8–6.1 contrasts the structure of a crystal with that of an amorphous solid of the same composition. Atoms in both have the same first neighbors, i.e., each oxygen atom is between two silicon atoms, and *bridges* two tetrahedra. Each silicon is among four oxygens (the fourth is envisioned as being above *or* below the plane of the paper). However, in glass, differences arise in more-distant neighbors because the glass, unlike the crystalline silica, does not have a regular long-range pattern.

When only silica (SiO_2) is present, and every oxygen atom serves as a bridge, the glass is very rigid. Fused silica, for example, is extremely viscous even when it gets into the temperature range in which it is a true liquid. In polymer terms, its *network* structure is highly cross-linked. Fused silica is very useful in some applications because it has a low thermal expansion. However, its high viscosity makes it extremely difficult to shape.

Network modifiers. Most silicate glasses contain *network modifiers*. These are oxides such as CaO and Na_2O which supply cations (positive ions) to the structure. As shown in Fig. 8–6.2, the addition of Na_2O to a silica glass introduces (1) two Na^+ ions, and (2) *nonbridging oxygens*. These are oxygens which are attached to only one silicon. As in the chain structure of Fig. 8–5.5(b), these oxygens carry a negative charge.

Commercial glasses contain these network modifiers because their presence markedly lowers the high-temperature viscosity of the glass by reducing the effective cross-linking, to make the glass more "thermoplastic." Therefore this glass is much easier to shape into the desired product.

The choice of a glass depends on striking a balance between the requirements of processing and the properties of the finished product. For example, a fused-silica glass has both a very low thermal expansion and a very high viscosity. Thus, even though there are many applications in which the low thermal expansion would be desirable, costs of processing fused silica may exclude its use because of the difficulty involved in shaping it into the desired product.

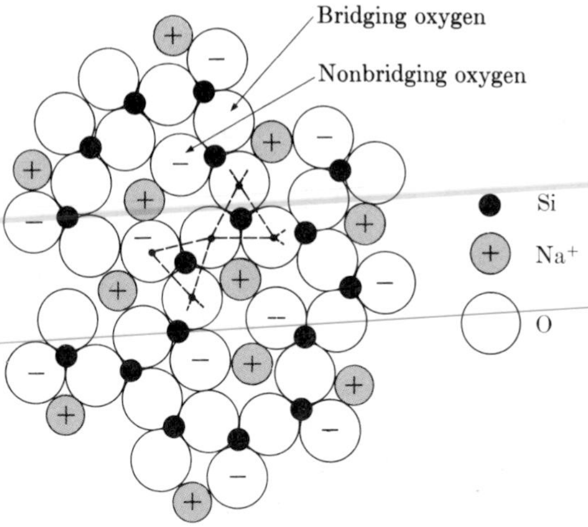

Fig. 8–6.2. Structure of soda–silica glass. The addition of Na_2O to a silica glass decreases the number of bridging oxygen atoms. Each nonbridging oxygen carries a charge. (As in Fig. 8–6.1, the fourth oxygen of each tetrahedron has been removed for clarity.) This glass is less viscous at high temperatures than the silica glass of Fig. 8–6.1(a).

Example 8–6.1 A glass contains 80 w/o SiO_2 and 20 w/o Na_2O. What fraction of the oxygens is nonbridging?

Solution: Basis: 100 g = 80 g SiO_2 + 20 g Na_2O.

$$80 \text{ g } SiO_2/[(28.1 + (2 \times 16.0)) \text{ g/mole}] = 1.33 \text{ mole } SiO_2 \quad \text{or} \quad 80.6 \text{ m/o } SiO_2;$$
$$20 \text{ g } Na_2O/[((2 \times 23.0) + 16.0) \text{ g/mole}] = 0.32 \text{ mole } Na_2O \quad \text{or} \quad 19.4 \text{ m/o } Na_2O.$$

New basis: 100 moles.

$$\begin{aligned} 80.6\ SiO_2 &= 80.6\ Si + 161.2\ O \\ 19.4\ Na_2O &= \underline{\qquad\qquad\quad 19.4\ O + 38.8\ Na^+} \\ &\quad\ 80.6\ Si + 180.6\ O + 38.8\ Na^+ \end{aligned}$$

Note (from Fig. 8–6.2) that there is one nonbridging oxygen for each Na^+ added.

$$\text{Fraction nonbridging oxygens} = 38.8/180.6 = 0.215.$$

Comments. Calcium forms a double charge and provides electrons to two nonbridging oxygens. Therefore each Ca^{2+} plus the accompanying oxygen of CaO breaks open a bridging oxygen and forms two nonbridging, charge-carrying oxygens. ◀

8–7 ELECTROMAGNETIC BEHAVIOR OF CERAMICS

A broad-scope generalization categorizes ceramics as being nonelectrical; that is, they have low conductivities and therefore are *insulators.* Many ceramics are used for insulating purposes. In addition, certain ceramics have electromagnetic capabilities ranging from capacitors with high dielectric constants, to piezoelectric* responses, to magnetic behavior. Finally, oxides of most transition elements are semiconductors.

Ceramic dielectrics. Ceramics are well known for their use as electrical insulators (Fig. 8–7.1). One of the better ways to tie up valence electrons so they can't carry charge is to transfer them from the metal atoms to the oxygen atoms, forming oxygen ions O^{2-}. Let us once more refer to the spark-plug insulator of Fig. 2–7.3. The Al^{3+}

Fig. 8–7.1. Surface breakdown. Adsorbed moisture and contaminants provide a surface path for electrical shorting. (R. Russell, *Brick and Clay Record.*)

---

* Pressure-electric interactions.

Table 8–7.1
PROPERTIES OF CERAMIC DIELECTRICS

Material	Resistivity (volume), ohm · cm	Dielectric strength, volts/mil*	Relative dielectric constant, κ		Loss factor, $\tan\delta$	
			60 Hz	10^6 Hz	60 Hz	10^6 Hz
Electrical porcelain	10^{13}–10^{15}	40–200	6	—	0.010	—
Steatite insulators	$>10^{14}$	200–400	6	6	0.005	0.003
Zircon insulators	$\sim 10^{15}$	250–400	9	8	0.035	0.001
Alumina insulators	$>10^{14}$	250	—	9	—	<0.0005
Soda-lime glass	10^{14}	250	7	7	0.1	0.01
E-glass	$>10^{17}$	—	—	4	—	0.0006
Fused silica (SiO_2)	$\sim 10^{20}$	250	4	3.8	0.001	0.0001

* Volts per 0.001 in. This value varies somewhat with service conditions.

ions of Al_2O_3 have been stripped of the valence electrons that would carry charge in metallic aluminum. Those electrons are held firmly by the oxygen ions. In other insulating materials, Mg^{2+} ions lose their electrons to O^{2-} ions in MgO; and silicon and oxygen rigidly share electrons within the SiO_4 tetrahedron (Fig. 8–5.1). As a result, compositions of MgO–Al_2O_3–SiO_2 form some of our best insulators.

Even materials which are commonly considered to be insulators can break down under high electrical voltages. Usually the breakdown is a *surface* phenomenon. For example, the spark plugs of an automobile may short out on a damp morning because condensed moisture on the surface of the ceramic insulators permits the current to short-circuit the spark gap. Insulators are designed with lengthened surface paths (see Fig. 8–7.1) to decrease the possibility of surface shorting, and since internal pores and cracks provide opportunity for additional "surface" failure, the insulators are usually glazed to make them nonabsorbent. *Volume breakdown* occurs only when extremely high voltage gradients are encountered. A very strong electric field can be sufficient to disrupt the induced dipoles in the insulator, and when the strength of the field exceeds the strength of the dipole, rupture may occur.

Dielectric properties are listed in Table 8–7.1 for selected ceramic insulators. The *dielectric strength* is the voltage gradient which produces electrical breakdown *through* the insulator. The *dielectric constant* κ may be compared with similar data for plastics in Table 7–6.1. In general, these values are slightly higher in ceramic materials, since ions, rather than induced dipoles, respond to the electric field. As with polymeric insulators, the dielectric constant is sensitive to frequency. However, in the normal temperature range, there is less variation among ceramic insulators. The electrical engineer also pays attention to the *loss factor*, $\tan\delta$, when he designs electrical circuits. The product of the dielectric constant and the loss factor, $\kappa\tan\delta$, is a measure of the electrical energy consumed by a capacitor in an a.c. circuit. This directs specific attention to both alumina and the glasses designed for electrical purposes (E-glasses) when designing for the megahertz range.*

Ceramic semiconductors. Although ceramic compounds are nominally insulators, they may become semiconductors if they contain multivalent transition elements. This was illustrated in Fig. 5–5.4(a), where electron holes carry charges by moving from one nickel ion to another. Magnetite (Fe_3O_4 or $Fe^{2+}Fe_2^{3+}O_4$) is a ceramic semiconductor with a resistivity of 10^{-2} ohm · cm, which is comparable to graphite and gray tin (Fig. 5–1.1). The origin of the conductivity is identical to that of NiO in Fig. 5–5.4(a); however, the number of electron holes for carrying the charge is much higher because the fraction of Fe^{3+} ions is greater.

The resistivity can be increased by solid solutions which replace the multivalent iron ions with other ions. This is shown in Table 8–7.2 for solid solutions of $MgCr_2O_4$ and $FeFe_2O_4$. Neither Mg^{2+} nor Cr^{3+} ions can react with the electrons or electron holes. Thus the resistivity can be adjusted to selected levels. The temperature coefficient of resistivity of these semiconductors is equally interesting to the engineer. As noted in Table 8–7.2, the resistance change is more than 1%/°C, and in other solid solutions may be as high as 4%/°C. This sensitivity is sufficient for accurate temperature measurement and has led to devices called *thermistors* which are used for thermometric purposes. Because thermistors usually have a negative temperature coefficient of resistivity, they may also be used to compensate for positive resistance changes in the metallic components of a circuit.

Table 8–7.2
RESISTIVITIES OF CERAMIC SEMICONDUCTORS*

Composition, mole percent		Resistivity, ohm · cm		$\Delta\rho/\Delta T$
$FeFe_2O_4$	$MgCr_2O_4$	25°C	60°C	%/°C
100	0	0.005	0.0045	−0.3
75	25	0.7	0.45	−1.0
50	50	1.8×10^2	75.0	−1.6
25	75	4.0×10^4	1.2×10^4	−2.0
0	100	$>10^{12}$	$>10^{12}$	—

* Adapted from E. J. Verwey, P. W. Haayman, and F. C. Romeijn, *J. Chem. Phys.*

Piezoelectric ceramics. Certain ceramic crystals lack a center of symmetry. We will not consider their crystallographic features but will note an important consequence when ionic crystals are involved. The centers of positive and negative charges are not identical. As a result each unit cell acts as a small electric dipole with a positive and negative end. This is illustrated by $BaTiO_3$. Recall from Fig. 8–4.1 that $BaTiO_3$ is

* As indicated in Section 8–6, fused silica is difficult to process because of its high viscosity.

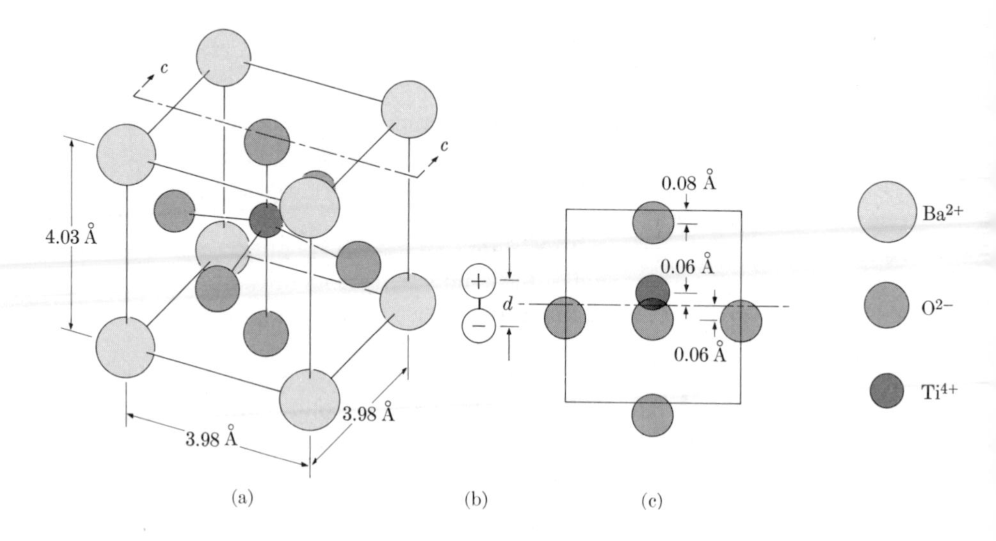

Fig. 8–7.2. Tetragonal $BaTiO_3$. Above 120°C (250°F), $BaTiO_3$ is cubic (Fig. 8–4.1). Below that temperature, the ions shift with respect to the corner Ba^{2+} ions. Since the Ti^{4+} and the O^{2-} ions shift in opposite directions, the centers of positive and negative charges are not identical. The unit cell becomes noncubic.

cubic—above 120°C. Below that temperature, called the *Curie point*, there is a slight but important shift in the ions. The central Ti^{4+} ion shifts about 0.06 Å with respect to the corner Ba^{2+} ions. The O^{2-} ions shift in the opposite direction, as indicated in Fig. 8–7.2.* The center of positive charge and the center of negative charge are separated by the dipole length d.

This type of material changes dimensions in an electric field because the dipole length can be changed by the field. This provides a means of changing electrical energy into mechanical energy and *vice versa.* To understand this, refer first to Fig. 8–7.3(a). The cooperative alignment of the internal dipoles in many unit cells builds up a negative charge on one end of the crystal and a positive charge on the other. The amount of charge on the electrodes depends on the *polarization* $\mathscr{P}$, which, in turn, is related to the dipole length d (Eq. 7–6.1). Now consider two alternatives shown in Fig. 8–7.3.

1. Compress (or pull) the crystal with a stress σ. The length of d is changed; therefore the charge difference between the two ends is decreased if there is a conductive path (Fig. 8–7.3(b)). If not, a voltage differential is established.

2. Alternatively, place a voltage across this material (Fig. 8–7.3(c)). The dipoles respond to the field and lengthen their dipole distance d. This changes the dimension of the crystal. We have a voltage-strain, or electromechanical *transducer*.

* The shift of the Ti^{4+} in Fig. 8–7.2 is upward as drawn. Actually, the shift could be in any one of the six coordinate directions. In any event, the O^{2-} ions shift in the opposite direction.

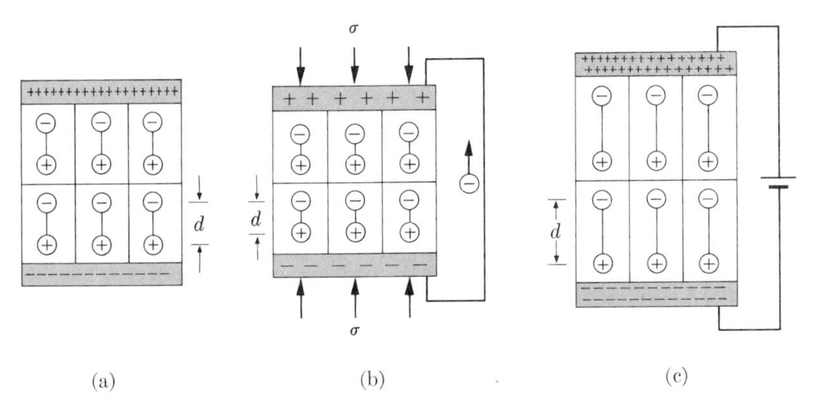

Fig. 8–7.3. Piezoelectric material. An electric field induces dimensional changes, shown in part (c). Conversely, strain from pressure induces a change in the charge density (b).

Materials with the above characteristics are called *piezoelectric*, i.e., pressure-electric. We use crystals of the $BaTiO_3$ type* for pressure gages, for phonograph cartridges, and for high-frequency sound generators.

Fig. 8–7.4. Piezoelectric crystals (quartz). These crystals are made artificially to be used for frequency control. As they vibrate elastically, the electric field responds at the same frequency, and vice versa. (Courtesy Western Electric Co.)

* The most widely used piezoelectric ceramics are $PbZrO_3$–$PbTiO_3$ solid solutions called PZT's. They have the structure of $BaTiO_3$ in Fig. 8–7.2. Their Curie points are higher than that of $BaTiO_3$.

Quartz crystals (SiO_2) are also piezoelectric. They are produced (Fig. 8–7.4) for special applications in circuits requiring frequency control. Once cut to a selected geometry, the elastic vibrational frequency is constant to one part in 10^8! Thus, in resonance, the crystal can control the frequency of an electronic circuit to that same accuracy. Fine watch circuits and control circuits for radio broadcasting are but two applications which utilize quartz crystals for their piezoelectric characteristics.

• **Ferroelectric ceramics.** The electric dipole in quartz is permanently oriented. One end of the crystal is always the positive end, and the other end of the crystal is always the negative end. In contrast, the $BaTiO_3$ of Fig. 8–7.2 is reversible. The Ti^{4+} ion can be made to shift downward and the O^{2-} ions to shift upwards to give a mirror image of the present arrangement, if an electric field is applied with the negative electrode below the unit cell (and the positive electrode above). This reversibility with the applied field is called ferroelectricity.* The electric field need not be an external field but may be a field from the dipoles in the next unit cells. Thus it is common to find adjacent unit cells spontaneously developing parallel electrical polarity. Furthermore, the polarity of a group of unit cells, called a *domain* (Fig. 8–7.5) can be maintained for a period of time, because external energy is required to invert the structure.

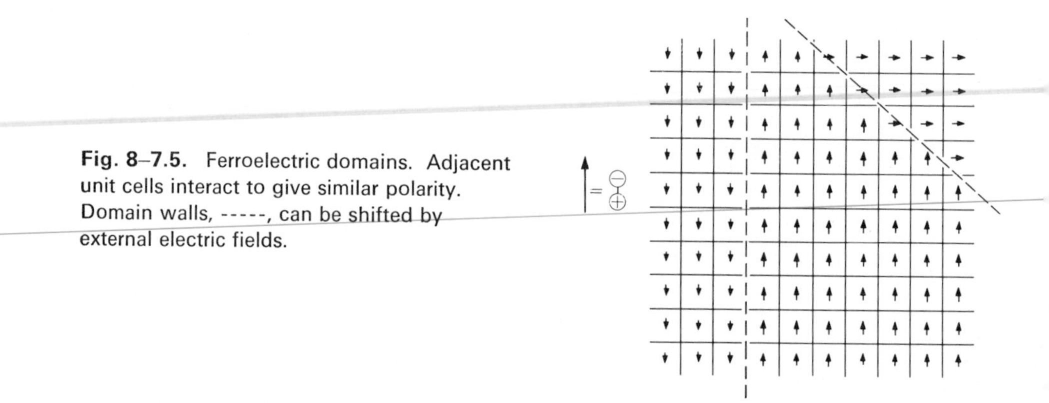

Fig. 8–7.5. Ferroelectric domains. Adjacent unit cells interact to give similar polarity. Domain walls, -----, can be shifted by external electric fields.

Consider a ferroelectric material containing many domains of many different alignments, so that none is generally preferred. If an external electric field is applied, the boundaries between the domains will move in such a way that the more favorably

* All ferroelectrics are piezoelectric; not all piezoelectrics are ferroelectric. Quartz, for example, will not reverse with the field. The name ferroelectric originated because the hysteresis loop, to be presented shortly (Fig. 8–7.6), "resembles" a ferromagnetic hysteresis loop, but originates from an electric field. Iron ions are not common among the ferroelectric materials.

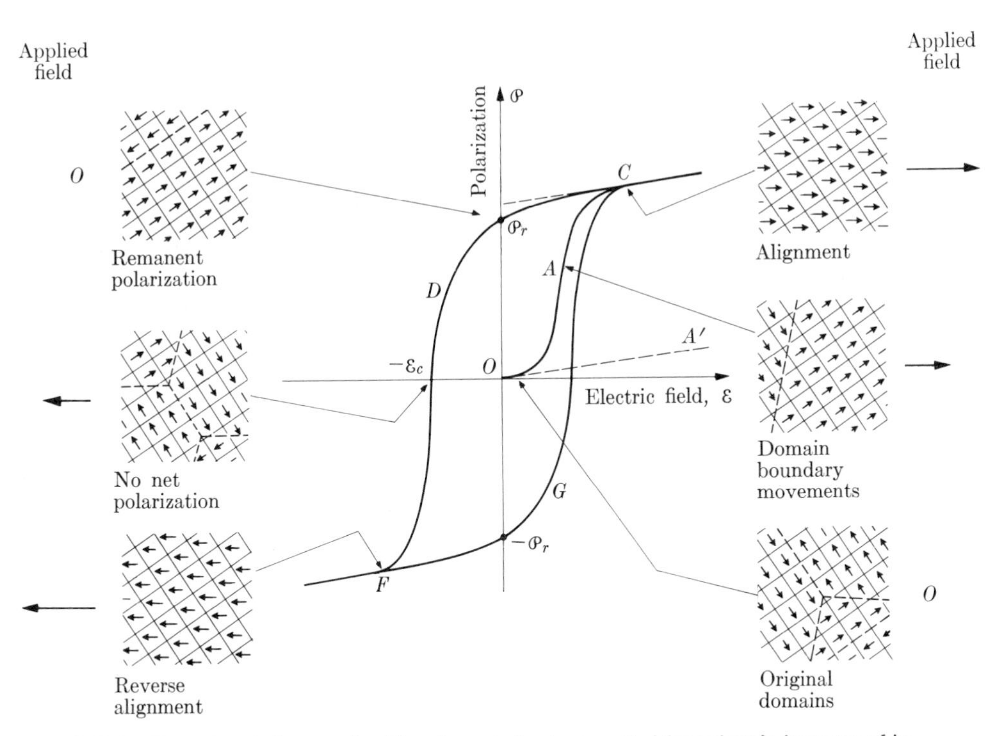

Fig. 8–7.6. Ferroelectric hysteresis. The above behavior is called ferroelectric because of its similarity to ferromagnetic behavior. However, this behavior arises from electric fields rather than magnetic fields. (See the text for an explanation. Also compare with Fig. 8–7.8.)

oriented domains expand and the less favorably oriented domains contract in volume. This gives a net polarization which increases rapidly, as shown in the *O*-to-*A* part of the curve in the $\mathcal{P}$–$\mathcal{E}$ diagram of Fig. 8–7.6. Eventually, alignment approaches a maximum and a further increase in the electric field makes only slight increases in preferential orientation, as in the *A*-to-*C* part of the curve in Fig. 8–7.6. A removal of that external electric field does not remove the polarization, so that some remanent polarization, $\mathcal{P}_r$, is maintained; not until a coercive field $\mathcal{E}_c$, of opposite polarity is applied does the material lose its net polarization. Cyclic fields produce a *hysteresis loop*, as indicated by the completed *CDFGC* path of the $\mathcal{P}$–$\mathcal{E}$ curves.

A *ferroelectric Curie point* exists at 120°C in $BaTiO_3$ because the spontaneous polarization of adjacent unit cells is lost when the crystal changes to a cubic structure (Fig. 8–4.1). Although new domains form spontaneously as $BaTiO_3$ is cooled below 120°C, these do not have any preferred alignment until a new external field is applied.

• **Magnetic ceramics.** Metallic magnets possess a major disadvantage in UHF and VHF circuits. The high frequency induces currents and therefore introduces power losses within the metallic cores. In fact the metallurgist capitalizes on this to melt high-quality alloys. Such power loss into heat cannot be tolerated in an electronic circuit.* Fortunately, ceramic compounds containing magnetic atoms can be magnetic, *and* nonconducting.

If we consider the individual iron atom (Eq. 2–1.1), we observe that there are six 3*d* electrons. The same is true for a ferrous (Fe^{2+}) ion (Eq. 2–1.2). A ferric (Fe^{3+}) ion has five 3*d* electrons (Eq. 2–1.3). These electrons have electron spin alignments as follows in their five 3*d* orbitals:

$$Fe^{2+}: \quad \underline{\uparrow\downarrow} \quad \underline{\uparrow} \quad \underline{\uparrow} \quad \underline{\uparrow} \quad \underline{\uparrow}; \tag{8–7.1}$$

$$Fe^{3+}: \quad \underline{\uparrow} \quad \underline{\uparrow} \quad \underline{\uparrow} \quad \underline{\uparrow} \quad \underline{\uparrow}. \tag{8–7.2}$$

This means that each Fe^{2+} has a magnetic imbalance of four unpaired electron spins (called four *Bohr magnetons*). Since the magnetic moment of each electron spin is 9.27×10^{-24} amp · m², each Fe^{2+} ion has a total magnetic moment of 37×10^{-24} amp · m². Each Fe^{3+} ion has an imbalance of five Bohr magnetons and magnetic moment of 46.3×10^{-24} amp · m².

Many ceramic magnetic phases fall in the $A_mB_nX_p$ category. Examples are $Ni_8Fe_{16}O_{32}$ and magnetite (Fe_3O_4, or $Fe_8^{2+}Fe_{16}^{3+}O_{32}$). Their unit cell is shown in Fig. 8–7.7. The structure is complex, but may be simplified by noting that the O^{2-} ions assume an fcc arrangement. As described in Fig. 8–4.2, the unit cell is double the dimensions of what we considered previously in Figs. 3–1.1 and 8–2.1(b). The iron cations reside in both four-fold and six-fold sites. Those in the four-fold sites (CN = 4) point their magnetic moment in one ⟨111⟩ direction; those in six-fold sites (CN = 6) point their magnetic moment in the *opposite* ⟨111⟩ direction. Detailed analyses reveal that the 16 Fe^{3+} ions per unit cell are almost equally divided between four-fold and six-fold sites. Therefore there is no *net* magnetic moment from the Fe^{3+} ions:

$$(8\ Fe^{3+}\ \text{ions})(5\uparrow\ \text{per ion}) + (8\ Fe^{3+}\ \text{ions})(5\downarrow\ \text{per ion}) = 0. \tag{8–7.3}$$

However, almost all of the Fe^{2+} ions reside in the six-fold sites. Since each of the 8 Fe^{2+} ions has four Bohr magnetons, β, the unit cell has a magnetic moment of 32β, or a magnetic moment of about 0.3×10^{-21} amp · m². Inasmuch as the unit cell has a lattice content of 8.37 Å, and *saturation magnetization* M_s is equal to the maximum magnetic moment per unit volume, we can calculate the saturation magnetization of Fe_3O_4 as:

$$M = (0.3 \times 10^{-21}\ \text{amp} \cdot \text{m}^2)/(8.37 \times 10^{-10}\ \text{m})^3 = 0.5 \times 10^6\ \text{amp/m}.$$

This compares with 0.53×10^6 amp/m by experiment.

---

* The electrical engineer overcomes this trouble in 60-Hz circuits by using laminates rather than solid cores in transformers, armatures, etc. This increases the effective resistance. In a high-frequency circuit, however, the laminates would have to be microscopically thin.

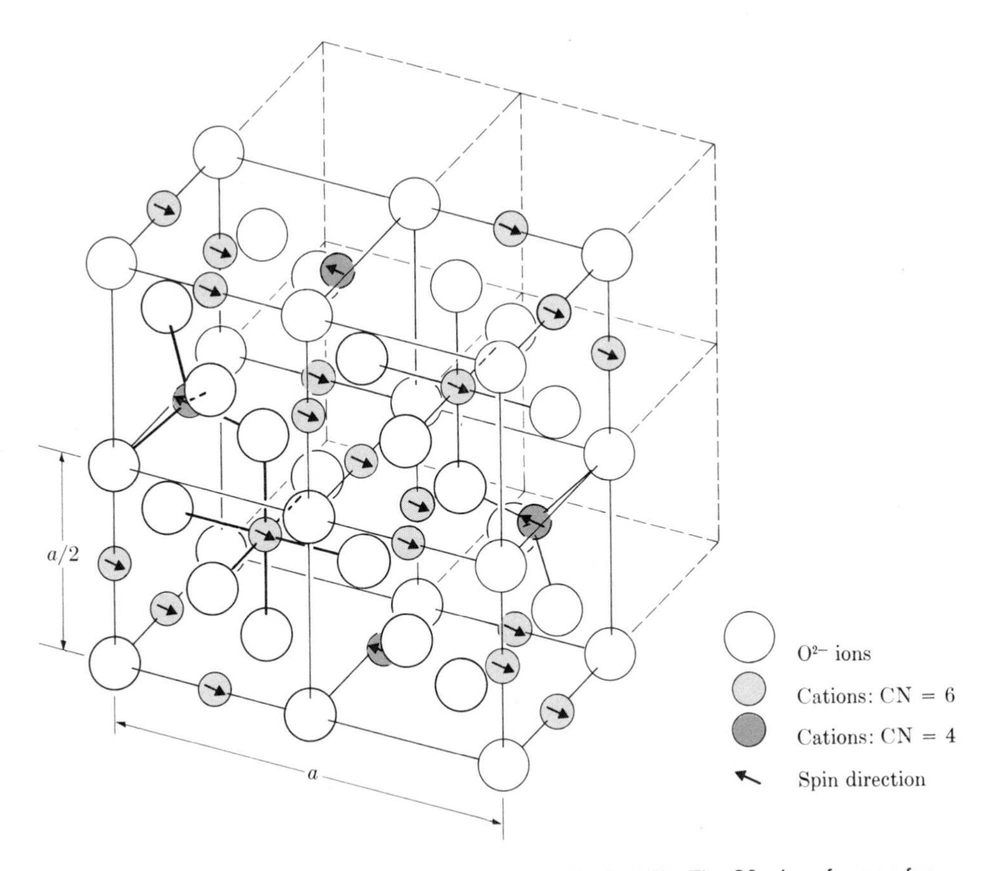

Fig. 8–7.7. Ferrospinel structure (more detail than in Fig. 8–4.2). The O^{2-} ions form an fcc lattice, with cations in the indicated interstitial sites. (Only the front half of the unit cell is sketched.) The cell dimensions are double those of the NaCl- and ZnS-type structures (Fig. 8–2.1), because adjacent octants are not identical. The magnetic moments of the two spin directions do not balance. Therefore each unit cell has a net magnetic moment. (See the text for elaboration.)

Since Fe_3O_4 is not a "hard" magnetic material, there is a deorientation when the external field is removed. However, domains remain (cf. Fig. 8–7.5). A *B–H* hysteresis loop forms as the magnetic field is cycled. This is similar in principle to the $\mathcal{P}$–$\mathcal{E}$ loop in Fig. 8–7.6.

The most common permanent ceramic magnets have a hexagonal variant in the crystal structure of Fig. 8–7.7. We will not consider these; however, we will observe in Fig. 8–7.8 that the *B–H* loop has a magnetic remanence B_r which remains when the field is removed. Further it requires a reversed field, called the coercive field H_c, to

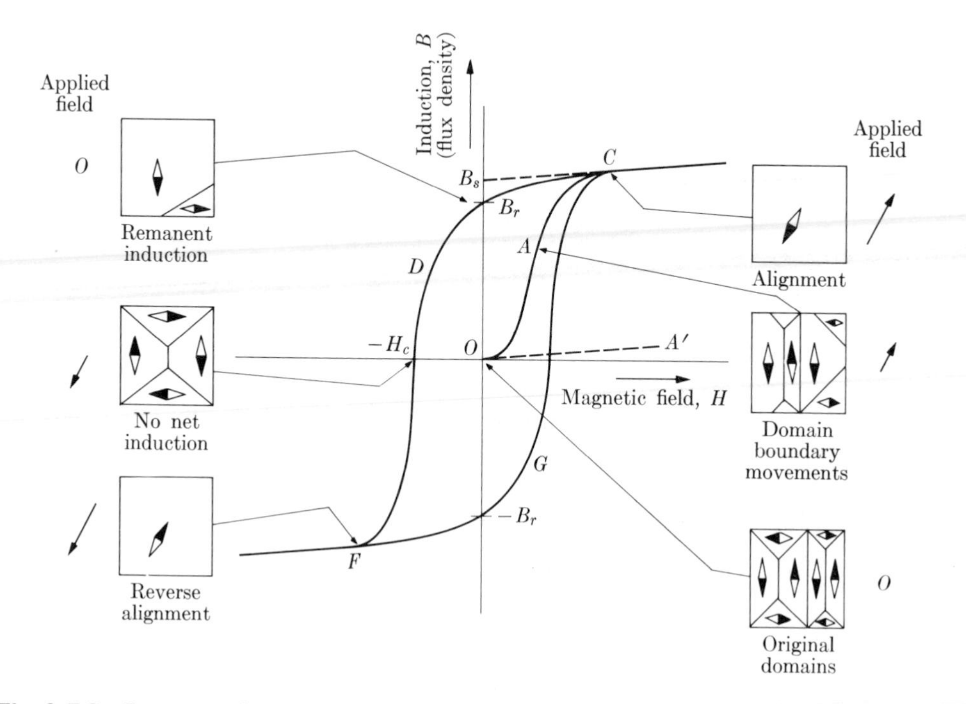

Fig. 8–7.8. Ferromagnetic hysteresis. Magnetic saturation, B_s, occurs when most of the domains are aligned in a common direction. When the magnetic field is removed, a magnetic remanence, B_r, remains. A hard magnet has a high remanence; a soft magnet has negligible remanence. A coercive force, $-H_c$, is required to eliminate the residual magnetization. (Cf. Fig. 8–7.6 for ferroelectric hysteresis.)

remove the initial magnetization. (Compare B_r with $\mathscr{P}_r$ and $-H_c$ with $-\mathscr{E}_c$ in Figs. 8–7.8 and 8–7.6.)

Example 8–7.1 The surface of silicon is oxidized to give a 25-μm silica (SiO_2) glass surface layer, and then coated with aluminum. How does the capacitance of this layer compare at 10^6 Hz with that which is produced when a 2-mil (50-μm) film of polyethylene is vapor-coated with aluminum?

Solution: Using Eq. (1–4.6), and data from Tables 7–6.1 and 8–7.1,

$$\frac{C_{PE}}{C_{SiO_2}} = \frac{2.3(8.85 \times 10^{-12}\ \mathrm{C/V \cdot m})A/(50 \times 10^{-6}\ \mathrm{m})}{3.8(8.85 \times 10^{-12}\ \mathrm{C/V \cdot m})A/(25 \times 10^{-6}\ \mathrm{m})}$$
$$= 0.3. \blacktriangleleft$$

Example 8–7.2 Calculate the polarization $\mathscr{P}$ of $BaTiO_3$ based on Fig. 8–7.2 and the information of Section 7–6.

Solution: With reference to the center of a cell cornered on the Ba^{2+} ions:

Ion		Q, coul	d, m	Qd, coul · m
Ba^{2+}		$+2(0.16 \times 10^{-18})$	0	0
Ti^{4+}		$+4($,, $)$	$+0.06(10^{-10})$	3.84×10^{-30}
$2\ O^{2-}$	(side of cell)	$-4($,, $)$	$-0.06(10^{-10})$	3.84 ,,
O^{2-}	(top and bottom)	$-2($,, $)$	$-0.08(10^{-10})$	2.56 ,,
				$\Sigma = 10.24 \times 10^{-30}$

From Eq. (7–6.1),

$$\begin{aligned}\mathscr{P} &= \sum Qd/V \\ &= (10.24 \times 10^{-30}\ \mathrm{C \cdot m})/(4.03 \times 3.98^2 \times 10^{-30}\ \mathrm{m^3}) \\ &= 0.16\ \mathrm{coul/m^2}. \blacktriangleleft\end{aligned}$$

• **Example 8–7.3** By substituting $(4Li^{+} + 4Fe^{3+})$ for $8Fe^{2+}$ magnetite is altered from (a) $Fe_8^{2+}Fe_{16}^{3+}O_{32}$ to (b) $Li_4^{+}Fe_{20}^{3+}O_{32}$. Assume that there is negligible change in unit-cell size and that the Li^{+} ions enter the four-fold sites and six-fold sites equally. What is the percent change in saturation magnetization?

Solution:

	6-f↑	4-f↓	
Mag:	$8Fe^{3+} + 8Fe^{2+}$	$8Fe^{3+}$	$+32\ O^{2-}$
LiMag:	$14Fe^{3+} + 2Li^{+}$	$6Fe^{3+} + 2Li^{+}$	$+32\ O^{2-}$

(O^{2-} and Li^{+} are nonmagnetic; Fe^{2+} has 4β/ion; Fe^{3+} has 5β/ion)

$$\begin{aligned}\beta_{\mathrm{Mag}} &= (8)(5\uparrow) + (8)(4\uparrow) + (8)(5\downarrow) = 32\ \beta/\text{unit cell}; \\ \beta_{\mathrm{LiMag}} &= (14)(5\uparrow) + (6)(5\downarrow) = 40\ \beta/\text{unit cell}. \\ \Delta M_s &= [(40-32)(9.27 \times 10^{-24})/V]/[32(9.27 \times 10^{-24})/V] = +25\% \blacktriangleleft\end{aligned}$$

8–8 MECHANICAL BEHAVIOR OF CERAMICS

With the exception of a few materials such as plasticized clay, ceramic materials are characterized by high shear strengths; thus, they are not ductile. We will see that this leads (1) to high hardnesses and compressive strengths and (2) to notch sensitivity and low fracture strengths.

All compounds have ordered arrangements of dissimilar atoms (or ions). This was discussed for intermetallic compounds as an optional topic in Section 6–4. In ceramic phases, cations are coordinated with anions; and anions are coordinated with cations (Fig. 8–2.1). This is natural, since unlike ions attract and like ions repel. However this interferes with slip, as shown in Fig. 8–8.1. The slip process in the horizontal direction (as drawn) is precluded in NiO, MgO, and other NaCl-type crystals. The displaced arrangement would have to be achieved by first breaking strong positive-to-negative bonds of the Ni^{2+} and O^{2-} ions. Next, like charges would have to pass adjacently as the Ni^{2+} ion moves to join the next O^{2-} ions. This consideration does not exist in the metal.

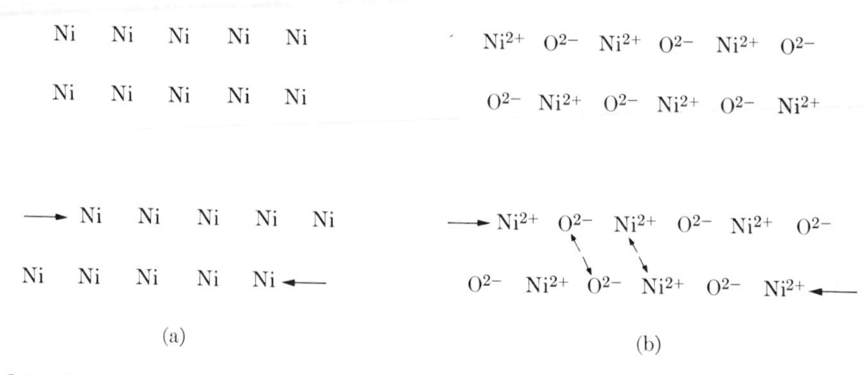

Fig. 8–8.1. Comparison of slip processes (metallic nickel and nickel oxide). More force is required to displace the ions in NiO than the atoms in nickel. The strong repulsive forces between like ions becomes significant. Nickel also has more slip systems than does nickel oxide.

In NiO, slip is possible in certain other directions without the above restriction, e.g., at a 45° direction. This would be slip in a $\langle 1\bar{1}0 \rangle$ direction on a $\{110\}$ plane. There are *six* combinations possible. This compares with twelve combinations for $\{111\}\langle 1\bar{1}0 \rangle$ slip in fcc nickel (Table 6–4.1). Furthermore, the displacement distance **b** in NiO is $(r + R)\sqrt{2}$, or 3 Å, which is greater than the 2.5 Å in nickel. Everything considered—(a) coulombic repulsion, (2) fewer slip systems, and (3) longer displacement distances*—adds up to a much greater resistance to plastic deformation in crystalline ceramics than in their metallic counterparts. This difference is accentuated even further when more complex ceramic phases are considered, e.g., silicates, the spinel of Fig. 8–7.6, or portland cement.†

Hardness. Ceramic phases are hard because they generally cannot undergo plastic slip. We see commercial utilization of hard ceramics in abrasive materials. Emery, a very widely used ceramic abrasive, is Al_2O_3 (usually with some Fe_2O_3 and/or TiO_2

---

* The energy for dislocation movements increases with the square of the displacement distance, **b** (Eq. 6–4.2).

† Some chemists consider the chief hydrated phase to be

$$Ca_2[SiO_2(OH)_2]_2 \cdot (CaO)_{y-1} \cdot xH_2O!!$$

in solid solution). Silicon carbide (SiC) and TiC are equally important for grinding and cutting metals, and similar manufacturing processes. They have extensive covalent bonding but are still considered to be ceramic products. One form of SiC has the ZnS-type structure (Fig. 8–2.1c); TiC has the NaCl-type structure (Fig. 8–2.1(b)).

Exceptions to the general rule that ceramics are relatively hard can be related directly to structures. Talc, clays, and mica are soft because they have sheet structures (Fig. 8–5.6). They are strong within the sheets. However, sheets are held together by weak secondary bonds (Section 2–4).

Notch sensitivity. A notch or crack is a *stress-raiser*. The effect is much greater than would be anticipated simply on the basis of the decreased cross-sectional area. The stress, σ_c, at the tip of the notch or flaw (Fig. 8–8.2) is related to the nominal tensile stress, σ_n, the depth of the crack, c, and the radius of curvature, r, at the tip of the notch:

$$\sigma_c = \sigma_n 2\sqrt{c/r}. \tag{8–8.1}$$

If σ_c exceeds the yield strength in a *ductile* material, the tip of the notch will deform to a larger radius, and the stress concentration will be reduced. If the notch is a crack in a *nonductile* material, the radius of curvature may be of atomic dimensions (say 1 Å). Thus a crack which reaches only 0.1 to 10 μm (10^{-5} to 10^{-3} cm) into the surface will give a *stress concentration factor*, σ_c/σ_n, of 10^2 or 10^3. Deeper cracks are more severe. Even though strong in shear, we find stress concentrations σ_c which exceed the bond strength between the atoms. Thus the crack may propagate; this accentuates the stress concentration of Eq. (8–8.1) still further, until catastrophic failure occurs.

Ceramic materials (and intermetallic compounds, Section 6–4) are generally weak in tension because of their nonductility just described. The same factor makes them strong in compression. A compressive load can be supported across a microcrack without extending the flaw.

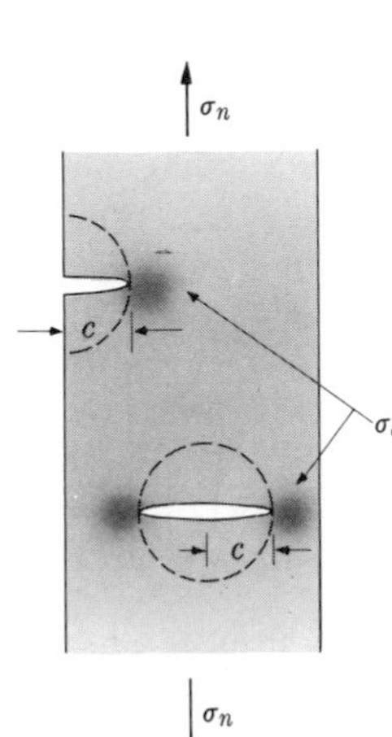

Fig. 8–8.2. Stress concentration. The concentration factor depends upon the crack depth and the radius of curvature at the tip of the crack (Eq. 8–8.1).

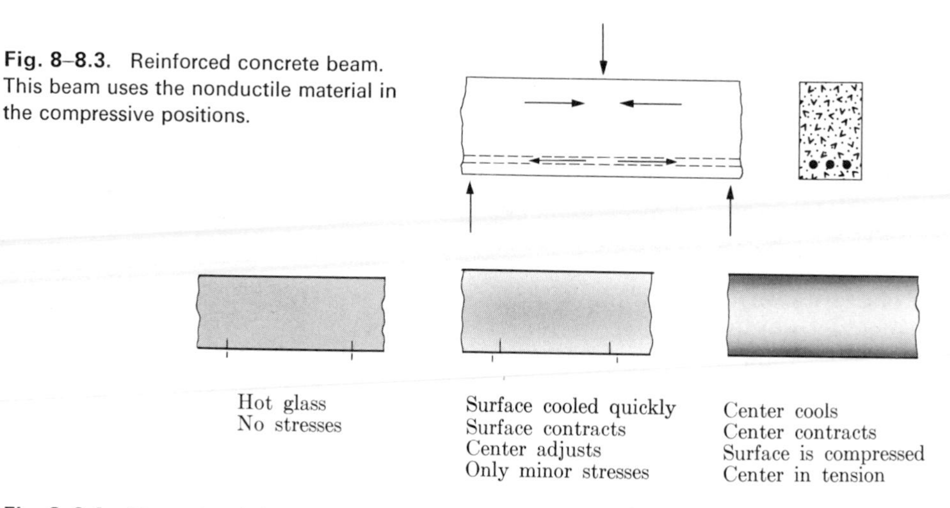

Fig. 8–8.3. Reinforced concrete beam. This beam uses the nonductile material in the compressive positions.

Fig. 8–8.4. Dimensional changes in "tempered" glass.

Use of nonductile materials. This relationship of high compressive strength and lower tensile strength is important to the design engineer. Concrete, brick, and other ceramics are used primarily in compressive locations (Fig. 8–8.3). When it is necessary to subject such materials as glass to bending (and therefore tensile loading), it is usually necessary to increase some dimensions. For example, the viewing glass of a television picture tube may be as much as 1.5 cm thick.

Since ceramic materials are stronger in compression than in tension, "tempered" glass is used for glass doors, rear windows of cars, and similar high-strength applications. To produce tempered glass, the glass plate is heated to a temperature high enough to permit adjustments to stresses among the atoms, and is then quickly cooled by an air blast or oil quench (Fig. 8–8.4). The surface contracts because of the drop in temperature and becomes rigid, while the center is still hot and can adjust its dimensions to the surface contractions. When the center cools and contracts slightly later, compressive stresses are produced at the surface (and tensile stresses in the center). The stresses which remain in the cross section of the glass are diagrammed in Fig. 8–8.5. A considerable load must be applied to the glass before tensile stresses can be developed in the surface of the glass where cracks start.* In effect, since the compressive stresses must be overcome first, the overall strength of the glass is greatly enhanced.

• **Viscous deformation of glass.** Ceramic glasses are generally siliceous materials and have many of the characteristics of thermoplastic polymers. In fact, the initial recognition of the *glass temperature* T_g was for siliceous glasses (Fig. 4–5.3). The

* If a crack penetrates through the compressive skin (e.g., by scratching) into the tension zone shown in Fig. 8–8.5, the crack may become rapidly self-propagating. The aftermath of this effect can be observed in a broken rear window of a car.

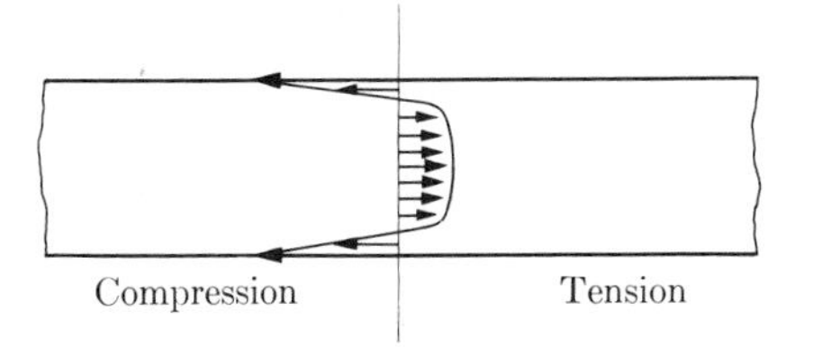

Fig. 8–8.5. Surface compression of "tempered" glass. These compressive stresses must be overcome before the surface can be broken in tension.

viscosity η of a glass is temperature-sensitive:

$$\log_{10} \eta = C + \frac{B}{T}, \tag{8–8.2}$$

where C and B are constants* for a glass.

In manufacturing, close attention is paid to viscosity. First, the glass must have a low viscosity (<300 poises) in the *melting range* so that bubbles can escape and homogenization can occur. The *working range* covers several orders of magnitude. Fast operations, such as the production of light bulbs, require a rather fluid glass ($\sim 10^{3.5}$ poises); the drawing operation on heavy glass sheet requires a more viscous product ($\sim 10^7$ poises).

Two other viscosity temperature points are important: (1) the *annealing point*, with $\eta = 10^{13}$ poises, and (2) the *strain point*, with $\eta = 10^{14.5}$ poises. At the temperature of the annealing point, atoms can move sufficiently so that residual thermal stresses may be relieved within about 15 minutes, a time compatible with production schedules. Below the strain point, the glass is sufficiently rigid so that it may be handled without the generation of new residual stresses.

Example 8–8.1 Glass has a theoretical strength in excess of 7000 MPa (10^6 psi). Strengths approaching this value have been reached for pristine glass. A piece of plate glass fails with a bending tension of 60 MPa (8,600 psi). What crack depth would be responsible for this low-stress fracture, if we assume the tip of the crack has the dimension of an oxygen ion?

Solution: From Eq. (8–8.1), and assuming $r = 1.4$ Å,

$$c = \frac{r(\sigma_c/\sigma_n)^2}{4} = \frac{(1.4 \text{ Å})(7000/60)^2}{4} = 4800 \text{ Å}.$$

Comments. Cracks of this size are observed in the surfaces of glass which is exposed to normal environments; however, special techniques are required to detect them, since that dimension is near the limit of resolution by visible light.

Tempered plate glass can support an added load, since the compressive stress of the skin (Fig. 8–8.5) would have to be exceeded before stresses start to concentrate from tension. ◀

---

* The flow of glass requires the movement of atoms past one another under stress. A glass with a high fluidity f (low viscosity, since $\eta = 1/f$) corresponds to a glass which has fast diffusion. Thus, as in Eqs. (4–6.6) and (4–7.4),

$$\log_{10} f = \log f_0 - \frac{Q}{4.575T} = -\log_{10} \eta. \tag{8–8.3}$$

This leads to Eq. (8–8.2), where $C = -\log f_0$, and $B = Q/4.575$. Viscosity is expressed here as poises (1 poise $= 0.1$ Pa · sec).

• **Example 8–8.2** An electrical glass has a working range of 870°C ($\eta = 10^7$ poises) to 1300°C ($\eta = 10^{3.5}$ poises). Estimate its annealing point ($\eta = 10^{13}$ poises).

Solution: From Eq. (8–8.2),

$$7 = C + B/(870 + 273),$$
$$3.5 = C + B/(1300 + 273).$$

Solving simultaneously,

$$B = 14{,}600 \qquad \text{and} \qquad C = -5.8$$

For $\eta = 10^{13}$ poises,

$$13 = -5.8 + 14{,}600/T,$$
$$T = 775°\text{K} \quad (\simeq 500°\text{C}).$$

Comment. The annealing point corresponds approximately to the glass temperature, T_g. ◀

REVIEW AND STUDY

Since ceramic phases are compounds of metallic and nonmetallic elements, we find that it is not as easy to generalize about structure and properties as for metals or nonmetals alone. Metals, for example, are always electrical and thermal conductors; most ceramics are insulators but some possess semiconductivity that has technical uses. Organic polymers always transmit light in thin sections; ceramics may have the transparency of optical glass or may be opaque in magnetic spinels. Structural ceramics are very strong in compression, but must be considered weak in tension. Fiberglass may have a tensile strength greater than steel and therefore can be used as a reinforcement; glass is also recognized as a brittle, friable material, so that special care must be taken when it is shipped or handled.

The examples above are just a few of those we could cite to indicate that ceramic materials have a variety of characteristics. Many of these are indispensable for current-day technology and societal needs. However, they are more complex than other materials and therefore require greater technical familiarity and understanding.

SUMMARY: STRUCTURAL

The simpler ceramic compounds may be viewed in terms of a basic lattice of anions with cations in selected interstitial sites. We used CsCl-, NaCl-, and ZnS-type structures as our prototypes. Cations reside in eight-fold, six-fold, and four-fold interstitial sites, respectively. Table 8–2.1 provides a review summary of these structural types. More complex structures such as the one possessed by magnetic spinels can be analyzed by observing how the cations are ordered among the four-fold and six-fold sites.

Silicates involve more covalent bonding than do the majority of other ceramic compounds. Therefore we find useful analogies with the polymer structures studied in Chapter 7. The SiO_4 tetrahedron is polyfunctional. Depending upon the relative amount of silica present, we may observe bifunctional SiO_4 units in chain structures, trifunctional bonding in sheet structures (clays, talc, mica, etc.), and tetrafunctional bonding in network silicates (quartz, feldspar, etc.). Commercial glasses are amorphous network silicates. Additions of network modifiers reduce the number of bridging oxygens.

SUMMARY: PROPERTIES AND BEHAVIOR

Ceramic insulators are designed to be electrically inert. The constituent ions, however, can respond to electric fields. This provides a polarization inside the dielectric. The resulting dielectric constant is used advantageously in capacitors; but it produces power losses if the ion displacements lag the applied field. Permanent polarization produces piezoelectricity and leads to electromechanical transducers. Reversible polarization produces ferroelectric effects. Semiconducting ceramics owe their existence to atoms with multiple valence states. Magnetic ceramics require ions with unfilled subvalence shells. Each atom, therefore, acts as a small magnet; these couple together to give magnetic domains, which respond to magnetic fields.

Compounds, be they ceramic or intermetallic, strongly resist shear deformation. As a result they are notch- (or crack-)sensitive. This produces low tensile strengths. However, the resistance to shear produces high compressive strengths. Ceramic materials are thus selected for use under compression in structure and product designs. Also, ceramic materials are engineered to possess surface compression. The viscosity of glass follows the same laws as diffusion but with an applied force.

TERMS FOR REVIEW

Abrasives

$A_mB_nX_p$ compounds

• Annealing point

AX structures

  CsCl-type
  NaCl-type
  ZnS-type

A_mX_p structures

• Bohr magneton, β

Bridging oxygen

• Coercive field

  electric, $\mathscr{E}_c$
  magnetic, H_c

• Curie point

Dielectric strength

• Domain

• Ferroelectric

Glass

Glass temperature, T_g

• Hysteresis loop

  ferroelectric
  magnetic

Insulator

Interstitial sites

  four-, six-, eight-fold

Network modifier

Nonbridging oxygen

Piezoelectric

Polarization, $\mathscr{P}$

Relative dielectric constant, κ

• Remanence

  ferroelectric, $\mathscr{P}_r$
  magnetic, B_r

TERMS FOR REVIEW—CONTINUED

- •Silicate structures
  - chain
  - network
  - sheet
- •SiO_4 tetrahedra
- •Spinel
- •Strain point
- •Stress concentration factor, σ_c/σ_n
- Tempered glass
- Thermistors
- Transducer
- •Viscosity, η

FOR DISCUSSION

A_8 Cite examples of (a) intermetallic compounds, (b) organic compounds, and (c) ceramics.

B_8 Distinguish between CsCl-type, NaCl-type, and ZnS-type structures with regard to coordination number, minimum ion-size ratio, and the number of ions per unit cell.

C_8 Make a sketch of a hexagonal array of atom nuclei. Now dot in a second layer which could lie in close packing above the first layer. Indicate the locations of six-fold sites between the two layers; of four-fold sites.

D_8 Sketch the atom arrangements on the $(1\bar{1}0)$ plane of ZnS. (Cf. Fig. 8–3.2 for CaF_2.) Repeat for the (110) plane.

E_8 Refer to cubic $BaTiO_3$ (Fig. 8–4.1). Which cubic lattice of Fig. 3–2.2 does it possess?

F_8 Figure 4–3.1 with Mg^{2+}, Fe^{2+}, and O^{2-} ions, and Fig. 8–4.1 with Ba^{2+}, Ti^{4+}, and O^{2-} ions, both have three types of ions. We call one a solid solution and the other an $A_mB_nX_p$ compound. Why the difference?

G_8 Both MnO and MnS have NaCl-type structures. A limited amount ($\sim$1 a/o) of solid solution occurs. Suggest the mechanism. Why is the extent of solid solution limited?

•H_8 Use electron dots (Cf. Fig. 2–2.3(a)) to show covalent bonding in an SiO_4^{4-} ion; to show ionic bonding. Repeat for $Si_2O_7^{6-}$ of Fig. 8–5.2.

•I_8 What is the Si/O ratio in a chain silicate? In a sheet silicate? In a network silicate?

•J_8 A common silicone is a condensation polymer of silicic acid, $Si(OH)_4$. By a sketch show how this can form a silicone.

•K_8 Compare and contrast the structure of cristobalite (Fig. 8–5.7) with that of ZnS (Fig. 8–2.1(c)), and with silicon, which has the structure of diamond (Fig. 5–4.3).

•L_8 Cristobalite (Fig. 8–5.7) has almost no expansion when it melts to fused silicon. Suggest why.

M_8 Distinguish between bridging oxygens and nonbridging oxygens, and their effect on glass deformation above the glass temperature.

N_8 Boric oxide (B_2O_3) is a good glass-former: Compare its structure and viscous behavior with fused silica (SiO_2).

O_8 a) Discuss two ways in which cracks affect an electrical insulator. b) Why are insulators glazed before use on high-voltage lines?

P_8 Distinguish between dielectric constant and dielectric strength.

•Q_8 Why are ceramic semiconductors composed of oxides of transition metals?

R_8 Explain to someone who is not taking this course how a ceramic cartridge on a phonograph provides a signal which can then be amplified.

S_8 Suggest a concept for a piezoelectric "spark plug."

•T_8 Suggest a concept for a "memory unit" to be made from a thin ferroelectric sheet.

•U_8 From Appendix B indicate the number of Bohr magnetons (unpaired $3d$ electrons per atom) possessed by the elements with atomic numbers between 20 and 30.

•V_8 Normally we do not observe magnetism in MnO; yet each cation possesses five Bohr magnetons, and experiments show that MnO responds to changes in a magnetic field. Suggest a reason that we do not observe magnetization. (MnO has an NaCl-type structure.)

W_8 Why do tensile test data for ceramics show more scatter than for metals?

X_8 Glass can be treated to replace Na^+ ions in the surface zone with K^+ ions. This "tempers" the glass. Explain.

Y_8 A ceramic (Al_2O_3) cutting tool has to be mounted in a lathe differently from a tool made of "high-speed" steel. Why?

Z_8 A glaze for dinnerware is chosen so that it has a slightly lower thermal expansion coefficient than the underlying porcelain. Explain why this makes the product stronger.

STUDY PROBLEMS

8–2.1 Periclase (MgO) has an fcc structure of O^{2-} ions with Mg^{2+} ions in all the six-fold sites. a) The radii are 1.40 Å and 0.66 Å, respectively; what is the atomic packing factor? b) What would this factor be if $r/R = 0.414$?

Answer: a) 0.73 b) 0.79

8–2.2 CsCl has a simple cubic structure of Cl^- ions with Cs^+ ions in all the eight-fold sites. a) The radii are 1.87 Å and 1.72 Å, respectively, for CN = 8; what is the atomic packing factor? b) What would this factor be if $r/R = 0.732$?

8–2.3 With the data in the previous problem and in the appendix, calculate the density of CsCl.

Answer: 3.94 g/cm^3 (vs. 3.97 g/cm^3 by experiment)

8–2.4 X-ray data show that the unit-cell dimensions of cubic MgO are 4.2 Å. It has a density of 3.6 g/cm^3. How many Mg^{2+} ions and O^{2-} ions are there per unit cell?

8–2.5 Estimate the densities of the two polymorphs of MnS in Example 8–2.2.

Answer: a) 3.9 g/cm^3; b) 3.4 g/cm^3

8–2.6 It is hypothesized that at high pressures, NaCl can be forced into a CsCl-type structure. What would be the percent volume change? (See the footnote of Appendix B for differences in radii with CN = 6 and CN = 8.)

8–3.1 With CN = 4, $R_{O^{2-}} = 1.28$ Å (Table 2–6.1). With CN = 8, $r_{Zr^{4+}} \simeq 0.85$ Å. Estimate the size of the unit cell of cubic ZrO_2, which has the CaF_2-type structure.

Answer: 4.9 Å (vs. 5.07 Å by experiment)

8–3.2 Can MgF_2 have the same structure as CaF_2? Explain.

8–3.3 The uranium oxide used as a nuclear fuel material has the structure of CaF_2. As expected, there are 8 O^{2-} ions per unit cell; however, the average uranium-ion valence is greater than 4+; therefore, there are, on the average only 3.5 metal ions per unit cell. The density of the oxide is 10.9 g/cm^3. a) What is the average distance $D_{U\text{–}O}$ between the centers of adjacent uranium and oxygen ions? b) Estimate the radius of the uranium ion.

Answer: a) 2.29 Å; b) 1.0 Å

8–4.1 The density of cubic $PbTiO_3$ is 7.5 g/cm^3. What is the lattice dimension of the unit cell?

Answer: 4.06 Å (vs. 3.97 Å by experiment)

8–4.2 Calculate the oxygen content (a) of Fe_3O_4; (b) of $NiFe_2O_4$.

8–4.3 A cubic form of ZrO_2 is possible when one Ca^{2+} ion is added in a solid solution for every six Zr^{4+} ions present. Thus the cations form an fcc structure, and O^{2-} ions are located in the four-fold sites. a) How many O^{2-} ions are there for every 100 cations? b) What fraction of the four-fold sites is occupied?

Answer: a) 185.7 O^{2-} ions; b) 92.9%

8–4.4 a) What type of vacancies, anion or cation, must be introduced with MgF_2, in order for it to dissolve in LiF? b) What type must be introduced with LiF, for it to dissolve in MgF_2?

8–4.5 A solid solution contains 30 mole percent MgO and 70 mole percent LiF. a) What are the weight percents of Li^+, Mg^{2+}, F^-, O^{2-}? b) What is the density?

Answer: a) Li^+, 16 w/o; Mg^{2+}, 24 w/o; F^-, 44 w/o; O^{2-}, 16 w/o; b) 3.0 g/cm^3

•**8–5.1** The average O—Si—O angle in silicon is 109.5°. a) What is the closest oxygen-to-oxygen distance? b) What is the closest silicon-to-silicon distance in Fig. 8–5.7? c) What is the size of the unit cell in Fig. 8–5.7. (With CN = 2, $R_O \approx 1.14$ Å. Determine center-to-center distances.)

Answer: a) 2.5 Å; b) 3 Å; c) 7 Å

•**8–5.2** What is the density of silica in the previous problem?

•**8–5.3** Coesite, the high-pressure polymorph of SiO_2, has a density of 2.9 g/cm^3. What is the ionic packing factor?

Answer: 0.36

•8–5.4 $AlPO_4$ can form the same crystal structure as SiO_2. Explain.

8–6.1 A glass contains 75 w/o SiO_2 and 25 w/o CaO. What fraction of the oxygens serve as bridges between pairs of silicon atoms?

Answer: 70 a/o

8–7.1 Refer to Example 8–7.2. What is the distance between the centers of positive and negative charges of each unit cell?

Answer: 0.107 Å

8–7.2 Refer to Example 8–7.2. A one-centimeter cube of $BaTiO_3$ is compressed 1%. The two ends receiving pressure are connected electrically. How many electrons travel from the negative end to the positive end?

8–7.3 A piezoelectric crystal has a Young's modulus of 80,000 MPa (11,600,000 psi). What stress must be applied to reduce its polarization from 80×10^{-3} to 78.8×10^{-3} coul · cm/cm^3?

Answer: 1200 MPa (175,000 psi)

8–7.4 Lithium niobate ($LiNbO_3$) has the same structure as $BaTiO_3$ but with Li^+ replacing Ba^{2+}, and Nb^{5+} replacing Ti^{4+}. Assume the dimensions of the $LiNbO_3$ unit cell are approximately the same as described for $BaTiO_3$ in Fig. 8–7.2. a) What is the distance between the centers of positive and negative charge in the unit cell? b) What is the dipole moment, $\mu = Qd$, of the unit cell? c) What polarization is possible?

•8–7.5 A ceramic magnetic material *nickel ferrite* has eight [$NiFe_2O_4$]'s per unit cell, which is cubic, with $a = 8.34$ Å. a) Assume all the unit cells have the same magnetic orientation; what is the saturation magnetization? b) Assume the N–S polarity is 180° reversed in 40% of the unit cells when the external field is removed. How much magnetization remains?

Answer: a) 250,000 amp/m; b) 50,000 amp/m

•8–7.6 Refer to Example 8–4.1. In this particular ferrospinel, the Zn^{2+} ions preferentially choose four-fold sites and force a corresponding number of Fe^{3+} ions into six-fold sites. a) How many Bohr magnetons does this produce per unit cell of $(Zn_3, Ni_5)Fe_{16}O_{32}$? b) How many are there in $[NiFe_2O_4]_8$ where all of the divalent ions are in six-fold sites?

8–8.1 A 1-mm round rod with the composition of plate glass (Appendix C) is coated with 0.1 mm of borosilicate glass, so that the rod is now 1.2 mm in diameter. Assuming no initial stresses at 200°C, what surface stress is developed when the composite rod is cooled to 20°C?

Answer: $\sigma_{boro} = -8000$ psi (55 MPa), $\sigma_{pl} = +3500$ psi (25 MPa)

•8–8.2 A large glass plate contains a hole 1 cm in diameter. a) What is the stress-concentration factor adjacent to the hole from a tensile stress which is applied parallel to the plate? b) When the hole was drilled, the side developed 10-μm cracks. What total stress concentration results if we assume, as we did in Example 8–8.1, that this type of crack has the dimensions of an oxygen ion?

•8–8.3 Assume that the flow rate of a glass is inversely proportional to the viscosity η. How much faster will a molten electrical glass (Example 8–8.2) flow from a furnace at 920°C than at 900°C?

Answer: $\eta_{920}/\eta_{900} = 0.618$; $\dot{F}_{920} = 1.6\, \dot{F}_{900}$

CHAPTER NINE

MULTIPHASE MATERIALS: EQUILIBRIA

PREVIEW

We now shift from single-phase materials to materials with two or more phases. Our prime attention in Chapter 9 will be directed at phase diagrams, because they give us the basis for Chapters 10 and 11.

We can learn to use the phase diagrams (1) to predict *what phases* are in equilibrium for selected alloy compositions at desired temperatures; (2) to determine the *chemical composition* of each phase; and (3) to calculate the *quantity* of each phase which is present. Phase diagrams are powerful tools in the hands of scientists and engineers who design materials for specific applications, as well as others who must anticipate the stability of specific materials when designing products for service environments.

Throughout this chapter, we shall assume that equilibrium is attained; i.e., there is no continuing reaction between the phases that are present. In Chapter 10, we shall look at the microstructures that phases develop within a material, and some of the resulting properties. In Chapter 11, we shall see how we can manipulate these equilibria and microstructures to our advantage.

9–1 QUALITATIVE PHASE RELATIONSHIPS

The three preceding chapters have been concerned successively with metallic, organic, and ceramic phases, and with the dependence of their properties on phase structure. In each chapter only single-phase materials were considered. However, although many useful engineering materials consist predominantly of one phase,* a greater number are *mixtures of phases*; for example, steel, solder, portland cement, grinding wheels, paints, and glass-reinforced plastics. The mixture of two or more phases in one material permits interaction among the phases, and the resulting properties are usually different from the properties of individual phases. It is frequently possible, also, to modify these properties by changing either the shape or the distribution of the phases (see Chapter 10).

Solutions versus mixtures. Different *components* can be combined into a single material by means of *solutions* or of *mixtures*.† Solid solutions have already been discussed in Sections 4–2ff, 6–4, and 8–4, and we are all familiar with liquid solutions. The composition of solutions can vary widely, because (1) one atom may be substituted for another at lattice points of the phase structure, or (2) atoms may be placed in the interstices of the structure. The solute does not change the structural pattern of the solvent. A mixture, on the other hand, contains more than one phase (structural pattern). Sand and water, rubber with a carbon filler, and tungsten carbide with a cobalt binder are examples of mixtures. In each of these aggregates there are two different phases, each with its own atomic arrangement.

It is, of course, possible to have a mixture of two different solutions. For example, in a lead–tin solder, one phase is a solid solution in which tin has replaced some of the lead in the fcc structure, and the other phase has the structure of tin (body-centered tetragonal). At elevated temperatures, lead atoms may replace a limited number of atoms in the tin structure. Thus an ordinary 60–40 solder (60% Sn–40% Pb) contains two structures, each a solid solution.

Liquid solubility limits. Figure 9–1.1 shows the *solubility limit* of ordinary sugar in water; the curve is a *solubility curve*. All compositions shown to the left of the curve will form only one phase, because all the sugar is dissolved in the liquid phase. With the higher percentages of sugar shown to the right of the curve, however, it is impossible to dissolve the sugar completely, with the result that we have a mixture of two phases, solid sugar and liquid "syrup." This example shows the change of solubility with temperature, and also demonstrates a simple method for plotting temperature (or any other variable) as a function of composition. From left to right, the abscissa of Fig. 9–1.1 indicates the percentage of sugar. The percentage of water may be read directly from right to left, since the total of the components must, of course, equal 100%.

---

* See the end of Section 4–5 for our definition of phases.

† A *solution* is a phase with more than one component; a *mixture* is a material with more than one phase.

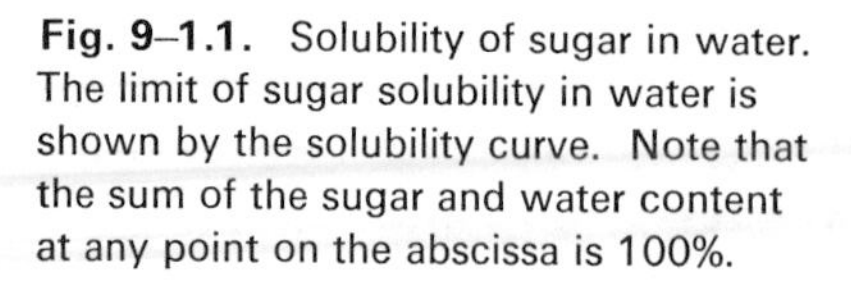
Fig. 9–1.1. Solubility of sugar in water. The limit of sugar solubility in water is shown by the solubility curve. Note that the sum of the sugar and water content at any point on the abscissa is 100%.

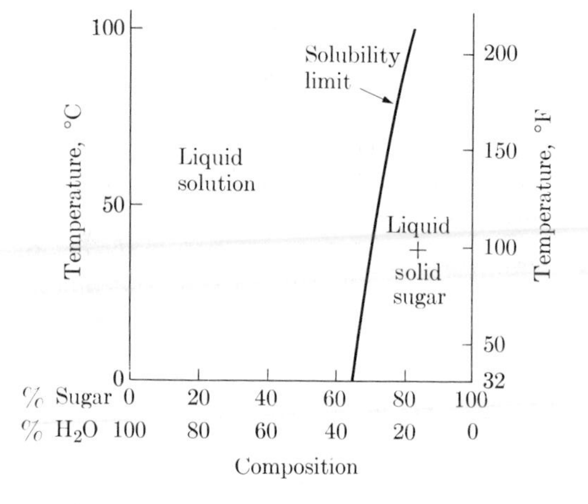

Figure 9–1.2 shows a two-component system which has more engineering importance than our first example. Here the extremes of the abscissa indicate 100% H_2O and 30% NaCl. Note from the figure that (1) the solubility limit of NaCl in a brine solution decreases with decreasing temperature, (2) the solubility limit of H_2O in a brine solution also decreases with decreasing temperature, and (3) intermediate compositions have melting temperatures lower than that of either pure ice (0°C or 32°F), or pure salt (800.4°C, or 1473°F). (1) and (3) are well-known facts, and (2), the less familiar limited solubility of ice in the aqueous liquid can be verified by a simple experiment. A salt and water solution, e.g., sea water with 1.7% NaCl, can be cooled to less than 0°C (<32°F) and, according to Fig. 9–1.2, it will still be entirely liquid

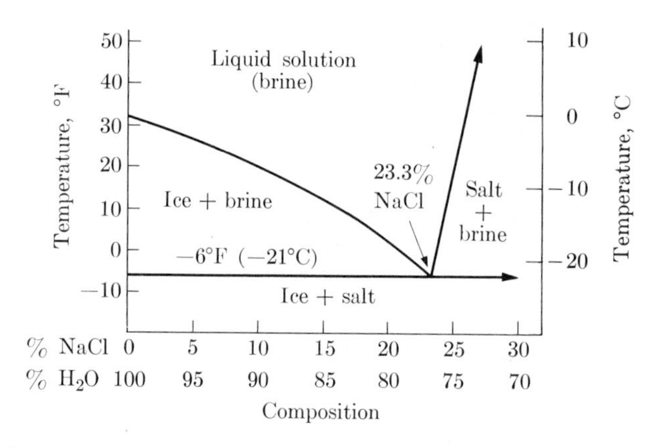

Fig. 9–1.2. Solubility of NaCl salt in brine (right upward-sloping line) and solubility of ice in brine (left curve).

until minus 1°C (30.2°F) is reached. This is in agreement with observations of any arctic saline sea.* When such a salty liquid is cooled below −1°C, ice crystals will form, and because the solution cannot contain more than 98.3% H_2O at that temperature, these ice crystals must separate from the liquid. At minus 20°C (−4°F), the maximum H_2O content possible in a brine solution is 77%, as can be verified by making a slush at that temperature and separating the ice from this liquid; the ice will be nearly pure H_2O and the remaining liquid will be saltier (i.e., lower in H_2O) than the original brine solution.

Eutectic temperatures and compositions. Another example with reduced melting temperatures is a mixture of H_2O and $CaCl_2$ (Fig. 9–1.3). Calcium chloride, rather than NaCl, is used to remove ice from highways in very cold climates. The reason is apparent from a comparison of Figs. 9–1.3 and 9–1.2. An aqueous solution of calcium chloride remains liquid at temperatures as low as minus 51°C (−60°F), while a similar solution of NaCl will freeze at minus 21°C (−6°F). The lowest temperature at which a solution will remain completely liquid is called the *eutectic temperature*, and the solution which possesses this lowest freezing point is called the *eutectic composition*. The eutectic composition for the H_2O–NaCl system of Fig. 9–1.2 is 76.7% H_2O and 23.3% NaCl. From Fig. 9–1.3, the eutectic composition for H_2O–$CaCl_2$ is 71% H_2O and 29% $CaCl_2$. The intersection of the solubility curves in such diagrams marks the eutectic composition for the two components in a liquid solution.

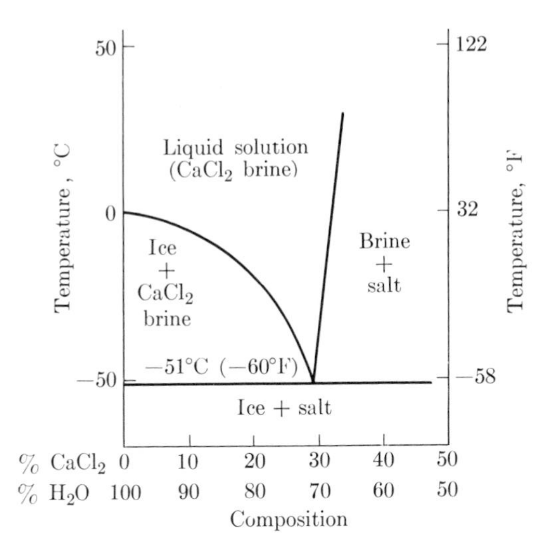

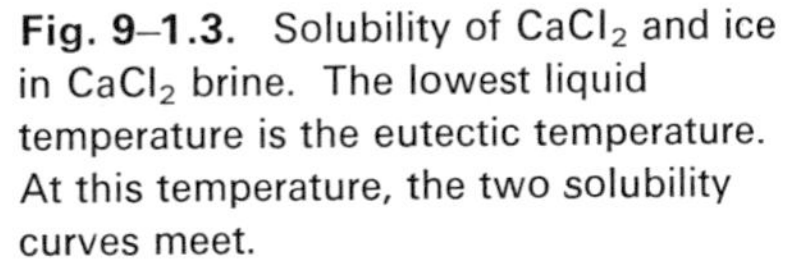
Fig. 9–1.3. Solubility of $CaCl_2$ and ice in $CaCl_2$ brine. The lowest liquid temperature is the eutectic temperature. At this temperature, the two solubility curves meet.

* There can be a slight variation if the salt content is not exactly 1.7%.

These melting–freezing relationships are quite common in all types of two-component combinations. Figure 9–1.4 shows the solubility curves for lead and tin. The low-melting "60–40" alloy is used in many *solders* because this eutectic composition permits the formation of metal-bonded joints with a minimum of heating. If the solder contained more lead (say 80% lead and 20% tin), during cooling the liquid metal would become saturated with lead at 280°C which is above the eutectic temperature and a lead-rich phase would precipitate from the liquid metal solution.

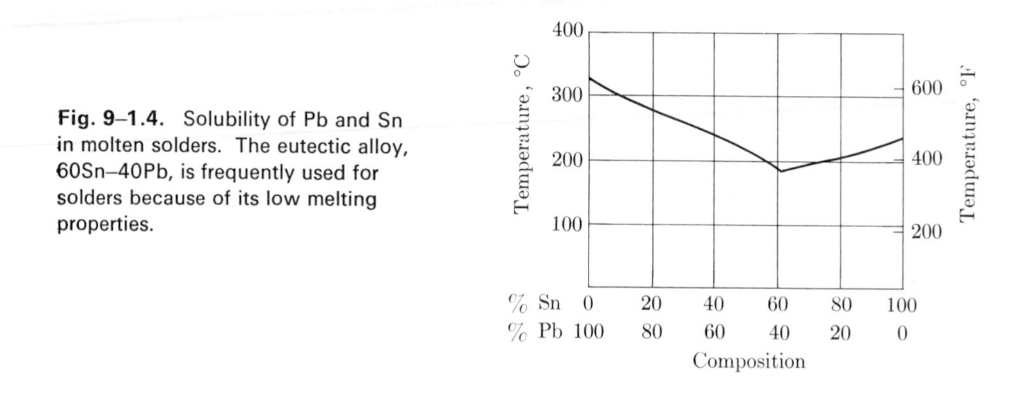

Fig. 9–1.4. Solubility of Pb and Sn in molten solders. The eutectic alloy, 60Sn–40Pb, is frequently used for solders because of its low melting properties.

Solid solubility limits. Experiments have demonstrated that at 280°C (536°F), 13% (8 w/o) of the lead atoms in the solid phase of this alloy may be replaced by tin atoms. Figure 9–1.5 shows the solubility curves for tin in the *solid* lead-rich structure and for lead in the *solid* tin-rich structure. In these particular alloys, the temperature at 183°C (361°F) is the eutectic temperature and represents (1) the lowest temperature at which any liquid can exist, (2) the temperature with the greatest solid solubility, and (3) the temperature above which any excess over the solid solubility limit is liquid, and below which any excess over the solid solubility limit is solid.

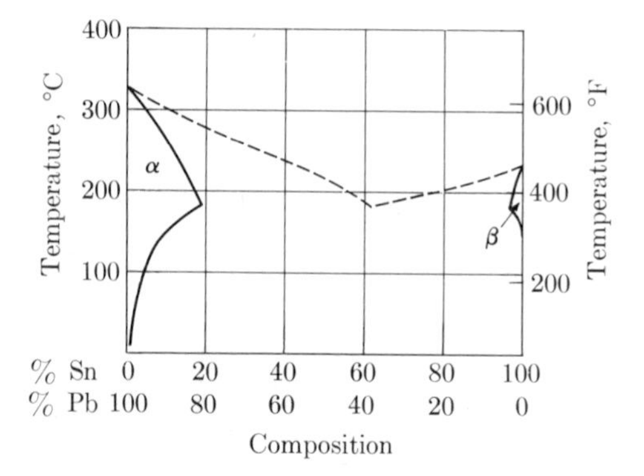

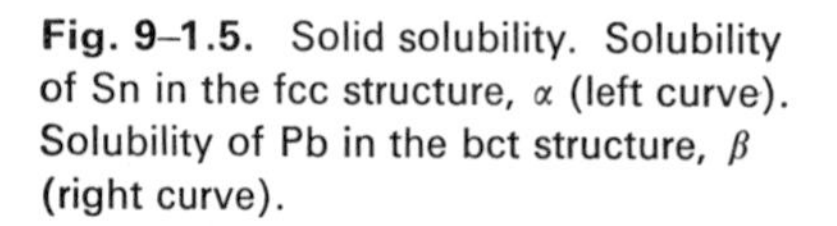

Fig. 9–1.5. Solid solubility. Solubility of Sn in the fcc structure, α (left curve). Solubility of Pb in the bct structure, β (right curve).

Phase names and/or labels. We have already given names to certain solid-solution phases. *Brass* is an fcc phase of zinc in copper, and *sterling silver* is a solid-solution phase of copper in fcc silver. Later we will give the name *ferrite* to the phase of bcc iron when it contains solutes; likewise, *austenite* is the common name for fcc iron when it contains added elements.

Greek-letter labels are used more commonly than names, for simplicity's sake. The solid, lead-rich fcc phase in the previous section, which contains tin, is called α; and the solid tin-rich bct phase, which contains lead, is called β.* Thus, α and β are *phases* in the Pb–Sn *system*, while lead and tin are *components* of the Pb–Sn system. Although pure lead has the structure of the α phase, α is not necessarily pure lead.

Example 9–1.1 A syrup may contain only 67% sugar (33% water) at 20°C (68°F), but 83% sugar at 100°C (212°F). One hundred grams of sugar and 25 grams of water are mixed and boiled until all the sugar is dissolved. During cooling, the solubility limit is exceeded, so that (with time) excess sugar separates from the syrup. If equilibrium is attained, what is the weight ratio of syrup to excess sugar at 20°C?

Solution

Basis: 100 g sugar + 25 g H_2O = 125 g

At 20°C, amount of syrup = 25 g H_2O/(33 g H_2O/100 g syrup)

= 75.7 g syrup

125 g − 75.7 g = 49.3 g excess sugar.

Comment. All the sugar is dissolved at the higher temperature, to give a single phase of syrup. When the mixture cools, the *solubility limit* for an 80 sugar–20 water syrup is reached at 87°C (189°F). Typically, however, some *supercooling* is encountered before the excess phase (in this case, sugar) starts to separate. In fact, in this case, supercooling can proceed to room temperature, so that the start of separation may be delayed considerably. Similar supersaturation commonly occurs in metals and other materials. ◀

9–2 PHASE DIAGRAMS

Figure 9–2.1 is a completed *phase* (equilibrium) *diagram* for the lead–tin system. This diagram can be used as a "map" from which the phases present at any particular temperature or composition can be read if the alloy is at equilibrium.

For example, at 50% tin and 100°C, the "map" indicates two solid phases: α is a lead-rich solid solution with some dissolved tin; β is almost pure tin with very little dissolved lead. At 200°C, an alloy of 10% tin and 90% lead lies in an area which is entirely in the α phase. It is a solid solution of lead with some tin dissolved in it. At the same temperature, but for 30% tin and 70% lead, the "map" indicates a mixture

---

* Admittedly, the label α appears in almost all systems as the first-named phase. However, this does not prove to be a problem. (The reader is aware that x is repeatedly used in algebraic calculations without special labelling confusion.)

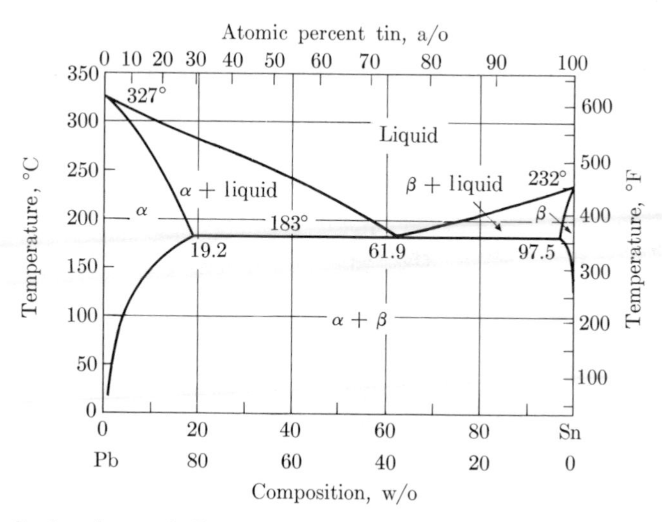

Fig. 9–2.1. Pb–Sn diagram. Such a diagram indicates the phase compositions and permits the calculation of phase quantities for any lead-tin mixture at any temperature. (Adapted from *Metals Handbook*, American Society for Metals.)

of two phases—liquid and the α solid solution; if this latter alloy were heated to a temperature of 300°C, it would become all liquid.

The phase fields in equilibrium diagrams, of course, depend on the particular alloy systems being depicted. When copper and nickel are mixed, the phase diagram is as shown in Fig. 9–2.2. This "map" is comparatively simple, since only two phases

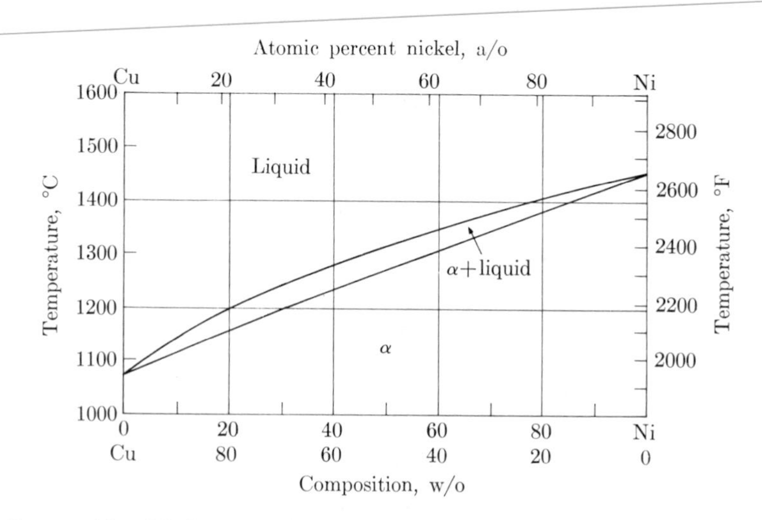

Fig. 9–2.2. Cu-Ni diagram. All solid alloys contain only one phase. This phase is fcc. (Adapted from *Metals Handbook*, American Society for Metals.)

are present. In the lower part of the diagram, all alloys form only one solid solution and therefore only one crystal structure. Both the nickel and the copper have face-centered cubic structures. Since the atoms of each are nearly the same size, it is possible for nickel and copper atoms to replace each other in the crystal structure in any proportion at 1000°C. When an alloy containing 60% copper and 40% nickel is heated, the solid phase exists until the temperature of about 1235°C (2255°F) is reached. Above this temperature and up to 1275°C (2330°F) the solid and liquid solutions coexist. Above 1275°C only a liquid phase remains.

Freezing ranges. As shown in the foregoing phase diagrams, the range of temperatures over which freezing occurs varies with the composition of the alloy. This situation influences the plumber, for example, to select a high-lead alloy as a "wiping" solder when he needs a solder which will not freeze completely at one temperature. If he chooses an 80Pb–20Sn solder, the freezing range is from 280 to 183°C as compared with 188 to 183°C (370 to 360°F) for a 60Sn–40Pb solder.

The terms *liquidus*, the locus of temperatures above which all compositions are liquid, and *solidus*, the locus of temperatures below which all compositions are solid, are applied in this connection. Every phase diagram for two or more components must show a liquidus and solidus boundary and an intervening freezing range. Whether the components are metals or nonmetals (Fig. 9–2.3), there are certain locations on the phase diagram where the liquidus and solidus meet. For a pure component, this point lies at the edge of the diagram. When it is heated, a pure material will remain solid until its melting point is reached, and will then change entirely to liquid before it can be raised to a higher temperature.

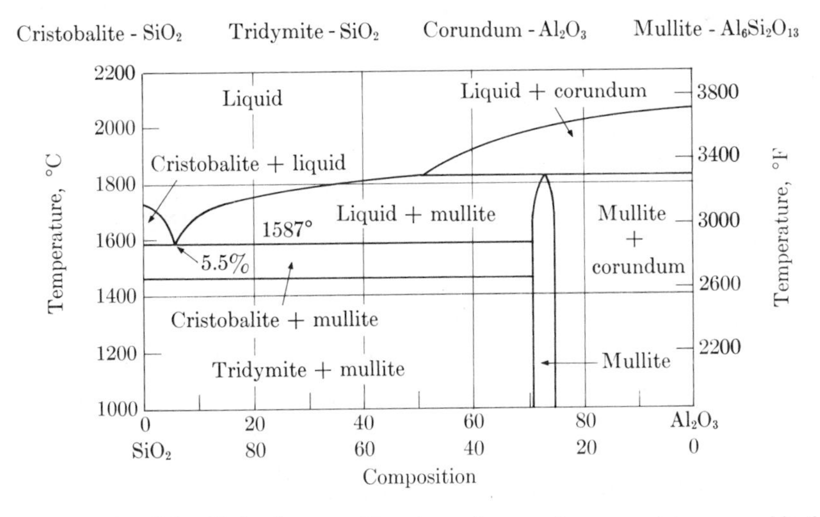

Fig. 9–2.3. SiO_2–Al_2O_3 diagram. The phase diagrams for nonmetals are used in the same manner as those for metals. The only difference is the longer time required to establish equilibrium. (Adapted from Askay and Pask, *Science*, 1974.)

The solidus and liquidus must also meet at the eutectic. In Fig. 9–2.1 the liquid solder composed of 61.9% tin and 38.1% lead is entirely solid below the eutectic temperature and entirely liquid above it. At the eutectic temperature three phases can coexist [(α + Liq + β) in the Pb–Sn system].

Isothermal cuts. A traverse across the phase diagram at a constant temperature (*isotherm*) provides a simple sequence of alternating 1- and 2-phase fields. Consider the SiO_2–Al_2O_3 diagram at 1650°C (3000°F), and you will see that the sequence is 1-2-1-2-1-2-1. With pure SiO_2, only *one* phase exists (named cristobalite). It holds negligible amounts of Al_2O_3 in solid solution.* Therefore, a second phase (liquid) appears with the addition of Al_2O_3. The *two*-phase region contains cristobalite and liquid. Between 4% Al_2O_3 (96% SiO_2) and 8% Al_2O_3 (92% SiO_2), the liquid can dissolve all of the SiO_2 and Al_2O_3 that is present, so just *one* phase exists. Beyond 8% Al_2O_3 (<92% SiO_2), the solubility limit of the liquid for Al_2O_3 is exceeded, and solid mullite precipitates. The *two* phases liquid and mullite coexist.† The solid-solution range of mullite is from 71% Al_2O_3 (29% SiO_2) to 75% Al_2O_3 (25% SiO_2). Only *one* phase is stable in this range, because it can accommodate both the SiO_2 and Al_2O_3 that are present. A *two*-phase field, mullite and corundum (Al_2O_3) follows, and extends to within a line's width of the right side of the phase diagram. With only Al_2O_3, this *one* phase is called corundum.

Example 9–2.1 Sterling silver, an alloy containing approximately 92.5% silver and 7.5% copper (Fig. 9–2.4), is heated slowly from room temperature to 1000°C (1830°F). What phase(s) will be present as heating progresses?

Answer:

Room temperature to 760°C	$\alpha + \beta$
760°C to 800°C	Only α
800°C to 900°C	α + liquid
900°C to 1000°C	Only liquid ◀

Example 9–2.2 A combination of 90% SiO_2 and 10% Al_2O_3 is melted at 1800°C, and then cooled extremely slowly to 1400°C. What phase(s) will be present in the cooling process?

Answer:

1800°C to 1700°C	Only liquid
1700°C to 1587°C	Liquid + mullite ($Al_6Si_2O_{13}$)
1587°C to 1470°C	Mullite + cristobalite (SiO_2)
< 1470°C	Mullite + tridymite (SiO_2)

---

* There is a slight solubility, but it is so low that we are not able to show it on the phase diagram.

† When 8% Al_2O_3 is just slightly exceeded there will be very little mullite. When the right side of the 2-phase field is approached, very little liquid remains.

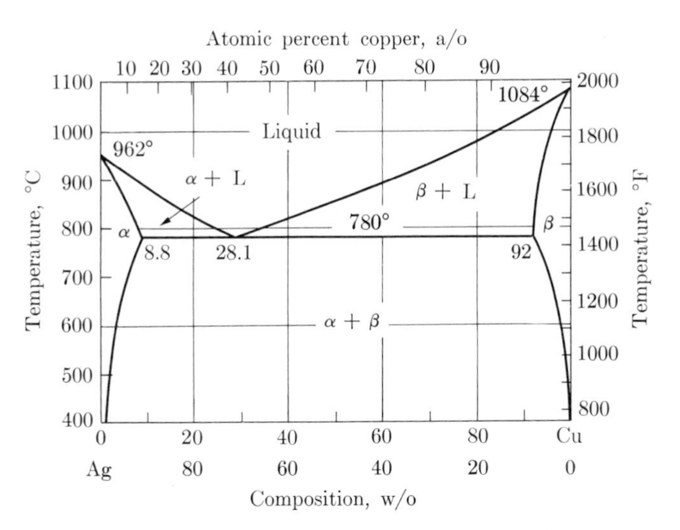

Fig. 9–2.4. Ag–Cu diagram. (Adapted from *Metals Handbook*, American Society for Metals.)

Comments: The cooling would have to be extremely slow, because the process of changing the strong Si—O bonds from one structure to another is very slow.

Pure silica has three common polymorphs: cristobalite and tridymite at high temperatures, and quartz at low temperatures. ◀

Example 9–2.3 Refer to the Al–Mg diagram (Fig. 9–5.4) later in the chapter. Apply the 1–2–1–2–··· rule (a) at 500°C; (b) at 200°C.

Answer:

a) α — $(\alpha + \text{L})$ — Liq — $(\text{L} + \epsilon)$ — ϵ.

		α		Liq		ϵ
%Mg	0	11	29	76	92	100
%Al	100	89	71	24	8	0

b) α — $(\alpha + \beta)$ — β — $(\beta + \beta')$ — β' — $(\beta' + \gamma)$ — γ — $(\gamma + \epsilon)$ — ϵ.

%Mg 3 35 37 ~41 ~42 49 56 96

Comment. Dashed lines are the best estimates based on present information. ◀

9–3 CHEMICAL COMPOSITIONS OF PHASES

In addition to serving as a "map," a phase diagram provides the chemical *compositions* of the phases which are present under conditions of equilibrium. This information, along with information on the amount of each phase (Section 9–4), constitutes very useful data for the scientists and engineers who are involved with materials development, selection, and application in product design.

One-phase areas. The determination of the chemical composition of a single phase is automatic. *It has the same composition as the alloy.* This is to be expected; since only liquid is present in a 60Sn–40Pb alloy at 225°C (Fig. 9–3.1), the liquid has to have the same 60–40 composition. This also holds when the location in the phase diagram involves a single-phase solid solution.

You will observe that we reported the chemical composition of the individual phase (and of the total alloy) in terms of the *components*—in this case, lead and tin.

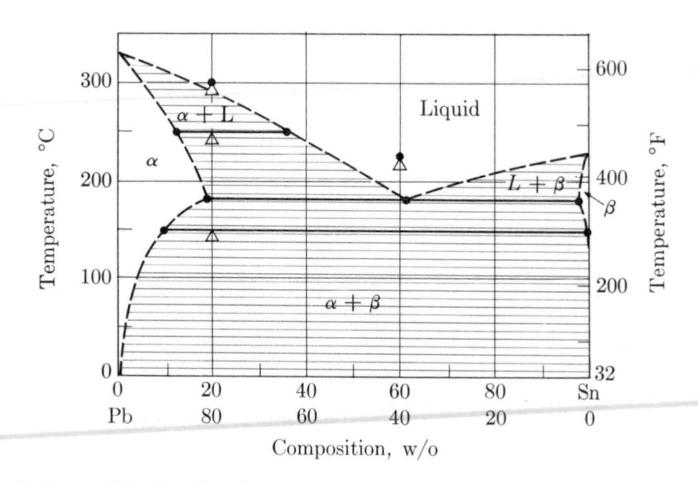

Fig. 9–3.1. Compositions of phases (Pb–Sn alloys). At 150°C, an 80Pb–20Sn alloy contains α and β. The chemical composition of α is dictated by the solubility curve. At this temperature, the solubility limit is 10% Sn (and 90% Pb) in the fcc α phase. See the text.

Two-phase areas. The determination of the chemical compositions of two phases can be handled on a rote basis. We will do this first; however, the rationale will follow in the next paragraph.

The chemical compositions of the two phases are located *at the two ends of the isotherm across the two-phase area.* To illustrate, take an 80Pb–20Sn solder at 150°C. As indicated in Fig. 9–3.1, α has a chemical composition of 10 w/o tin (and therefore 90 w/o lead). The composition of the β is nearly 100 w/o tin. Other isotherms on Fig. 9–3.1 permit us to read the chemical compositions of the two phases of any Pb–Sn alloy at any temperature.

The basis for the above procedure is simply that the *solubility limit* for tin in α at 150°C is 10 w/o. Our alloy exceeds this with its composition of 20 w/o tin. Therefore, α is saturated with tin and the excess tin is present in β. Conversely, the solubility limit for lead in β is <1 w/o; therefore, almost all of the lead must be in a phase other than β—specifically, in α.

Three-phase temperatures and eutectic reactions. A liquid which has the analysis of the eutectic composition (38.1Pb–61.9Sn, when we consider the Pb–Sn system) separates into two solid phases (α and β) at the eutectic temperature (183°C). Thus, at this temperature *only*, three phases can be in equilibrium. If this alloy is heated, the two solid phases of this solder melt into a one-phase liquid. We can write this reversible reaction as

$$L(61.9\%\ Sn) \overset{183°C}{\rightleftarrows} \alpha(19.2\%\ Sn) + \beta(97.5\%\ Sn). \qquad (9\text{–}3.1)$$

The tin analyses are shown for the three phases that are in equilibrium at 183°C.*

The reaction of Eq. (9–3.1) is called an *eutectic reaction* and involves a liquid and two solids. The more general form would be:

$$L_2 \underset{\text{Heating}}{\overset{\text{Cooling}}{\rightleftarrows}} S_1 + S_3, \qquad (9\text{–}3.2)$$

where the 1,2,3 subscripts refer to progressively increasing contents of one of the two components.

Example 9–3.1 In the Cu–Ni system (Fig. 9–2.2) and at 1300°C: a) What is the solubility limit of copper in solid α? b) Of nickel in the liquid? c) What are the chemical compositions of the phases in a 45Cu–55Ni alloy at 1300°C?

Answer: a) 42 w/o Cu b) 45 w/o Ni c) α – (42Cu–58Ni), L – (55Cu–45Ni).

Comment. It is the solubility limit which determines the chemical compositions of the phases in a two-phase area. ◀

Example 9–3.2 An alloy of 40Ag–60Cu (Fig. 9–2.4) is cooled slowly from 1000°C (1832°F) to room temperature. a) What phase(s) will be present as cooling progresses? b) Indicate their compositions. c) Write the eutectic reaction.

Answer:

a), b)

Temperature	α	Liquid	β
1000°C	—	40Ag–60Cu	—
800	—	66Ag–34Cu	8Ag–92Cu
780	91.2Ag–8.8Cu	71.9Ag–28.1Cu	8Ag–92Cu
600	96.5Ag–3.5Cu	—	2Ag–98Cu
400	99Ag–1Cu	—	Near 100Cu
20 (extrapolated)	Near 100Ag	—	Near 100Cu

---

* We could have used 38.1% Pb, 80.8% Pb, and 2.5% Pb for L, α, and β, respectively, rather than the tin values.

c) Liquid (71.9%Ag) $\xrightleftharpoons{780°C}$ α(91.2%Ag) + β(8%Ag).

Comment. The liquid becomes saturated with copper at about 890°C. The first β to separate at this temperature has the composition of 7Ag–93Cu. ◀

9–4 QUANTITIES OF PHASES

In addition to (1) identifying the stable, or equilibrium phases (Section 9–2) and (2) obtaining their chemical compositions from the phase diagrams (Section 9–3), we can (3) determine the *quantities* of each phase which is present under equilibrium conditions. This will be useful to us later when we consider the properties of multiphase materials in Chapter 10.

One-phase areas. Again, as in Section 9–3, we have an automatic situation. With 225 g of a 80Pb–20Sn alloy at 300°C, only liquid is present, and its quantity is 225 g. It is equally valid to state that all (or 100%) of the alloy is liquid. In that manner we do not have to specify the exact weight of the alloy which is present.

Two-phase areas. We can use a rote basis to indicate the quantities (or the *quantity fractions*) of the phases which are present in the two-phase areas of phase diagrams. We will do this first; the rationale will follow. The quantities of the two phases are determined by *interpolating the composition of the alloy between the compositions of the two phases.* To illustrate, we will again take an 80Pb–20Sn solder at 150°C. As indicated in Fig. 9–4.1, the chemical composition of the alloy (80Pb–20Sn) is at a position which is 0.11 of the distance between the chemical composition of α (90Pb–10Sn) at this temperature, and the chemical composition of β (<1Pb and ~100Sn).

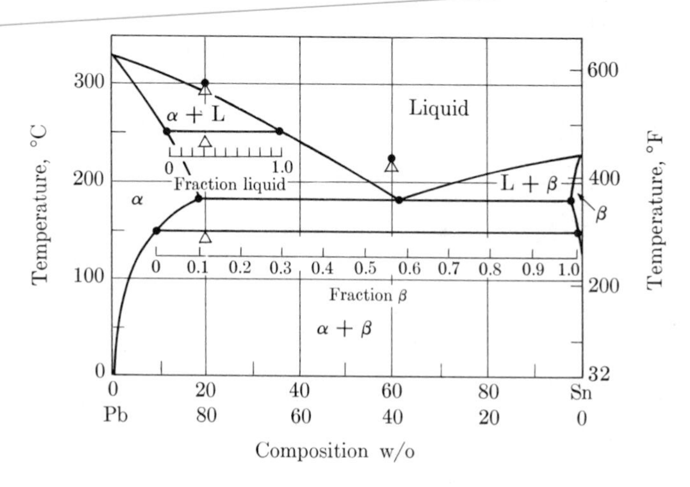

Fig. 9–4.1. Quantities of phases (Pb–Sn alloys). As observed in Fig. 9–3.1, an 80Pb–20Sn alloy contains α and β at 150°C. By interpolation, the fraction of β is 0.11. This same alloy contains 0.33 liquid at 250°C (and 0.67 α). See text.

Therefore, of the total amount of solder, the quantity fraction of β is 0.11 (and 0.89 of α) at 150°C.* It is equally appropriate to report this as 89% α and 11% β.

As another example, consider the same alloy at 250°C. On the basis of Fig. 9–4.1, we have α (88Pb–12Sn) and L (64Pb–36Sn). The chemical composition of this alloy as a whole (80Pb–20Sn) is $\frac{1}{3}$ of the distance between the chemical composition of α and the chemical composition of the liquid. Therefore, of the total amount of solder at 250°C, the quantity fraction of liquid is $\frac{1}{3}$ and of α is $\frac{2}{3}$.

You will observe that we have reported these quantity figures in terms of *phases*—in these cases, α and β, or α and L. This is in contrast to the chemical compositions which were reported in terms of the *components*—Pb and Sn.

To follow the above procedure, we would need to have a stretchable scale, because the length of the isotherm between the two solubility limit curves continuously varies. Interpolation by using a millimeter scale offers a simple alternative. The alert student will also be quick to suggest interpolating on the basis of the numbers on the abscissa of the graph.

Materials balance. The rationale for the above interpolation procedure originates as a materials balance. By this we mean that the whole is equal to the sum of the parts. At 250°C, and with the above example, the lead in the total alloy is equal to the lead in the α plus the lead in liquid.

To illustrate, let us use 600 g of our same 80Pb–20Sn solder at 250°C. There are 480 g of Pb (and 120 g of Sn). We can consider that there are A grams of α (88Pb–12Sn) and L grams of liquid (64Pb–36Sn). Those compositions are located in Figs. 9–3.1 and 9–4.1. Of course

$$A + L = 600 \text{ g}.$$

Thus, on the basis of the chemical compositions of the two phases and the total alloy, the lead balance is

$$0.88A + 0.64L = 0.80(A + L), \tag{9–4.1}$$

or, in this case, $A = 2L$. This means that we have 200 g of liquid and 400 g of α—the same $\frac{1}{3}$ and $\frac{2}{3}$ values we calculated previously.

A generalization of Eq. (9–4.1) is

$$C_x(X) + C_y(Y) = C(X + Y), \tag{9–4.2}$$

where C_x and C_y are the chemical *compositions* of phases x and y respectively; C is the analysis of the total alloy. Likewise, X and Y are the quantities of the phases x and y. Algebraic rearrangement of Eq. (9–4.2) gives either:

$$\frac{X}{X + Y} = \frac{C_y - C}{C_y - C_x}, \tag{9–4.3(a)}$$

* Thus if we have 225 grams of solder, there would be ~200 grams of α and ~25 grams of β.

or

$$\frac{Y}{X + Y} = \frac{C_x - C}{C_x - C_y}. \tag{9–4.3(b)}$$

The value of $X/(X + Y)$ is the quantity fraction of x; and, by like token, $Y/(X + Y)$ is the quantity fraction of y. Equation (9–4.3) is the *inverse lever rule*, which is commonly presented in other materials texts as a means for calculating the *quantity fractions* of the phases.*

• **The three-phase special case.** Lead–tin solders have three phases when they are equilibrated at 183°C. In this special case, we *cannot* calculate exactly the quantity fractions of α, β, and the liquid that are present. We can calculate that a 70Pb–30Sn alloy contains 0.86 α at 182°C (and 14% β). We also can determine that the same solder contains $\frac{1}{4}$ liquid (and 75% α) at 184°C. At the eutectic temperature, between 182 and 184°C, the β and some of the α react on heating to give an eutectic liquid. In the process, 1000 grams of solder would have 110 g of α (i.e., 860 g − 750 g) and 140 g of β consumed to form 250 g of eutectic liquid (61.9Pb–38.1Sn). Therefore, with three phases at 183°C, we can only indicate that the quantity of α is between 860 and 750 grams; the quantity of β, between 140 and 0 grams; and of liquid, between 0 and 250 grams.

Example 9–4.1 Consider a 90Cu–10Sn bronze. a) What phases are present at 300°C? What are their chemical compositions? What is the fraction of each? b) Repeat for 600°C. c) At what temperature is there $\frac{1}{3}$ liquid? $\frac{2}{3}$ liquid?

Solution: Refer to Fig. 9–5.2.

a) α: 93Cu–7Sn (37–10)/(37–7) = 0.9,

ϵ: 63Cu–37Sn (10–7)/(37–7) = 0.1;

b) α: 90Cu–10Sn 1.0;

c) By using a mm scale: $\frac{1}{3}$ L @ 900°C; $\frac{2}{3}$ L @ 975°C. ◀

Example 9–4.2 Using lead, make a material balance which shows A, the quantity of α, and B, the quantity of β, in 75 grams of a 70Pb–30Sn solder at 182°C.

Solution: From Eq. (9–4.2), and using the lead analyses of Fig. 9–2.1,

$$0.808A + 0.025B = 0.70(A + B).$$

Also,

$$A + B = 75 \text{ grams}.$$

---

* The student may use this lever rule in his calculations of the quantity fraction of the phases, or if simpler, one may interpolate according to Fig. 9–4.1. If one uses the abscissa as a scale for interpolation, Eq. (9–4.3) is followed indirectly. In any event, observe that the chemical composition of the alloy is at the "center of gravity" between the compositions of the two contributing phases. This is the fulcrum of the lever; the majority phase is on the *shorter* arm of the lever.

Solving simultaneously,

$$A = 64.7 \text{ g}, \qquad B = 10.3 \text{ g},$$

$$A/(A + B) = 0.86; \qquad B/(A + B) = 0.14.$$

Comment. A material balance using tin, $0.192A + 0.975B = 0.30(A + B)$, gives the identical answer. ◀

Example 9–4.3 As a function of temperature, plot the quantity of β in 10.0 grams of a 40Ag–60Cu alloy.

Solution: The chemical compositions of the phases were previously determined for Example 9–3.2. At 890°C and above, all liquid; therefore: 0 grams of β. By interpolating:

At 800°C	0.45(10 g)	= 4.5 g
780°+	0.50(10)	= 5.0
780°−	0.615(10)	= 6.15
600°	0.60(10)	= 6.0
400°	0.595(10)	= 5.95
20°	0.60(10)	= 6.0.

This is plotted as the curve over the dark shaded area of Fig. 9–4.2.

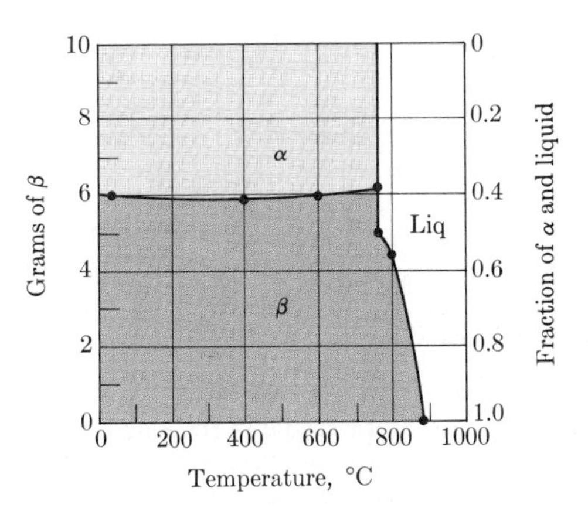

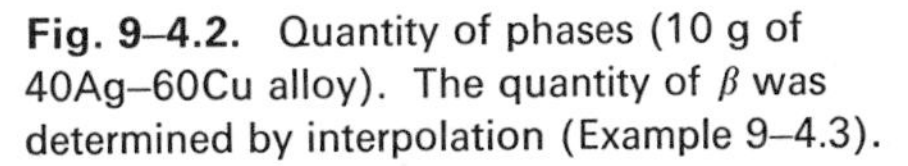

Fig. 9–4.2. Quantity of phases (10 g of 40Ag–60Cu alloy). The quantity of β was determined by interpolation (Example 9–4.3).

Comments. We can also indicate the number of grams (or the quantity fraction) of α and liquid in Fig. 9–4.2, because the total must add up to 10.0 grams at all temperatures. We have done this with modified shading.

Note the discontinuity in quantities at the eutectic temperature (780°C), which is the special case where three phases are present simultaneously. ◀

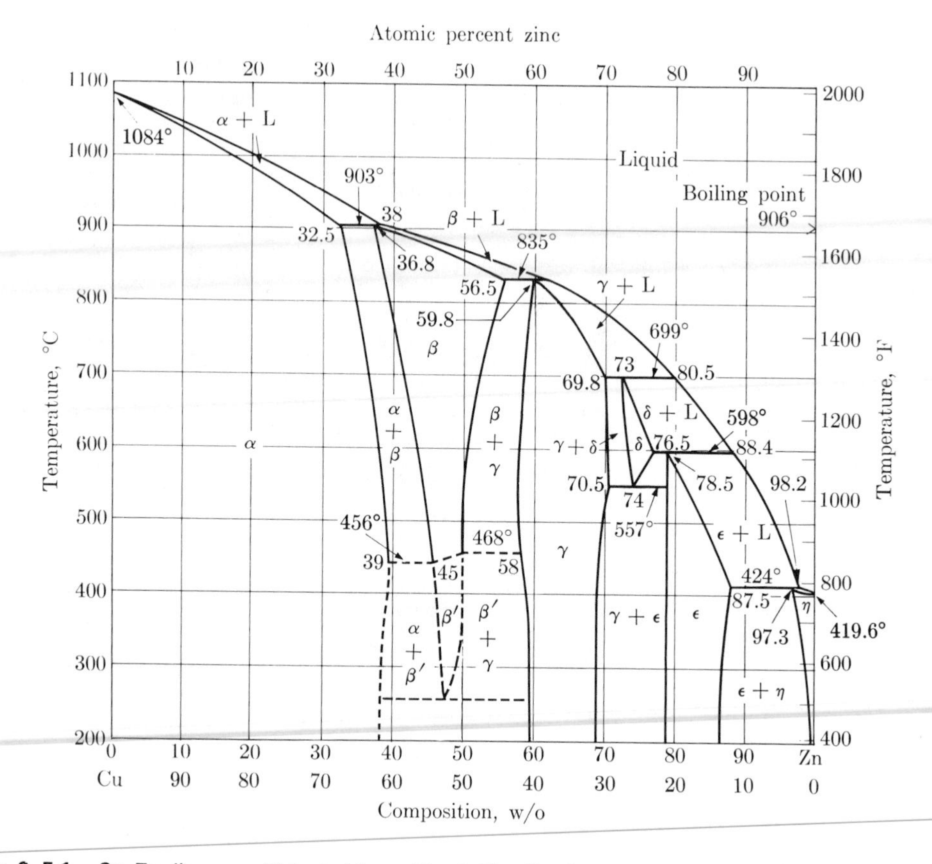

Fig. 9–5.1. Cu–Zn diagram. (Adapted from *Metals Handbook*, American Society for Metals).

9–5 COMMERCIAL ALLOYS AND CERAMICS

The reader may be relieved to learn that most commercial alloys lie in the simpler parts of the phase diagrams. For example, 99% of the brasses lie in the single-phase α area of Fig. 9–5.1. Likewise, the common bronzes possess less than 10% tin. There is little commercial interest in the more complex-appearing central areas* of the Cu–Sn system (Fig. 9–5.2). In Chapter 11, we will give special attention to alloys

* Although more complex in appearance, all areas are either 1-phase or 2-phase. We have already considered these. Thus, for any alloy we can determine: (1) *what* phase(s), (2) their chemical *compositions*, and (3) the *quantity* of each phase.

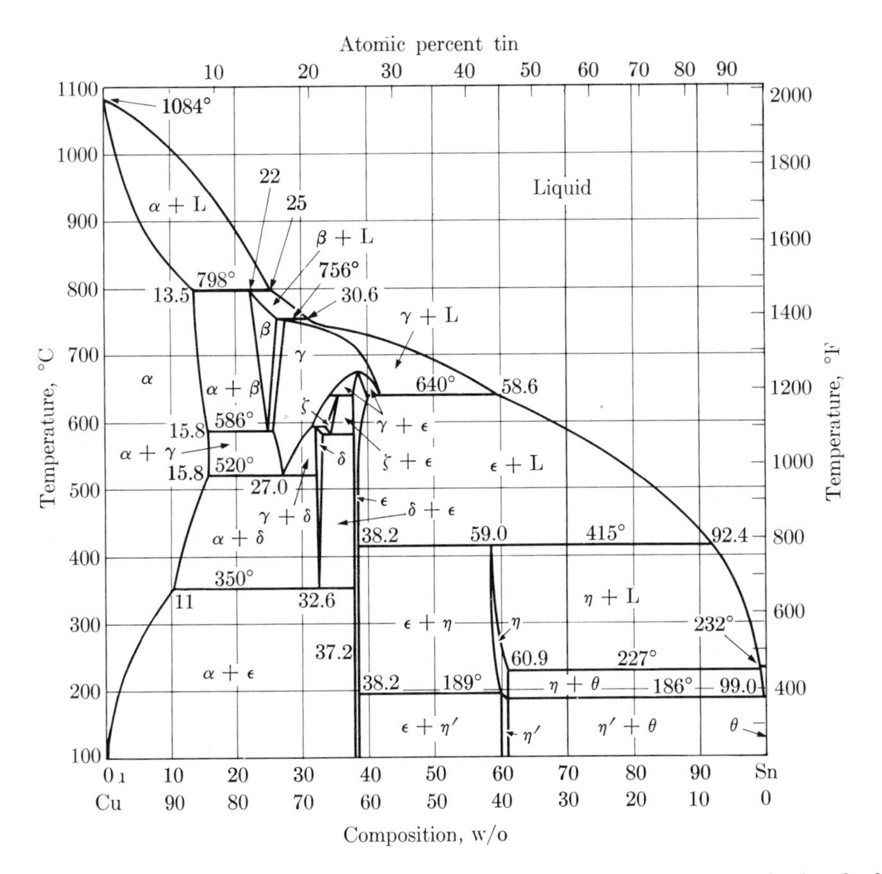

Fig. 9–5.2. Cu–Sn diagram. (Adapted from *Metals Handbook*, American Society for Metals.)

such as 95Al–5Cu, 90Al–10Mg, 90Mg–10Al (Figs. 9–5.3 and 9–5.4) because each forms one phase at elevated temperatures, but crosses a solubility limit curve during cooling. By controlling the rate of separation of the second phase, the engineer can greatly increase the strength of the alloy. This, of course, is of major interest.

The Al–Si system (Fig. 9–5.7) reveals the commercial basis for purifying semiconducting and related materials (Ex. 9–5.1).

In the next two sections we shall give detailed attention to iron–carbon alloys for two reasons. First, steels constitute very prominent alloys in any technical civilization. Secondly, steels can serve as a prototype for heat-treating procedures. It is through the application of phase diagrams to heat treatment that the microstructures of the alloys, and therefore their properties, are controlled.

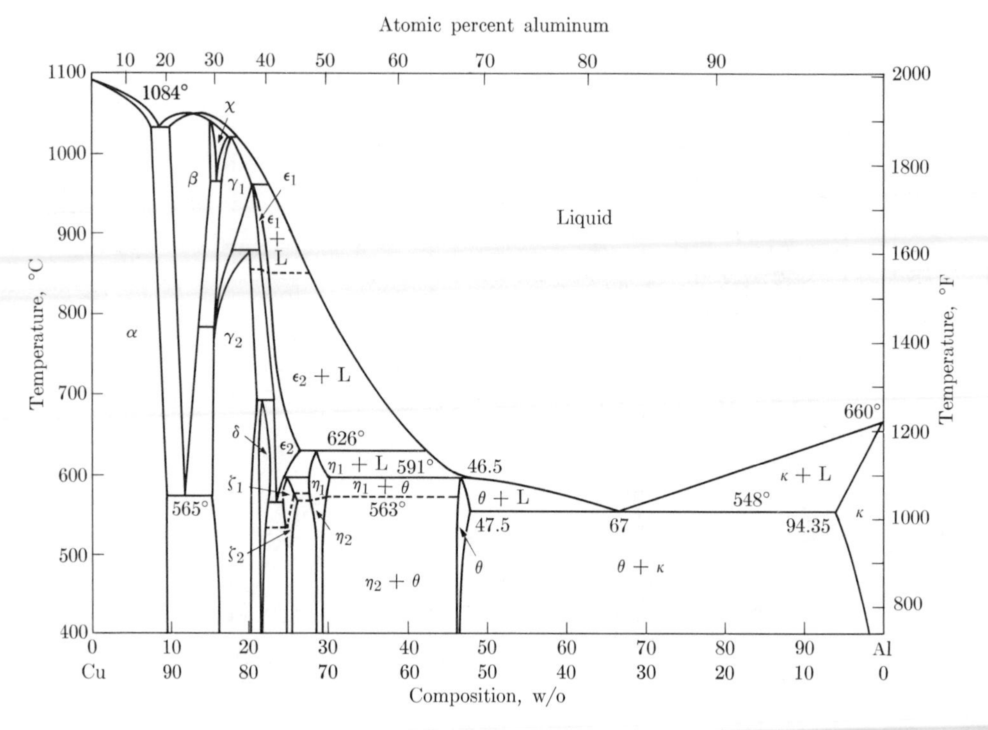

Fig. 9–5.3. Al–Cu diagram. (Adapted from *Metals Handbook*, American Society for Metals.)

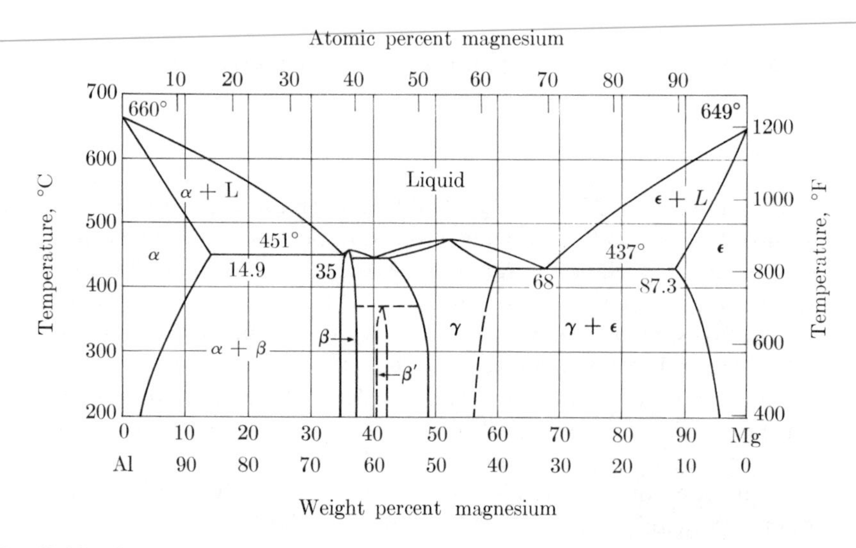

Fig. 9–5.4. Al–Mg diagram. (Adapted from *Metals Handbook*, American Society for Metals.)

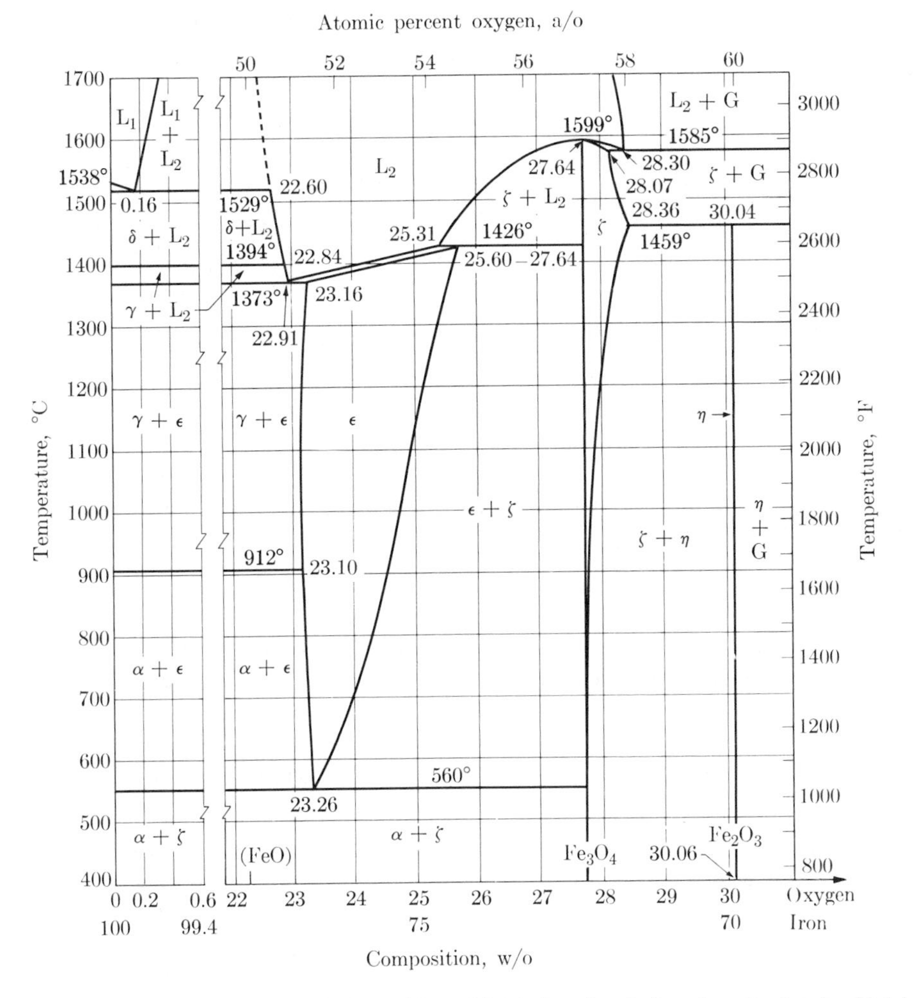

Fig. 9–5.5. Fe–O diagram. (Adapted from *Metals Handbook,* American Society for Metals.)

Ceramists pay as much attention to phase diagrams as do metallurgists.* However, within this text, we shall observe only three: Al_2O_3–SiO_2 (Fig. 9–2.3), Fe–O (Fig. 9–5.5) and FeO–MgO (Fig. 9–5.6).† The first is pertinent to clay-base ceramics

* See, for example, the 3-volume compendium, *Phase Diagrams for Ceramists,* American Ceramic Society.

† Commonly, ceramic products involve three or more components, for example, K_2O–Al_2O_3–SiO_2 in electrical porcelains. Although important, this text will not consider the more complex ternary-phase diagrams.

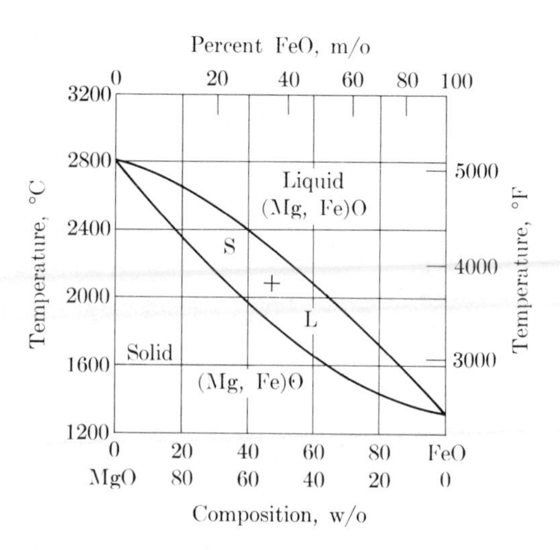

Fig. 9–5.6. MgO–FeO diagram. Comparable to the Cu–Ni diagram (Fig. 9–2.2); both the liquid and the solid phases possess full solubility.

since the better-grade clays are approximately $40Al_2O_3$–$60SiO_2$ after processing. The Fe–O diagram shows the nonstoichiometric range of $Fe_{1-x}O$ (ϵ of Fig. 9–5.5) which was discussed in Chapters 4, 5, and 8. The FeO–MgO diagram reveals that the solid solution between FeO and MgO (Fig. 4–3.1) is a complete series below the solidus temperature. The Cu–Ni system (Fig. 9–2.2) is comparable, but for metallic materials.

•**Example 9–5.1** The Al–Si diagram (Fig. 9–5.7) has the solubility limit exaggerated for Al in β because, at the eutectic temperature (577°C), β will hold a maximum of only 0.17 w/o (99.83% Si). a) Assume a linear solubility curve; what is the aluminum analysis of β at 1300°C?

An alloy containing 98Si–2Al is equilibrated at 1300°C, and the liquid removed. b) What fraction of the aluminum is removed?

The remaining solid is remelted at 1450°C. c) At what temperature will the new solid start to form? d) What is its chemical analysis?

Solution:

a) Being linear,

$$\frac{0.17\%}{(1414 - 577°\text{C})} = \frac{C_\beta}{(1414 - 1300°\text{C})},$$

$$C_\beta = 0.023 \text{ w/o Al.}$$

b) From Fig. 9–5.7, $C_L = 21$ w/o Al. Using 100 g,

$$L = [100\text{ g}][(2 - 0.023)/(21 - 0.023)] = 9.4 \text{ g Liq};$$

$$\beta = 90.6 \text{ g.}$$

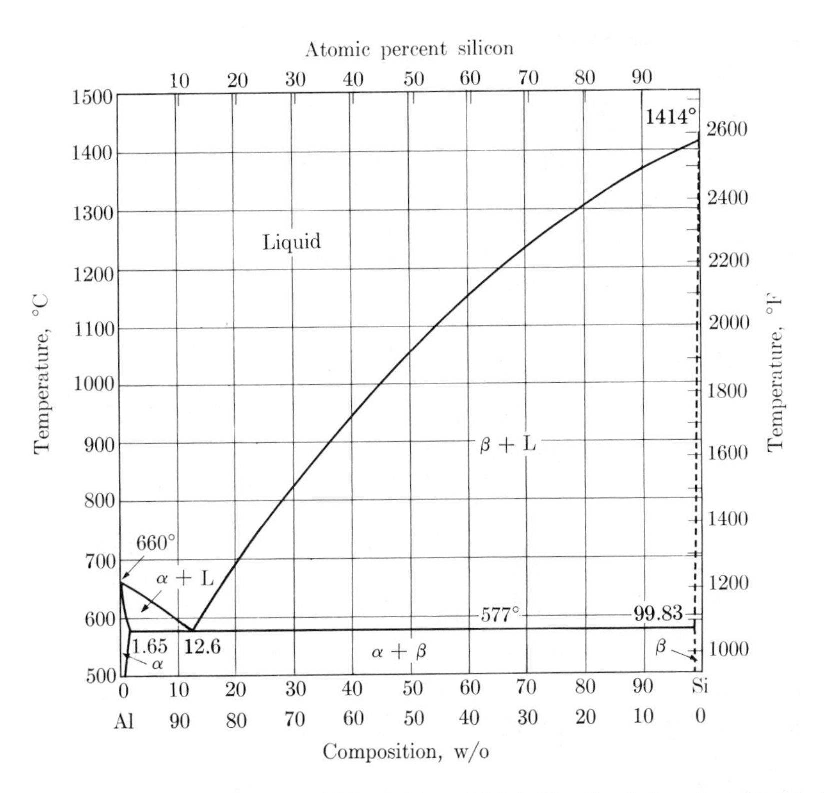

Fig. 9–5.7. Al-Si diagram. (Adapted from *Metals Handbook*, American Society for Metals.)

Fraction aluminum removed:

$$\frac{(9.4\text{ g})(0.21)}{(100\text{ g})(0.02)} = 0.99.$$

c) The upper end of the liquidus curve of Fig. 9–5.7 is approximately linear. Therefore,

$$\frac{21}{(1414 - 1300°\text{C})} = \frac{0.023}{(1414 - T°\text{C})};$$

$$T = 1414 - 0.124°\text{C}$$

d) Repeating (a),

$$\frac{0.17\%}{(1414 - 577°\text{C})} = \frac{C'_\beta}{0.124°\text{C}},$$

$$C'_\beta = 0.00003 \text{ w/o Al}.$$

Comment. By repeating this type of procedure several times, the metallurgist is able to produce silicon with $< 1/10^9$ aluminum for controlled-purity semiconductor materials. ◀

9–6 PHASES OF THE IRON–CARBON SYSTEM

Steels, which are primarily alloys of iron and carbon, offer illustrations of the majority of reactions and microstructures available to the engineer for adjusting material properties. Also, the iron–carbon alloys are among the most prominent structural engineering materials.

The versatility of the steels as engineering materials is evidenced by the many kinds of steel that are manufactured. At one extreme are the very soft steels used for deep-drawing applications such as automobile fenders and stove panels. At the other are the extremely hard and tough steels used for gears and bulldozer blades. Some steels must have abnormally high resistance to corrosion. Steels for such electrical purposes as transformer sheets must have special magnetic characteristics so that they may be magnetized and demagnetized many times each second with low power losses. Other steels must be completely nonmagnetic, for such applications as wrist watches and minesweepers. Phase diagrams can be used to help explain each characteristic described above.

Pure iron changes its crystal structure twice before it melts. As discussed in Section 3–5, iron changes from bcc to fcc at 912°C (1673°F). This change is reversed at 1394°C (2540°F) to form the bcc structure again. The bcc polymorph then remains stable until iron melts at 1538°C (2800°F).

Ferrite, or α-iron. The structural modification of pure iron at room temperature is called either *α-iron* or *ferrite*. Ferrite is quite soft and ductile; in the purity which is encountered commercially, its tensile strength is less than 45,000 psi (310MPa). It is a ferromagnetic material at temperatures under 770°C (1418°F).

Since ferrite has a body-centered cubic structure, the interatomic spaces are small and pronouncedly oblate, and cannot readily accommodate even a small spherical carbon atom. Therefore, solubility of carbon in ferrite is very low. The carbon atom is too small for substitutional solid solution, and too large for extensive interstitial solid solution (Section 4–2).

Austenite, or γ-iron. The face-centered modification of iron is called *austenite*, or *γ-iron*. It is the stable form of pure iron at temperatures between 912°C and 1394°C. Making a direct comparison between the mechanical properties of austenite and ferrite is difficult because they must be compared at different temperatures. However, at its stable temperatures, austenite is soft and ductile and consequently is well suited to fabrication processes. Most steel forging and rolling operations are performed at 1100°C (2000°F) or above, when the iron is face-centered cubic. Austenite is not ferromagnetic at any temperature.

The face-centered cubic structure of iron has larger interatomic spacings than does ferrite. Even so, in the fcc structure the holes are barely large enough to crowd the carbon atoms into the interstices, and this crowding introduces strains into the structure. As a result, not all the holes can be filled at any one time. The maximum

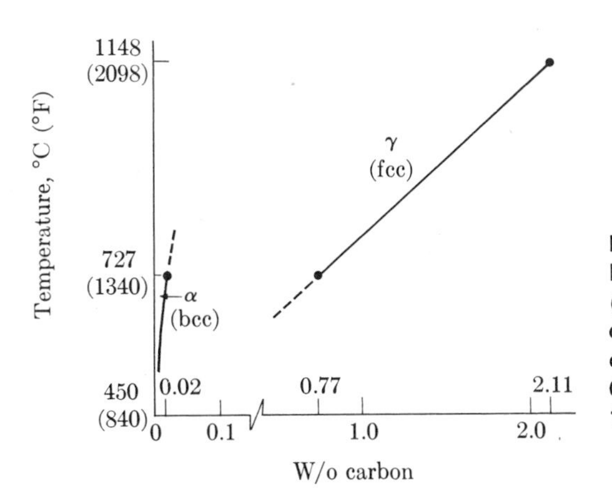

Fig. 9–6.1. Carbon solubility in iron. Negligible carbon dissolves in ferrite (α) at 20°C. The solubility increases to only 0.02 w/o at 727°C. The solubility of carbon in austenite (γ) increases from 0.77 w/o at 727°C to 2.11 w/o at 1148°C. (Cf. Fig. 9–7.1.)

solubility is only 2.11% (9 a/o) carbon (Fig. 9–6.1). By definition, steels contain less than 1.2% carbon; thus steels may have their carbon completely dissolved in austenite at high temperatures.

δ-iron. Above 1394°C (2540°F), austenite is no longer the most stable form of iron, since then the crystal structure changes back to a body-centered-cubic phase called *δ-iron.* δ-iron is the same as α-iron except for its temperature range, and so it is commonly called δ-ferrite. The solubility of carbon in δ-ferrite is small, but it is measurably larger than in α-ferrite, because of the higher temperature.

Iron carbide. In iron–carbon alloys, carbon in excess of the solubility limit must form a second phase, which is most commonly iron carbide (*cementite*). Iron carbide has the chemical composition of Fe_3C. This does not mean that iron carbide forms molecules of Fe_3C, but simply that the crystal lattice contains iron and carbon atoms in a three-to-one ratio. Fe_3C has an orthorhombic unit cell with 12 iron atoms and 4 carbon atoms per cell, and thus has a carbon content of 6.7%.

As compared with austenite and ferrite, cementite is very hard. The presence of iron carbide with ferrite in steel greatly increases the strength of the steel (Section 10–4). However, because pure iron carbide is nonductile, it cannot adjust to stress concentrations; therefore it is relatively weak by itself. (Compare with nonductile ceramic materials in Section 8–8.)

Example 9–6.1 Carbon atoms sit in the $\frac{1}{2}, \frac{1}{2}, 0$ positions in bcc iron. a) Using $r_C = 0.77$ Å and $R_\alpha = 1.24$ Å, how much must the nearest iron atoms be displaced to accommodate the carbon?

A carbon atom sits in the $\frac{1}{2}, \frac{1}{2}, \frac{1}{2}$ position in fcc iron. b) Using $R_\gamma = 1.27$ Å (Table 2–6.1) for the radius of γ-iron, how much crowding is present?

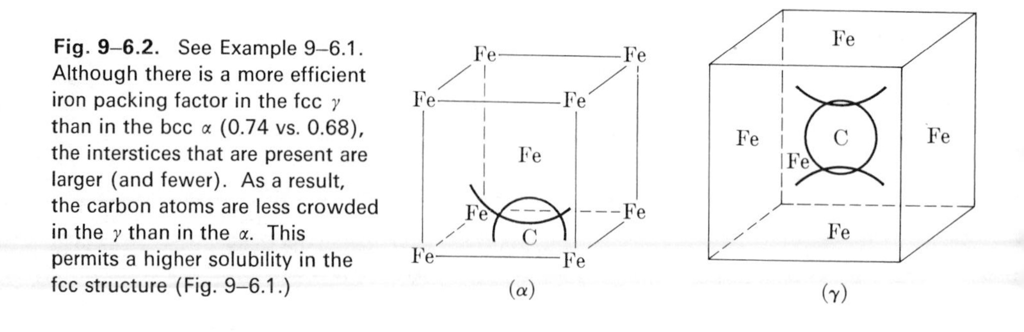

Fig. 9–6.2. See Example 9–6.1. Although there is a more efficient iron packing factor in the fcc γ than in the bcc α (0.74 vs. 0.68), the interstices that are present are larger (and fewer). As a result, the carbon atoms are less crowded in the γ than in the α. This permits a higher solubility in the fcc structure (Fig. 9–6.1.)

Solution:

a) Refer to Fig. 9–6.2(a). The nearest iron atoms are at $\frac{1}{2}, \frac{1}{2}, \frac{1}{2}$ and $\frac{1}{2}, \frac{1}{2}, -\frac{1}{2}$.

$$a = 4(1.24\ \text{Å})/\sqrt{3} = 2.86\ \text{Å},$$
$$\Delta = (1.24\ \text{Å} + 0.77\ \text{Å}) - (2.86\ \text{Å})/2 = 0.6\ \text{Å}.$$

b) Refer to Fig. 9–6.2(b). The nearest iron atoms are at the fcc positions.

$$a = 4(1.27\ \text{Å})/\sqrt{2} = 3.59\ \text{Å}$$
$$\Delta = (1.27\ \text{Å} + 0.77\ \text{Å}) - (3.59\ \text{Å})/2 = 0.25\ \text{Å}.$$

Comments. Although bcc iron has a lower packing factor than fcc, the interstices do not accommodate a carbon atom without crowding. [The lower APF of a bcc metal (0.68) than of an fcc metal (0.74) occurs because there are 12 of the $\frac{1}{2}, \frac{1}{2}, 0$ (and $0, 0, \frac{1}{2}$) interstices per bcc unit cell of two atoms. In fcc there are only four interstices per unit cell of four atoms.] ◀

9–7 THE Fe–Fe_3C PHASE DIAGRAM

Figure 9–7.1 presents the phase diagram between iron and iron carbide, Fe_3C.* This phase diagram is the basis for the heat-treating of the majority of our steels.

If you cover the 0–1% carbon region of Fig. 9–7.1 by your hand, you will see a close resemblance to previous phase diagrams. The eutectic composition is at 4.3 w/o (17 a/o) carbon; the eutectic temperature is 1148°C (2100°F). *Cast irons* are based on this eutectic region, since they typically contain 2.5–4% carbon. They have relatively low melting points with certain desirable processing and mold-filling characteristics.

The iron-rich γ can dissolve up to 9 a/o (2.1 w/o) carbon. As discussed in the last section, the carbon atoms dissolve interstitially in fcc iron. *Steels* are based on this

* It will not be necessary for us to give consideration to higher carbon contents because graphite develops and this does not give usable metal products. Even at lower carbon contents, graphite will form in lieu of iron carbide if enough time is available, particularly when certain other elements such as Si or Te are present.

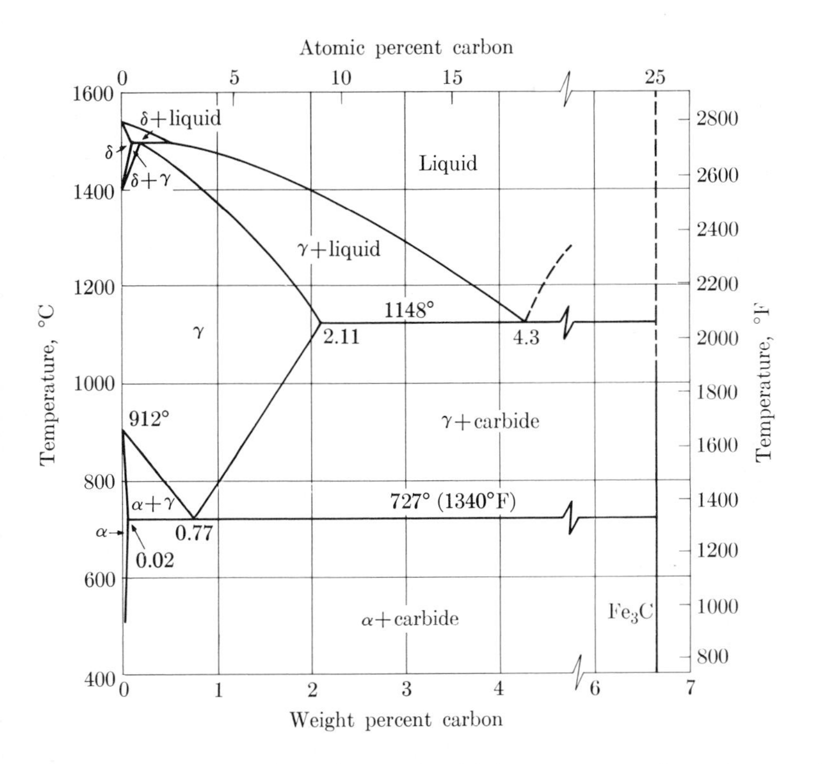

Fig. 9–7.1. Fe–Fe_3C phase diagram. The left lower corner receives prime attention in heat-treating steels (Fig. 9–7.3).

solid-solution phase. Since steels contain less than 1.2% carbon, they can be single-phase at forging and other hot-working temperatures, which are commonly in the region of 1100 to 1250°C (2000 to 2300°F).

In the iron-rich region (>99% Fe and <1% C), the Fe–Fe_3C diagram differs from any of the diagrams we have examined previously. This difference arises from the fact that iron is polymorphic with bcc and fcc phases. Since we are not studying steel melting and solidification in this text, we will skip the features of this low-carbon region that are found above 1400°C. We will pay considerable attention to the features of the diagram in the 700–900°C (1300–1650°F) temperature range and the 0–1% carbon range, because it is here that the engineer can develop within the steel those microstructures that are required for the desired properties.

The eutectoid reaction. In Fig. 9–7.2 a comparison is made between the addition of common salt to water (Fig. 9–1.2) and the addition of carbon to austenite. In each case, the addition of the solute lowers the stable temperature range of the solvent phase. These two examples differ in only one respect: in the ice–salt system, a *liquid*

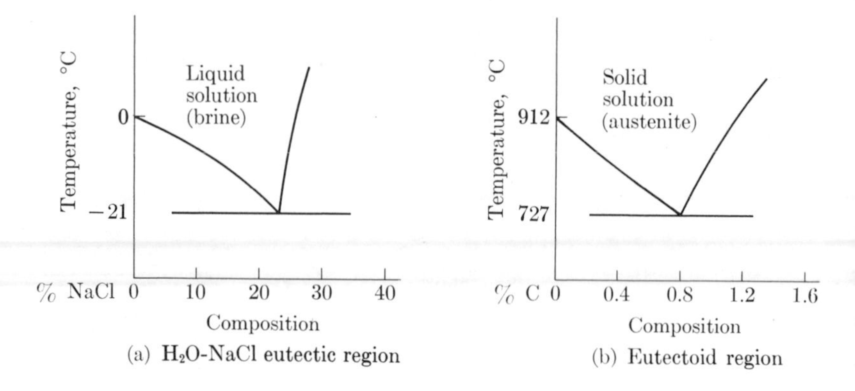

Fig. 9–7.2. Eutectic and eutectic-like (eutectoid) regions of phase diagrams.

solution exists above the eutectic temperature; in the iron–carbon system, a *solid solution* exists, so that a true eutectic reaction does not occur upon cooling. However, because of the analogy of this reaction to the eutectic reaction, it is called *eutectoid* (literally, eutectic–like).

$$\text{Eutectic:} \qquad L_2 \underset{\text{Heating}}{\overset{\text{Cooling}}{\rightleftharpoons}} S_1 + S_3; \tag{9–3.2}$$

$$\text{Eutectoid:} \qquad S_2 \underset{\text{Heating}}{\overset{\text{Cooling}}{\rightleftharpoons}} S_1 + S_3. \tag{9–7.1}$$

The *eutectoid temperature* for iron–carbon alloys is 727°C (1340°F). The corresponding *eutectoid composition* is approximately 0.8% carbon. The specific eutectoid reaction for Fe–C alloys is:

$$\gamma(0.77\%\text{C}) \overset{727°\text{C}}{\rightleftharpoons} \alpha(0.02\%\text{C}) + \text{Fe}_3\text{C}(6.7\%\text{C}). \tag{9–7.2}$$

Figure 9–7.3 shows the eutectoid region in greater detail than in Fig. 9–7.1.

Example 9–7.1 An eutectoid steel (~0.8% carbon) is heated to 800°C (1472°F) and cooled slowly through the eutectoid temperature (727°C). Calculate the number of grams of carbide which form per 100 g of steel.

Solution: At 726°C (1338°F),

$$\begin{aligned} \text{Carbide} &= [(0.8 - 0.02)/(6.7 - 0.02)](100\text{ g}) \\ &= 12\text{ g} \qquad (\text{and } 88\text{ g } \alpha). \end{aligned}$$

Comment. Careful measurements indicate that the eutectoid composition is 0.77% carbon. The figure, 0.8% C, is generally used, however. ◀

Example 9–7.2 Plot the fraction of ferrite, austenite, and carbide in an alloy of 0.60% carbon, 99.40% iron as a function of temperature.

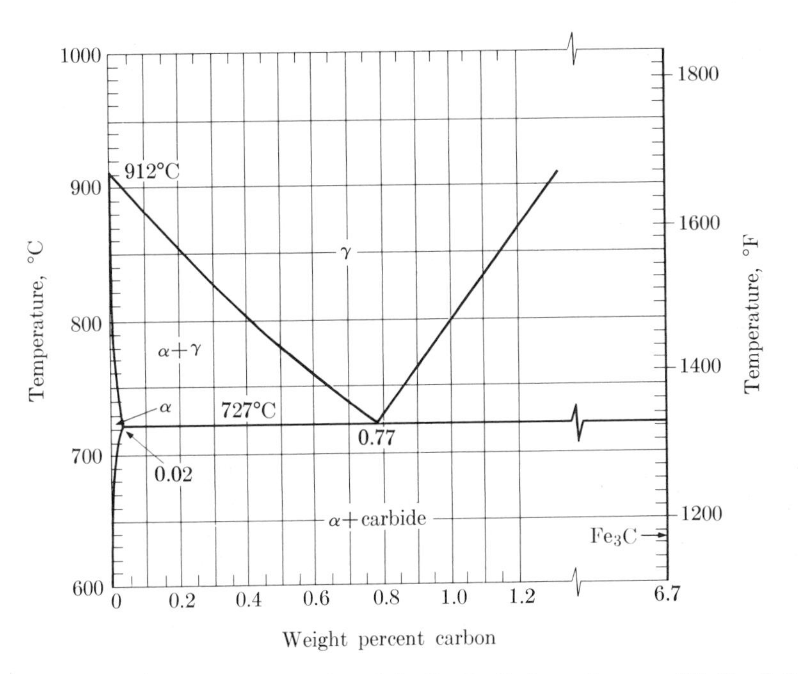

Fig. 9–7.3. The eutectoid region of the Fe–Fe_3C phase diagram. (Cf. Fig. 9–7.1.)

Solution: At 728°C (1342°F),

$$\text{Ferrite} = \frac{0.8 - 0.60}{0.8 - 0.02} = 0.26;$$

At 726°C (1338°F),

$$\text{Ferrite} = \frac{6.7 - 0.60}{6.7 - 0.02} = 0.91.$$

Results are shown in Fig. 9–7.4 for other temperatures and for other phases. ◀

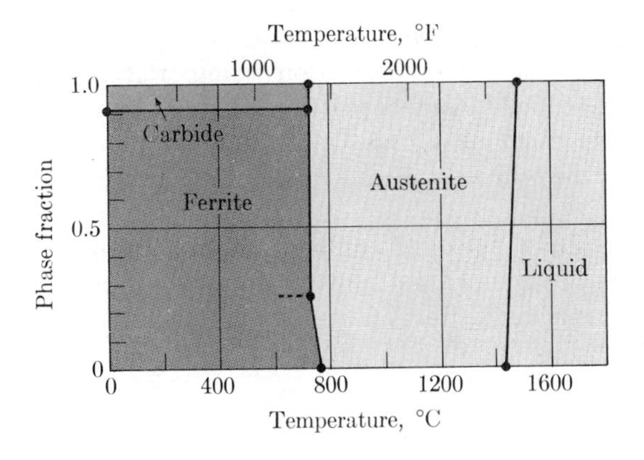

Fig. 9–7.4. Phase fractions (99.4Fe–0.6C). (See Example 9–7.2.) Almost three-fourths of this steel is austenite at 728°C (1342°F). (Since the α has an eutectoid composition at that temperature, we will observe in the next section that the steel will contain almost 75% pearlite (and about one fourth proeutectoid ferrite).)

9–8 AUSTENITE DECOMPOSITION

During cooling, the Fe–C eutectoid reaction involves the simultaneous formation of ferrite and carbide from the decomposition of austenite of eutectoid composition:

$$\gamma(\sim 0.8\%\mathrm{C}) \rightarrow \alpha + \bar{\mathrm{C}}. \qquad (9\text{–}8.1)^*$$

There is nearly 12% carbide and more than 88% ferrite in the resulting mixture. Since the carbide and ferrite form simultaneously, they are intimately mixed. Characteristically, the mixture is *lamellar*; i.e., it is composed of alternate layers of ferrite and carbide (Fig. 9–8.1). The resulting microstructure, called *pearlite*, is very important in iron and steel technology, because it may be formed in almost all steels by means of suitable heat treatments.

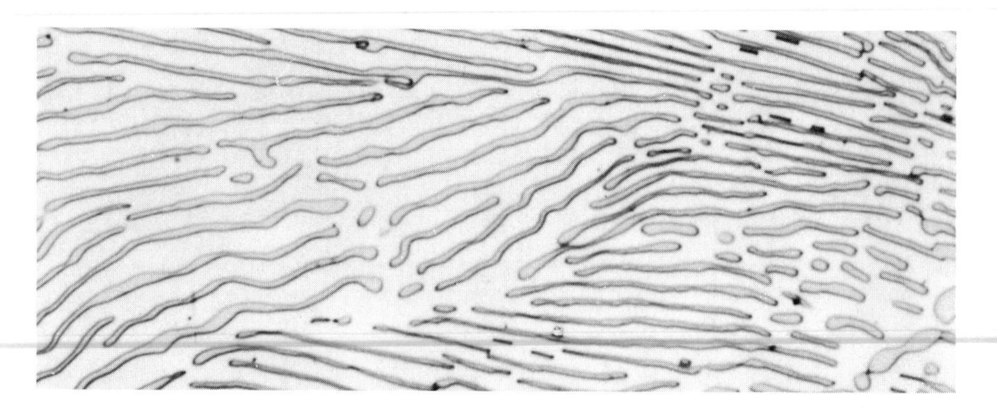

Fig. 9–8.1. Pearlite, X2500. This microstructure is a lamellar mixture of ferrite (lighter matrix) and carbide (darker). Pearlite forms from austenite of eutectoid composition. Therefore the amount and composition of pearlite is the same as the amount and composition of eutectoid. (U.S. Steel Corp.)

Pearlite is a specific mixture of two phases formed by transforming austenite of eutectoid composition to ferrite and carbide. This distinction is important, since mixtures of ferrite and carbide may be formed by other reactions as well. However, the microstructure resulting from other reactions will not be lamellar (compare Figs. 10–4.7(b) and 9–8.1) and consequently the properties of such mixtures will be different. (See Section 10–4.)

Since pearlite comes from austenite of eutectoid composition, the amount of pearlite present is equal to the amount of eutectoid austenite transformed (Fig. 9–8.2). We can determine this amount by measuring the fraction quantity of γ just above the eutectoid temperature.

---

* We will use $\bar{\mathrm{C}}$ to denote *carbide*, to avoid confusion with C for elemental carbon.

Example 9–8.1 Determine the amount of pearlite in a 99.6% Fe–0.4% C alloy which is cooled slowly from 870°C (1600°F). Basis: 100 g of alloy.

Solution: From 870 to 805°C: 100 g austenite with 0.4% C.

From 805 to 727°C(+): ferrite separates from austenite and the carbon content of the austenite increases to 0.8% C.

At 727° C(+): composition of ferrite = 0.02% C,
amount of ferrite = 51 g;
composition of austenite = 0.8% C,
amount of austenite = 49 g.

At 727°C(−): amount of pearlite = 49 g. (It came from, and replaced, the austenite with an eutectoid composition.)

Comments. Each of the above calculations assumes sufficient time for equilibrium to be attained. Ferrite which formed above 727°C (i.e., before the eutectoid reaction) is called *proeutectoid ferrite*. That which is part of the pearlite, having been formed from austenite with the eutectoid composition is called *eutectoid ferrite*. (See Fig. 9–8.2(b–d).) ◀

Example 9–8.2 From the results of the example above, determine the amount of ferrite and carbide present in the specified alloy (a) at 725°C, and (b) at room temperature. Basis: 100 g of alloy. (Some data come from Example 9–8.1.)

Solution: (a) at 725°C:

$$\text{Amount of carbide:}\quad 49\,\frac{0.8 - 0.02}{6.7 - 0.02} = 5.7\,\frac{\text{g carbide}}{\text{100 g steel}};$$

Amount of ferrite: 49 − 5.7 = 43.3 g ferrite formed with the pearlite (eutectoid),

51 g ferrite formed before the pearlite (proeutectoid),

94.3 g ferrite total/100 g steel

Alternative calculations:

$$\text{Amount of carbide:}\quad \frac{0.4 - 0.02}{6.7 - 0.02} = 5.7\,\frac{\text{g carbide}}{\text{100 g steel}};$$

$$\text{Amount of ferrite:}\quad \frac{6.7 - 0.4}{6.7 - 0.02} = 94.3\,\frac{\text{g ferrite}}{\text{100 g steel}}.$$

b) At room temperature (the solubility of carbon in ferrite at room temperature may be considered zero for these calculations),

$$\frac{0.4 - 0}{6.7 - 0} = 6.0\,\frac{\text{g carbide}}{\text{100 g steel}},$$

$$\frac{6.7 - 0.4}{6.7 - 0} = 94.0\,\frac{\text{g ferrite}}{\text{100 g steel}}.$$

Comments. Additional carbide is precipitated from the ferrite below the eutectoid point because the solubility of carbon in ferrite decreases to nearly zero. This additional carbide is not part of the pearlite. (Each of these calculations assumes that equilibrium prevails.) ◀

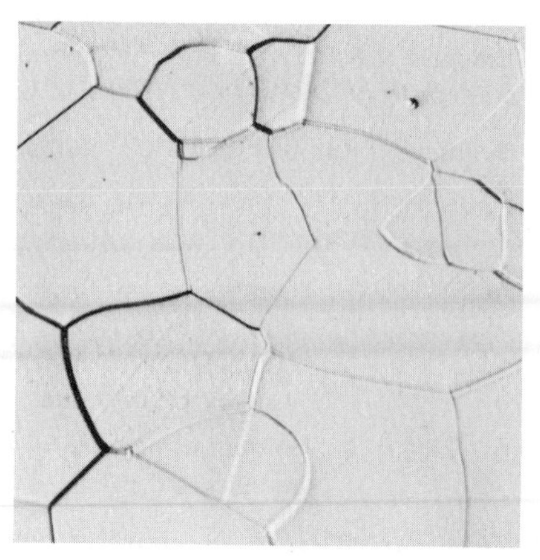

(a) 0% C

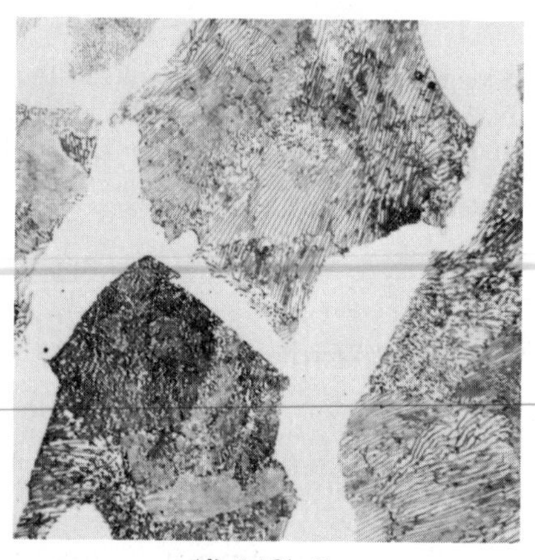

(d) 0.5% C

Fig. 9–8.2. Annealed iron–carbon alloys (X500). The sequence (a) through (f) increases in carbon content; therefore there is a corresponding decrease in the amount of proeutectoid ferrite (white areas) which separated from the austenite prior to the eutectoid reaction. The gray areas are pearlite, a lamellar mixture of eutectoid ferrite and carbide. The network between the former grains of austenite in (f) is proeutectoid carbide since the carbon content of that alloy is greater than 0.8 w/o. (U.S. Steel Corp.)

9–9 PLAIN-CARBON AND LOW-ALLOY STEELS

An important category of steels is designed for heat-treating into the γ range (*austenitizing*), followed by cooling and decomposition of the austenite either directly (or indirectly) to ferrite plus carbide ($\alpha + \bar{C}$). If a steel contains primarily iron and carbon,* the alloy is called a *plain-carbon steel.* We use the term *low-alloy steel* if we add up to five percent of alloying elements such as nickel, chromium, molybdenum, manganese, silicon, etc. Alloying elements are added primarily to reduce the decomposition rate of austenite to ($\alpha + \bar{C}$) during heat treatment. The result is a much harder steel, as we will see in Chapter 10.

* Plus approximately 0.5% Mn for processing purposes.

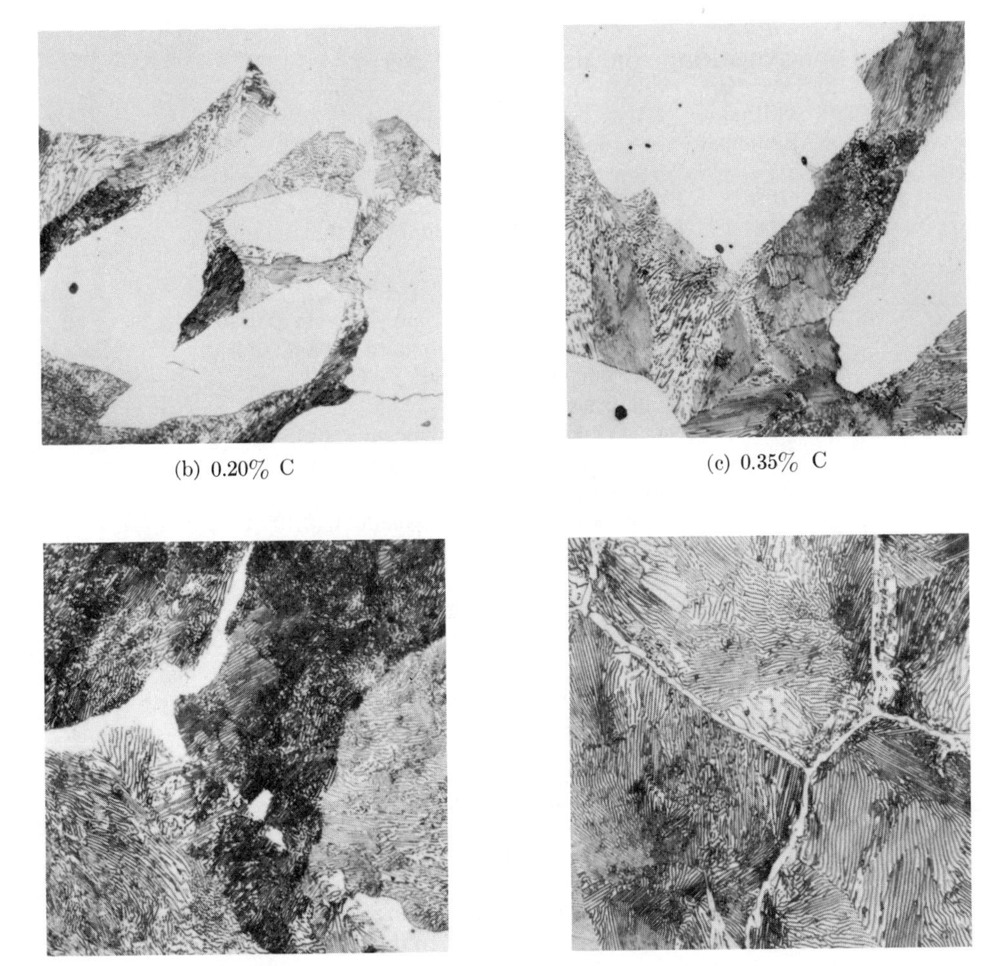

(b) 0.20% C

(c) 0.35% C

(e) 0.7% C

(f) 1.2% C

The proeutectoid ferrite (Figs. 9–8.2(b) through (e)) deforms more readily than the ferrite within the pearlite since the former is more massive and is not reinforced by the harder, more rigid carbide. A metallurgist refers to *hypoeutectoid* steels (literally, lower carbon than the eutectoid composition) when he speaks of steels with microstructures containing the separate ferrite areas. In a like manner, an *eutectoid steel* is predominantly pearlite if it is cooled slowly; and a *hypereutectoid steel* contains carbon *above* the eutectoid composition, thus producing proeutectoid carbide (Fig. 9–8.2(f)). Hypoeutectoid steels are more common than hypereutectoid steels.

Nomenclature for steels. The importance of carbon in steel has made it desirable to indicate the carbon content in the identification scheme of steel types. A four-digit numbering scheme is used, in which the last two digits designate the number of

Table 9–9.1
NOMENCLATURE FOR AISI AND SAE STEELS

AISI or SAE number	Composition
10xx	Plain-carbon steels*
11xx	Plain-carbon (resulfurized for machinability)
15xx	Manganese (1.0–2.0%)
40xx	Molybdenum (0.20–0.30%)
41xx	Chromium (0.40–1.20%), molybdenum (0.08–0.25%)
43xx	Nickel (1.65–2.00%), chromium (0.40–0.90%) molybdenum (0.20–0.30%)
44xx	Molybdenum (0.5%)
46xx	Nickel (1.40–2.00%), molybdenum (0.15–0.30%)
48xx	Nickel (3.25–3.75%), molybdenum (0.20–0.30%)
51xx	Chromium (0.70–1.20%)
61xx	Chromium (0.70–1.10%), vanadium (0.10%)
81xx	Nickel (0.20–0.40%), chromium (0.30–0.55%), molybdenum (0.08–0.15%)
86xx	Nickel (0.30–0.70%), chromium (0.40–0.85%), molybdenum (0.08–0.25%)
87xx	Nickel (0.40–0.70%), chromium (0.40–0.60%), molybdenum (0.20–0.30%)
92xx	Silicon (1.80–2.20%)

xx—carbon content, 0.xx%.
* All plain-carbon steels contain 0.50% ± manganese, and residual amounts (<0.05 w/o) of other elements.

hundredths of percent of carbon content (Table 9–9.1). For example, a 1040 steel has 0.40% carbon (plus or minus a small workable range). The first two digits indicate the type of alloying element that has been added to the iron and carbon. The classification (10xx) is reserved for plain-carbon steels with only a minimum amount of other alloying elements.

These designations for the steels are accepted as standard by both the American Iron and Steel Institute and the Society of Automotive Engineers (AISI and SAE). Many commercial steels are not included in this classification scheme because of larger additions or more subtle variations in alloy contents. Usually, however, such steels have more specialized applications and may not be stocked as regular warehouse items.

•**Eutectoid shifts.** Figure 9–7.3 shows the eutectoid region for the Fe–Fe_3C phase diagram, i.e., with only iron and iron carbide present. The eutectoid temperature is 727°C (1340°F); the eutectoid composition is just under 0.8% carbon.

In alloy steels, the carbon atoms and the iron atoms coordinate with atoms of other types, as well as with each other. Therefore, we should not be surprised that the eutectoid carbon content and temperature are altered from the above values when

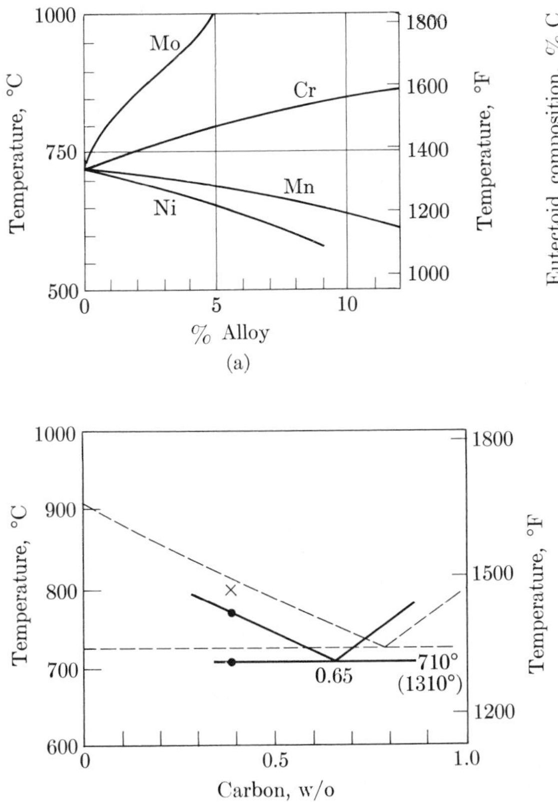

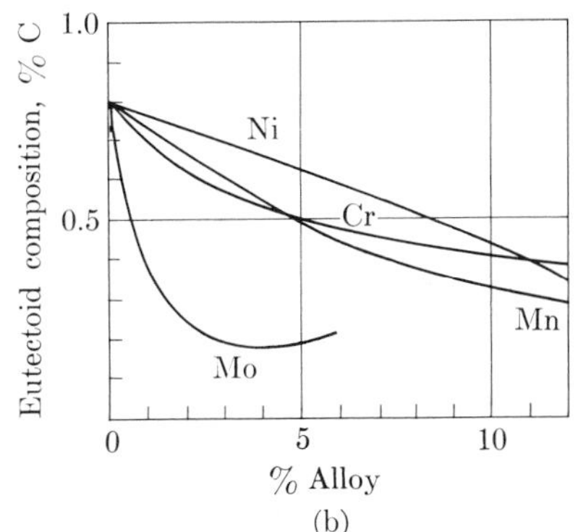

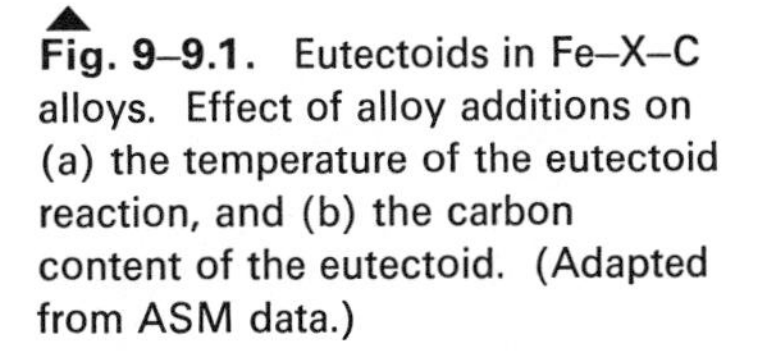

Fig. 9–9.1. Eutectoids in Fe–X–C alloys. Effect of alloy additions on (a) the temperature of the eutectoid reaction, and (b) the carbon content of the eutectoid. (Adapted from ASM data.)

Fig. 9–9.2. Eutectoid shift (Example 9–9.1). With nearly 2% Mn, the eutectoid analysis drops to 0.65% carbon, and the eutectoid temperature drops to 710°C (1310°F). Therefore, there is more γ from which pearlite forms.

other elements are present. Figure 9–9.1(a) shows the shift in eutectoid temperature as nickel, manganese, chromium and molybdenum are added. The first two elements lower the eutectoid temperature because they preferentially form fcc solid solutions with iron (austenite-formers); Cr and Mo raise the eutectoid temperature because they are bcc themselves, and are thus ferrite-formers in steels. Figure 9–9.1(b) shows that each of the alloying elements lowers the carbon content of the eutectoid analysis. These curves are valuable because they let us predict the temperature of austenitization for heat treatment of low-alloy steels.

• **Example 9–9.1** An SAE 1540 steel (C = 0.39%; Mn = 1.97%) is to be austenitized. a) What temperature is required if it must be heated 30°C into the austenite range? b) How much pearlite will it form during cooling?

Solution: From Fig. 9–9.1, the eutectoid is shifted to C ~ 0.65%, and T_e = 710°C. At 0.39% C, the lower side of the γ-range is 770°C (Fig. 9–9.2).

a) 770 + 30°C = 800°C (1470°F);

b) Fraction γ at 711°C = 0.6 (by interpolation on Fig. 9–9.2).
Fraction pearlite below 710°C = 0.6.

Comment. For our purposes we can assume that solubility lines which outline the γ field do not change slope as small amounts of alloying elements are added.

In the absence of 1.97% Mn, 810°C + 30° would have been required to austenitize; and half of the steel would be pearlite. ◀

REVIEW AND STUDY

The purpose of this chapter was two-fold:

1. One must understand how to extract necessary data from phase diagrams if one is to select, control, and modify multiphase materials.

2. The Fe–C system was examined more closely so we can use it as a prototype for solid-state microstructures in Chapter 10, and for the control of microstructures in Chapter 11.

SUMMARY: PHASE DIAGRAMS

We can obtain three types of data from a phase diagram, which is a graph of solubility-limit curves:

1. The phase diagram shows *what phase(s)* are present under equilibrium conditions for various temperatures and compositions.

2. The phase diagram also provides the *chemical compositions* for all equilibrated phases: (a) In a single-phase area the composition is simply the same as the alloy. (b) In a two-phase area, the compositions are located where the isotherm crosses the solubility-limit curves.

3. Finally, we can calculate the *quantity fractions* of the phases in a two-phase alloy by interpolation along the isotherm. This is the same as the "lever rule."

SUMMARY: Fe–Fe_3C DIAGRAM

The iron–carbon system has an eutectic region which is the basis for cast irons. Steels, however, are based on the fact that iron becomes fcc at elevated temperatures and can dissolve the carbon which is present. Thus steels are normally forged and hot-worked as a single-phase material.

The fcc solid solution (austenite, or γ) decomposes during cooling to bcc ferrite (α) and carbide ($\bar{C}$). The decomposition is an eutectoid reaction:

$$\gamma(\sim 0.8\%\mathrm{C}) \rightarrow \alpha(\sim 0.02\%\mathrm{C}) + \bar{\mathrm{C}}(6.7\%\mathrm{C}). \qquad (9\text{–}10.1)$$

Heat treatment of steel involves forming austenite and then decomposing the austenite directly or indirectly (Chapter 10) to form a variety of microstructures. The microstructure which is obtained controls the properties of the steel (Chapters 10 and 11).

Two prominent categories of steels are (1) the plain-carbon steels, and (2) the low-alloy steels. The first are essentially Fe–Fe_3C alloys; the latter may contain up to five percent of other alloying elements, such as Mn, Ni, Cr, Mo, W, Si, etc.

TERMS FOR REVIEW

Austenite, γ
Austenite decomposition
Austenitization
Carbide, $\bar{C}$
Cast iron
Component
Equilibrium
Eutectic composition
Eutectic reaction
Eutectic temperature
Eutectoid composition
Eutectoid reaction
•Eutectoid shift
Eutectoid temperature
Ferrite, α
  eutectoid
  proeutectoid
Isotherm
Lever rule (inverse)
Liquidus
Materials balance
Mixture
Nomenclature, (AISI–SAE)
Pearlite, $\alpha + \bar{C}$
Phase diagram
  isothermal cut
  one-phase area
  two-phase area
  three-phase temperature
Phases
  chemical compositions of
  quantities of
Quantity fraction, $X/(X + Y)$
Solder, Pb–Sn
Solidus
Solubility limit
Solution
Steel
  eutectoid
  hypereutectoid
  hypoeutectoid
  low-alloy
  plain-carbon

PHASE-DIAGRAM INDEX

FOR DISCUSSION

A_9 From Fig. 9–1.2, suggest a method for purifying sea water.

B_9 Arctic oceans (−1.5°C) cease melting snow when the NaCl content drops to 2.5%. Why?

C_9 Explain why honey will "sugar" if it is stored in an open jar.

D_9 Check the 1–2–1–2–1–··· sequences at various isotherms in a complex phase diagram such as the Cu–Sn system (Fig. 9–5.2). (Note that there is still a one-phase δ field at 400°C. Admittedly it is narrow.)

E_9 Distinguish between the chemical composition of a phase, and the quantity of a phase.

F_9 Why does the composition of an eutectic liquid, L_2, always lie between the compositions of the two solids, S_1, and S_3?

G_9 Why are the equilibrium compositions of two phases always at the limits of where the isotherm passes through the two-phase area?

H_9 We can use interpolation in the horizontal isothermal direction to get the quantity fractions of the two phases. We can *not* obtain a valid answer by interpolating along a vertical composition line. By using a 70Pb–30Sn solder, explain why the latter doesn't work.

I_9 Perform the algebraic rearrangement which we mentioned just before Eq. (9–4.3). [*Hint:* Solve Eq. (9–4.2) for X, to use in Eq. (9–4.3).]

J_9 Devise a "center-of-gravity" calculation to determine the quantity fraction of a phase. Show that this is identical to the "lever rule" and the "interpolation procedure."

•K_9 Explain to a classmate why the quantity of α in an 80Ag–20Cu alloy is indeterminate at the eutectic temperature. However, if you have 50 g of α in 100 g of alloy, the temperature would have to be 780°C, if equilibrium exists.

L_9 Take the 475 to 575°C and 10 to 33% Sn region of the Cu–Sn phase diagram (Fig. 9–5.2) and sketch it separately with an enlarged scale. Prepare a problem involving phase compositions and quantities to be solved by a classmate.

M_9 A long rod of silicon is "zone-refined" by passing it through a short heater so that a molten zone starts at one end and moves toward its other end. It solidifies as it leaves the heater. Explain how this could be repeated to remove impurities (cf. Example 9–5.1).

N_9 Why can austenite dissolve more carbon than ferrite?

O_9 Based on the comments of Example 9–6.1, explain why $D_{\text{C in }\alpha} > D_{\text{C in }\gamma}$, even though austenite can dissolve many times more carbon atoms than can ferrite.

P_9 The data of Fig. 9–7.3 will be used on numerous occasions during the balance of the text. They would be worth remembering so you don't have to look them up on each occasion.

Q_9 Pearlite is a 2-phase microstructure containing α and $\bar{C}$. Distinguish between phases and microstructures.

R_9 The solution to Example 9–8.2(a) mentions eutectoid ferrite and proeutectoid ferrite. Based on Fig. 9–8.2(d) and Fig. 9–7.3, explain the difference.

S_9 Explain the AISI–SAE nomenclature pattern to a freshman who has had chemistry but who is not in this course.

•T_9 Predict whether tungsten (bcc) will raise or lower the temperature of austenite decomposition.

STUDY PROBLEMS

9–1.1 Which will melt more ice at −5°C (23°F), (a) 1 kg of NaCl or (b) 1 kg of $CaCl_2$?

Answer: a) ~13 kg b) ~9 kg

9–1.2 A syrup contains equal quantities of water and sugar. How much more sugar can be dissolved into 100 g of the syrup at 80°C?

9–2.1 A 90Cu–10Sn bronze is cooled slowly from 1100°C to 20°C. What phase(s) will be present as the cooling progresses?

Answer: 1100 → 1020°C, L; 1020 → 830°C, (L + α); 830°C → 340°C, α; 340 → 20°C, (α + ϵ)

9–2.2 A 65Cu–35Zn brass is heated from 300°C to 1000°C. What phase(s) are present at each 100°C interval?

9–2.3 Show the sequential changes in phases when the composition of an alloy is changed from 100% Cu to 100% Al (a) at 700°C, (b) at 450°C, (c) at 900°C. (See Fig. 9–5.3.)

Answer: c) At 900°C: α, $\alpha + \beta$, β, $\beta + \gamma_1$, γ_1, $\gamma_1 + \epsilon_1$, ϵ_1, ϵ_1 + L, L

9–2.4 Show the sequential changes in phases when the composition of an alloy is changed from 100% Mg to 100% Al (a) at 300°C, (b) at 400°C. (See Fig. 9–5.4.)

9–3.1 What are the chemical compositions of the phase(s) in Study Problem 9–2.2?

Answer: 300°, 400°, 500°, 600°, 700°: α(65Cu–35Zn); 800°C: α(66Cu–34Zn) and β(61Cu–39Zn); 900°C: α(67.5Cu–32.5Zn) and β(~63Cu–37Zn); 1000°C: Liq(65Cu–35Zn)

9–3.2 What are the chemical compositions of the phase(s) in Study Problem 9–2.1?

9–4.1 Assuming 1500 g of bronze in Study Problem 9–3.2, what is the weight of solid phase(s) at each 100°C interval?

Answer: 1100°C, 0 g; 1000°C, ~150 g α; 800°C, 600°C, 400°C, 1500 g α; 200°C, ~1350 g α + 150 g ϵ

9–4.2 With 200 g of 65Cu–35Zn, what weight of α is present at each temperature of Study Problem 9–3.1?

Answer: 300 → 700°C, 200 g α; 800°C, 160 g α; 900°C, ~90 g α; 1000°C, 0 g α

9–4.3 The solubility of tin in solid lead at 200°C is 18% Sn. The solubility of lead in the molten metal at the same temperature is 44% Pb. What is the composition of an alloy containing 40% liquid and 60% solid α at 200°C?

Answer: 66.8% Pb, 33.2% Sn

9–4.4 (a) At what temperature would a monel-metal alloy (70% nickel, 30% copper) contain $\frac{2}{3}$ liquid and $\frac{1}{3}$ solid, and (b) what would be the composition of the liquid and of the solid?

9–5.1 An alloy of 50 g Cu and 30 g Zn is melted and cooled slowly. (a) At what temperature will there be 40 g α and 40 g β? (b) 50 g α and 30 g β? (c) 30 g α and 50 g β?

Answer: a) 780°C b) 750°C c) 800°C

9–5.2 (a) What are the chemical compositions of the phases in a 10% magnesium–90% aluminum alloy at 600°C, 400°C, 200°C? (b) What are the quantities of these phases at each of the temperatures in part (a)? (c) Make a material balance for the distribution of the magnesium and aluminum in the above alloy at 600°C.

9–5.3 Make a material balance for *P* grams of a 92–8 Ag–Cu alloy at 500°C (equilibrium conditions).

Answer: 0.938*P* g α, 98% Ag–2% Cu; 0.062*P* g β, 1% Ag–99% Cu

9–5.4 Make a material balance for 100 g of a 90 Mg–10 Al alloy at 200°C (assume equilibrium).

Answer: 84.6 g ϵ, 96% Mg–4% Al; 15.4 g γ, 57% Mg–43% Al

9–5.5 A die-casting alloy of 95 Al–5 Si is cooled so that the metal contains primary (pro-eutectic) α and an eutectic mixture of ($\alpha + \beta$). What fraction of the casting is primary α?

Answer: 0.7 primary α

9–5.6 How much mullite would be present in a 60% SiO_2–40% Al_2O_3 brick at the following temperatures under equilibrium conditions: (a) 1400°C, (b) 1580°C, (c) 1600°C?

9–5.7 (a) Determine the compositions of Al–Si alloys that would contain $\frac{1}{3}$ liquid and $\frac{2}{3}$ solid when brought to equilibrium at 600°C. (b) Give the chemical compositions of the liquids.

Answer: a) 96 Al–4 Si, 29 Al–71 Si b) 91 Al–9 Si, 86 Al–14 Si

9–5.8 Based on Figs. 9–2.1 and 9–7.4, make a graph for an alloy containing 80% Pb and 20% Sn showing (a) the fraction of liquid versus temperatures, (b) the fraction of α versus temperature, and (c) the fraction of β versus temperature.

9–5.9 Draw on cross-section paper the equilibrium diagram for alloys of metals A and B from the following data. Label all areas of your diagram.

Melting point of A	700°C
Melting point of B	1000°C
Eutectic temperature	500°C
Composition of liquid in equilibrium at the eutectic temperature:	30 w/o A 70 w/o B

Solubilities at 500°C:	B in A = 15 w/o
	A in B = 20 w/o
Solubilities at 70°C:	B in A = 15 w/o
	A in B = 8 w/o

9–6.1 How many unit cells are there per carbon atom in austenite of eutectoid composition?

Answer: ~7 unit cells

9–6.2 Based on the data of Appendix C and Example 3–5.1, plot an approximate volume–temperature curve for iron between 20°C and T_m. If you have to make estimates, state your assumptions.

9–7.1 Making use of the iron–carbon diagram, calculate the fraction of α and the fraction of carbide present at 700°C in a metal containing 2% carbon and 98% iron.

Answer: 0.30 carbide, 0.70 ferrite

9–7.2 Describe the phase changes which occur on heating a 0.20% carbon steel from room temperature to 1200°C.

9–7.3 On the basis of this chapter, would you choose a high- or low-carbon steel for an automobile fender? Give reasons.

9–8.1 Calculate the percent ferrite, carbide, and pearlite, at room temperature, in iron–carbon alloys containing (a) 0.5% carbon, (b) 0.8% carbon, (c) 1.5% carbon.

Answer: a) 7.5% carbide, 92.5% α, 62% pearlite b) 12% carbide, 88% α, 100% pearlite c) 22.5% carbide, 77.5% α, 88% pearlite.

9–8.2 (a) Determine the phases present, the composition of each of these phases, and the relative amounts of each phase for 1.2% carbon steel at 870°C, 760°C, 700°C. (Assume equilibrium.) (b) How much pearlite is present at each of the above temperatures?

9–9.1 A steel has the following composition. Give it an AISI–SAE number: C 0.21, Mn 0.69, Cr 0.62, Mo 0.13, Ni 0.61.

Answer: AISI 8620

9–9.2 A steel has the following composition. Give it an AISI–SAE number: C 0.38, Mn 0.75, Cr 0.87, Mo 0.18, Ni 0.03.

9–9.3 Calculate the fractional quantities of the phase(s) at 700°C, 750°C, 800°C, and 900°C for the following steels: (a) 0.8C–99.2Fe; (b) 1.2C–98.8Fe; •(c) 0.6C–0.6Mo–98.8Fe.

Answer: a) 700°C, 0.88α–$0.12\overline{C}$; 750°C, 800°C, 900°C, 1.0γ b) 700°C, 0.82α–$0.18\overline{C}$; 750°C, 0.95γ–$0.05\overline{C}$; 800°C, 0.965γ–$0.035\overline{C}$; 900°C, 1.0γ c) 700°C, 750°C, 0.91α–$0.09\overline{C}$; 800°C, 900°C, 1.0γ ($\overline{C}$ = carbide)

•**9–9.4** Modify Fig. 9–7.3 for a steel containing (a) 1% Mn, (b) 1% Cr, (c) 1% Ni. (The new solubility curves remain essentially parallel to the previous ones.)

CHAPTER TEN

MULTIPHASE MATERIALS: MICRO-STRUCTURES AND PROPERTIES

PREVIEW

Reactions which produce specific phases and microstructures within solids have important engineering consequences because they permit the attainment of desired, and even enhanced, properties. We will categorize these reactions in Section 10–1.

Recall from Chapter 6 that the microstructures of single-phase metals can be adjusted (1) by plastic deformation and (2) by recrystallization, which in turn modify the properties. Further choices of properties were available for single-phase materials (3) through the appropriate selection of solid solutions and copolymers and (4) by crystal and molecular orientations.

Multiple-phase materials can have their properties modified and controlled through the same procedures that are available for single-phase materials. There are additional means of control of microstructures in materials with more than one phase present:

1. The *relative amounts of the phases* may be varied.

2. The *size of the phase grains* within the microstructure may be increased.

3. The *shape and distribution of the phases* can be modified.

Each of these three microstructural variations provides means of modifying the properties of materials. The resulting microstructures and their relationship to properties will be the subject of this chapter.

10–1 SOLID-PHASE REACTIONS

We encountered our first solid-phase reaction in Section 6–7 when we considered *recrystallization*. This is a relatively simple reaction because there is no compositional change. In fact, both the *reactant* (deformed metal) and the *product* (annealed metal) have the same crystal structure. However, the atoms rearranged themselves to form new, more perfect crystals. This takes time, as we will see later.

We shall consider four other types of reactions which produce new phases.

1. *Polymorphic reactions* have changes in crystal structure, but not in composition.
2. *Solution* and *precipitation reactions* involve composition changes at the solubility limit, so two phases are involved.
3. The *eutectoid reaction* which we will consider is austenite decomposition (Eq. 9–8.1).
4. *Martensitic* reactions are special shear reactions that have particular significance for steel properties.

This list is not all-inclusive, but will provide us with the bases for understanding selected heat-treatment processes which are used to control microstructure and properties.

Polymorphic reactions. Pure materials may undergo phase transformations from one polymorphic form to another.

$$\text{Polymorphic reaction:} \qquad \text{Phase } A \rightleftarrows \text{Phase } \mathscr{A}. \qquad (10\text{–}1.1)$$

Our prototype of this reaction, discussed in Section 9–6, was the $\alpha \rightleftarrows \gamma$ transformation of iron at 912°C (1673°F). Other familiar polymorphic changes (out of many) include the α-to-β transformation in titanium and the quartz-to-tridymite-to-cristobalite changes in SiO_2. These transformations commonly result from temperature changes. The polymorphic transformation between graphite and diamond is a consequence of pressure variations.

Density and volume changes accompany phase transformations because the polymorphic structures do not have identical atomic packing factors. These packing factors are 0.68 and 0.74, respectively, for the bcc and fcc phases of iron; hence, there is a noticeable dimensional change at 912°C (Fig. 10–1.1(a)). However, the heating or cooling must be quite slow in order for the change to occur precisely at the transformation temperature without a lag.

The transformation temperature can be defined as the temperature at which the two phases have identical amounts of energy available for chemical reaction. Thus, below 912°C austenite is more reactive than ferrite because it has more free energy; consequently, austenite is unstable and ferrite is stable. Immediately above 912°C, the opposite is true; then austenite is the stable phase of pure iron because the bcc iron reacts to form fcc iron.

Polymorphic reactions usually require only minor atom movements because the composition of the reactant and the product are identical (Eq. 10–1.1). However, it

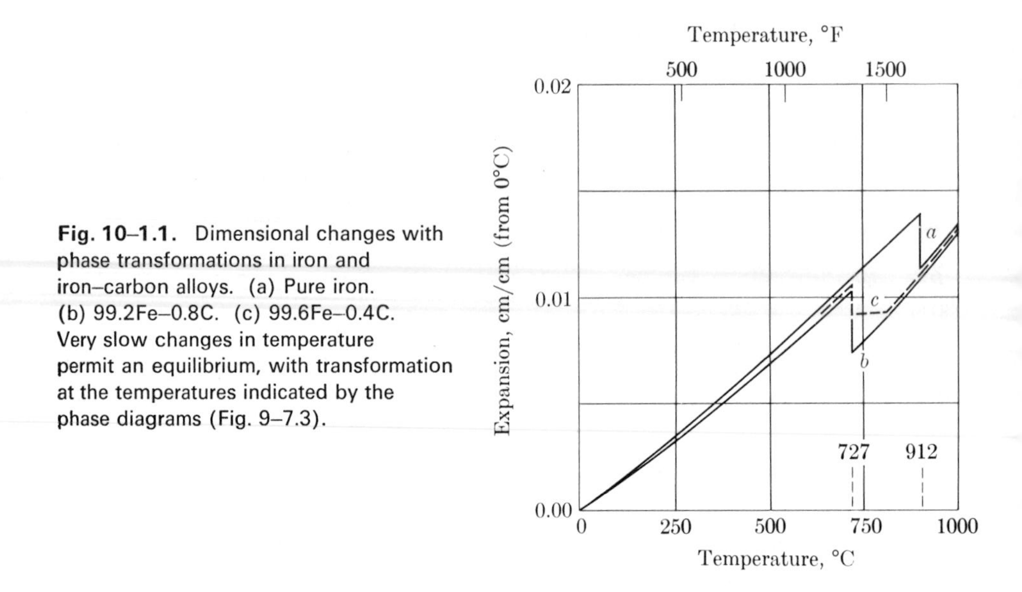

Fig. 10–1.1. Dimensional changes with phase transformations in iron and iron–carbon alloys. (a) Pure iron. (b) 99.2Fe–0.8C. (c) 99.6Fe–0.4C. Very slow changes in temperature permit an equilibrium, with transformation at the temperatures indicated by the phase diagrams (Fig. 9–7.3).

is still necessary to break the existing bonds and rearrange the atoms into the new structure.

Solution and precipitation within solids. These two reactions are opposite to each other and may be illustrated in Fig. 10–1.2. At temperatures below 370°C (700°F), an alloy with 90% Al and 10% Mg contains two phases, α and β. Above 370°C, all the magnesium may be contained in fcc α; consequently, as the metal is heated, solid β is *dissolved* in the solid α. *Solid precipitation* occurs when this alloy is cooled into the two-phase temperature range after being *solution-treated* above 370°C. Such precipitation is useful for the age-hardening of alloys (Chapter 11). The following equations are appropriate for these two reactions in binary alloys. Note that only

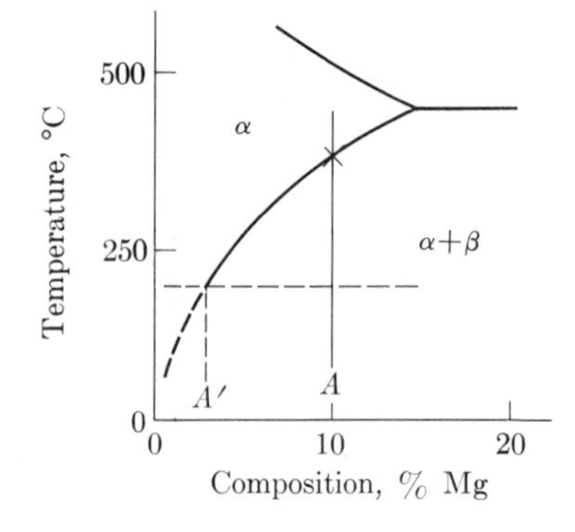

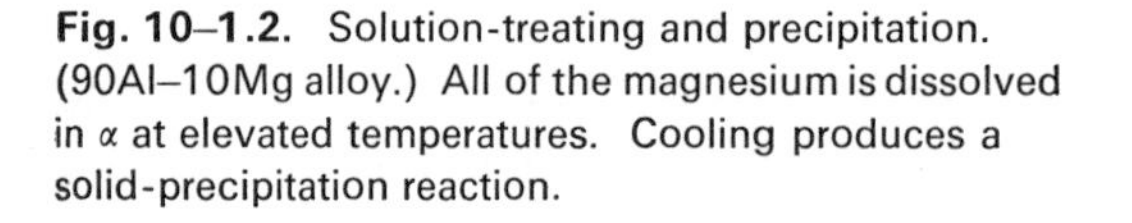

Fig. 10–1.2. Solution-treating and precipitation. (90Al–10Mg alloy.) All of the magnesium is dissolved in α at elevated temperatures. Cooling produces a solid-precipitation reaction.

two phases are involved in each case; however, the composition of the solution changes from A to another value, A′, as the solubility limit is exceeded (Fig. 10–1.2).

$$\text{Solution-treating:} \qquad \text{Solid A}' + \text{Solid B} \xrightarrow{\text{Heating}} \text{Solid A}. \qquad (10\text{–}1.2)$$

$$\text{Precipitation:} \qquad \text{Solid A} \xrightarrow{\text{Cooling}} \text{Solid A}' + \text{Solid B}. \qquad (10\text{–}1.3)$$

The above reactions require diffusion. Consider, for example, the composition of 95.5 w/o Al and 4.5 w/o Cu, which is a widely used aluminum alloy. As shown in Fig. 9–5.3, there are two phases, θ and κ, in equilibrium below 500°C (930°F). $CuAl_2$ comprises the θ phase, so that it contains an appreciably larger fraction of copper atoms than the κ phase. During solution-treatment these copper atoms must move through the κ structure to take on random substitutional positions; during precipitation, the copper atoms must be collected onto the growing $CuAl_2$ (that is, θ) particles.

Precipitation, like phase transformations, is nucleated at boundaries and other imperfections within a material. Grain-boundary precipitation is predominant at those temperatures just below the solubility limit where the atoms can diffuse easily, and intragrain precipitation is common at lower temperatures where diffusion rates are restricted (Fig. 10–1.3(a)).

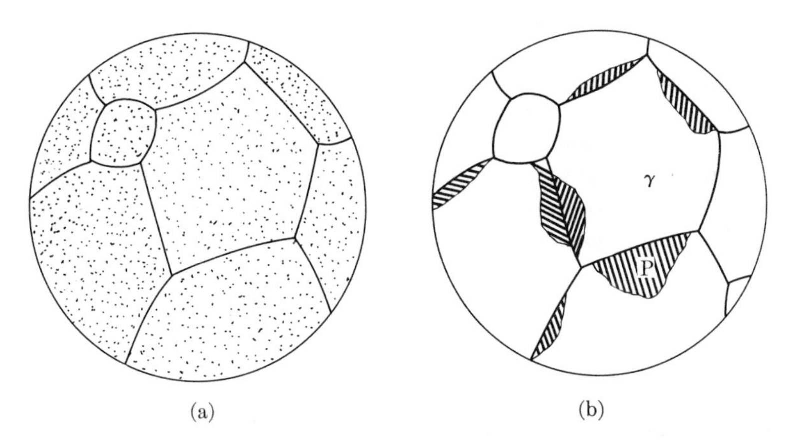

Fig. 10–1.3. Solid-phase reactions. (a) Nucleation at imperfections within the grains ($\kappa \rightarrow \kappa' + \theta$ in an Al–Cu alloy). (b) Nucleation at grain boundaries ($\gamma \rightarrow \alpha$ + carbide in steel).

Eutectoid reactions. Again let us use the ferrite–austenite transformation as an example, and refer to Fig. 9–7.1. For reasons cited in Section 9–6, carbon has greater solubility in austenite than in ferrite. As a result, the stability range of austenite is extended to higher and lower temperatures when carbon is present with iron. Figures 9–7.1 and 9–7.3 indicate that the lower end of the temperature range drops from

912°C (1673°F) for pure iron to 727°C (1340°F) for an alloy with approximately 0.8 w/o carbon. Presumably this temperature would be still lower for higher percentages of carbon, except for the fact that carbide, $\bar{C}$, is formed. In Chapter 9 we referred to this lowest temperature as the eutectoid temperature, and discussed the fact that it involved (in a binary alloy) an eutectoid reaction of three phases:

$$\text{Eutectoid:} \qquad \text{Solid 2} \underset{\text{Heating}}{\overset{\text{Cooling}}{\rightleftarrows}} \text{Solid 1} + \text{Solid 3}. \tag{10–1.4}$$

Materials of eutectoid composition may have their transformation completed at a single temperature (Fig. 10–1.1(b)) if sufficient time is given during heating or cooling. Other compositions (Fig. 10–1.1(c)) require a range of temperatures even under equilibrium conditions.

The *decomposition of austenite* to pearlite containing ferrite and carbide is typical of many reactions within solids, in that it starts at the grain boundaries and proceeds into the former grains (Fig. 10–1.3(b)). This sequence of reactions is to be expected, since it will be recalled from Section 4–4 and Fig. 4–4.7 that the atoms at the grain boundaries have higher energies than the atoms within the grains. Therefore, the atoms along the boundary require less additional energy to break away from their existing neighbors and to form a new structure.

However, grain boundaries are not the only locations of more energetic atoms. Those atoms around point or line defects (Section 4–4) have extra energy and can serve as locations for the nucleation of reactions (Fig. 10–1.3(a)). These locations increase in importance as the temperature is lowered.

In contrast to polymorphic reactions, and similar to precipitation reactions, an eutectoid reaction requires diffusion because the product phases are not identical to the reactant phases. This is illustrated in Fig. 10–1.4 for pearlite formation, where carbon must diffuse from areas forming ferrite into areas forming carbide. Of course, this requires time for completion.

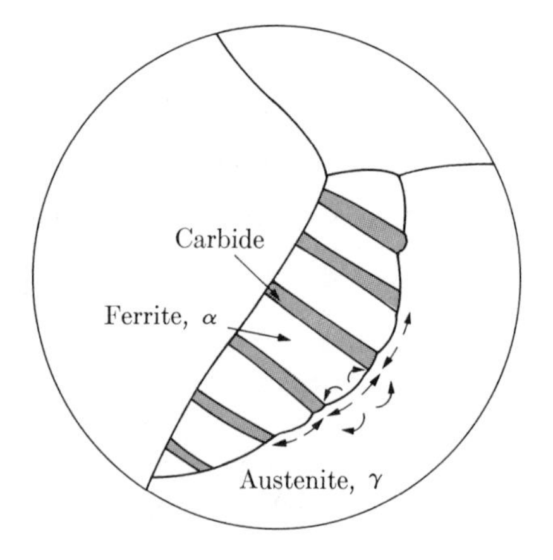

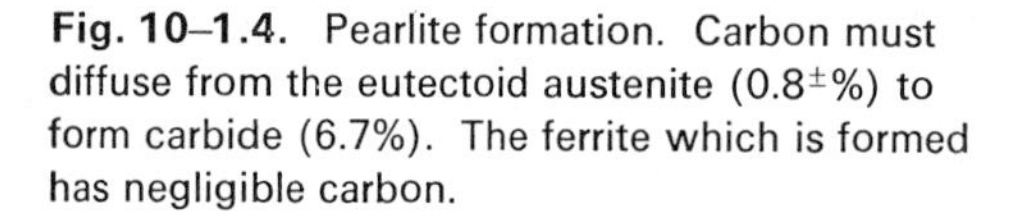

Fig. 10–1.4. Pearlite formation. Carbon must diffuse from the eutectoid austenite ($0.8^{\pm}$%) to form carbide (6.7%). The ferrite which is formed has negligible carbon.

Martensite formation. As indicated by Eq. (9–8.1), austenite decomposes to ferrite plus carbide ($\alpha + \bar{C}$). This assumes that there is time for the carbon to diffuse and relocate itself into the carbide phase and deplete itself from the ferrite. If we quench austenite very rapidly, Eq. (9–8.1) can be detoured. We can indicate this as follows:

$$\gamma(\text{fcc}) \xrightarrow[\text{Cooling}]{\text{Slow}} \alpha(\text{bcc}) + \text{carbide}. \qquad (10\text{–}1.5)$$

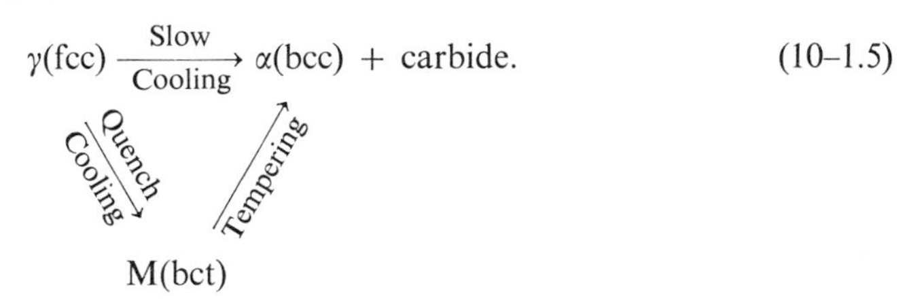

This alternate route of forming (α + carbide) involves a *transition phase* of *martensite*, M (Fig. 10–1.5). This polymorphic phase of austenite is not stable because, given an opportunity, martensite will proceed to form ($\alpha + \bar{C}$). However, it is a very important phase.

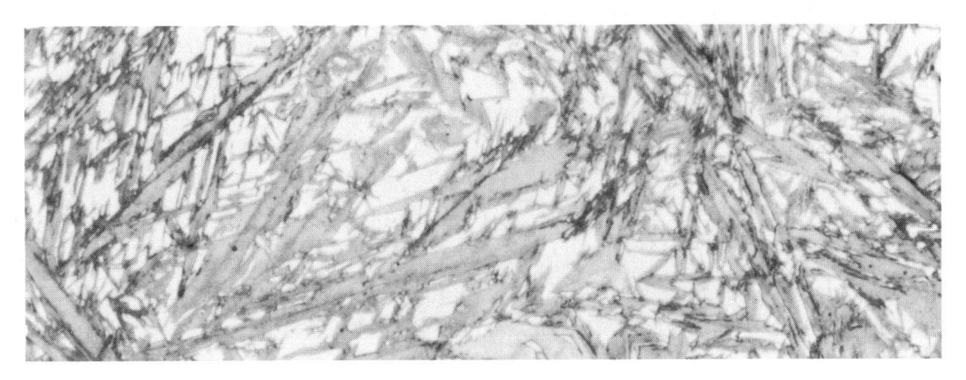

Fig. 10–1.5. Martensite (X1000). The individual grains are platelike crystals with the same composition as that of the grains in the original austenite. (U.S. Steel Corp.)

Martensite forms at lower temperatures (but still above room temperature) because the fcc structure of austenite becomes sufficiently unstable so that it changes spontaneously to a body-centered structure in a special way. It does not involve diffusion, but results from a shearing action. All the atoms shift in concert, and no individual atom departs more than a couple of angstroms from its previous neighbors. As a result, the change is very rapid. All of the carbon which was present remains in solid solution. The resulting body-centered structure is tetragonal (bct) rather than cubic; it is definitely different from ferrite.

Since martensite has a noncubic structure, and since carbon is trapped in the lattice, slip does not occur readily; and therefore martensite is hard, strong, and

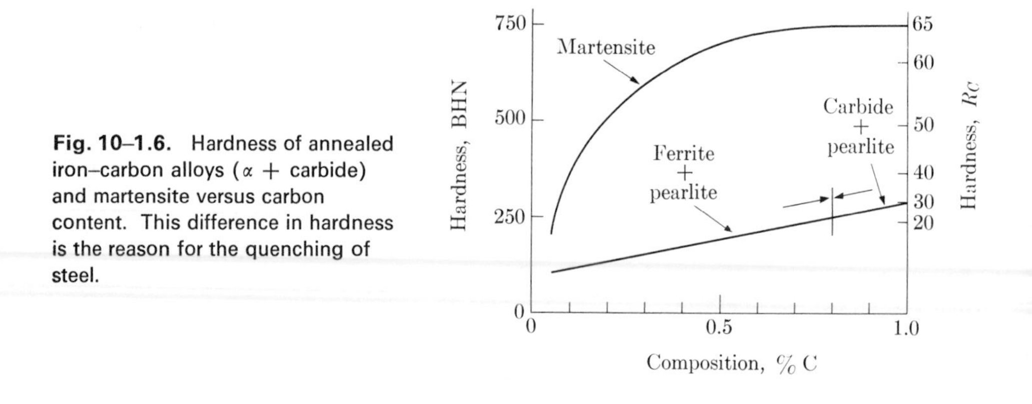

Fig. 10–1.6. Hardness of annealed iron–carbon alloys (α + carbide) and martensite versus carbon content. This difference in hardness is the reason for the quenching of steel.

brittle. Figure 10–1.6 shows a comparison of the hardness of martensite with that of pearlite-containing steels as a function of carbon content. This enhanced hardness is of major engineering importance, since it provides a steel which is extremely resistant to abrasion and deformation.

The existence of martensite as a *metastable* phase which contains carbon in solid solution in a bct structure does not alter the iron–carbide phase diagram (Fig. 9–7.1). With sufficient time at temperatures below the eutectoid temperature, the supersaturated solution of carbon in iron continues its progress to the more stable ferrite and carbide (Eq. 10–1.5). This process is known commercially as *tempering*;

$$\underset{\text{(Martensite)}}{\text{M}} \longrightarrow \underset{\text{(Tempered martensite)}}{\alpha + \text{carbide.}} \qquad (10\text{–}1.6)$$

The resulting (α + $\bar{C}$) microstructure is not lamellar like that of the pearlite we have observed, but contains many dispersed carbide particles (Fig. 10–1.7) because there

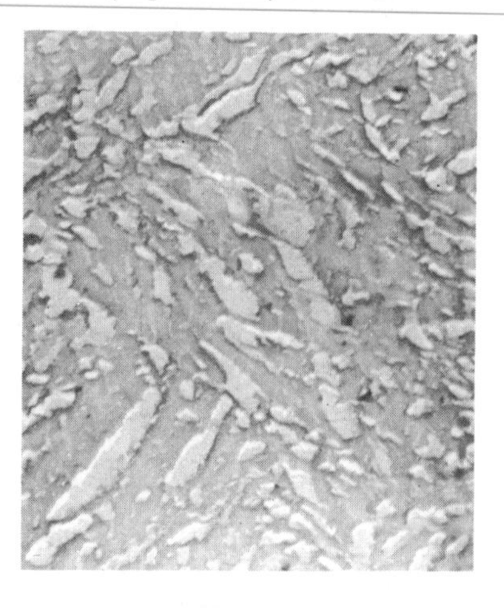

800°F, 1 hr, 44 R_C
(Chrysler Corp.)

Fig. 10–1.7. Tempered martensite (eutectoid steel, X11,000). The steel was previously quenched to form martensite, which is a body-centered tetragonal (bct) phase and was 65 R_C. It was then tempered for 1 hour at 425°C (800°F). The tempered martensite is a two-phase microstructure containing carbide particles (light) in a matrix of ferrite (dark). Initially the tempered martensite was very hard and brittle; however, with the above heating, it is now 44 R_C, but much tougher. (Chrysler Corp.)

are numerous nucleation sites within the martensitic steel. This *tempered martensite** is much tougher than the metastable martensite, making it a more desirable product although it may be slightly softer.

Example 10–1.1 A 95.4Al–4.6Cu alloy has 2 atom percent copper in solid solution κ at 550°C. It is quenched, then reheated to 100°C, where θ precipitates (Fig. 9–5.3). The θ ($CuAl_2$) develops many *very small* particles throughout the alloy so that the average interparticle distance is only 50 Å. (a) Approximately how many particles form per cm³? (b) If, by extrapolating Fig. 9–5.3, we assume that negligible copper remains in κ at 100°C, how many copper atoms are there per θ particle?

Solution:

a) Assume one particle per 50 Å cube.

$$\text{Count} \sim 1/(50 \times 10^{-8}\ \text{cm})^3 = \sim 8 \times 10^{18}/\text{cm}^3.$$

b)

$$a_\kappa = 4(\sim 1.43\ \text{Å})/\sqrt{2} = \sim 4.04\ \text{Å};$$

$$\text{Cu/cm}^3 = 0.02(4\ \text{atoms/u.c.})/(4.04 \times 10^{-8}\ \text{cm})^3$$

$$= 1.2 \times 10^{21}/\text{cm}^3.$$

$$\text{Cu/particle} = (1.2 \times 10^{21}/\text{cm}^3)/(\sim 8 \times 10^{18}/\text{cm}^3)$$

$$= \sim 150\ \text{Cu/particle}.$$

Comment. This microstructure is approximately what we will encounter in precipitation hardening (Section 11–3). ◀

10–2 MULTIPHASE MICROSTRUCTURES

Specific attention has already been given to pearlite as a multiphase microstructure (Fig. 9–8.1). In this example the microstructure contained a *mixture* of ferrite and carbide, which was lamellar in shape and distribution because growth proceeded inward from the boundaries of the former austenite grains (Fig. 10–1.3(b)). However, not all microstructures which contain ferrite and carbide are the same. This is observed in Fig. 10–2.1(a), where the ferrite is present in two generations. The *proeutectoid ferrite* formed before the austenite of eutectoid composition changed to pearlite. In contrast, the *eutectoid ferrite* is within the pearlite. (See Example 9–8.1.) The steel of Fig. 10–2.1(b) contains only the latter.

A second microstructural modification involves the thicknesses of the ferrite and carbide lamellae in pearlite. The diffusion shown in Fig. 10–1.4 can proceed over greater distances if the transformation temperature is increased or the cooling rate decreased. These factors will produce thicker lamellae in the pearlite. The effect of this variable on properties will be discussed later in this chapter.

---

* Note that tempered martensite does not have the crystal structure of martensite. Rather, it is a two-phase microstructure containing *ferrite* and *carbide*.

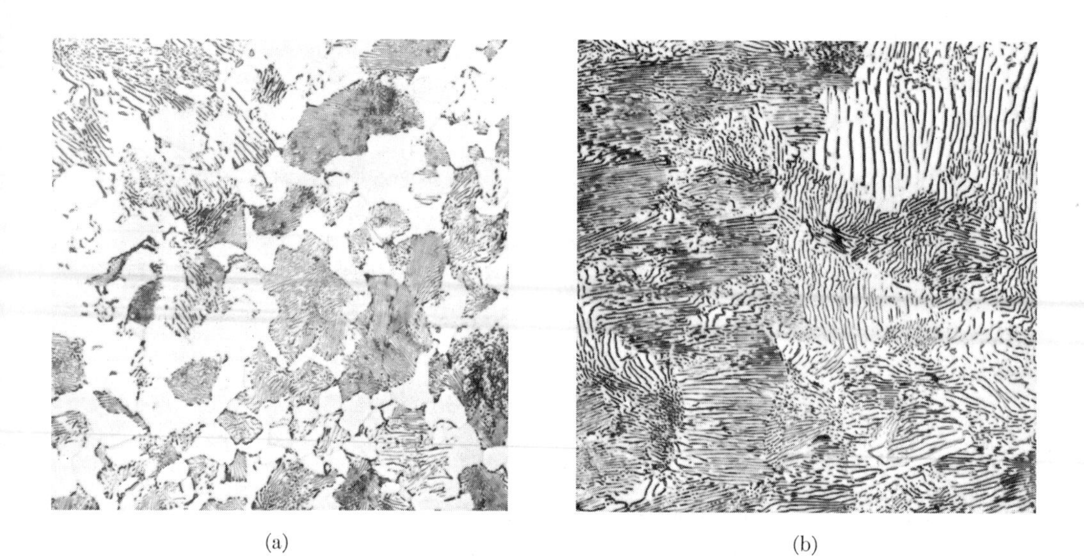

Fig. 10–2.1. Microstructure of pearlitic steels, X500. (a) 0.40% carbon. (b) 0.80% carbon. The larger (white) ferrite areas of the 0.40% carbon steel formed before the eutectoid reaction produced the lamellar ferrite in the pearlite (gray). The 0.80% carbon steel contains only eutectoid ferrite. (Compare with Figs. 9–7.3 and 9–8.1) (U.S. Steel Corp.)

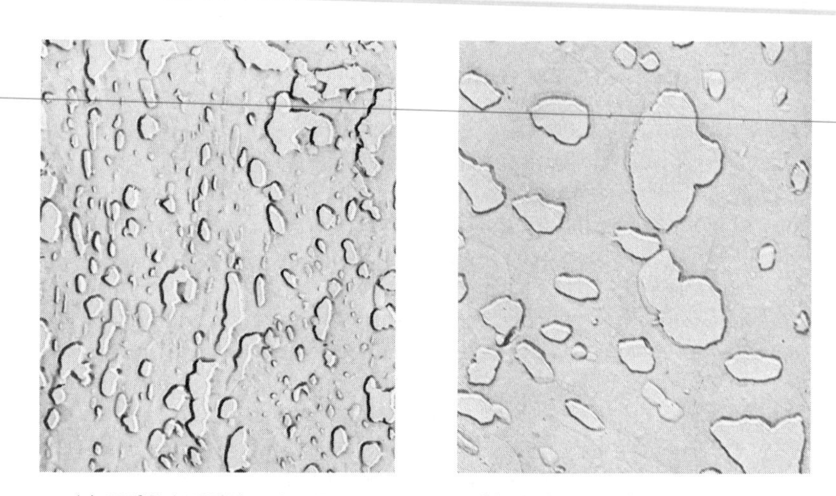

(a) 595°C (1100°F), 1 hr, R_C 33 (b) 675°C (1250°F), 12 hr, R_C 20

Fig. 10–2.2. Electron micrographs of tempered martensite (X11,000). Each of these is eutectoid steel which was previously quenched to maximum hardness (65 R_C). (*Electron Microstructure of Steel,* American Society for Testing and Materials, and General Motors.)

The microstructure of tempered martensite (Figs. 10–1.7 and 10–2.2) has a dispersion of carbide particles within a matrix of ferrite. Such a dispersion originates because the microstructure of the parent martensite (Fig. 10–1.5) provided many nucleating sites on which transformation could start. It is important to note that increased temperatures permit a growth of the carbide particles, with an accompanying decrease in hardness (Fig. 10–2.2). This growth, which does not alter the relative amounts of the ferrite and carbide phases, occurs through a precipitation of the very small amounts of carbon that are in the ferrite onto the larger carbide particles, and an accompanying solution of the smaller carbide particles into the ferrite.

Other materials besides steel have a variety of microstructures; for example, certain aluminum alloys contain a dispersion of small $CuAl_2$ particles with a soft aluminum matrix. In gray cast iron, graphite is present as flakes within a ferrite-plus-carbide matrix. Likewise, the microstructures of nonmetallic materials range from plastics, with glass-fiber reinforcement, and rubber, containing colloidal carbon, to clays which have been plasticized by interparticle water.

Pores are a very important microstructural feature in many materials and may be considered as a phase of zero composition. In electrical and magnetic ceramics, they are undesirable and are to be avoided. However, they are necessary in other materials, such as thermal insulations and powdered metal bearings. Even when they are desired, their amount, size, shape, and distribution must be controlled if the material is to have optimum properties.

Example 10–2.1 An eutectoid (1080) steel contains ferrite ($\rho = 7.86$ g/cm^3) and carbide ($\rho = 7.69$ g/cm^3). (a) What is the density of the steel? (b) What is the approximate ratio of ferrite lamellae thickness d_α to carbide lamellae thickness $d_{\overline{C}}$?

Solution: Basis: 100 g steel = 0.8 g carbon = 12 g Fe_3C + 88 g α

$$88 \text{ g } \alpha/(7.86 \text{ g/cm}^3) = 11.20 \text{ cm}^3\ \alpha,$$
$$12 \text{ g } \overline{C}/(7.69 \text{ g/cm}^3) = \underline{1.56 \text{ cm}^3\ \overline{C}}.$$
$$12.76 \text{ cm}^3$$

a) $\rho = 100 \text{ g}/12.76 \text{ cm}^3 = 7.84 \text{ g/cm}^3$.

b) Since the long dimensions of the lamallae of α and $\overline{C}$ are approximately identical,

$$d_\alpha/d_{\overline{C}} = V_\alpha/V_{\overline{C}} = 11.20/1.56 = \sim 7.$$

Comment. An examination of Fig. 9–8.1 reveals that the carbide is less continuous than the ferrite; thus, their thickness ratios will be somewhat less than the above estimate. ◀

10–3 ADDITIVE PROPERTIES (MIXTURE RULES)

Two phases never have completely identical properties because, as we have seen, each is structurally different. This generalization also applies to the properties of multiphase materials. Certain of these properties are additive and may be determined by suitably weighted averages of the properties of each of the individual phases.

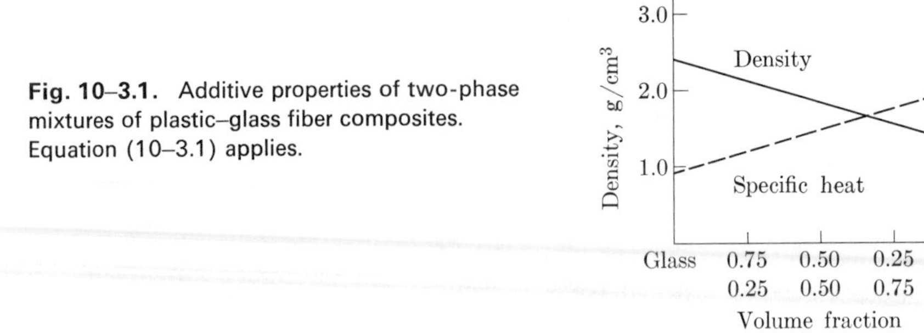

Fig. 10–3.1. Additive properties of two-phase mixtures of plastic–glass fiber composites. Equation (10–3.1) applies.

Other properties are interactive, because the behavior of each phase depends on the nature of the adjacent phase.

The *density* of a multiphase microstructure may be calculated directly from the density ρ of each phase and the corresponding volume fraction f:

$$\rho_{\text{material}} = f_1\rho_1 + f_2\rho_2 + f_3\rho_3 + \cdots \tag{10–3.1}$$

With only two phases present, the density varies linearly with the fraction which is present (Fig. 10–3.1). When there is pore space, the $f\rho$ product is zero because the phase has no density.

The *specific heat* of a multiphase microstructure has a similar additive relationship which is also proportional to the volume fraction. The averages of both these properties, density and specific heat, depend on volume because the contribution of one phase does not affect the contribution of the adjacent phase. Likewise, neither property is influenced by the size of the grains in the phase mixture.

The *thermal* and *electrical conductivities* of multiphase microstructures are also additive. However, a more complex weighting procedure is necessary because the shape and distribution of each phase is important. The three simplified examples which may be cited are shown in Fig. 10–3.2; although these examples apply equally to either thermal or electrical conductivity, thermal-conductivity notations will be incorporated into the following equations. *Parallel conduction* applies in the first case; therefore,

$$k_m = f_1k_1 + f_2k_2 + \cdots, \tag{10–3.2}$$

and the average conductivity for the material k_m is a summation of the volume (or cross-sectional area) contribution of all phases. *Series conduction* applies in Fig. 10–3.2(b); therefore

$$\frac{1}{k_m} = \frac{f_A}{k_A} + \frac{f_B}{k_B} + \cdots, \tag{10–3.3}$$

or, rewriting for a two-phase microstructure,

$$k_m = \frac{k_Ak_B}{f_Ak_B + f_Bk_A}. \tag{10–3.4}$$

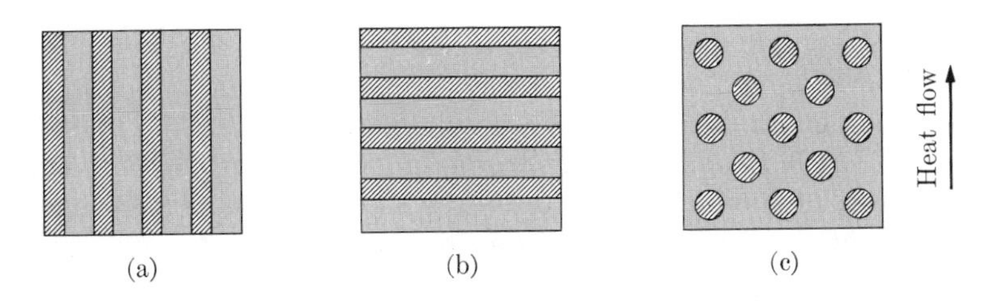

Fig. 10–3.2. Conductivity versus phase distribution (idealized). (a) Parallel conductivity (Eq. (10–3.2)). (b) Series conductivity (Eq. (10–3.4)). (c) Conductivity through a material with a dispersed phase (Eq. (10–3.5)).

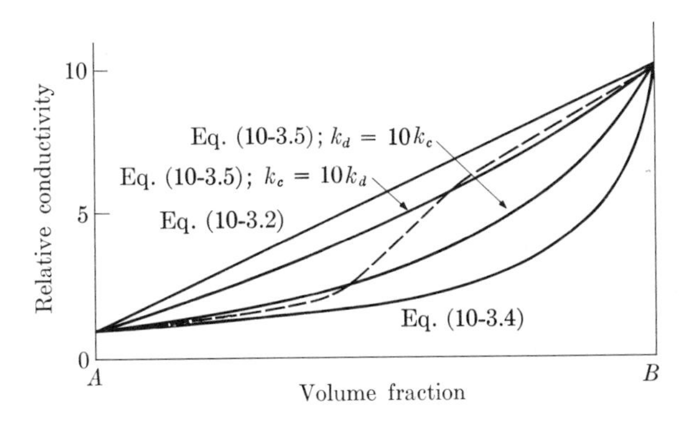

Fig. 10–3.3. Conductivity versus volume fraction. (Adapted from Kingery, W.D., *Introduction to Ceramics*, Wiley.)

In this case the conductivity of the material is less than that obtained by a linear interpolation from the phase volumes (Fig. 10–3.3).

If one phase is dispersed in another, as is common (Fig. 10–3.2(c)), a more complicated interpolation results:

$$k_m = k_c \frac{1 + 2f_d \dfrac{1 - k_c/k_d}{2k_c/k_d + 1}}{1 - f_d \dfrac{1 - k_c/k_d}{2k_c/k_d + 1}}. \qquad \bullet(10\text{–}3.5)$$

In this equation the subscript c refers to the continuous phase and d to the dispersed phase. If $k_c \ll k_d$, the equation simplifies to

$$k_m = k_c(1 + 2f_d)/(1 - f_d). \qquad (10\text{–}3.6)$$

Example 10–3.1 reveals the validity of this simplification. If $k_d \ll k_c$, Eq. (10–3.5) simplifies to

$$k_m = k_c(1 - f_d)/(1 + f_d/2). \qquad (10\text{–}3.7)$$

Such an equation could also be applicable to the conductivity of a solid containing a dispersion of small pores.

•Equation 10–3.5 is plotted in Fig. 10–3.3 for two examples: when $k_c/k_d = 10$, and when $k_d/k_c = 10$. The results fall between the two curves for Eqs. (10–3.2) and (10–3.4). Data* for thermal conductivity of MgO–Mg_2SiO_4 mixtures (dashed line) find substantial agreement with Eq. (10–3.5). However, cognizance had to be taken of the fact that the continuous and dispersed phases were interchanged as the volume fraction of either phase increased from zero to 100 percent. Therefore, the experimental curve switches from one dispersion curve to the other.

Example 10–3.1 Fifty w/o SiO_2 flour (i.e., quartz powder) is added to a phenol-formaldehyde resin as a filler. (a) What is the density of the mixture? •(b) What is the thermal conductivity?

Answer: From Appendix C:

$$\rho_{SiO_2} = 2.65\ \text{g/cm}^3, \qquad \rho_{pf} = 1.3\ \text{g/cm}^3,$$

$$k_{SiO_2} = 0.12\ \text{J} \cdot \text{cm/°C} \cdot \text{cm}^2 \cdot \text{sec},$$

$$k_{pf} = 0.0016\ \text{J} \cdot \text{cm/°C} \cdot \text{cm}^2 \cdot \text{sec}.$$

(Basis: 100 g.)

a)

$$\begin{array}{llll} 50\ \text{g}\ \ SiO_2 = & 18.8\ \text{cm}^3 & SiO_2; & f_{SiO_2} = 0.33 \\ 50\ \text{g}\ \ \text{pf} = & \underline{38.4\ \text{cm}^3} & \text{pf}; & f_{pf} = \underline{0.67} \\ & 57.2\ \text{cm}^3 & & \qquad\ \ 1.0 \end{array}$$

By Eq. (10–3.1),

$$\rho_m = (0.33)(2.65) + (0.67)(1.3) = 1.75\ \text{g/cm}^3.$$

• b) By Eq. (10–3.5), and setting $k_c/k_d = 0.0016/0.12 = 0.0133$,

$$k_m = 0.0016 \frac{1 + 2(0.33)\dfrac{1 - 0.0133}{2(0.0133) + 1}}{1 - (0.33)\dfrac{1 - 0.0133}{2(0.0133) + 1}}$$

$$= 0.004\ \text{J} \cdot \text{cm/cm}^2 \cdot \text{°C} \cdot \text{sec}.$$

b) Alternative: Since $k_{pf} \ll k_{SiO_2}$,

$$k_m \simeq k_c \left(\frac{1 + 2f_d}{1 - f_d}\right),$$

$$k_m \simeq 0.0016 \left[\frac{1 + 2(0.33)}{1 - 0.33}\right] = 0.004 \frac{\text{cal} \cdot \text{cm}}{\text{°C} \cdot \text{cm}^2 \cdot \text{sec}}.$$

---

* Kingery, W. D., *Introduction to Ceramics*, Wiley.

10–4 INTERACTIVE PROPERTIES

Properties such as hardness and strength cannot be approximated by simple mixture rules of properties of the contributing phases. For example, the steels of Fig. 10–2.2 have tensile strengths in excess of 100,000 psi (>700 MPa). However, the strength of the matrix phase (ferrite) is less than one-third of those values. Since the ferrite is the continuous phase and therefore all of the load must be carried through it, we may conclude that "the chain is stronger than its weakest link!" While there is a mathematical rationale for this, our qualitative explanation will simply be that the rigid carbide particles inhibit slip and prevent shear of the matrix.*

Effects of phase quantities. Although we have not presented a rigorous explanation for these interactive properties, we make use of them, because this interdependence of phases for the mechanical properties makes it possible to strengthen materials by addition of "fillers." For example, the addition (1) of carbon to rubber, (2) of sand to clay, (3) of sand to tar or asphalt, (4) of wood flour to plastics, or (5) of intermetallic precipitates to soft metal, increases the resistance to deformation or flow. The effect on strength in the fourth example is shown graphically in Fig. 10–4.1. Although a phenol-formaldehyde resin has considerable strength alone, it is subject to eventual shear failure under stress, and the incorporation of a second phase produces added resistance to deformation. At the other end of the composition range, the strength of wood flour (fine sawdust) alone is nil; there are no forces which hold the individual particles of cellulose into a coherent mass. The addition of the resin

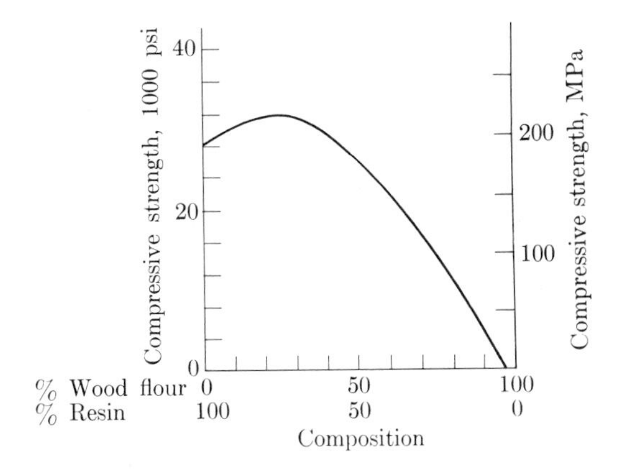

Fig. 10–4.1. Strength of mixtures (wood-flour filler in phenol-formaldehyde resin). The mixture of wood flour and resin is stronger than either alone. The wood flour prevents slip in the resin; the resin bonds the particles of wood flour.

---

* This can be demonstrated dramatically by using some solder (T.S. $\simeq$ 5000 psi, or ~35 MPa), to join two pieces of steel together. If the soldered joint is thin, this joint can support a stress over 25,000 psi (>170 MPa)! Although weak, the solder does not deform because it is rigidly held by the stronger adjacent steel. With thinner and thinner joints, there is less and less chance for plastic deformation within the weak solder, so the joint will become even stronger.

to the fine particles serves to cement the wood flour together. Maximum strength is developed with intermediate compositions, with the two phases strengthening each other.

Highway construction is another example of the use of phase mixtures. For obvious reasons, roadbeds composed solely of clay would be unsatisfactory, as would those made completely of gravel or sand. However, an appropriate combination of clay and gravel produces a practical, stabilized roadbed (Fig. 10–4.2). The clay is strengthened by the hard gravel, and the gravel is bound by the clay into a coherent mass that resists concentrated loads.

Fig. 10–4.2. Stabilized roadbed. The mixture of gravel and clay is more durable than either gravel or clay alone.

Mixtures of ferrite and carbide in steel are less obvious but equally common Carbide is harder than the accompanying ferrite, and so it increases the resistance of steel to deformation. Figure 10–4.3 shows graphically the *hardnesses* of carbon steels which have been annealed (i.e., cooled slowly from austenitic temperatures to ensure coarse pearlite). The microstructures are shown in Fig. 10–2.1. The 0.4-percent carbon steel contains approximately 50% pearlite (6% carbide), and 50% proeutectoid ferrite; the 0.8-percent carbon steel is all pearlite (12% carbide). Slip occurs more readily in the former than in the latter.

In addition to being harder, steels with higher carbon contents are stronger. Both the *yield strength* and the *tensile strength* are included in Fig. 10–4.3. Note, however, that the curves in this figure cannot be extrapolated to very high values at 100% iron carbide. Because iron carbide is lacking in ductility and cannot adjust to stress concentrations, an extension of Fig. 10–4.3 to 100% iron carbide will show the curves dropping to approximately 5000 psi (35 MPa).

Figure 10–4.4 shows the effect of iron carbide on *ductility*. As expected, ductility decreases with increased carbon content. Therefore for automobile fenders the manufacturer chooses a very low-carbon steel that can be rolled into a thin sheet and

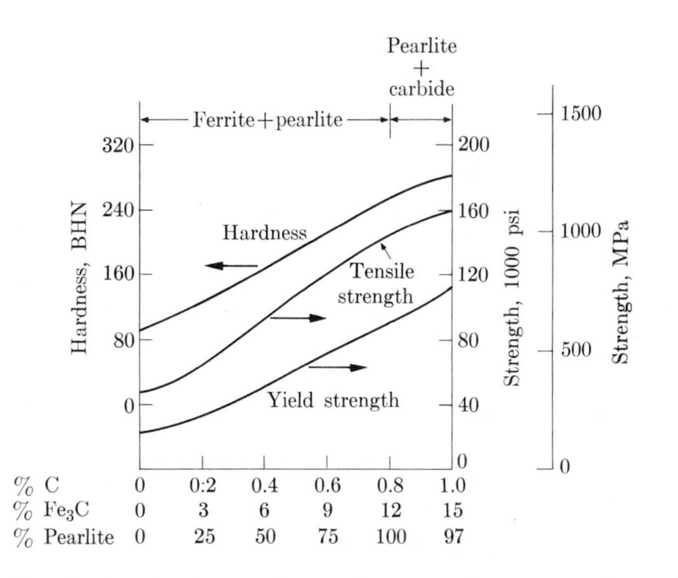

Fig. 10–4.3. Hardness and strength versus carbon content of annealed plain-carbon steels. The steels contain microstructures similar to those shown in Fig. 10–2.1. Hardness: read left. Strength: read right.

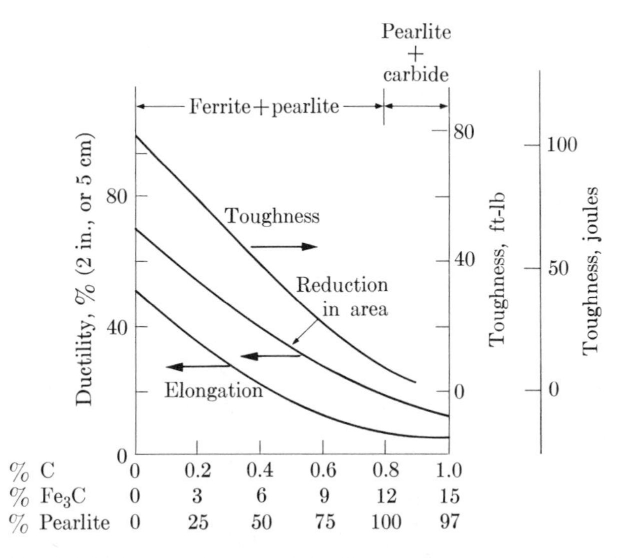

Fig. 10–4.4. Toughness and ductility versus carbon content of annealed plain-carbon steels. (Cf. microstructure of Fig. 10–2.1.) Those steels with more proeutectoid ferrite are both more ductile and tougher. Ductility: read left. Toughness: read right.

deep-drawn to take the sharp curvatures required both for styling and for single-step production. By contrast, the dies and shears used in forming and cutting these fenders are made of high-carbon steels which may be made hard and strong.

The *toughness* of a steel (Fig. 10–4.4) is also important if the engineer is designing or using equipment which is subject to impact, since a brittle (opposite of tough) steel will break under comparatively light blows. Fractures are more readily propagated in high-carbon steels because there are more brittle carbide paths through which cracks may progress.

Effects of phase size. The coarseness of the microstructure of a material directly affects its mechanical properties. The addition of very fine sand to asphalt will produce a more viscous mixture than will the addition of an equal weight (or volume) percent of gravel. Similarly, steel with a very finely mixed microstructure of ferrite and carbide will be much harder and stronger than will steel of the same carbon content but with a much coarser microstructure. A comparison is made in Fig. 10–4.5. The lower set of data describes a steel which was cooled slowly to produce a coarse pearlite structure; the upper set is for a steel that was cooled more rapidly, to form a much finer pearlite.

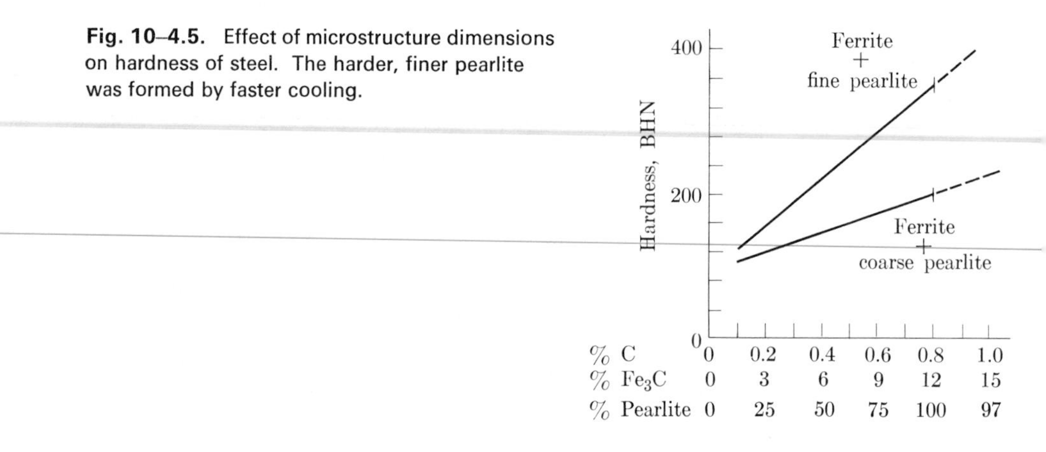

Fig. 10–4.5. Effect of microstructure dimensions on hardness of steel. The harder, finer pearlite was formed by faster cooling.

The same variation in properties is observable in martensitic steels tempered at different temperatures, or for increasing periods of time (Fig. 10–4.6). Initial tempering produces a fine mixture of carbide particles in a ferrite matrix (Fig. 10–2.2(a)). Figure 10–2.2(b) shows a sample tempered longer at a higher temperature, which permits the carbide particles to grow. The hardness values in the legend indicate the degree of softening. When the harder particles are agglomerated, there are larger areas of the soft ferrite matrix in which slip can occur without restriction.

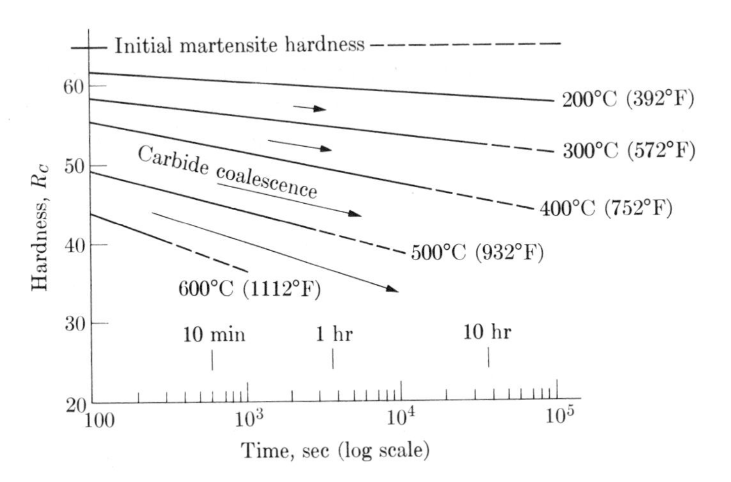

Fig. 10–4.6. Hardness of tempered martensite (1080 steel quenched to 65 R_C). Softening occurs as the carbide particles coalesce, giving greater intervening ferrite distances (Fig. 10–2.2). An Arrhenius pattern (Eq. 4–6.6) is followed in the early stages of softening.

Effects of phase shape and distribution. The shape and distribution of microstructure phases also affect the properties of a material. For example, in pearlite the carbide is lamellar (Fig. 10–4.7(a)). However, if the pearlitic steel were held just under the

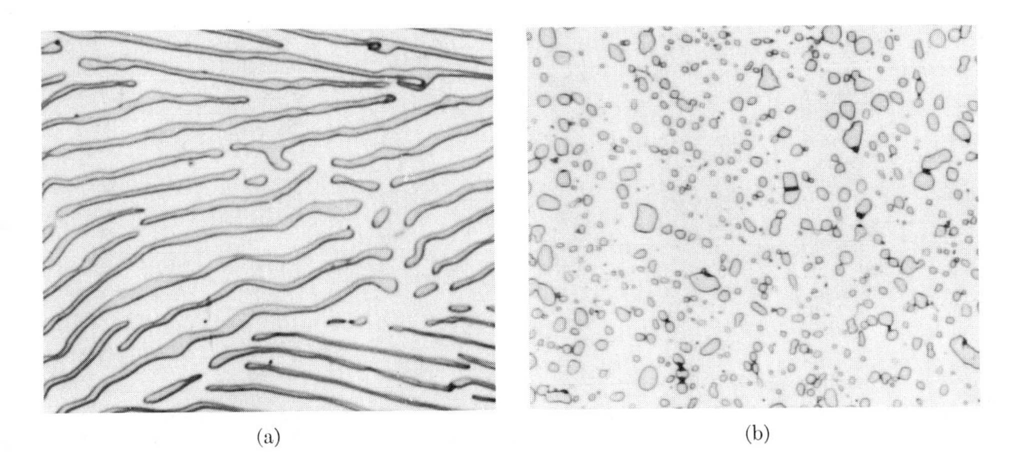

Fig. 10–4.7. Microstructures of eutectoid steels (0.80% C) (X2500). (a) Pearlite formed by transforming austenite (γ) of eutectoid composition. (b) Spheroidite formed by tempering at 700°C (1300°F). (Courtesy U.S. Steel Corp.)

eutectoid temperature for a long period of time, the carbide would *spheroidize** and develop the structure shown in Fig. 10–4.7(b) and called *spheroidite*. The effect of spheroidization on hardness is shown in Fig. 10–4.8(a); the hardness decreases because the spheroidized structure reinforces the metal less. The shape of the phases also

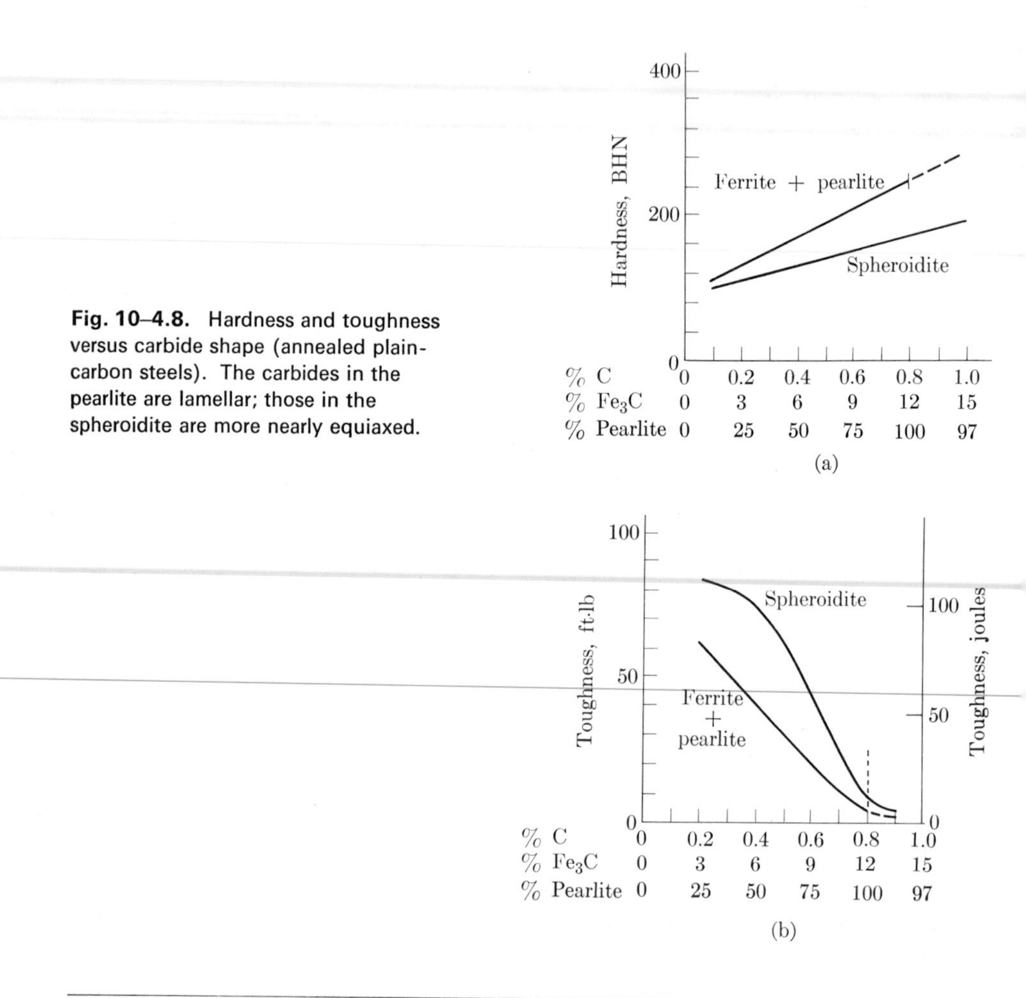

Fig. 10–4.8. Hardness and toughness versus carbide shape (annealed plain-carbon steels). The carbides in the pearlite are lamellar; those in the spheroidite are more nearly equiaxed.

---

* The boundaries between two phases (or grains) are sites of higher energy than the lattice itself because the boundary atoms are not as closely aligned as the atoms within a crystal (Fig. 4–4.7). Since the boundary atoms have higher energy, the boundaries will be spontaneously eliminated if the atomic mobility is sufficient. Boundary area may be reduced by grain growth (Fig. 10–2.2) or by spheroidization (Fig. 10–4.7). High temperatures facilitate either of these reactions.

rather markedly affects the toughness of the mixture (Fig. 10–4.8(b)). A spheroidized structure is tougher because cracks in the carbide cannot propagate very far before entering tough ferrite.

•**Example 10–4.1** The data of Fig. 10–4.6 reveal that it takes 10^4 sec to temper martensite from 65 R_C to 50 R_C at 360°C and 100 sec at 490°C. Assuming an Arrhenius relationship (Eq. 4–6.6), how long will it take at 300°C?

Solution: Using $\log t = C + B/T$,

$$\log 10^4 = 4 = C + B/(360 + 273),$$
$$\log 100 = 2 = C + B/(490 + 273).$$

Solving simultaneously,

$$B = 7430,$$
$$C = -7.74;$$
$$\log t = -7.74 + 7430/573$$
$$t = 170{,}000 \text{ sec} \quad (\sim 2 \text{ days}).$$

Comment. This calculation is appropriate for hardness values greater than about 35 R_C, while the carbide particles are still small and numerous. ◀

REVIEW AND STUDY

SUMMARY

We depend on solid-phase reactions to change a material from one microstructure to another. Polymorphic reactions, like recrystallization reactions, do not require diffusion. Solution and precipitation involve two phases of different compositions. In solution treatments, the atoms of the minor component must diffuse to become distributed throughout the solvent phase; in precipitation reactions, the excess atoms of the solute must segregate into a new phase. The new phase will collect at grain boundaries if diffusion permits. At lower temperatures where diffusion is more restricted, the precipitate is more commonly distributed as fine particles throughout the grain of the original solvent phase.

The eutectoid reaction involves three phases. During cooling, it is common to have the two product phases grow from the phase boundary into the disappearing reaction phase. This is the way pearlite is formed.

Martensite is a transition phase which forms intermediately between austenite, and ferrite plus carbide ($\alpha + \bar{C}$). The quenching must be sufficiently fast to cool the austenite to low temperatures before normal decomposition can occur. This hard phase is of major interest in steel processing. When it finally proceeds to the more stable ($\alpha + \bar{C}$), the carbides form a very fine dispersion in the ferrite matrix, to give a microstructure called tempered martensite.

Properties related primarily to the relative amounts of the phases can be described by mixture rules. These are sometimes considered to be additive. Other properties are more complex because the behavior of one phase interacts with its neighboring phases. We are able to observe the qualitative effects of phase amounts, phase size, and phase shape and distribution by examining the properties of various annealed steels (Fig. 10–4.3 through 10–4.8).

TERMS FOR REVIEW

Additive properties
Austenite decomposition
Dispersed phases
Eutectoid reaction
Interactive properties
Microstructural effects
  phase distribution
  phase quantities
  phase shape
  phase size
Martensite, M
Metastability

Mixture rules
Multiphase microstructures
Phase, transition
Polymorphic reaction
Precipitation in solids
Solid-phase reactions
Solution treatment
Spheroidite, ($\alpha + \bar{C}$)
Tempered martensite, ($\alpha + \bar{C}$)
Tempering

FOR DISCUSSION

A_{10} Which of the following polymorphic reactions will require the least activation energy? The most? Why? a) Bcc Fe ⇄ fcc Fe; b) Graphite ⇄ diamond; c) Cubic $BaTiO_3$ ⇄ tetragonal $BaTiO_3$.

B_{10} Explain solid precipitation to a student who has had college chemistry, but who has not had this course.

C_{10} Highly deformed copper (CW > 80%) recrystallizes at lower temperatures than slightly deformed copper (CW < 20%). Why?

D_{10} Why is time required for pearlite formation from austenite?

E_{10} Explain the three curves of Fig. 10–1.1 to a classmate.

F_{10} Why do eutectoid reactions generally require longer times than polymorphic reactions?

G_{10} Why does martensite resist slip?

H_{10} In your own words, explain why the formation of martensite is not time-dependent.

I_{10} The bct structure of martensite contains two iron atoms per unit cell and has $c = 2.95$ Å and $a = 2.85$ Å when the w/o carbon is 0.8. Point out the locations within the unit cell where the carbon atoms can reside with least strain. (See Fig. 9–6.2(a).)

J_{10} Equation 10–1.5 is irreversible as written. Under what conditions could austenite reform from ($\alpha + \bar{C}$) or from M? [*Note:* The direct reversal of ($\alpha + \bar{C}$) to M has never been observed, nor is it expected. Why?]

K_{10} Compare and contrast: a) pearlite b) martensite, and c) tempered martensite. Distinguish between a phase and a microstructure.

L_{10} "Oilless" bronze bearings are made by sintering bronze powders to have a porosity of 10–20 v/o. What mixture rule would best describe their density? Their thermal conductivity?

M_{10} Parallel resistance of an electrical circuit may be determined by:

$$1/\bar{R} = 1/R_1 + 1/R_2 + 1/R_3 + \cdots \qquad (10\text{–}5.1)$$

Relate this equation to those in Section 10–3.

N_{10} Why does Fe_3C strengthen ferrite even though the carbide is so brittle that "it breaks like glass"?

O_{10} Why is fine pearlite stronger than coarse pearlite?

P_{10} Cite those properties of pearlite and spheroidite (Fig. 10–4.7) which are the same; those which are different. Explain your choices.

Q_{10} Explain to a freshman chemistry student why the curves of Fig. 10–4.6 have a negative slope.

•R_{10} Give a basis for using the Arrhenius assumption in Example 10–4.1.

S_{10} On the basis of the information in *this* chapter, suggest a way of preparing spheroidite.

•T_{10} The 90Al–10Mg alloy of Fig. 10–1.2 is cooled slowly so β forms a network at the α grain boundaries. Based on Fig. 9–5.4 and Sections 6–4 and 8–8, predict various mechanical properties.

STUDY PROBLEMS

10–1.1 A 90Pb–10Sn solder is held at 185°C until equilibrium is established. The average grain size is 10^6 μm^3. The alloy is then *rapidly* cooled to room temperature where a β precipitate forms within the former α grains. The resulting matrix within those former α grains is 99Pb–1Sn (ρ = 13.3 g/cm^3), and the β particles ($\sim$100% Sn) are separated by an average distance of 0.1 μm. a) How many particles are there per original grain? b) What is the volume fraction of β? c) Approximately how many tin atoms are there per β particle?

Answer: a) $\sim 10^9$ b) 15 v/o c) $\sim 5 \times 10^6$ Sn

10–1.2 A sterling silver (92.5Ag–7.5Cu) is quenched after solution treatment, then reheated to 400°C until equilibrium is attained; at equilibrium it contains a β precipitate (approximately spherical in shape) in an α matrix. a) If the representative diameter of the β particle is 0.1 μm, how many will there be per cm^3? If d = 0.05 μm? b) What is the average distance between particles?

10–1.3 a) Cite the eutectoid reaction in the Fe–O system (Fig. 9–5.5). b) What is the Fe^{3+}:Fe^{2+} ratio in the oxide of eutectoid composition? c) How much metallic iron is in the two-phase product below the eutectoid temperature?

Answer: a) ϵ(23.26% O) $\underset{}{\overset{560°C}{\rightleftharpoons}}$ α(0% O) + ζ(27.64% O) b) 0.13 c) 16%

10–1.4 Refer to the Cu–Sn diagram in Chapter 9. Locate and state the reactions (on cooling) for four eutectoid transformations.

10–1.5 Refer to the data in Discussion Item I_{10}. Calculate the density of this martensite.

Answer: 7.9 g/cm^3

10–1.6 An oxide containing 75 w/o iron and 25 w/o oxygen is solution-treated at 1300°C (2370°F). It is then cooled slowly to room temperature. Compare and contrast the reactions and possible microstructures with those found after slow cooling a 1060 steel from 900°C (1650°F).

10–2.1 The ferrite lamellae of pearlite may be as thick as 1 μm. As a 1080 steel transforms, how many carbon atoms must have diffused from each μm^3 of iron that becomes ferrite?

Answer: $3 \times 10^9/\mu m^3$

10–3.1 Calculate the density of a glass-reinforced phenol-formaldehyde plastic, in which the glass content is 15 w/o. (A borosilicate glass is used for the longitudinal glass fibers.)

Answer: 1.4 g/cm^3

10–3.2 At room temperature the α of sterling silver contains much less than 1% Cu (extrapolate Fig. 9–2.4). If pure metals have heat capacities of 6 cal/mole (25 J/mole), what is the heat capacity of sterling silver if the β has completely precipitated from α?

10–3.3 Estimate the thermal conductivity (longitudinal) of the reinforced plastic of Study Problem 10–3.1.

Answer: 0.002 J · cm/°C · cm^2 · sec

• **10–3.4** Repeat Study Problem 10–3.3 for a 60Pb–40Sn solder equilibrated at 20°C. (The β is randomly distributed within the α. The thermal conductivity of tin is 0.63 J · cm/°C · cm^2 · sec.)

10–3.5 A cube (2.5 cm along each side) is made by laminating alternate sheets of aluminum and vulcanized rubber (0.5 mm and 0.75 mm thick, respectively). What is the thermal conductivity of the laminate: a) parallel to and b) perpendicular to the sheets? (Use the data of Appendix C.)

Answer: a) 0.9 $J \cdot cm/°C \cdot cm^2 \cdot sec$ b) 0.002 $J \cdot cm/°C \cdot cm^2 \cdot sec$

10–4.1 Austenite (0.45% C) is quenched rapidly to form 100% martensite. The martensite is tempered at 300°C to form ($\alpha + \overline{C}$). What is the weight fraction of carbide?

Answer: 0.067

CHAPTER ELEVEN

THERMAL PROCESSING OF MULTIPHASE MATERIALS

PREVIEW

It is already evident that solid-phase reactions do not occur simultaneously. We will observe that time is required for the nucleation of new grains and phases. We will also learn that when the reactant and product phases differ in composition, time must be allowed for diffusion and relocation of the component atoms. Finally, diffusional time is required for the growth of particles and grains within the microstructure, to give us the microstructural variables of Chapter 10.

We will use an isothermal reaction procedure to examine reaction rates. This provides us with a method of anticipating the time necessary to complete a reaction, or equally important, the time available to avoid an unwanted reaction. For example, when we quenched steel in Section 10–1, it was to avoid the normal reaction and to obtain a very hard martensite as a substitute.

This chapter cites different thermal-processing treatments that are used to engineer materials. With each, it points out the role of composition, nucleants, and/or solid-phase reactions for manipulating the microstructures to the selected geometries for the desired properties.

11–1 REACTION RATES

Rates for various reactions range from those which are almost instantaneous to those which are so slow that for practical purposes the reactions may be considered non-existent. No reaction is *instantaneous*, and even those we call "instantaneous" actually require a finite time to occur. The burning of the gas–air mixture in an internal combustion engine is typical of a fast reaction, but if there were not a delay of a few milliseconds during combustion, the engine would knock.

The crystallization of glass is an example of a very *slow* reaction. Glass has the structure of a liquid and, like other materials, it should crystallize below its melting point. However, the rate of crystallization at room temperature is so slow that we still have samples of noncrystalline glass made in early historic times (Fig. 11–1.1). Thus, glass is considered to be a *metastable phase*.

The time requirement for reaction arises because (1) existing bonds must be broken, (2) atoms must be moved to or from the reaction site, and (3) a new boundary is required whenever a new phase is to be nucleated. Each of the above steps requires a supply of energy; therefore, it should be anticipated that each will be sensitive to the temperature of the material.

Fig. 11–1.1. Early historical glass (1400 B.C.). This glass has not crystallized during this long period even though it is a supercooled liquid. (Corning Glass Works.)

Isothermal reactions. Reactions do not progress linearly with time, even when the temperature is constant, i.e., *isothermal.* As revealed in Fig. 11–1.2, even recrystallization requires "incubation" time to nucleate new grains. Then the rate accelerates, and is maximum when the reaction is about 50% complete. The rate slows down as the reaction nears completion; the final few percent take considerable time, particularly when one considers that the abscissa of Fig. 11–1.2 is plotted on a logarithmic scale.

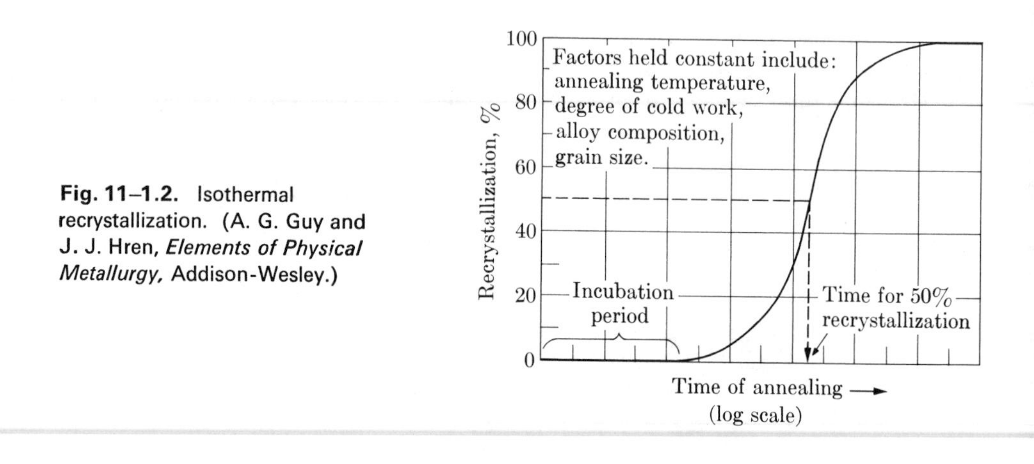

Fig. 11–1.2. Isothermal recrystallization. (A. G. Guy and J. J. Hren, *Elements of Physical Metallurgy,* Addison-Wesley.)

Isothermal recrystallization versus temperature. We observed in Section 6–7 that in hot-working, recrystallization occurs as fast as the metal is deformed. The time is longer at lower temperatures; at room temperature the time is sufficiently long so we may consider the deformed metal to be metastable. Figure 11–1.3 shows the time required to produce 50% recrystallization of high-purity copper. This percentage is used since it is most sharply defined and most easily measured (Fig. 11–1.4). Since the atomic rearrangements must be thermally activated, the time has an Arrhenius relationship with the absolute temperature (Eq. 4–6.6);

$$\log t = C + B/T, \tag{11–1.1}$$

where C and B are constants* for each reaction.

* A fast reaction rate, R (short time, since $R = 1/t$), corresponds to easy diffusion. As in Eqs. (4–7.4) and (8–8.3),

$$\log_{10} R = \log_{10} R_0 - Q/4.575T = -\log_{10} t. \tag{11–1.2}$$

This leads to Eq. (11–1.1), where $C = -\log_{10} R_0$, and $B = Q/4.575$.

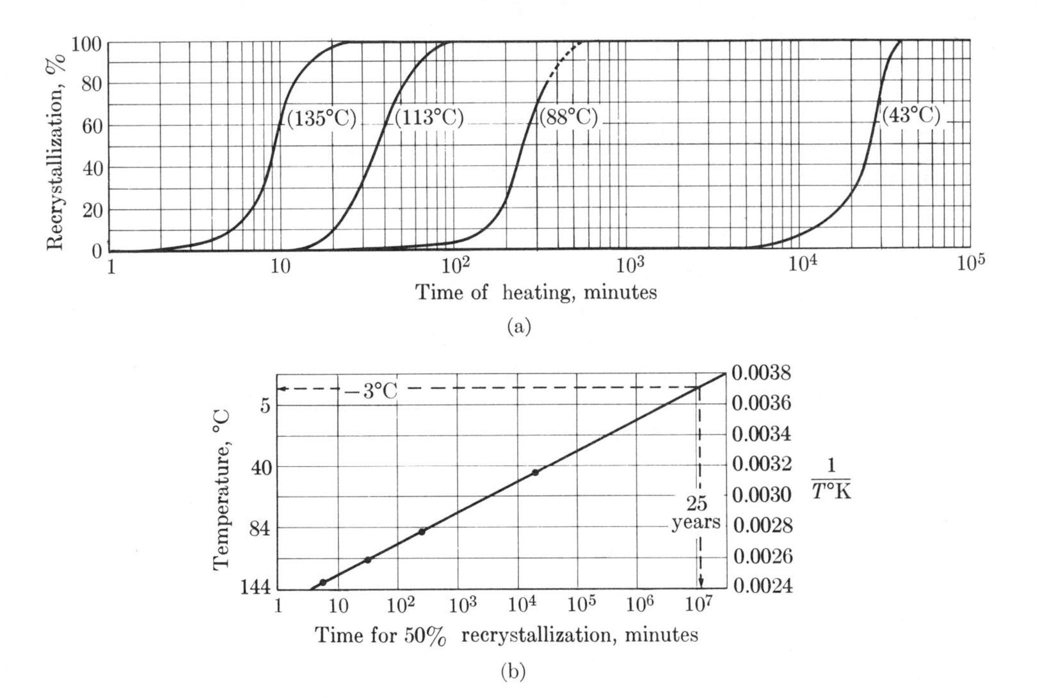

▲ **Fig. 11–1.3.** Isothermal recrystallization of 99.999% pure copper. (After Decker and Harker.) (a) Isothermal recrystallization of pure copper cold-rolled 98%. (b) Plot for extrapolation from the data given by the four curves of (a). (Adapted from A. G. Guy and J. J. Hren, *Elements of Physical Metallurgy*, Addison-Wesley.)

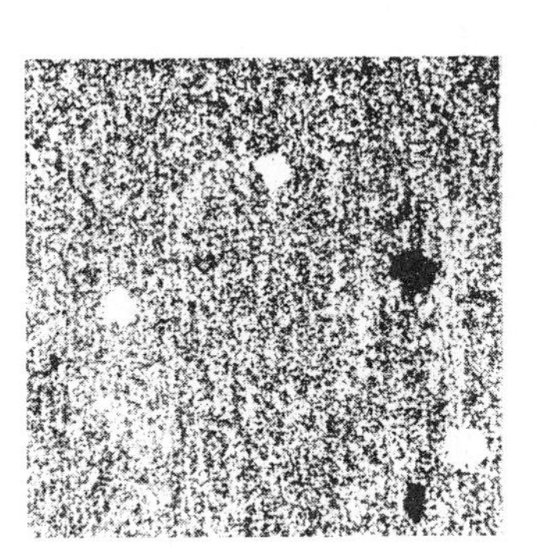

(a) Heated for 50 hours

Fig. 11–1.4. The progress of isothermal recrystallization at 310°C (590°F) in 99.95% aluminum that has been cold-worked 5%. (a) Heated for 50 hours, (b) 70 hours, (c) 80 hours, (d) 100 hours. (×7.5) ◀ (Courtesy W. A. Anderson, Alcoa.) ▼

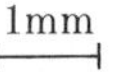

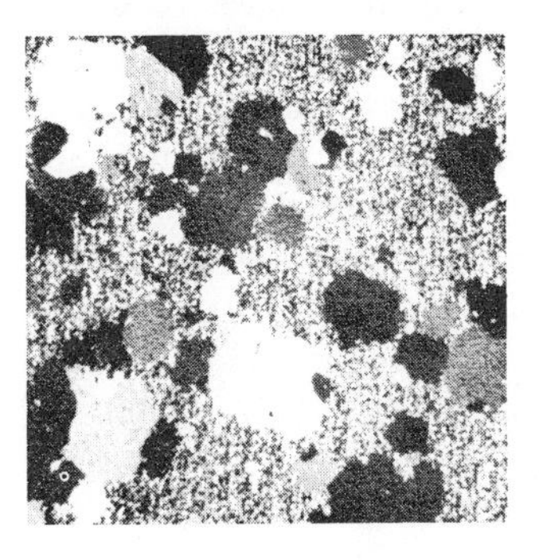

(b) 70 hours

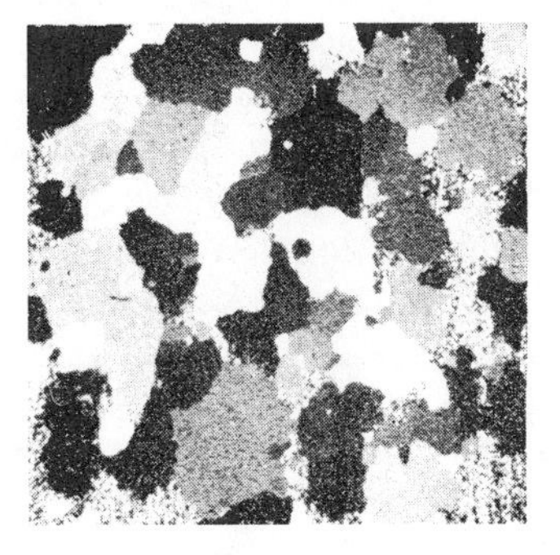

(c) 80 hours

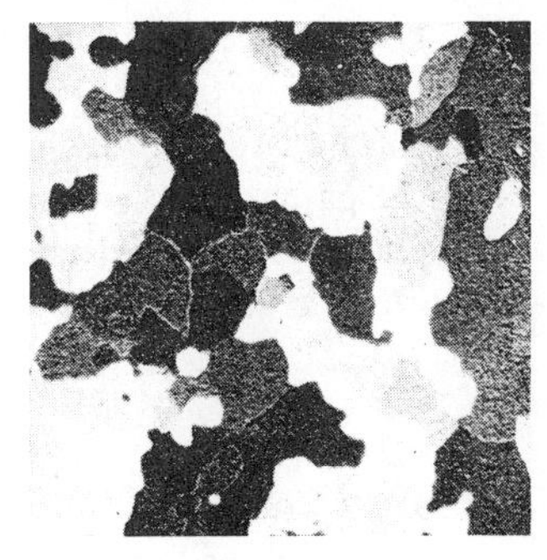

(d) 100 hours

Isothermal precipitation versus temperature. Precipitation is more complex than recrystallization because a new phase must form. Consider a 90Pb–10Sn alloy (Fig. 9–2.1) that has a solubility limit curve similar to the one encountered for a 90Al–10Mg alloy in Fig. 10–1.2. Above 155°C, only α (Pb-rich) is stable; below 155°C, some β (Sn-rich) will separate from supersaturated α if time is available. Figure 11–1.5 shows the time required for 50% completion as a function of temperature. We will not be able to present an Arrhenius-type equation for this data. However, we will observe that the required time decreases as the temperature is raised from −50°C to room temperature. At about 50°C, the pattern is reversed and eventually runs off the long-time end of the graph just below the solubility limit of 155°C. This reversal will be of interest to us since we find the same *C*-type curve in other reactions too. The explanation lies with the time required for nucleation of the new β phase.*

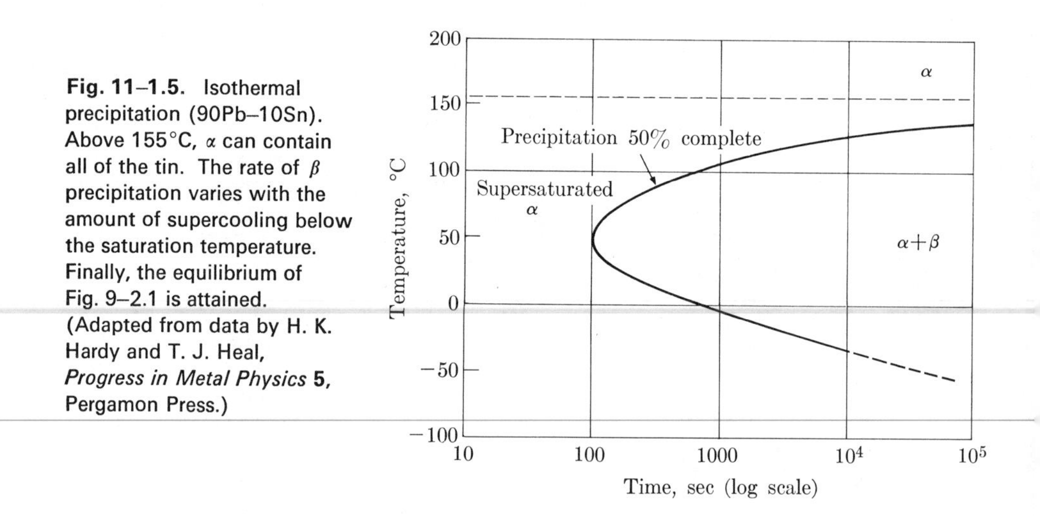

Fig. 11–1.5. Isothermal precipitation (90Pb–10Sn). Above 155°C, α can contain all of the tin. The rate of β precipitation varies with the amount of supercooling below the saturation temperature. Finally, the equilibrium of Fig. 9–2.1 is attained. (Adapted from data by H. K. Hardy and T. J. Heal, *Progress in Metal Physics* **5**, Pergamon Press.)

•**Homogeneous nucleation.** Extra energy is associated with any phase boundary because the atoms along the interface are less completely coordinated than those in the interior of the crystal (Sections 4–4 and 6–2). When a new phase must be nucleated, extra energy is therefore required to form the interface between the new particle and the parent phase. Since this extra boundary energy γ is not available at the saturation temperature, reaction is infinitely slow. With supercooling, however, energy can be made available. We shall use water and ice to illustrate.

---

* The reader will observe that Fig. 11–1.5 is *not* a phase diagram; rather, it is a transformation diagram, where the phases change with time until equilibrium is attained at the far right. The eventual results after long times should correspond to the data available from phase (or equilibrium) diagrams.

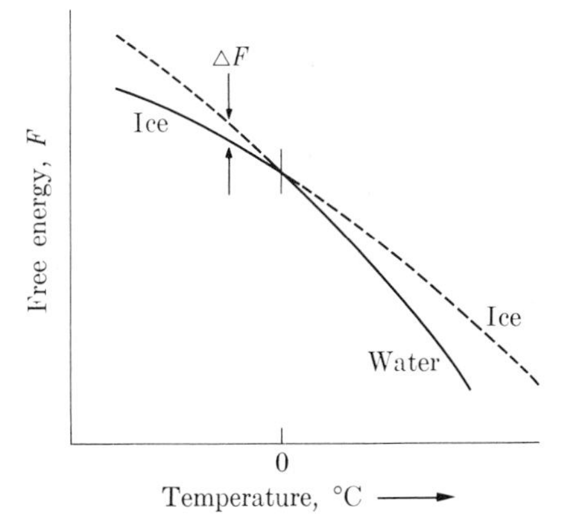

Fig. 11–1.6. Free energy of water and ice. Above 0°C, ice has a higher free energy than water. It gives up this energy by changing to the lower-free-energy form of water. The driving force is equal to ΔF.

At 0°C, the free energy* of ice and water are equal, since they are in equilibrium. Below 0°C, ice is more stable than water because it has less free energy (Fig. 11–1.6). The free energy difference ΔF becomes available for nucleation if the water is supercooled below 0°C.

The energy released during the formation of a new phase of radius r is $\frac{4}{3}\pi r^3 \Delta F_V$, where ΔF_V is the decrease ($-$) in free energy per unit volume. However, the energy required to produce the new surface between the nucleated phase and the parent phase is equal to $4\pi r^2 \gamma$, where γ is the interface energy per unit area. Thus, the total change in energy ΔF_r for a precipitate of radius r is

$$\Delta F_r = 4\pi r^2 \gamma + \tfrac{4}{3}\pi r^3 \Delta F_V. \tag{11–1.3}$$

Although ΔF_V is negative for any spontaneous reaction, γ is always positive; therefore, we can plot Eq. (11–1.3) as shown in Fig. 11–1.7. Below a critical nucleus radius r_c,

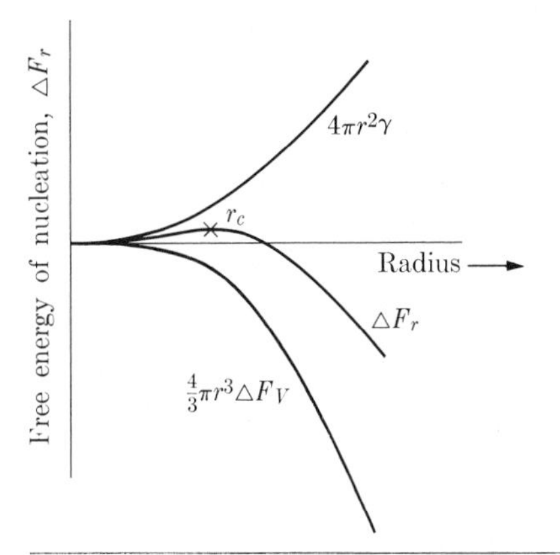

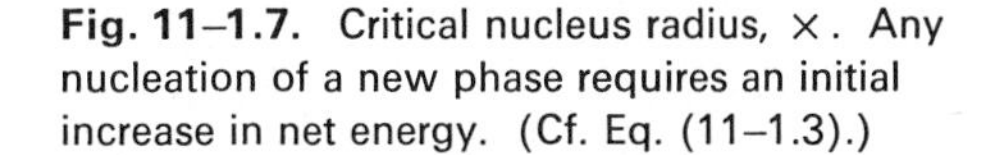

Fig. 11–1.7. Critical nucleus radius, ×. Any nucleation of a new phase requires an initial increase in net energy. (Cf. Eq. (11–1.3).)

* The energy available for a chemical reaction.

net energy is required for the nucleation of a new phase and nucleation cannot be spontaneous. Cooling beyond the equilibrium temperature, i.e., *supercooling*, increases the negative value of ΔF and therefore markedly reduces both the critical radius and the maximum amount of nucleation energy (Fig. 11–1.8). Thus there is a much greater probability that the necessary number of atoms can receive the appropriate amount of energy from some local energy source, such as a grain boundary or crystal imperfections. As a result, a reaction which takes infinitely long at the equilibrium temperature occurs in much less time with supercooling. Of course, extremely low temperatures restrict atom movements so that the time for reaction is extended even in the presence of nuclei (Fig. 11–1.5).

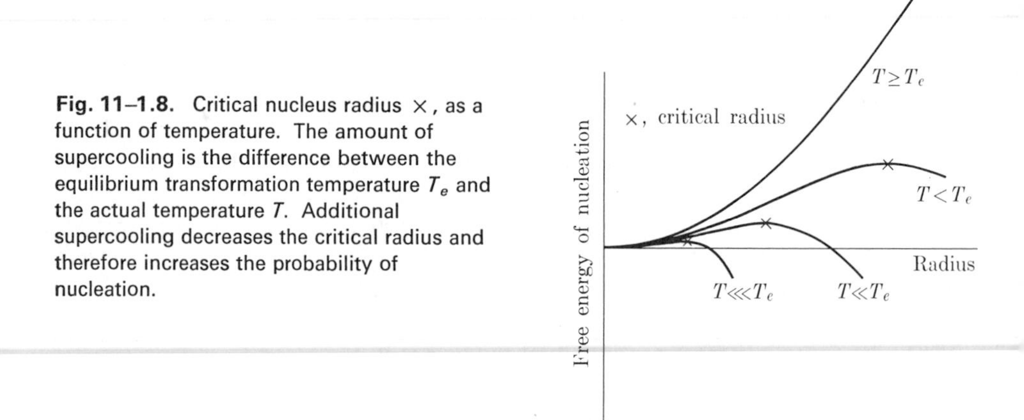

Fig. 11–1.8. Critical nucleus radius ×, as a function of temperature. The amount of supercooling is the difference between the equilibrium transformation temperature T_e and the actual temperature T. Additional supercooling decreases the critical radius and therefore increases the probability of nucleation.

Less interface energy is required if an existing surface is available. If, for example, a plane surface of the product phase is already present, transformations can proceed without any increased surface area. Very often impurity particles may be present (by chance or by intent) and their surfaces may serve as a nucleating surface. If the impurity and the product phase have similar structures, and thus comparable interfaces of low energy, the impurity may serve as a base for the transformation. Less surface energy is required under such conditions and the transformation can proceed more rapidly. Essentially all solid-state reactions are initiated in this manner.

Example 11–1.1 A copper alloy, which received 35% cold work, had recrystallized (50% completion) after 1 hour at 295°C (563°F). At 325°C (617°F), the time required was only 15 min. What minimum temperature is required for hot-working where this recrystallization must be 50% complete in 6 sec?

Solution: Use the Arrhenius relationship of Eq. (11–1.1).

$$\log 60 = C + B/(295 + 273),$$
$$\log 15 = C + B/(325 + 273).$$

Solving simultaneously,

$$C = -10.2, \qquad B = 6800;$$
$$\log 0.1 = -10.2 + 6800/T$$
$$T = 739°K = 466°C \quad \text{(or 871°F)}.$$

Comment. A calculation of the time required reveals that this strain-hardened alloy will remain in the metastable condition for a long time at room temperature. However, it may also be noted that softening occurs ahead of 50% recrystallization. These data are for a 35% cold-worked copper alloy, and not for a 98% cold-worked, high-purity copper; therefore they should not be expected to match Fig. 11–1.3(b). ◀

•**Example 11–1.2** Consider the polymorphic transformation, $A \rightleftarrows \mathscr{A}$. The interface energy between A and $\mathscr{A}$ is 0.5 J/m², and the values of ΔF_V for $A \to \mathscr{A}$ are -0.4×10^9 J/m³ at 1000°C and -2×10^9 J/m³ at 900°C. a) Determine the critical nucleus radius for the nucleation of $\mathscr{A}$ within A at each temperature. b) Calculate the energy that must be supplied for reaction to proceed in each case.

Solution

a)

$$r_c = \text{radius when } \frac{d\,\Delta F_r}{dr} = 0,$$

$$\Delta F_r = 4\pi r^2 \gamma + \tfrac{4}{3}\pi r^3\,\Delta F_V,$$

$$\frac{d\,\Delta F_r}{dr} = 0 = 8\pi r\gamma + 4\pi r^2\,\Delta F_V,$$

$$r_c = \frac{-2\gamma}{\Delta F_V}.$$

At 1000°C,

$$r_c = \frac{-2(0.5\ \text{J/m}^2)}{-0.4 \times 10^9\ \text{J/m}^3} = 25\ \text{Å}.$$

At 900°C,

$$r_c = \frac{-2(0.5\ \text{J/m}^2)}{-2 \times 10^9\ \text{J/m}^3} = 5\ \text{Å}.$$

b)

$$\Delta F_r = 4\pi r^2\left(\gamma + r\frac{\Delta F_V}{3}\right).$$

At 1000°C,

$$\Delta F_r = 4\pi(25 \times 10^{-10})^2\left[0.5 + \frac{(25 \times 10^{-10})(-0.4 \times 10^9)}{3}\right]$$
$$= 13 \times 10^{-18}\ \text{J}.$$

At 900°C,

$$\Delta F_r = 4\pi(5 \times 10^{-10})^2\left[0.5 + \frac{(5 \times 10^{-10})(-2 \times 10^9)}{3}\right]$$
$$= 0.5 \times 10^{-18}\ \text{J}. \quad ◀$$

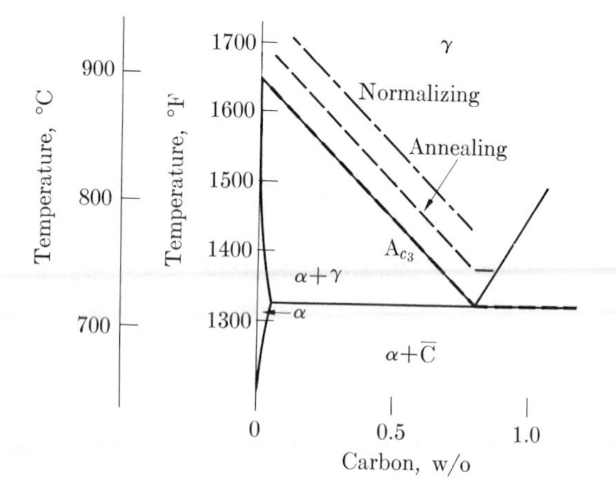

Fig. 11–2.1. Annealing and normalizing (plain-carbon steels). The heat-treating temperature varies with the carbon content. For annealing, the temperature is sufficiently high to ensure complete austenization. The steel is then cooled slowly to form a coarse pearlite and soft product. For homogenization, the steel is heated somewhat higher to promote atom diffusion and compositional uniformity. Excessive heating, however, would permit undesirable grain growth. Following homogenization, the normalized steel is air-cooled because low hardness values are not a primary objective. (See Fig. 11–2.3(a).)

11–2 ANNEALING AND HOMOGENIZATION TREATMENTS

The term *annealing* is used generally to denote either a softening or a toughening process. In Section 6–7, the word *annealing* was used to describe the softening which accompanies recrystallization of strain-hardened metals. We saw in Chapter 8 that the glass technologist uses *annealing* to denote a heat-treatment for removing residual stresses and thereby reducing the probability of the development of cracks in a brittle glass. *Annealing* has special connotations when it describes a heat treatment of steel. (We shall discuss these connotations shortly.) In the meantime, we shall simply note that *annealing* entails heating in each case to a temperature at which the individual atoms have added freedom for movement and rearrangement into a more stable structure, i.e., a structure with less energy.

The following outline will point out the purposes of several common annealing treatments, together with some of their technical requirements. The "shop name" of the treatment is given in boldface type.

Recrystallization (or anneal)

Materials: Nonferrous, cold-worked metals

Purpose: To soften and remove strain hardening

Temperature: $T_m/3$ to $T_m/2$—chosen to give a time compatible with the production process.

Process anneal

Materials: Steel wire and sheet products

Purpose: To soften and remove strain hardening

Temperature: Immediately under the eutectoid temperature to permit the process to be as fast as possible (and even be in tandem with the cold-working process). However, austenite must not be formed, or hard martensite could develop as these small-dimensional products are cooled.

Full anneal

Material: Steel

Purpose: To soften prior to machining

Temperature: *Austenization* 25–30°C (45–55°F) above the stability of the last ferrite (Fig. 11–2.1). This is followed by furnace cooling, so that the austenite decomposes to a coarse pearlite. This is a relatively expensive process because it ties up furnace time.

Anneal

Material: Glass

Purpose: To remove residual stresses, and avoid thermal cracking

Temperature: Varies with the glass composition; requires a viscosity η of 10^{13} poises. The glass must be cooled slowly past the strain point ($\eta = 10^{14.5}$ poises) in order to avoid the reintroduction of residual stresses (Section 8–8).

Stress relief

Materials: All metals, and steels in particular, since they have a discontinuous volume change during transformation. This is particularly important if machining is to follow a previous heat treatment, because when an area with residual stresses is removed, the remainder of the product is subject to warpage.

Temperature: A few minutes at ~600°C (~1100°F) for steels. Allowance must, of course, be made for the center of large masses to attain this temperature.

Spheroidization

Materials: High-carbon steels, e.g., ball-bearings

Purpose: To toughen an otherwise brittle steel. It is necessary to have high carbon content for wear resistance; however, in the pearlitic form, it would have very little ductility (Fig. 10–4.8).

Temperature: If initially pearlite, 15–16 hours just below the eutectoid temperature; if initially martensite, 1 hour at that temperature (Fig. 10–2.2).

Homogenization (soaking)

Materials: Cast metals

Purpose: To make the composition more uniform. During solidification, the first solid to form is not the same as the overall composition. This is illustrated in Fig. 11–2.2(a) for a 96Al–4Cu alloy. The appropriate phase diagram (Fig. 9–5.3) reveals that κ, the very first solid to form at $\sim$650°C (1200°F) has only 1% copper; the copper content of the solid gradually increases as the temperature falls. At the eutectic temperature, the liquid contains 33% copper.

Temperature: As high as possible short of forming liquid or producing excessive grain growth. Some atoms must diffuse long distances compared to the atomic scale ($\sim 10^{-2}$ cm, or $\sim 10^{6}$ unit-cell distances). It is therefore necessary to heat the metal considerably above the recrystallization temperatures so that homogenization may occur in a reasonable length of time.

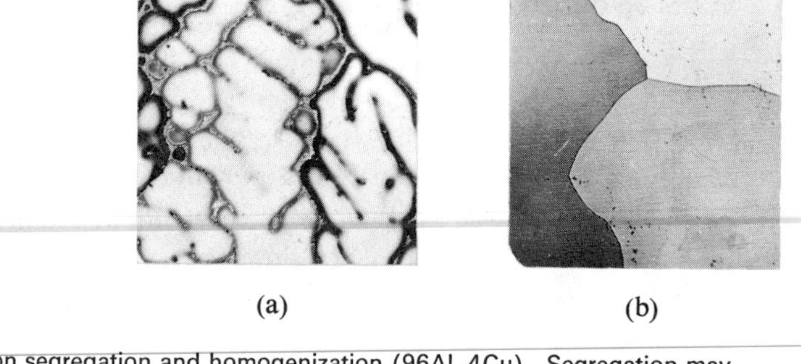

Fig. 11–2.2. Solidification segregation and homogenization (96Al–4Cu). Segregation may arise from rapid solidification, during which there has been insufficient time for diffusion to provide equilibrium (Fig. 9–5.3). (a) The cast alloy has a copper-rich residual (dark) between the growing crystals. (b) The same alloy after "soaking." The copper has diffused until it is uniform throughout each grain. (Courtesy Alcoa Research Laboratories.)

Normalizing

Material: Steel

Purpose: Homogenization of alloy additions for more uniform austenite decomposition

Temperature: *Austenization*, 50–60°C (90–110°F) into the range of complete austenite, followed by air-cooling. Normalizing uses higher temperatures than the full anneal in order to accelerate diffusion (Fig. 11–2.1). Still higher temperatures are precluded because of excessive austenitic grain growth. Since low hardnesses are not required, the product is removed from the furnace and air-cooled, permitting the furnace to receive new products.

These processes as they pertain to steels are summarized in Fig. 11–2.3. The reference heat-treating temperature varies with the steel composition (Figs. 9–7.3 and 9–9.1).

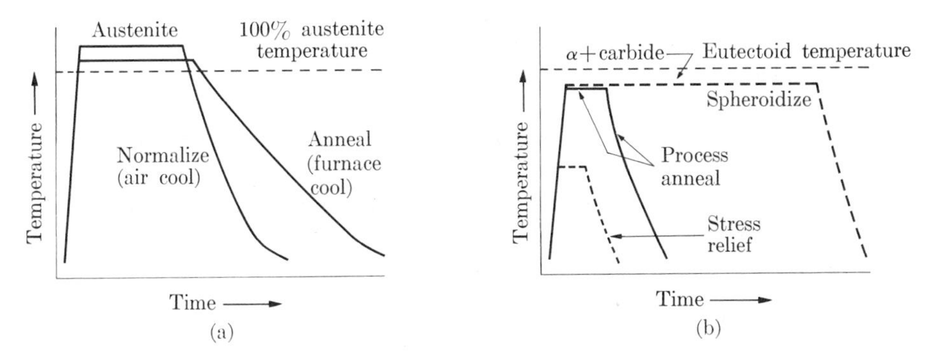

Fig. 11–2.3. Steel heat-treatment processes (schematic). (a) Austenization processes. (b) Subeutectoid processes.

Example 11–2.1 Let 100 g of a 96Al–4Cu alloy equilibrate at 620°C (1150°F), forming the κ and liquid, as shown by the phase diagram. The alloy is then cooled rapidly to 550°C (1020°F) with no chance for the initial solid to react. A liquid phase will still be present. a) What will its composition be? b) How many grams will there be of this final liquid?

Solution: At 620°C,

Liq: 88Al–12Cu, κ: 98Al–2Cu.

$$\text{Grams liquid} = 100\text{ g}\left(\frac{98 - 96}{98 - 88}\right) = 20\text{ g liquid.}$$

At 550°C, and with only the 20 g of liquid reacting,

κ: 94.4Al–5.6Cu.

a) Liq: 67Al–33Cu.

$$\text{b) Grams liquid} = 20\text{ g}\left(\frac{94.4 - 88}{94.4 - 67}\right) = 4.6\text{ g liquid.}$$

Comments. In the final 4.6 g of liquid, there will be local areas with high copper segregation (33%). There will also be nearly 80 g of metal with only 2% Cu.

Cooling may be considered to take place in a series of small steps, in which the solute in the residual liquid is progressively concentrated. ◀

11–3 PRECIPITATION PROCESSES

Age-hardening (precipitation-hardening). A very noticeable increase in hardness may develop during the *initial stages of precipitation* from a supersaturated solid solution (Section 10–1 and Eq. (10–1.3)). In fact, the start of the precipitation in Fig. 11–1.5 was detected by this increased hardness. This hardening is commonly called *age-hardening* because it develops with time. (It is also called *precipitation-hardening.*) The prime requirement for an alloy which is to be age-hardened is that solubility decreases with decreasing temperature, so that a supersaturated solid solution may be obtained. Numerous metal alloys have this characteristic.

The process of age-hardening involves a *solution treatment* (Eq. (10–1.2)) followed by a quench to supersaturate the solid solution. Usually the quenching is carried to a temperature where the precipitation rate is exceedingly slow. After the quench, the alloy is reheated to an intermediate temperature at which precipitation is initiated in a reasonable length of time. These are the two steps XA and AB in Fig. 11–3.1, and Table 11–3.1.

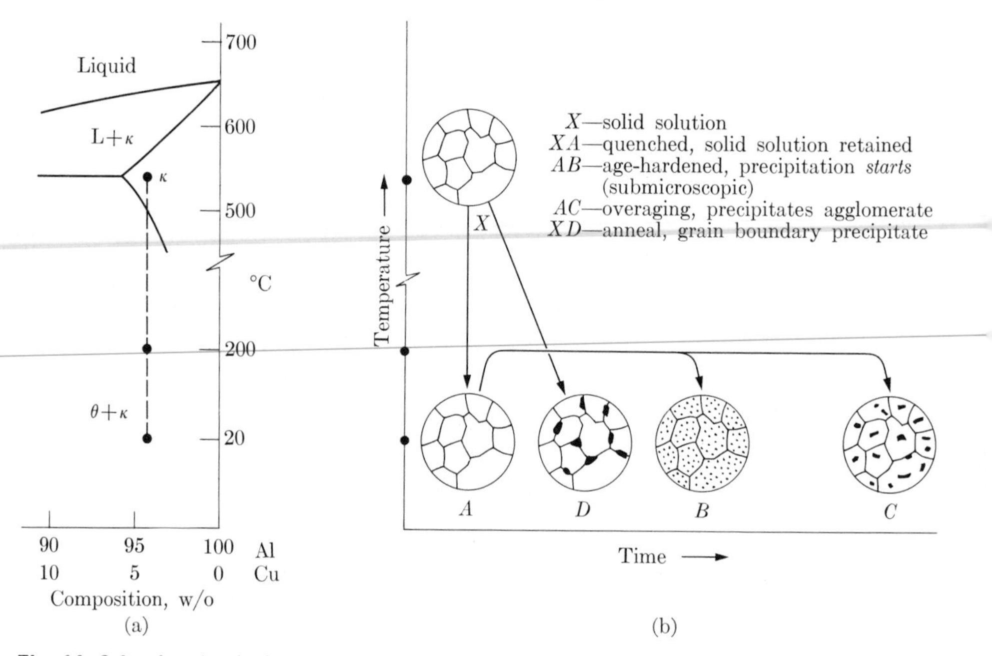

Fig. 11–3.1. Age-hardening process (96% Al–4% Cu alloy). See Table 11–3.1. The precipitates are still submicroscopic at the time of maximum hardness.

Fig. 11–3.2. Age-hardening mechanism: (a) κ solid solution. (b) Age-hardened; the θ precipitation has been initiated. Since the two structures are coherent at this stage, there is a stress field around the precipitate. (c) Overaged. There are two distinct and noncoherent phases, κ and θ. With limited numbers of solute atoms, maximum interference to dislocation movements occurs in (b). (Guy, A. G. and J. J. Hren, *Elements of Physical Metallurgy,* Addison-Wesley.)

Table 11–3.1
PROPERTIES OF AN AGE-HARDENABLE ALLOY (96% Al-4 % Cu)

	Treatment (See Fig. 11–3.1)	Tensile strength, psi (MPa)	Yield strength, psi (MPa)	Ductility, % in 2 in. (5 cm)
XA	Solution-treated and quenched	35,000 (240)	15,000 (105)	40
XAB	Age-hardened	60,000 (415)	45,000 (310)	20
XAC	Overaged	~25,000 (170)	~10,000 (70)	~20
XD	Annealed	25,000 (170)	10,000 (70)	15

An interesting example of the utility of the age-hardening process is the way it is used in airplane construction. Aluminum rivets are easier to drive and fit more tightly if they are soft and ductile, but in this condition they lack the desired strength. Therefore the manufacturer selects an aluminum alloy which can be quenched as a supersaturated solution, but which will age-harden at room temperature. The rivets are inserted while they are still relatively soft and ductile, and they harden further after they have been riveted in place. Since hardening sets in fairly rapidly at room temperature, there arises the practical problem of delaying the hardening process if the rivets are not to be used almost immediately after the solution treatment. Here advantage is taken of the known effects of temperature on the reaction rate. After the solution treatment the rivets are stored in a refrigerator, where the lower temperature will delay hardening for reasonable lengths of time.

Detailed studies have produced the following interpretation of the age-hardening phenomenon. The supersaturated atoms (Cu atoms in Example 10–1.1 and Fig. 11–3.1"*B*") tend to accumulate along specific crystal planes in the manner indicated in Fig. 11–3.2(b). The concentration of the copper (solute) atoms in these positions

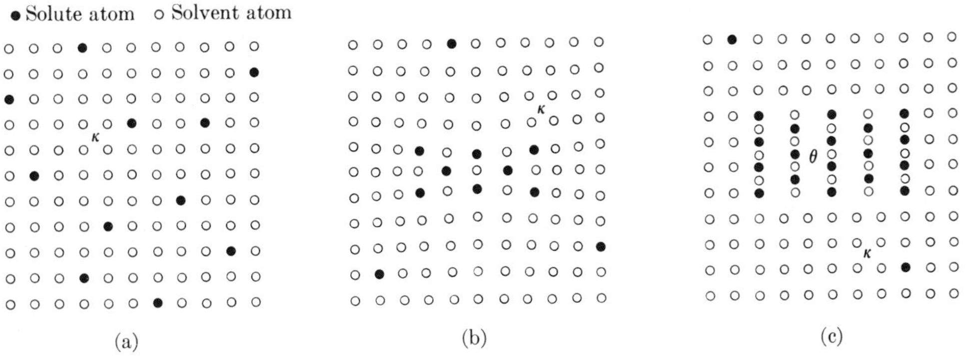

• Solute atom ○ Solvent atom

κ κ θ κ

(a) (b) (c)

lowers the concentrations in other locations, producing less supersaturation and therefore a more stable crystal structure. At this stage, the copper atoms have not formed a phase which is wholly distinct; a *coherency* of atom spacing exists across the boundary of the two structures. Dislocation movements occur with difficulty across these distorted regions, and consequently the metal becomes harder and more resistant to deformation under high stresses.

Overaging. A continuation of the local segregation process over long periods of time leads to true precipitation and *overaging*, or softening. For example, the development of a truly stable structure in an alloy of 96% aluminum and 4% copper involves an almost complete separation of the copper from the fcc aluminum at room temperature. Nearly all the copper forms $CuAl_2$ (θ in Fig. 11–3.2(c)). Because the growth of the second phase provides larger areas which have practically no means of slip resistance, a marked softening occurs.

Figure 11–3.3 shows data for the aging and overaging of a commercial aluminum alloy (2014). The initial hardening is followed by softening as the resulting precipitate is agglomerated. Two effects of the aging temperature may be observed: (1) precipitation, and therefore hardening, starts very quickly at higher temperatures; (2) overaging, and therefore softening, occurs more rapidly at higher temperatures. These two phenomena overlap to affect the maximum hardness that is attained. Lower temperatures permit greater increases in hardness, but longer times are required.

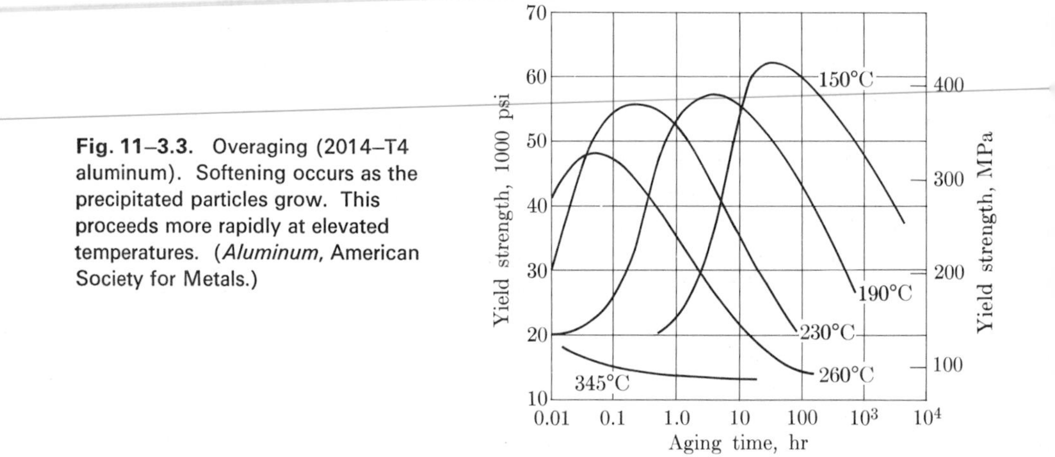

Fig. 11–3.3. Overaging (2014–T4 aluminum). Softening occurs as the precipitated particles grow. This proceeds more rapidly at elevated temperatures. (*Aluminum*, American Society for Metals.)

Combined hardening. Occasionally it is desirable to combine two methods of hardening. The cold-working of an alloy which has previously been age-hardened increases the hardness still further. However, there are some practical difficulties encountered in this process. Age-hardening increases resistance to slip and therefore increases the

Table 11–3.2
TENSILE STRENGTHS OF A STRAIN- AND AGE-HARDENED ALLOY (98% Cu–2% Be)

Annealed, 870°C (1600°F)	35,000 psi	240 MPa
Solution-treated, 870°C (1600°F) and cooled fast	72,000	500
Age-hardened only	175,000	1200
Cold-worked only (37%)	107,000	740
Age-hardened, then cold-worked*	200,000	1380
Cold-worked, then age-hardened	195,000	1340

* Cracked

energy required for cold-working, and it also decreases ductility so that rupture occurs more readily during cold-working. A possible alternative is to cold-work prior to the precipitation-hardening treatment. The metal is then cold-worked more readily and the age-hardening reaction occurs at a lower temperature because the slip planes serve as nuclei for the precipitation. However, the temperature of the aging process which follows cold-working may relieve some of the strain hardening and cause a slight loss in hardness. Although it does not produce hardnesses as great as those obtained from the reverse order, the final hardness is greater than that developed by using either method alone (Table 11–3.2).

• **High-strength, low-alloy (HSLA) steels.** Low-alloy steels have undergone some recent developments of significant importance that have led to high-strength products. Their development has provided the engineer with *structural* steels which have yield strengths of greater than 500 MPa (> 70,000 psi) in contrast to earlier structural steels of only half that figure.

The HSLA steels gain their strength from a very small grain size and finely distributed precipitates within the ferrite. In the latter respect, they may be compared to precipitation-hardened aluminum alloys which have just been discussed. The origin of the precipitate is significantly different, however. If "carbide-formers" such as vanadium or niobium are present, along with 0.05–0.2% carbon, we can still obtain a single phase (austenite) at hot-rolling temperatures. Thus beams, pipe, etc., can be readily processed. However, as the proeutectoid ferrite forms during austenite decomposition, there is a concurrent precipitation of alloy carbides. The precipitate forms, not because of the decreasing solubility curve for *one* phase, but rather because these alloy carbides are less soluble in α than in γ.

Unlike quenched and tempered steels (Section 11–5), these HSLA steels have the advantage of being weldable and of maintaining their mechanical properties without subsequent heat treatment. This has major implications for the engineer who is involved with major structures, or with pressurized pipelines.

Example 11–3.1 Based on Fig. 11–3.3, estimate the temperature required to reach the maximum hardness for this aluminum alloy in 10,000 hours (~ 14 months).

Solution: Assume that the Arrhenius relationship applies (Eq. 11–1.1), and use 150°C (423°K) and 260°C (533°K). At 150°C, with maximum hardness after 30 hrs:

$$\log t_{150} = 1.5 = C + B/423.$$

At 260°C, with maximum hardness after 0.05 hr:

$$\log t_{260} = -1.3 = C + B/533.$$

Solving simultaneously,

$$C = -12.1, \qquad B = 5750.$$

$$\log 10^4 \text{ hr} = 4 = -12.1 + 5750/T,$$

$$T = 357°\text{K} = 84°\text{C} \quad (\text{or } 183°\text{F}).$$

Comments. We may check our Arrhenius assumption at 230°C (503°K) and 190°C (463°K):

$$\log t_{230} = -12.1 + 5750/503 = -0.7, \qquad t = 0.2 \text{ hr};$$

$$\log t_{190} = -12.1 + 5750/463 = 0.3, \qquad t = 2 \text{ hr}.$$

The agreement with the experimental data in Fig. 11–3.3 is reasonable; however, an extrapolation to still longer times or lower temperatures is an approximation. ◀

11–4 $\gamma \rightarrow (\alpha + \overline{C})$ REACTION RATES

Isothermal transformation. The time requirement for austenite decomposition has been studied in considerable detail because of its practical importance. Figure 11–4.1

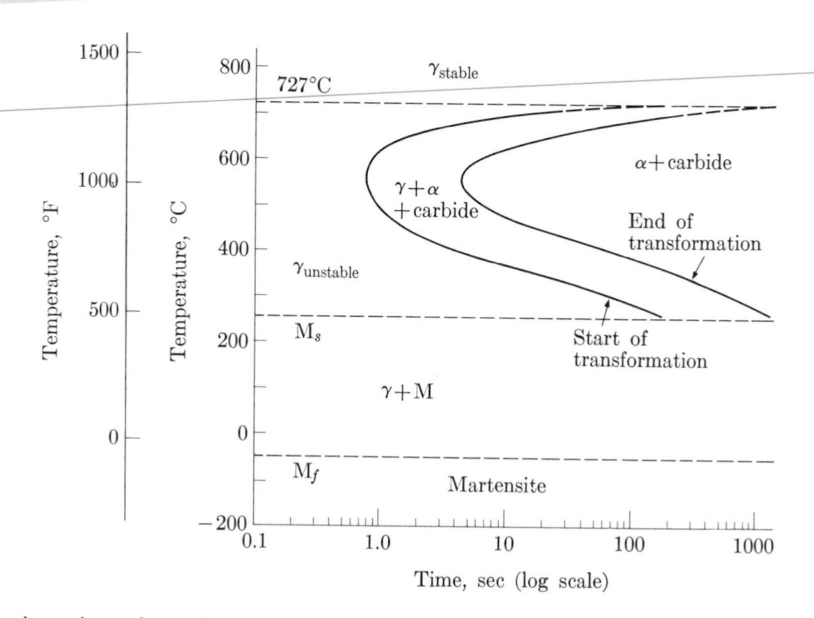

Fig. 11–4.1. Isothermal transformation curves for austenite decomposition (SAE 1080). (Adapted from U.S. Steel Corp. data.)

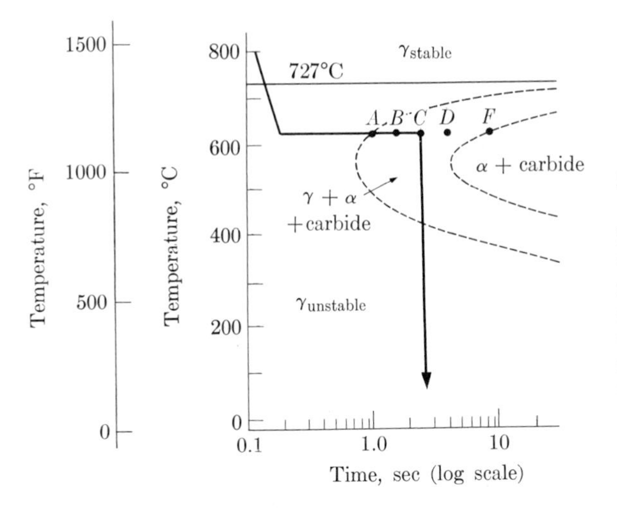

Fig. 11–4.2. Interrupted quench (eutectoid steel). This technique is used in establishing isothermal-transformation curves. The initial quench is made into a hot bath and it is held there for a prescribed time before the second quench to room temperature. (Cf. Figs. 11–4.3, 4, 5.)

shows time–temperature data for this reaction in an eutectoid steel (AISI–SAE 1080). The left solid curve is the time required for the decomposition to *start*. The right solid curve is the time for *completion* of the $\gamma \to (\alpha + \bar{C})$ reaction. These curves are commonly called isothermal transformation diagrams.* The data for Fig. 11–4.1 were obtained as follows. Small samples of eutectoid steel were heated into the austenite temperature range sufficiently long to assure complete transformation to austenite. These samples were then quenched to a lower temperature (e.g., 620°C) and held there for varying lengths of time before being quenched further to room temperatures (Fig. 11–4.2). The change $\gamma \to (\alpha + \bar{C})$ was not observed in samples held at 620°C for less than one second, and complete transformation to α + carbide was not observed until after more than 10 seconds had elapsed (Figs. 11–4.3 through 11–4.5). Similar data were obtained for other temperatures until the completed diagram shown in Fig. 11.4.1 was established.

The isothermal-transformation diagram shows that reactions occur slowly both at relatively low temperatures and at temperatures close to the transformation temperature. At intermediate temperatures, more rapid transformation occurs because (1) there is sufficient driving force [ΔF_V of Eq. (11–1.3)] to help nucleate the new phases, and (2) atomic diffusion is still quite rapid. With extremely fast cooling (severe quenching), it is often possible to miss the "knee" of the transformation curve for the beginning of the transformation, and to cool the steel to room temperature without transformation to ferrite and carbide, because the decomposition is detoured through the martensite (Eq. 10–1.5). In fact, this is the purpose of quenching steel in regular heat-treating operations.

The example just considered is the transformation of a eutectoid steel when there has been no advance separation of ferrite (or cementite) from the austenite

* They may also be called *C-curves* (because of their shape) or *T–T–T curves*, i.e., Temperature–Time–Transformation.

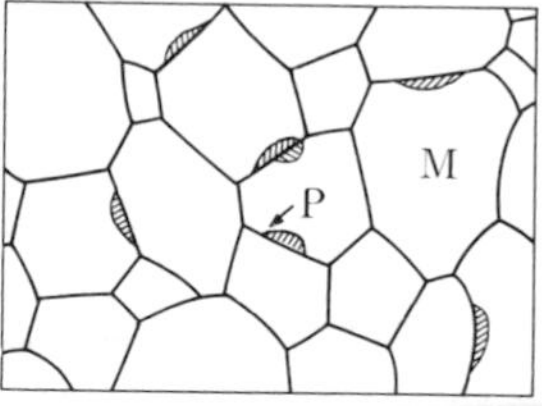

Fig. 11–4.3. Start of transformation at 620°C (1150°F) ($\gamma \rightarrow \alpha$ + carbide). M is steel which has not transformed to α + carbide. P is pearlite, i.e., lamellar (α + carbide). (Point *A* in Fig. 11–4.2.)

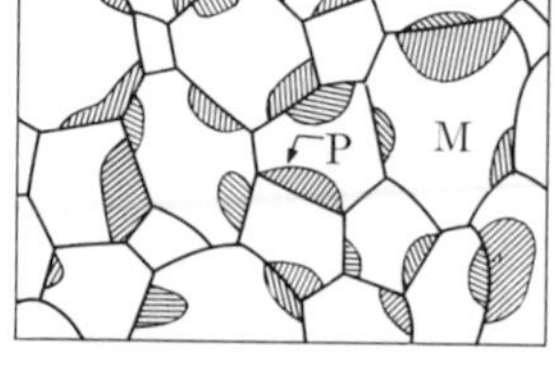

Fig. 11–4.4. Transformation 25% complete at 620°C (1150°F). (Point *B* in Fig. 11–4.2.)

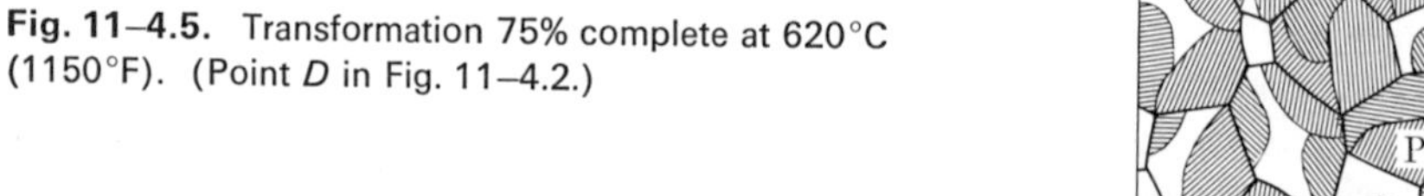

Fig. 11–4.5. Transformation 75% complete at 620°C (1150°F). (Point *D* in Fig. 11–4.2.)

prior to the formation of the pearlite ($\alpha + \bar{C}$). Figure 11–4.6(a) shows a similar transformation diagram for an SAE 1045 steel. Two features are different from those in Fig. 11–4.1.

1. Some ferrite may separate from the austenite above the eutectoid temperature. This could be predicted from the phase diagram which is shown at the right side of Fig. 11–4.6.

2. The isothermal transformation of 0.45 carbon steel occurs somewhat faster than the transformation of eutectoid steel. Comparison of the "knees" of the two curves shows this difference: the higher carbon steel starts to transform in about one second; in the 0.45% carbon steel, the reaction starts sooner. In fact, in the latter case the reaction occurs sufficiently fast so that we are not able to measure the rate at the "knee" of the curve with the interrupted-quench technique described above. A lower carbon content permits faster reaction, since part of the transformation delay is associated with the movement of the carbon atoms.

The isothermal-transformation curves which have been presented are *not equilibrium diagrams.* They indicate the changes which will occur with time as a variable. An equilibrium diagram shows no change with time. Of course, once equilibrium is established, the phases which eventually develop in an isothermal diagram (far right) must match the phase diagram.

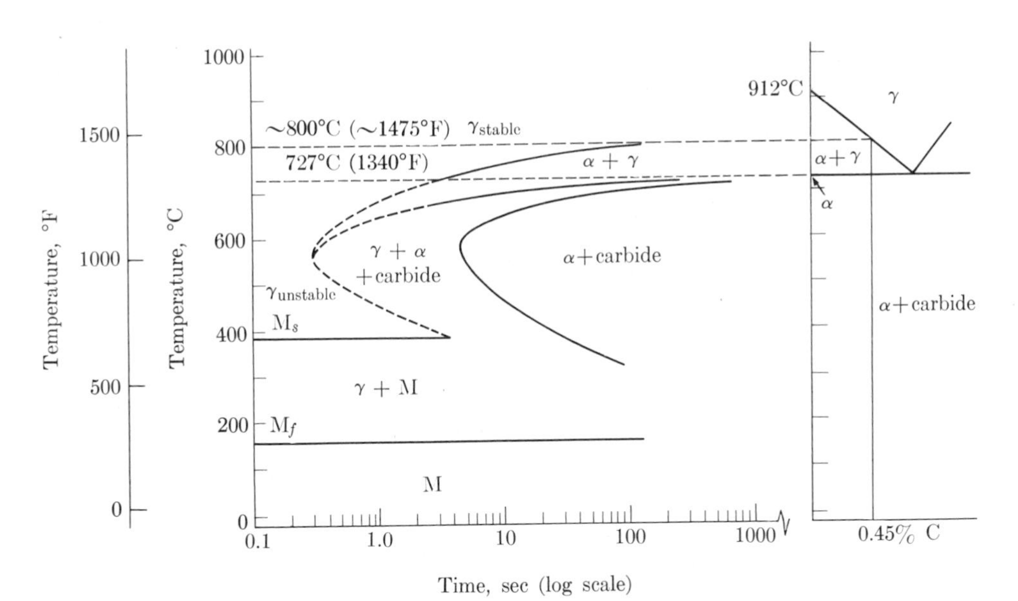

Fig. 11–4.6. Isothermal transformation diagram for SAE 1045 steel. The stable phases of the phase diagram (right) are not achieved immediately. However, in this steel, the austenite decomposition reaction is faster than in an eutectoid steel (Fig. 11–4.1).

Martensite formation. At the base of Figs. 11–4.1 and 11–4.6 are shown the temperatures at which the martensite reaction (Eq. 10–1.5) starts, M_s, and finishes, M_f, when the steel is quenched rapidly enough to miss the "knee" of the *C*-curve. Since diffusion is not involved,* the process of the martensite formation is not time-dependent. However, M_s and M_f depend on the amount of carbon which is present (Fig. 11–4.7). This has practical consequences in tool steels because a higher-carbon steel may be

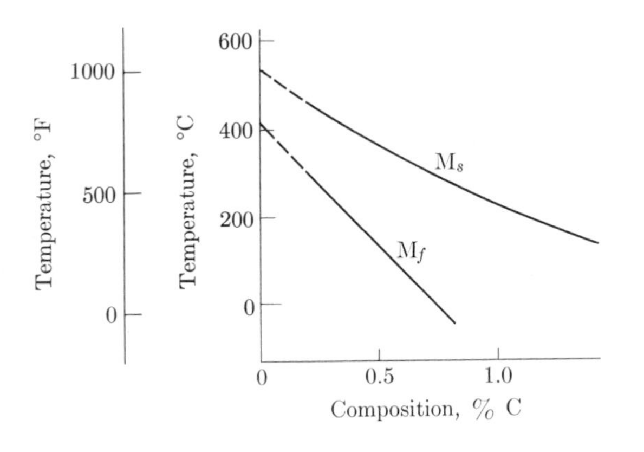

Fig. 11–4.7. Martensite transformation temperatures (plain-carbon steels). Transformation is first detected at M_s and is virtually complete at M_f. Between M_s and M_f, austenite is retained as a result of induced stresses.

* Austenite changes to martensite by shear (Section 10–1).

quenched to room temperature and still possess some *retained austenite*, i.e., the $\gamma \rightarrow$ M reaction is not complete.

Control of austenite decomposition rates. From the foregoing discussion, we see that temperature obviously affects the isothermal decomposition of austenite γ to ferrite plus carbide, $(\alpha + \bar{C})$. Other factors modify the time requirements, too. We will consider two: (1) *austenite grain size*, and (2) *alloy retardation.*

As sketched schematically in Figs. 11–4.3 to 11–4.5, the ferrite and carbide formation from the austenite starts at the grain boundary. A fine-grained steel offers more grain-boundary area per unit volume on which decomposition can be nucleated than does a coarse-grained steel. This is demonstrated in Fig. 11–4.8 where the two curves are for fine-grained (G.S. #8) steel and coarse-grained (G.S. #2) steel. The time requirements for the *start* of transformation differs by a factor of three at almost all temperatures.

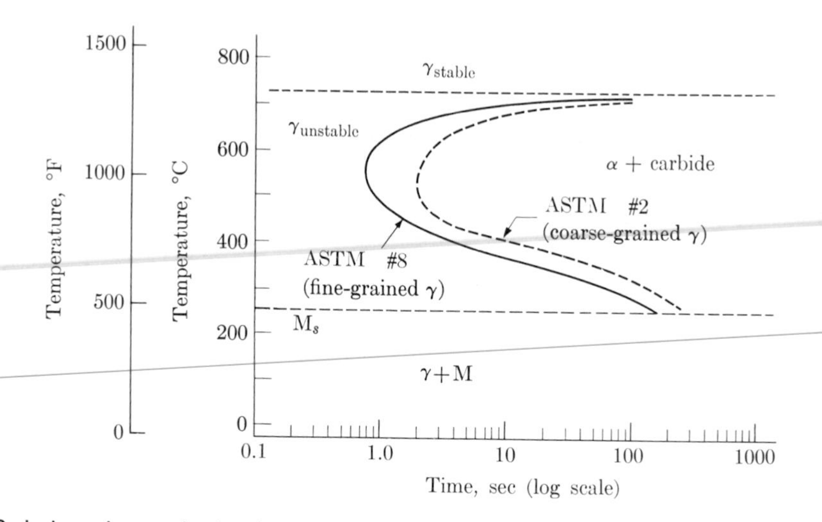

Fig. 11–4.8. Grain-boundary nucleation (eutectoid steel). The effect of austenite grain size on the *start* of austenite decomposition is shown for eutectoid steels. The steel with the fine-grained austenite has more grain-boundary area from which the $(\gamma \rightarrow \alpha + \bar{C})$ reaction can start.

The alloy retardation effect is equally pronounced. As an example, only 0.25% molybdenum (Fig. 11–4.9) delays the start of isothermal transformation by a factor of four at temperatures below 650°C (1200°F). This delay occurs because not only the carbon, but also the molybdenum, must be relocated when the austenite decomposes. Most of the Mo goes into the carbide. When silicon is the alloying element, most of it shifts into the ferrite. Since almost all alloying elements—Cr, Ni, Si, Mn, Mo, Ti, W, etc.,—diffuse more slowly in iron than carbon does (Fig. 4–7.3), we find

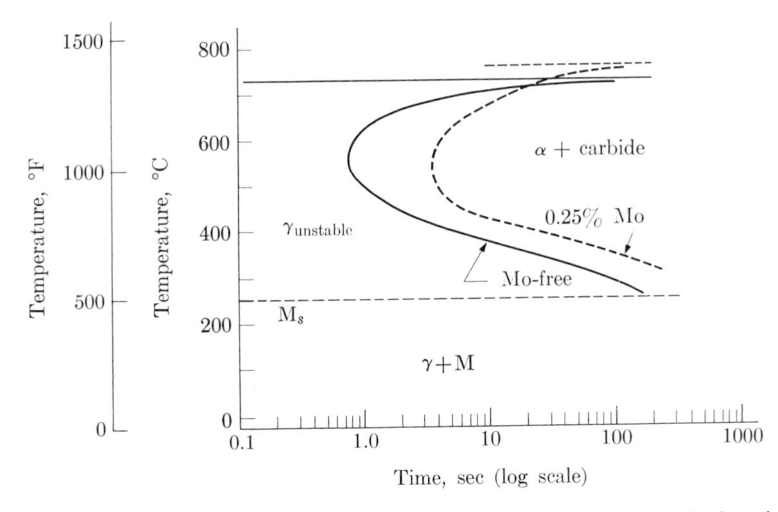

Fig. 11–4.9. Transformation retardation. Molybdenum, like other alloying elements, retards the *start* of transformation of austenite.

that all low-alloy steels (Table 9–9.1) have their isothermal transformation curves shifted to the right. In fact, it is easy to cool many low-alloy steels to room temperature and miss the "knee" of the transformation curve entirely, thus forming all martensite.

Continuous-cooling transformation. Isothermal transformation diagrams are convenient for interpretation, because we can hold the effect of temperature constant during the austenite decomposition. In practice, however, we more commonly encounter continuous cooling: (1) a hot piece of steel is removed from the furnace and air cooled, or (2) the steel is quenched in water. In neither is there an isothermal holding period while the $(\alpha + \bar{C})$ forms.

Let us consider the progress of austenite decomposition with different reaction rates of continuous cooling. A severe quench will miss the "knee" of the transformation curve, with the result that the austenite is changed to martensite rather than to pearlite (i.e., $\alpha + \bar{C}$). Slow cooling permits pearlite to form; however, the first decomposition occurs after a longer time (and therefore at a lower temperature) than for isothermal transformation, simply because part of the time was spent at higher temperatures where reaction rates were initiated more slowly. Thus the isothermal-transformation curves are displaced downward and to the right for *continuous-cooling* transformation (Fig. 11–4.10).

There are two important *critical cooling rates* for the continuous-cooling decomposition of austenite; these are included in Fig. 11–4.10. The first is the cooling rate, CR_M, which just misses the "knee" of the transformation curve; *more* rapid cooling rates to the left of this curve produce *only martensite*. The second is the cooling rate, CR_P, which produces no martensite; *less* rapid cooling rates to the right of the

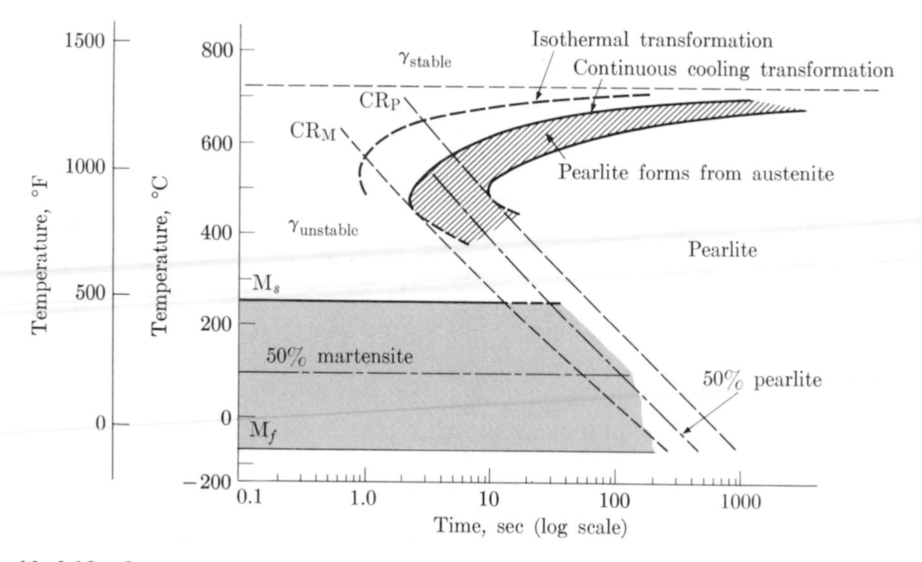

Fig. 11–4.10. Continuous-cooling transformation (eutectoid steel). Transformation temperatures and times are displaced from the isothermal-transformation curve for the same steel. (Cf. Fig. 11–4.1.) CR_M = minimum cooling rate for 100% martensite. CR_P = maximum cooling rate for 100% pearlite.

CR_P curve produce *only pearlite.* In a eutectoid steel (0.8C–99.2Fe), these two critical cooling rates are approximately 140°C/sec and 35°C/sec, respectively (~250°F/sec and ~65°F/sec), through the 750°C (1380°F) to 500°C (930°F) temperature range. Both of these rates are significantly slower in any steel which contains alloying elements because the alloy content slows down the $\gamma \rightarrow (\alpha + \bar{C})$ reaction (Fig. 11–4.9).

Example 11–4.1 Three AISI–SAE 1045 steel wires underwent the following thermal steps in the indicated sequences. Give the phases after each step and their approximate chemical analyses. (The symbol # means that the wire was held at that temperature until equilibrium was reached.)

		Time held
Wire a)	1) Heated to 820°C (1510°F):	#
	2) Quenched to 560°C (1040°F):	0
	3) Held at 560°C:	1 min
	4) Reheated to 820°C:	#
Wire b)	1) and 2) Same as (a), 1) and 2):	#
	3) Held at 560°C:	1 sec
	4) Quenched to 430°C (805°F):	0

Wire c) 1) Heated to 730°C (1345°F): #
2) Quenched to 430°C (805°F): 0
3) Quenched to 330°C (625°F): 10 sec
4) Held longer at 330°C: #

Solution and comments:

Wire a) 1) γ 0.45C–99.55Fe.
2) Same, but austenite is metastable (Fig. 11–4.6).
3) α Negligible carbon; $\bar{C}$ 6.7% carbon.
4) γ Same as (1).

Wire b) 3) γ(0.45% C) + [α(negligible carbon) + $\bar{C}$(6.7%C)].
4) Same as (3). ($\alpha + \bar{C}$) will *not* revert to austenite below the eutectoid temperature.

Wire c) 1) α(0.02% C) + γ(0.8%C). See the phase diagram.
2) Note that nothing is expected to happen to the ferrite as it cools. The austenite is of eutectoid composition; therefore, we should turn to Fig. 11–4.1 for its transformation. At zero time, γ is unstable.
3) Still, α(0.02% C) + unstable γ(0.8%C).
4) $\alpha + \bar{C}$(6.7%C). ◀

•**Example 11–4.2** By what factor was the grain size changed to decrease the start of transformation in Fig. 11–4.8 from the left curve (G.S. = #8) to the right curve (G.S. = #2)? Repeat for the boundary area.

Solution: Refer to Eq. (4–4.1). With ×100 and G.S. #8,

$$N = 2^{8-1} = 128/(0.01 \text{ in}^2);$$

With ×100 and G.S. #2,

$$N = 2^{2-1} = 2/(0.01 \text{ in}^2);$$

$$N_8/N_2 = 128/2 = 64.$$

Let δ be a representative linear dimension of a grain;

$$\delta_2/\delta_8 = \sqrt{64} = 8.$$

For the boundary area, assume either a cubelike or a spherelike grain, where

$$\text{Boundary area} = 3/\delta.$$

Therefore,

$$B_2/B_8 = (3/\delta_2)/(3/\delta_8) = \tfrac{1}{8}.$$

Comments. The surface area/volume ratio of a cube is $6a^2/a^3 = 6/a$; the same ratio of a sphere is $4\pi r^2/(4/3)\pi r^3 = 6/d$. The boundary area/volume is $3/\delta$ since each boundary is shared by two adjacent grains. Of course, the grains are neither spheres nor cubes, but something in between; thus the B/V ratio remains $3/\delta$. ◀

11-5 COMMERCIAL STEEL-TREATING PROCESSES

Since austenite can decompose in several different ways, the engineer has a choice of different microstructures. These are summarized in Table 11-5.1 and Fig. 11-5.1. The resulting properties vary significantly.

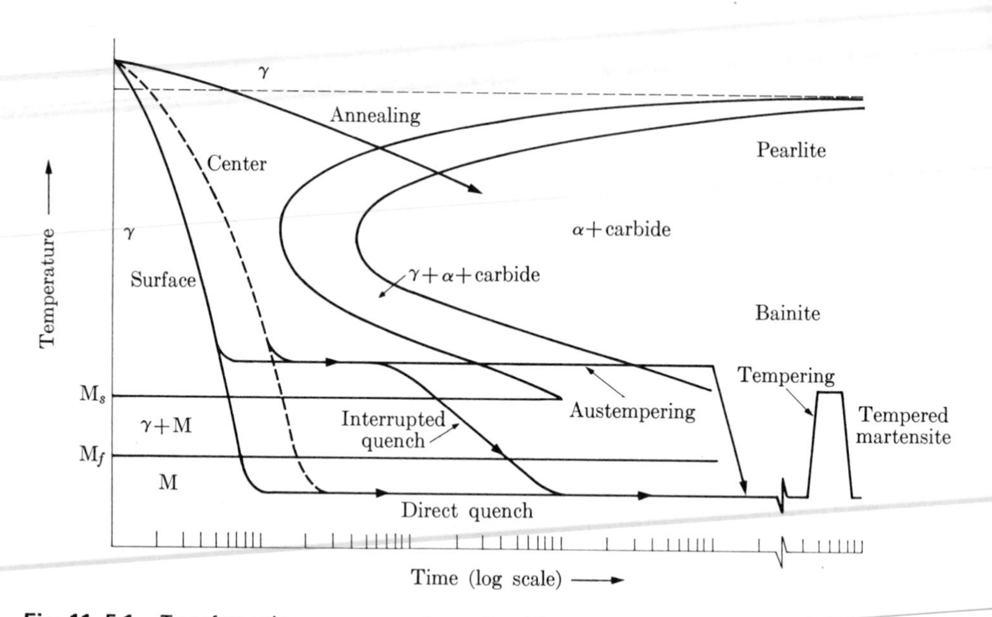

Fig. 11-5.1. Transformation processes. *Annealing:* The normal $\gamma \rightarrow \alpha$ + carbide transformation occurs. *Direct quench:* Martensite forms, first in the surface, then in the center. Severe stresses result. *Interrupted quench:* Time is available for the surface and center to transform nearly simultaneously, thus avoiding the quench cracking found in direct quenching. *Tempering:* Both the direct and the interrupted quench must be followed by a tempering process to complete the transformation. *Austempering:* Quenching avoids pearlite formation, but the $\gamma \rightarrow \alpha$ + carbide transformation may still occur above the M_s. The resulting microstructure is bainite.

Annealing. This has already been considered in Section 11-2 (also see Fig. 11-2.3(a)). The resulting properties are shown in Figs. 10-4.3 and 10-4.4 for plain-carbon steels.

Quenching. Cooling rates faster than CR_M of Fig. 11-4.10 give hard (and relatively brittle) martensite, by avoiding the transformation to $(\alpha + \overline{C})$. This is the purpose of quenching steels.

Austenite is denser than martensite (and also denser than ferrite plus carbide). This presents a problem with a direct austenite-to-martensite quench, because the slower-cooling center transforms and expands after the quickly cooled surface has

Table 11–5.1
TRANSFORMATION PROCESSES FOR STEELS*

Process	Purpose	Procedure	Phase(s)
Annealing	To soften	Slow cool from γ-stable range	α + carbide
Quenching	To harden	Quench more rapidly than CR_M	Martensite†
Interrupted quench	To harden without cracking	Quench, followed by slow cool from M_s to M_f	Martensite†
Austempering	To harden without forming brittle martensite	Quench, followed by isothermal transformation above the M_s	α + carbide
Tempering	To toughen (usually with minimum softening)	Reheating of martensite	α + carbide

* Cf. Fig. 11–5.1.
† Steels containing martensite must be toughened by the tempering process.

formed brittle martensite. Hence, cracking can occur if the steel is larger than sheet or wire dimensions, particularly if the carbon content is greater than 0.5%. This is an added reason for alloy additions to the steel. With more time available in which to form martensite (Fig. 11–4.9), an alloy steel can be cooled more slowly so that the surface and center transform more or less concurrently; thus it is possible to avoid the sharp volume change differentials and the resulting stresses that promote cracking.

Interrupted quench. By examining the isothermal transformation diagram, we can see an alternative to the direct quench just described. That is an interrupted quench (also called *martempering* or *marquenching*).

In this process, the steel is quenched rapidly past the "knee" of the transformation curve to avoid ($\alpha + \bar{C}$) formation, but the cooling is interrupted just above the M_s temperature. Cooling is then continued at a slow rate through the martensite range to ambient temperatures, so that the surface and the center of the steel may transform more or less simultaneously, thus avoiding quenching cracks. Slower cooling is possible at these lower temperatures because the ($\alpha + \bar{C}$) forms very slowly, while the martensite forms directly with the drop in temperature.

This process is more complicated from the production viewpoint because the cooling rate must be shifted from a quench to a "hold," and then to a slow cooling rate. As with the earlier direct quench, martensite is the product, and this must be tempered to secure toughness.

Tempering. Martensite has the attribute of being exceptionally hard. It is also very brittle, i.e., nonductile, when it contains carbon. This is not surprising since hardness and ductility are generally inversely related. (Compare Figs. 10–4.3 and 10–4.4, where we see that hardness and strength increase with carbon content while toughness and

Table 11–5.2
EFFECTS OF HEAT TREATMENTS (SAE 4140)

Microstructure	Tensile strength, psi (MPa)	Toughness, ft-lb (J)
Annealed, ($\alpha + \overline{C}$) Lamellar carbides	95,000 (655)	40 (55)
Spheroidite, ($\alpha + \overline{C}$) Large "spherical" carbides in ferrite matrix	70,000 (480)	80 (110)
Martensite, M	~200,000 (~1400)	<2 (<3)
Tempered martensite, ($\alpha + \overline{C}$) Dispersed carbides in ferrite matrix:		
500°C (930°F), 1 hr	185,000 (1275)	40 (55)
600°C (1110°F), 1 hr	150,000 (1035)	80 (110)

ductility decrease.) Fortunately, by tempering martensite, we can develop toughness faster than the hardness and strength decrease.*

Table 11–5.2 shows that a 4140 steel is tough in its annealed (or spheroidized) condition, but only moderately strong. Martensite, although very hard and basically strong, lacks toughness; therefore it may crack readily. However, tempered martensite can have a high strength *and* good toughness if it is heat-treated appropriately. This is achieved because the brittle martensite is replaced by a fine dispersion of rigid carbide particles within a tough ferrite matrix (Fig. 10–2.2). The carbide particles arrest dislocation movements and thereby prevent slip, and strengthen the alloy in a way similar to the precipitation hardening of Section 11–3. At the same time, the ductile ferrite can deform locally at points of stress concentration (Fig. 8–8.2) to blunt the tip of the cracks which start to form†.

* This fact was discovered by trial and error several centuries ago. Its governing principles have only become understood during the lifetime of the present-day metallurgist.

Tempering is certainly among the more important discoveries of nature that have affected the technological evolution of man. We can anneal steels so they can be machined. In turn, *these steels* can be hardened and tempered so they can machine *other steels*. A technological "bootstrap operation" such as this is not possible to the same degree of efficiency with other materials, because they lack the concurrent hardness and toughness. They either soften and become dull, or they break from brittleness.

† The metallurgist is aware that extended heating around 500°C (~900°F) will embrittle a steel because there are secondary reactions which can occur among the alloying elements. Therefore, he requires that the steel be cooled rapidly through this temperature region if tempering is performed at higher temperatures. He also knows that the toughness has a transition temperature which varies with the austenite grain size (Section 6–9).

Austempering. The final transformation process which we will consider for austenite decomposition is that of *austempering*. In this treatment, the austenite is allowed to transform *isothermally* to ferrite and carbide just above the M_s temperature. This requires a quench to avoid transformation to pearlite at higher temperatures. The advantage of austempering is that transformation occurs by a combination of shear and diffusion, to give a fine dispersion of carbides in ferrite; also, quench-cracking is avoided because the reaction (and volume change) takes place at constant temperature. In many respects the product, *bainite* (Fig. 11–5.2), is similar to tempered martensite, and the physical properties of the two microstructures are closely related.

A close examination of the isothermal-transformation process reveals that pearlite formation and bainite formation are different (Fig. 11–5.3). However, their combination gives the $\gamma \rightarrow (\alpha + \text{carbide})$ transformation curve of Fig. 11–4.1 for AISI–SAE

Fig. 11–5.2. Bainite at ×11,000. This SAE 1080 steel was austenitized, then transformed isothermally at 260°C (500°F) to form a fine dispersion of carbides in a ferrite matrix. The hardness is 57 R_C. (Courtesy General Motors and ASTM, *Electron Microstructure of Steel.*)

Fig. 11–5.3. Pearlite and bainite formation: (a) eutectoid steel, (b) 4340 steel. Pearlite forms in less time than bainite above 550°C, and bainite forms in less time than pearlite below 550°C (1000°F).

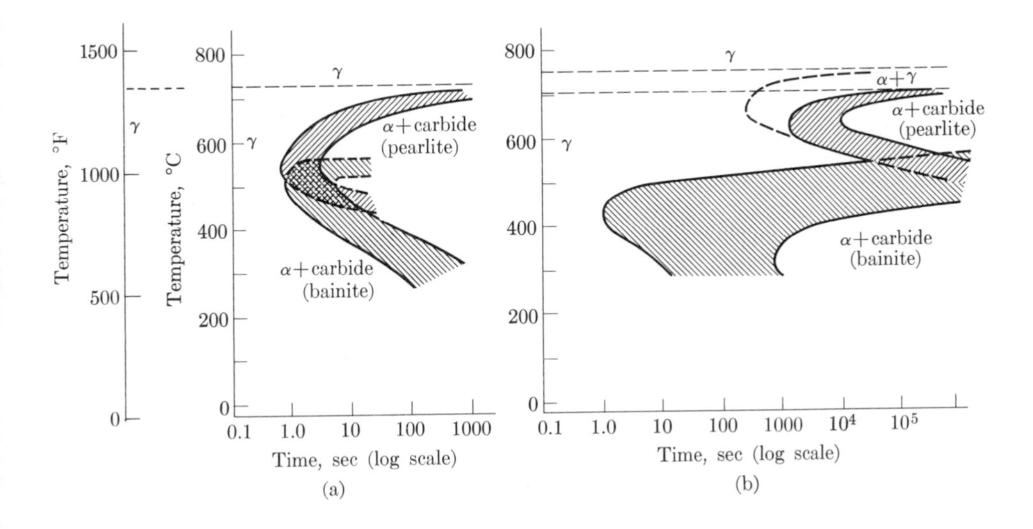

1080 steels. In some alloy steels, the two reactions do not blend together because their temperature ranges are separated quite distinctly (Fig. 11–5.3(b)). This difference allows bainite to be formed during the continuous cooling of those steels.

Example 11–5.1 a) A small piece of 1045 steel is heated to 850°C (1560°F), quenched to 650°C (1200°F), held for 5 sec, and then quenched to 20°C. What phases are present after each step? b) A small piece of 1080 steel is heated to 800°C (1470°F), quenched to 175°C (350°F), reheated to 290°C (550°F), and held 1 min. What phases are present after each step?

Solution

a) At 850°C (1560°F): γ_{stable}
After 5 sec at 650°C (1200°F): $\gamma_{\text{unstable}} + \alpha +$ carbide
After quenching to 20°C: $\text{M} + \alpha +$ carbide*

b) At 800°C (1470°F): γ_{stable}
After quenching to 175°C (350°F): $\gamma_{\text{unstable}} + \text{M}$
After 1 min at 290°C (550°F): $\gamma_{\text{unstable}} + \text{M}$*

Comment. With additional time at 290°C, both γ and M will transform to $\alpha +$ carbide. ◀

•11–6 HARDENABILITY

It is important to distinguish between *hardness* and *hardenability: Hardness* is a measure of resistance to plastic deformation. *Hardenability* is the ease with which hardness may be attained.

Figure 11–6.1 shows the maximum possible *hardnesses* for increasing amounts of carbon in steels; these maximum hardnesses are obtained only when 100% marten-

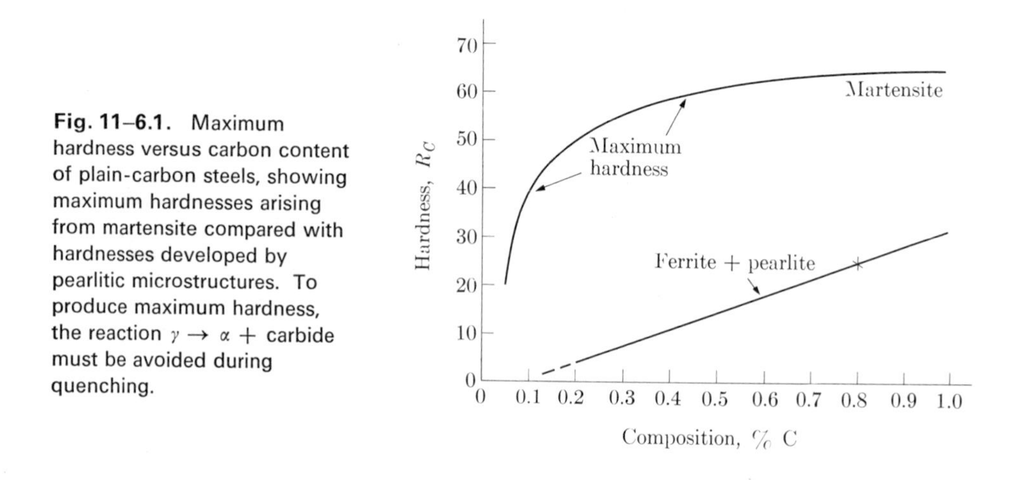

Fig. 11–6.1. Maximum hardness versus carbon content of plain-carbon steels, showing maximum hardnesses arising from martensite compared with hardnesses developed by pearlitic microstructures. To produce maximum hardness, the reaction $\gamma \rightarrow \alpha +$ carbide must be avoided during quenching.

---

* The reactions of Eq. (10–1.5) are *not* reversible below the eutectoid temperature.

site is formed. A steel that transforms rapidly from austenite to ferrite plus carbide has low *hardenability* because these high-temperature transformation products are formed at the expense of the martensite. Conversely, a steel that transforms very slowly from austenite to ferrite plus carbide has greater hardenability. Hardnesses nearer the maximum can be developed with less severe quenching in a steel of high hardenability, and greater hardnesses can be developed at the center of a piece of steel even though the cooling rate is slower there.

Hardenability curves. There is a standardized test, commonly called the *Jominy end-quench test*, to determine the hardenability of steel. In this test, a round bar of a specified size is heated to form austenite and is then end-quenched with a water stream of specified flow rate and pressure, as indicated in Fig. 11–6.2(a). Hardness

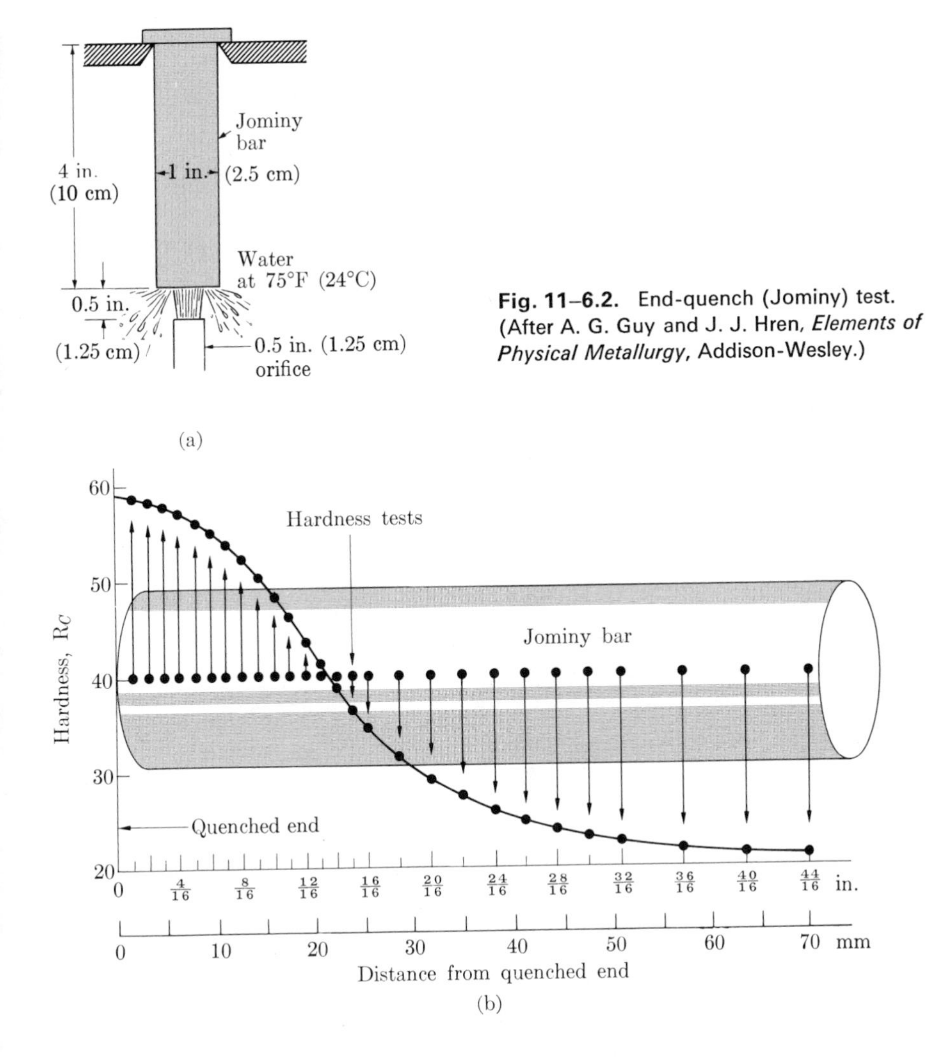

Fig. 11–6.2. End-quench (Jominy) test. (After A. G. Guy and J. J. Hren, *Elements of Physical Metallurgy*, Addison-Wesley.)

Table 11–6.1

COOLING RATES AT 700°C (1300°F) IN STEEL BARS
3-inch (76 mm) diameter

Position	Agitated water quench,		Agitated oil quench,	
	°C/sec	°F/sec	°C/sec	°F/sec
Surface	105	190	21	37
$\frac{3}{4}$ radius	30	54	16	29
Mid-radius	16	29	8	15
$\frac{1}{4}$ radius	14	25	7	12
Center	11	20	5.5	10

values along the cooling-rate gradient are determined on a Rockwell hardness tester, and a *hardenability curve* is plotted (Fig. 11–6.2(b)).

The quenched end is cooled very fast and therefore has the maximum possible hardness for the particular carbon content of the specimen. Since the steel behind the quenched surface is cooled at a slower rate, its hardness is less than the maximum possible for the particular steel being tested. The upper abscissa of Fig. 11–6.3 indicates rate of cooling as a function of distance from the quenched end. This curve holds for all *plain-carbon* or *low-alloy* steels* for which hardenability data are desired. Since the test is standardized as to temperature, specimen size, procedure, and other variables, the cooling rate at any specific location is almost completely independent of the type of steel.

Use of hardenability curves. End-quench hardenability curves are of great practical value because (1) if the cooling rate of a steel in any quench is known, the hardness may be read directly from the hardenability curve for that steel, and (2) if the hardness at any point can be measured, the cooling rate at that point may be obtained from the hardenability curve for that steel.

Figure 11–6.3 presents the end-quench hardenability curve for an AISI–SAE 1040 steel with the grain size and composition indicated.† The quenched end has nearly maximum hardness for 0.40% carbon steel because the cooling was very rapid and only martensite was formed. However, close behind the quenched end, the cooling

* The properties of steel which would alter the cooling rate are thermal conductivity, specific heat, and density. In *plain-carbon* and *low-alloy* steels, these vary so slightly that only very minor differences in cooling rates result. High-alloy, stainless-type steels would be exceptions. However, these steels are seldom quenched for hardness requirements.

† These data apply to this 1040 composition (and grain size). A slight variation is possible in the chemical specifications of any steel (e.g., in a 1040 steel, C = 0.37/0.44, Mn = 0.60/0.90, S = 0.05, P = 0.04, and Si = 0.15/0.25). As a result, two different 1040 steels may have somewhat different hardenability curves.

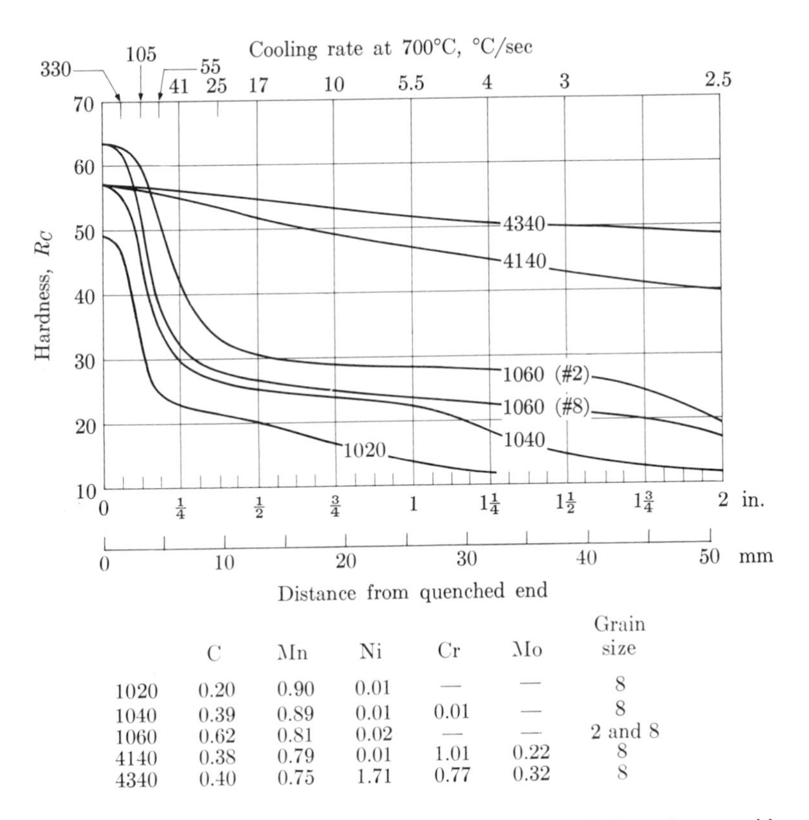

	C	Mn	Ni	Cr	Mo	Grain size
1020	0.20	0.90	0.01	—	—	8
1040	0.39	0.89	0.01	0.01	—	8
1060	0.62	0.81	0.02	—	—	2 and 8
4140	0.38	0.79	0.01	1.01	0.22	8
4340	0.40	0.75	1.71	0.77	0.32	8

Fig. 11–6.3. Hardenability curves for six steels with the indicated compositions and grain sizes. The steels were end-quenched as shown in Fig. 11–6.2(a). In commercial practice, the hardenability curve of each type of steel varies because of small variations in composition. (Adapted from U.S. Steel data.)

rate was not rapid enough to avoid some ferrite and carbide formation, and so maximum hardness was not attained at that point. (Compare the maximum hardness shown in Fig. 11–6.3 with the maximum hardness indicated for this steel in Fig. 11–6.1.)

In the laboratory, it is also possible to determine the cooling rates within bars of steel. Table 11–6.1, for example, shows the cooling rates at eutectoid temperatures for the surfaces, mid-radii, and centers of three-inch (76 mm) rounds quenched in mildly agitated water and oil. These cooling rates were determined by thermocouples embedded in the bars during the quenching operation. Similar data may be obtained for bars of other diameters. These data are summarized in Fig. 11–6.4.

By the use of the data of Fig. 11–6.4 and a hardenability curve, the *hardness traverse* which will exist in a steel after quenching may be predicted. For example, the center of the three-inch (76 mm) round quenched in oil has a cooling rate of 5.5°C per

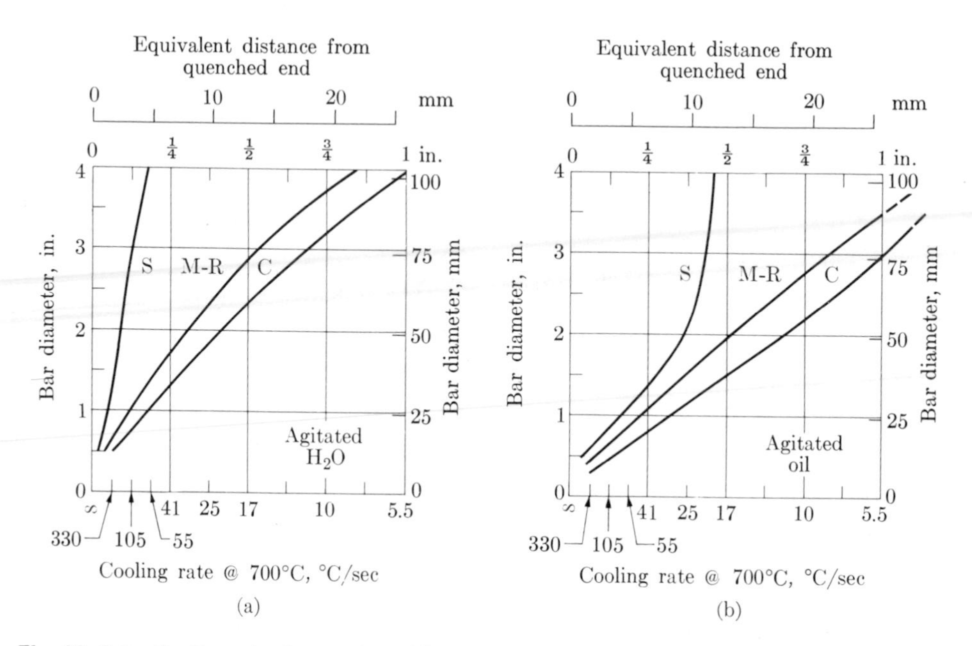

Fig. 11-6.4. Cooling rates in round steel bars quenched in (a) agitated water and (b) agitated oil. Bottom abscissa, cooling rates at 700°C; top abscissa, equivalent positions on an end-quench test bar. (C, center; M-R, mid-radius; S, surface.) The high heat of vaporization of water produces a severe quench in that quenching medium.

second (10°F/s). Since the center of this large round has the same cooling rate as a Jominy test bar at a point one inch (25 mm) from the quenched end, the hardnesses at the two positions will be the same. Thus if the bar is 1040 steel (Fig. 11-6.3), the center hardness will be 22 R_C. Figure 11-6.3 also shows that the following center hardnesses may be expected for bars of the following steels when cooled at 5.5°C per second:

AISI-SAE:	1040	4140	1020	4340	1060 (GS-8)	1060 (GS-2)
R_C:	22	47	13	52	24	28

Two determinations of quenched hardnesses are given in Examples 11-6.1 and 11-6.2 which follow shortly.

Tempered hardness. The results of Examples 11-6.1 and 11-6.2 are hardnesses of quenched steel. As indicated in Fig. 10-4.6, the hardness decreases with continued tempering because the carbide particles coalesce. The data of that figure are for a plain-carbon eutectoid steel (1080). Alloy steels temper more slowly.* Data for tempering rates are available in metallurgical books for use by the engineer who must specify heat-treating processes for specific steels.

Example 11–6.1 Sketch the *hardness traverses* for two steel rounds quenched in water; each is 1.5 inches (38 mm) in diameter, with AISI–SAE 1040 and 4140 compositions, respectively.

Solution

From Fig. 11–6.4(a)		From Fig. 11–6.3		
Position	Approx. cooling rate at 700°C (1300°F)	Cooling rate at 700°C (1300°F)	AISI–SAE 1040	AISI–SAE 4140
Surface	300°C/sec	300°C/sec	55 R_C	57 R_C
Mid-radius	55°C/sec	55°C/sec	35 R_C	55 R_C
Center	30°C/sec	30°C/sec	28 R_C	54 R_C

Comments. The hardness traverses for the two steels of Example 11–6.1 are shown in Fig. 11–6.5. Although the surface hardnesses of the two are practically identical, the difference in

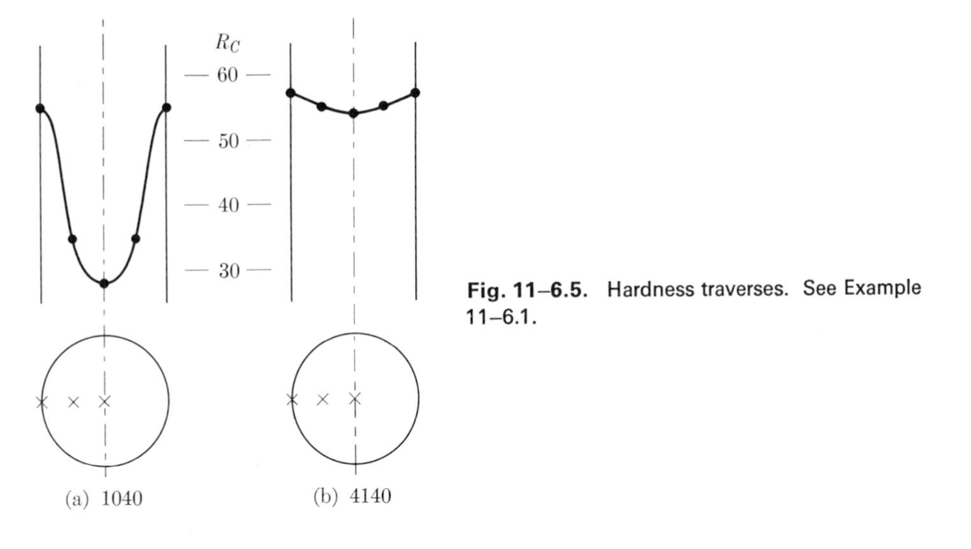

Fig. 11–6.5. Hardness traverses. See Example 11–6.1.

---

* A high-speed tool steel contains elements such as V, Cr, W, and Mo which make very stable carbides. These carbide particles coalesce much more slowly than Fe_3C does in a plain-carbon steel. Thus, tool steels can be used at higher temperatures (higher-speed operation) before overtempering, softening, and consequent destruction of the cutting edge becomes critical.

their hardenability produces a higher center hardness for the AISI–SAE 4140 steel. As indicated, this steel has a higher alloy content, which slows down the transformation of austenite to ferrite and carbide. Consequently, more martensite can form. ◀

Example 11–6.2 Figure 11–6.6 shows the points in the cross section of a V-bar of AISI–SAE 1060 steel (G.S.#2) in which the following hardness readings were obtained after oil quenching: A—40 R_C, B—36 R_C, C—33 R_C, D—32 R_C, E—31 R_C, F—63 R_C. What hardness values would be expected for an identically shaped bar of AISI–SAE 4068 steel?

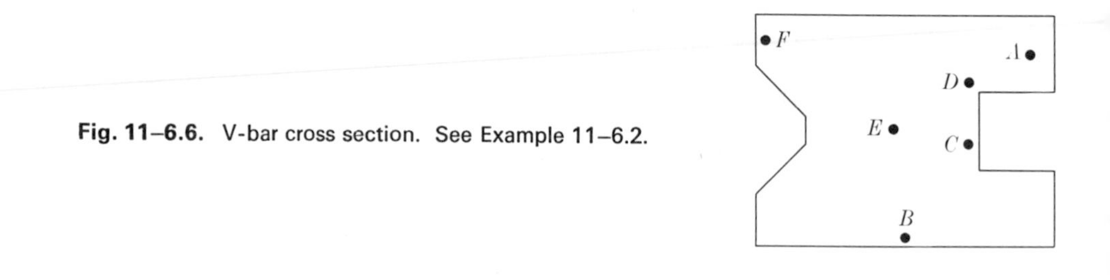

Fig. 11–6.6. V-bar cross section. See Example 11–6.2.

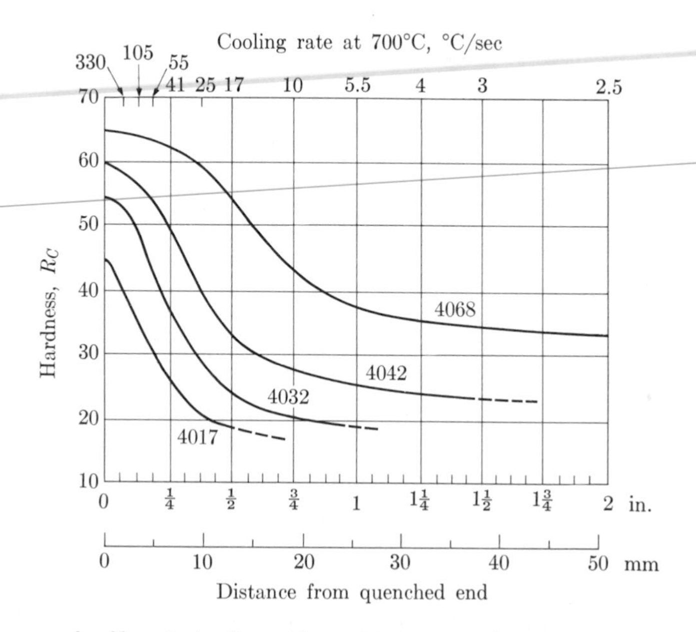

Fig. 11–6.7. Hardenability curves for 40xx steels. Except for carbon content, the composition is the same for each. Additional carbon gives harder martensite and harder $\alpha + \bar{C}$, according to Fig. 11–6.1.

Solution: For a given steel, the hardness is dependent on the cooling rate.

Point	AISI–SAE 1060 (G.S. #2) (From Fig. 11–6.3)		AISI–SAE 4068 (From Fig. 11–6.7)	
	Hardness	Approximate cooling rate	Cooling rate at 700°C	Hardness
A	40 R_C	39°C/sec	39°C/sec	62 R_C
B	36	33	33	61
C	33	25	25	59
D	32	20	20	56
E	31	17	17	54
F	63	330	330	64

REVIEW AND STUDY

To understand thermal processing, we first looked at some of the factors which govern reaction rates. Then we considered various heat treatments for annealing and homogenization, both of which bring the materials into a more stable form. Finally we looked at those microstructures which are obtained by manipulating the cooling rates and heating times to avoid the standard solid-phase reactions. It is here we are able to design high strengths and toughness into a material.

SUMMARY: REACTION RATES

Solid-phase reactions such as recrystallization can be related to temperature through the Arrhenius relationship, $\log t = C + B/T$. Even so, the progress of reactions is complicated by a slow start, followed by a more rapid change, and then a slow completion.

Solid-phase reactions which require nucleation proceed faster when supercooling is greater; however, the maximum rate (shortest time) is soon reached. At still lower temperatures, longer times are required because diffusion and compositional changes occur very slowly. This provides a typical *C*-curve for isothermal reactions, and gives a basis for studying the factors which affect reaction rates. Since many reactions start at grain boundaries, the *C*-curves are shifted to longer times in coarse-grained materials. Also, reactions are delayed when it is necessary to relocate solute atoms. For this reason, the *C*-curves for steels are shifted to longer times when alloying elements are present.

SUMMARY: EQUILIBRATION REACTIONS

One group of heat treatments is designed for stress relief, softening, and homogenization. These are outlined in Section 11–2. They capitalize on the driving force which causes a material to progress toward a lower energy state. In each case, the temperature is raised high enough so the atoms can rearrange themselves to eliminate strain energy, segregation, and/or grain boundaries. Each of these is characterized (1) by an extended heating time, or (2) by a slow cooling rate, so that atoms can move into the most stable positions.

SUMMARY: MICROSTRUCTURAL CONTROL

By fast cooling, we can modify reactions, and in some cases even avoid the normal reaction. This opens the door to new microstructures. The most favored microstructure for high strength and toughness is a fine dispersion of rigid particles within a tough ductile matrix.

The precipitation hardening (age-hardening) process achieves a fine dispersion by first forming a solid solution, followed by rapid cooling; then the supersaturated solution is heated at an intermediate temperature to initiate an extremely fine dispersion of many precipitated particles. These are highly effective in blocking slip, thereby strengthening the matrix material. Extended heating reduces the hardness level because as the precipitated particles grow, there are fewer of them, and they are separated by greater quantities of soft matrix.

A widely used process to heat-treat steels is to first form austenite, then to quench it to avoid the normal ($\alpha + \bar{C}$) reaction product. Martensite forms as a transition phase by a shear reaction. In turn, the martensite is tempered to produce an extremely fine dispersion of carbides. These are highly effective in blocking slip in the matrix of ferrite. The latter provides toughness. In practice the tempering process is stopped before the carbide particles grow significantly, so we have the favorable situation of high hardness and toughness simultaneously.

While hardness is a measure of resistance to plastic deformation, hardenability is an indication of the ease with which maximum hardness may be attained. The hardenability curve lets us predict the quenched hardness which may be obtained for a steel because it plots hardness versus cooling rate. A known cooling rate lets us predict the hardness.

TERMS FOR REVIEW

Alloy retardation
Annealing
Austenite grain size
Austenization
Austempering
Bainite ($\alpha + \bar{C}$)
C-curve
Continuous-cooling transformation
Critical cooling rates, CR_M and CR_P
• End-quench test (Jominy-bar)

TERMS FOR REVIEW—CONTINUED

- •Hardenability
- •Hardness traverse
- Homogenization
- Interrupted quench
- Isothermal precipitation
- Isothermal recrystallization
- Isothermal transformation
- Normalize
- Nucleation
- Overaging
- Precipitation-hardening (age-hardening)
- Process anneal
- Quenching
- Rate, $R = t^{-1}$
- Solution treatment
- Spheroidization
- Stress relief
- Tempering

FOR DISCUSSION

A_{11} Why do diffusion, recrystallization, stress relaxation (Section 7–5), and viscous flow (Section 8–8) have similar time–temperature relationships?

•B_{11} Explain the basis of cloud-seeding.

C_{11} Give two reasons why superheating occurs less readily in a solution reaction than supercooling occurs in a precipitation reaction.

•D_{11} How can a nucleus form if it "must be of critical size before the growth starts"?

•E_{11} Plaster of paris plus water forms gypsum when it "sets." The setting time of plaster can be shortened if some gypsum powder is mixed in with the plaster powder and water. Why?

F_{11} Point out the differences between normalizing steel and giving it a full anneal. What are the technical bases for these differences?

G_{11} On the basis of Section 11–4, explain why normalizing is commonly required for low-alloy steels.

H_{11} Explain the data variations in Table 11–3.1.

I_{11} As a practical production matter, why is it simpler to quench a 96Al–4Cu alloy to room temperature, then reheat to 100°C for aging, than to quench directly to 100°C?

J_{11} a) Refer to the phase diagrams of Chapter 9. What alloys are candidates for precipitation-hardening?
b) Cu–Zn brass is never age-hardened. Why?

K_{11} Steel used for food cans has less than 0.1% carbon, so that it is deformable. Therefore, it is not amenable to quenching; however, it hardens slightly during the food-packing process. Suggest an explanation.

L_{11} The lever rule is not applicable to an isothermal transformation diagram. Why?

M_{11} Why does the "start" curve in Fig. 11–4.6 have two branches on its upper arm?

N_{11} The start (or finish) of any reaction is generally very difficult to determine (compared to the midpoint). Give two reasons. [*Note:* Because of this, M_s is commonly considered to be ~1% reaction; and M_f is commonly considered to be 99% completion.]

O_{11} Why does the austenite in a fine-grained steel transform to (α + $\bar{C}$) faster than in a coarse-grained steel?

P_{11} Why does the austenite of a 2% Cr steel transform to (α + $\bar{C}$) less rapidly than the austenite in an 0.7% Cr steel?

Q_{11} Why did we *not* use Fig. 11–4.6 for wire (c), step 3, of Example 11–4.1?

R_{11} Assume CR_M and CR_P for a steel are 75°C/sec and 15°C/sec, respectively. What phases will be present if the steel is cooled at 10°C/sec? At 20°C/sec? At 50°C/sec? At 100°C/sec?

S_{11} Assume the same CR_M and CR_P as in R_{11}. If this steel is cooled at 10°C/sec, it is harder than if it is cooled at 5°C/sec. Why? (They both contain only α and $\bar{C}$.)

T_{11} Explain why low-alloy steels are commonly specified for steels which are to be heat-treated by quenching and tempering.

•U_{11} Why do the upper arms of the two curves in Fig. 11–4.9 cross? (Check the eutectoid temperature and composition.)

V_{11} Compare and contrast: tempered martensite, pearlite, bainite, martensite, spheroidite, ferrite, carbide.

W_{11} Bainite cannot be obtained by continuously cooling a 1080 steel. Why?

•X_{11} Explain the difference between hardness and hardenability.

•Y_{11} Why should a steel (or the water) be agitated when a gear is water-quenched?

•Z_{11} Why does the center of a 1-inch (2.5-cm) round steel bar quench to a greater hardness than the steel which is 1 inch behind the end of a Jominy end-quench bar? Which will quench faster?

STUDY PROBLEMS

11–1.1 The following data apply to the average times and temperatures for a certain solid-state reaction ($\alpha \rightarrow \beta$) which is diffusion-controlled. What temperature would accomplish the same reaction in 1 sec?

Transformation temperature	Time	Rate
380°C	10 sec	0.1 sec^{-1}
315°C	100 sec	0.01 sec^{-1}

Answer: 463°C

•**11–1.2** (Use the data of Example 11–1.2 and assume a linear relationship between ΔF_V and temperature.) Plot the values of r_c versus T and ΔF_c versus T at a) 1025°C, b) 1000°C, c) 900°C, and d) 850°C.

Answer: a) $r_c = \infty$, $\Delta F_r = \infty$ b) $r_c = 25$ Å, $\Delta F_r = 13 \times 10^{-18}$ J c) 5 Å, 0.5×10^{-18} J d) 35 Å, 0.25×10^{-18} J

11–2.1 In commercial practice, an annealed steel is heated an hour at 25–30°C above the highest temperature of α-ferrite. Indicate the annealing temperature for a 1040, a 1080, and a 1% C steel.

Answer: 835°C, 755°C, 755°C (1535°F, 1390°F, 1390°F)

11–2.2 In commercial practice, a normalized steel is heated an hour at 50–60°C above the lowest temperature of single-phase austenite. Indicate the normalizing temperature for a 1040, a 1080, and a 1% C steel.

11–3.1 Explain why a 92% copper, 8% nickel alloy can (or cannot) be age-hardened.

Answer: No supersaturation possible

11–3.2 Explain why the following alloys can (or cannot) be considered for age-hardening. a) 97% aluminum, 3% copper. b) 97% copper, 3% zinc. c) 97% nickel, 3% copper. d) 97% copper, 3% nickel. e) 97% aluminum, 3% magnesium. f) 97% magnesium, 3% aluminum.

11–3.3 A slight amount of age-hardening is realized when a steel (99.7 w/o Fe, 0.3 w/o C) is quenched from 700°C (1300°F) and reheated for 3 hr at 100°C. Account for the hardening.

Answer: Carbon solubility in ferrite decreases.

11–3.4 An aircraft manufacturer receives a shipment of aluminum alloy rivets that have already age-hardened. Can they be salvaged? Explain.

11–4.1 A small piece of 1080 steel is heated to 800°C, quenched to −60°C, reheated to 300°C, and held 10 sec. What phases are present at the end of this time?

Answer: Martensite (with some possible change to α + carbide).

11–4.2 Sketch an isothermal-transformation diagram for a steel of 1.0 w/o C and 99.0 w/o Fe.

11–4.3 Sketch an isothermal-transformation diagram for a 1020 steel.

Answer: M_s = 450°C, M_f = 290°C; α and γ are stable between 727°C and 855°C after long times; at 550°C the curve is farther to the left than in 1045 steel.

11–4.4 Different wire samples of an AISI–SAE 1045 steel received one of the following six heat-treating sequences. Indicate the phases which exist *immediately after* the completion of each sequence.

a) Heated to 825°C,# quenched to 550°C, held 10 sec;
b) Heated to 900°C,# quenched to 550°C, held 10 sec, quenched to 250°C;
c) Heated to 925°C,# quenched to 300°C, held;# *(Con't.)*

This symbol indicates that equilibrium was attained at this step of the heat treatment before proceeding to the other steps in the sequence.

d) Heated to 700°C,# quenched to 250°C;
e) Heated to 250°C,# heated to 425°C, held 1 sec;
f) Heated to 750°C,# quenched to 550°C.

11–4.5 One hundred grams of an AISI–SAE 1045 steel are heated and quenched in the ways indicated below. Indicate the phase(s) and grams of each phase at the *end* of each sequence.

a) Heated to 825°C,# quenched to 120°C;
b) Heated to 750°C,# quenched to 20°C;
c) Heated to 825°C;# quenched to 550°C, held 7 sec, quenched to 250°C;
d) Heated to 825°C,# quenched to 550°C, held 10 sec, heated to 750°C.#

Answer: a) 100 g M b) 70 g M, 30 g α c) 93 g α, 7 g carbide d) 70 g γ, 30 g α

11–4.6 A small wire of AISI–SAE 1045 steel is subjected to the following treatments as *successive* steps:
1) heated to 875°C, held there for 1 hr;
2) quenched to 250°C, held there 2 sec;
3) quenched to 20°C, held there 100 sec;
4) reheated to 550°C, held there 1 hr;
5) quenched to 20°C and held.
Describe the phases or structures present *after each step* of this heat-treatment sequence.

11–4.7 a) Repeat Problem 11–4.6 with steps (1), (2), (5).
b) Repeat Problem 11–4.6 with steps (1), (3), (4), (5).
c) Repeat Problem 11–4.6 with steps (1), (2), (4), (5).

11–5.1 Draw temperature (ordinate) and time (abscissa) plots for the following heat treatments. Indicate the important temperatures, relative times, and reasons for drawing the curves as you do. a) Normalizing a 1095 steel, contrasted with annealing the same steel. b) Solution-treating an aluminum 5% copper alloy, contrasted with aging the same alloy. c) Austempering a 1045 steel, contrasted with martempering (interrupted quench) the same steel. d) Spheroidizing 1080 steel, contrasted with spheroidizing 10105 steel.

•**11–6.1** How hard will the quenched end of an AISI–SAE 4620 steel test bar be?

Answer: Approximately 50 R_c

•**11–6.2** The quenched end of a Jominy bar has a hardness of 44 R_c. What is the carbon content of the steel? Explain.

•**11–6.3** What hardness would you expect the center of a 2-in. (5-cm) round bar of 1040 steel to have if it were quenched in a) agitated oil? b) Agitated water?

Answer: a) 24 R_c b) 26 R_c

•**11–6.4** A round bar of 1040 steel has a surface hardness of 41 R_c and a center hardness of 28 R_c. How fast were the surface and center cooled through 700°C?

•**11–6.5** A 2.5-in. (6.3-cm) diameter round of 1040 steel is quenched in agitated oil. Estimate the hardness 1 in. (2.5 cm) below the surface of the round bar. (Show reasoning.)

Answer: 24 R_c

•**11–6.6** How would the hardness traverse of the 1040 steel shown in Fig. 11–6.5 vary if a) it were quenched in still oil? b) it were quenched in still water? c) it had a coarser austenite grain size?

•**11–6.7** The center hardness of 6 bars of the same steel are indicated below. From these data, plot the hardenability curve for the steel. (*Hint:* There will be only *one* curve.)

Dia.	Quench	Hardness	Dia.	Quench	Hardness
1 inch	WQ	58 R_c	2 inch	OQ	47 R_c
1 inch	OQ	57 R_c	4 inch	WQ	34 R_c
2 inch	WQ	54 R_c	4 inch	OQ	30 R_c

Answer: EQ position $\frac{1}{4}$ in. (6.3 mm), 57 R_c; $\frac{1}{2}$ in. (12.7 mm), 53 R_c; $\frac{3}{4}$ in. (19 mm), 45 R_c; 1 in. (25.4 mm), 34 R_c; $1\frac{1}{4}$ in. (31.8 mm), 30 R_c.

•**11–6.8** Two 3-in. (76-mm) rounds were quenched, one in agitated water, the other in agitated oil. The following hardness traverses were made:

Distance below surface		Water, R_C	Oil, R_C
inches	cm		
0	0	57	39
$\frac{3}{8}$	0.95	46	35
$\frac{3}{4}$	1.90	36	32
$1\frac{1}{8}$	2.86	34	31
$1\frac{1}{2}$	3.81	33	30

Calculate and sketch the hardenability curve which would be obtained from an end-quenched test of the same steel. (*Hint:* There will be only *one* curve.)

•**11–6.9** A spline gear had a hardness of 45 R_c at its center when it was made of the 4068 steel shown in Fig. 11–6.7. What hardness would you expect the same gear to have if it were made of 1040 steel?

Answer: 24 R_c

•**11–6.10** The hardness of the surface of a round bar of 1040 steel quenched in agitated oil is 40 R_c. Determine the hardness of the center of a round bar of 4068 steel quenched in water if this bar is twice the diameter of the 1040 bar. (Indicate all steps in your solution.)

CHAPTER TWELVE

CORROSION OF METALS

PREVIEW

The reader is aware that corrosion degrades metals. The importance of corrosion is accentuated by the estimate that nearly 5% of every modern country's annual income goes directly or indirectly to maintain, repair, or replace deteriorated products. Most obvious to you, the reader, is the effect of corrosion on your automobile, particularly if you live in the northern latitudes where road salt is used. In addition, there are many other serious corrosion problems ranging from bridge maintenance to household plumbing, and from prosthetic implants (Fig. 12–1.1) to transatlantic cables.

If we own cars, or similar products, and do not simply store them in garages, we must expect to encounter corrosion. Corrosion cannot be eliminated by specifying gold-quality metals, because automobile design requires materials that are cheaper, stronger, and more available. However, corrosion can be minimized if technical designers become familiar with the causes of corrosion and how corrosion can be minimized.

We will observe that corrosion arises from an electrochemical reaction which is the reverse of electroplating. An electrolyte is required, and the rates can be modified by any factor that interferes with electrical conductivity.

When we examine means of reducing corrosion, we have the options (1) of avoiding galvanic cells, (2) of isolating the surface which is subject to corrosion, or (3) of reversing the electrochemical reaction by applying external potentials. We will examine each method in this chapter.

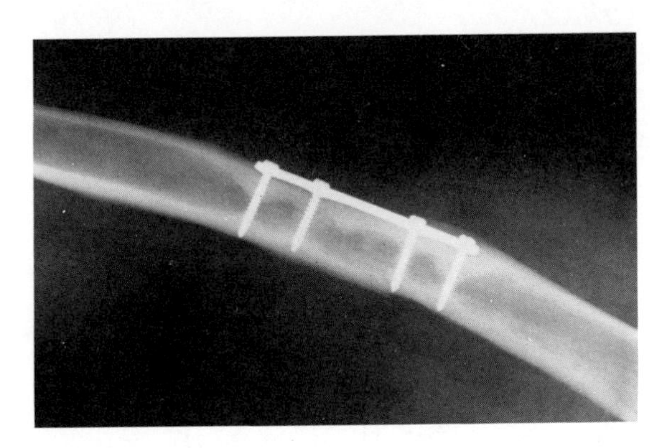

Fig. 12–1.1. Metallic prosthesis. This metal plate and screws, which are used for an internal fixation of a bone, must resist corrosion in the complex tissue environment: (1) without loss of strength, and (2) without introducing undesirable corrosion products into the body. (Courtesy S. F. Hulbert, Tulane University.)

12–1 ELECTROPLATING: CORROSION IN REVERSE

This common process is the reverse of corrosion. Metal ions are removed from an *electrolyte* by providing each with an electron to produce metallic atoms, M:

$$M^{n+} + ne^- \rightarrow M^0 \tag{12–1.1}$$

Since each electron carries a charge of 0.16×10^{-18} amp·sec, or coulombs, we can easily calculate the current required to plate metal at any given rate. One mole (0.6022×10^{24}) of monovalent ions requires (0.6022×10^{24} electrons) (0.1602×10^{-18} coul/electron), or 96,500 coul. This value is called a *faraday*, $\mathscr{F}$. Thus, to plate 107.87 g of silver from an Ag^+ solution, it would be necessary to use 1 amp for 96,500 sec (or some other combination of current and time to give 96,500 amp·sec). We may formalize the above calculation as Faraday's law;

$$w = ItM/n\mathscr{F}, \tag{12–1.2}$$

where w is the grams plated during time t with a current I for a metal of atomic weight M with a valence n.

The requirements for electroplating include two electrodes, an electrolyte, and a source of electrons. In commercial practice, the electrons are supplied by a d.c. current; however, in reality they come from a second electrode (Fig. 12–1.2). In this sketch the righthand electrode, or *cathode*, *receives electrons from the external circuit*.* Reaction (12–1.1) occurs, and plating proceeds. As this reaction continues, the left-hand electrode, or *anode*, *supplies the electrons to the external circuit** via the battery "pump." The electrons come from metal atoms of the left electrode, which are being

* These italicized definitions apply to anodes and cathodes in all electrical circuits, even in TV tubes which have chemically inert electrodes. In electrochemical reactions, it is always the *anode which undergoes corrosion.*

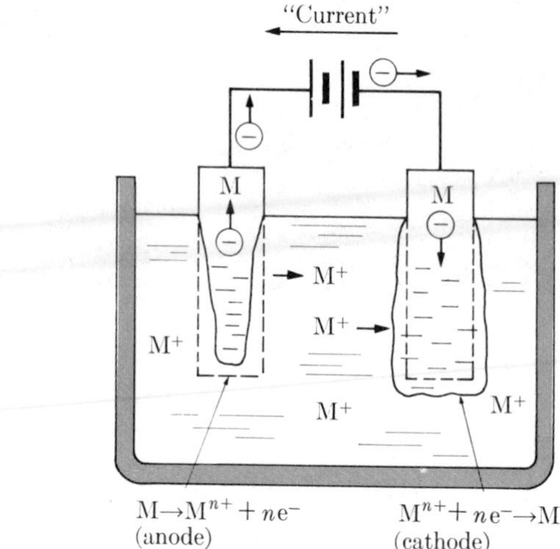

Fig. 12–1.2. Electroplating. The battery provides electrons to the cathode where the plating occurs (Eq. (12–1.1)). However, the ultimate source of the electrons is the metal in the anode which undergoes corrosion by Eq. (12–1.3).

oxidized to a higher valence level,

$$M^0 \rightarrow M^{n+} + ne^-. \tag{12–1.3}$$

The *anode undergoes corrosion.*

Example 12–1.1 An electroplating process (as in Fig. 12–1.2) must plate 100 g/hr of copper on a surface containing 25 cm². a) What current density, i, is required if the process is fully efficient? b) How much copper will be corroded from the anode?

Solution

a)

$$\text{Moles/cm}^2 = 100\text{ g}/[(63.54\text{ g/mole})(25\text{ cm}^2)] = 0.063;$$

$$\text{Electrons required} = (0.063\text{ moles/cm}^2)(0.6 \times 10^{24}\text{ Cu}^{++}\text{/mole})(2\text{ electrons/Cu}^{2+})$$
$$= 7.55 \times 10^{22}\text{ electrons/cm}^2;$$

$$\text{Coulombs required} = (7.55 \times 10^{22}\text{ electrons/cm}^2)(0.16 \times 10^{-18}\text{ coul/electron})$$
$$= 12{,}100\text{ coul/cm}^2;$$

$$\text{Current density} = (12{,}100\text{ amp}\cdot\text{sec/cm}^2)/(3600\text{ sec})$$
$$= 3.4\text{ amp/cm}^2;$$

Alternatively, rearranging Eq. (12–1.2),

$$I = n\mathscr{F}w/tM$$
$$= \frac{(2\text{ equiv./mole})(96{,}500\text{ coul/equiv.})(100\text{ g})}{(3600\text{ sec})(63.54\text{ g/mole})}$$
$$= 84.4\text{ amp};$$
$$i = 84.4\text{ amp}/(25\text{ cm}^2) = 3.4\text{ amp/cm}^2.$$

b) The same amount, 100 g of copper, will be corroded from the anode, and will appear in the electrolyte as Cu^{2+} ions.

Comments. Electroplating commonly has an inefficiency of a few percent because of side reactions. For example, some electrons may combine with H^+ ions at the anode to form H_2 gas. ◀

12–2 GALVANIC COUPLES (CELLS)

Although the anode (left electrode) of Fig. 12–1.2 exemplifies corrosion, it is atypical because it is assisted by a battery, and the two electrodes are identical. Corrosion, as we normally encounter it, involves dissimilar electrode conditions. We encounter a *galvanic couple*. To explain this, consider Fig. 12–2.1, where we have one electrode of zinc and the other of copper. Each metal is subject to oxidation:

$$Zn^0 \longrightarrow Zn^{2+} + 2e^-, \tag{12–2.1}$$

and

$$Cu^0 \rightarrow Cu^{2+} + 2e^-. \tag{12–2.2}$$

As indicated by the arrow lengths, we can envision that the zinc reaction produces a greater push to the right than does the copper reaction. Thus, if we connect the two metals through a voltmeter only, a voltage difference is registered. This difference is -1.1 volts when we use a standardized electrolyte at 25°C. This means that if we shunt the voltmeter, electrons will flow from the zinc electrode into the external circuit, and the copper will receive electrons from the external circuit. With electrons removed from the anode, Eq. (12–2.1) will progress to the right, and more zinc will be oxidized. The zinc is corroded. Meanwhile, electrons are supplied to the cathode in excess of the number released by Eq. (12–2.2). They will react with the copper ions which are available in the electrolyte, and Eq. (12–2.2) is *reversed* to give Eq. (12–1.1). Corrosion occurs at only one electrode of the galvanic couple.

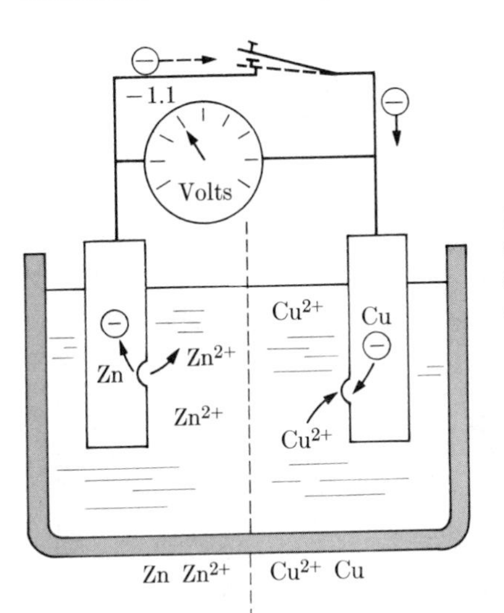

Fig. 12–2.1. Galvanic cell (Zn–Cu). With the switch closed, zinc provides electrons through the external circuit to copper. A 1.1-volt potential difference develops when the circuit is opened (Table 12–2.1).

Electrochemical potentials. The production of ions and electrons in reactions such as Eq. (12–2.1) builds up a potential called an *electrochemical potential*, which depends on (1) the nature of the metal and (2) the nature of the solution. Not all metals oxidize to ions and electrons with equal facility; this inequality is indicated by the first two equations of this section. In addition, atoms along a grain boundary are less stably located than those in the crystal lattice (Section 4–4), so they ionize more readily. Furthermore, the reaction in Eq. (12–2.1) will produce equilibrium with a greater electrode potential if the metal ions enter a solution in which they are relatively stable (e.g., the positive zinc ions are more stable in a uniformly Cl^- concentrated solution than in a uniformly dilute Cl^- solution).

To measure the electrode potential of any material (and therefore its corroding tendencies), we must first determine the voltage difference between the metal and a standard *hydrogen electrode*. According to Eqs. (12–1.1) and (12–1.3), iron has the

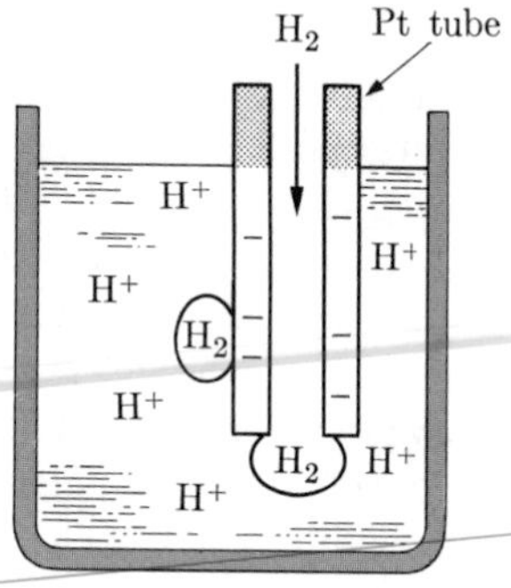

Fig. 12–2.2. Hydrogen electrode. Hydrogen establishes an equilibrium with its ions and electrons (Eq. (12–2.4)). The potential which is developed is the reference for electrochemical measurements (Table 12–2.1).

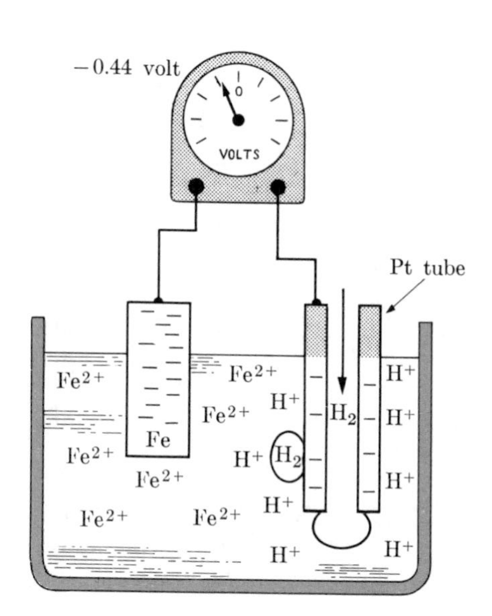

Fig. 12–2.3. Potential difference, Fe versus H_2. Iron produces a lower electron potential than does H_2 (see Table 12–2.1, center column (electrochemical notation)). Therefore iron is the anode and hydrogen the cathode.

Table 12–2.1
ELECTRODE POTENTIALS (25°C; MOLAR SOLUTIONS)

Anode half-cell reaction (the arrows are reversed for the cathode half-cell reaction)	Electrode potential used by electrochemists and corrosion engineers,* volts		Electrode potential used by physical chemists and thermodynamists,* volts
$Au \rightarrow Au^{3+} + 3e^-$	+1.50	↑ Cathodic (noble)	−1.50
$2H_2O \rightarrow O_2 + 4H^+ + 4e^-$	+1.23		−1.23
$Pt \rightarrow Pt^{4+} + 4e^-$	+1.20		−1.20
$Ag \rightarrow Ag^+ + e^-$	+0.80		−0.80
$Fe^{2+} \rightarrow Fe^{3+} + e^-$	+0.77		−0.77
$4(OH)^- \rightarrow O_2 + 2H_2O + 4e^-$	+0.40		−0.40
$Cu \rightarrow Cu^{2+} + 2e^-$	+0.34		−0.34
$H_2 \rightarrow 2H^+ + 2e^-$	0.000	Reference	0.000
$Pb \rightarrow Pb^{2+} + 2e^-$	−0.13	Anodic (active) ↓	+0.13
$Sn \rightarrow Sn^{2+} + 2e^-$	−0.14		+0.14
$Ni \rightarrow Ni^{2+} + 2e^-$	−0.25		+0.25
$Fe \rightarrow Fe^{2+} + 2e^-$	−0.44		+0.44
$Cr \rightarrow Cr^{2+} + 2e^-$	−0.74		+0.74
$Zn \rightarrow Zn^{2+} + 2e^-$	−0.76		+0.76
$Al \rightarrow Al^{3+} + 3e^-$	−1.66		+1.66
$Mg \rightarrow Mg^{2+} + 2e^-$	−2.36		+2.36
$Na \rightarrow Na^+ + e^-$	−2.71		+2.71
$K \rightarrow K^+ + e^-$	−2.92		+2.92
$Li \rightarrow Li^+ + e^-$	−2.96		+2.96

* The choice of signs is arbitrary. Since we are concerned with corrosion, we will use the middle column.

following electrochemical reaction:

$$Fe \rightleftharpoons Fe^{2+} + 2e^-. \tag{12–2.3}$$

With hydrogen (Fig. 12–2.2), equilibrium occurs by this reaction:

$$H_2 \rightleftharpoons 2H^+ + 2e^-. \tag{12–2.4}$$

The potential difference* (measured with a potentiometer) between the iron and hydrogen electrodes is 0.44 volt (Fig. 12–2.3).

Similar measurements for other metals yield the voltage comparisons listed in Table 12–2.1. The alkali and alkaline earth metals, which hold their outer-shell electrons rather loosely, show a greater potential difference with respect to hydrogen than does iron. Conversely, the noble metals, such as silver, platinum, and gold,

* Since the potential *difference* is the only voltage that can actually be measured, all comparisons are made with a standardized hydrogen electrode.

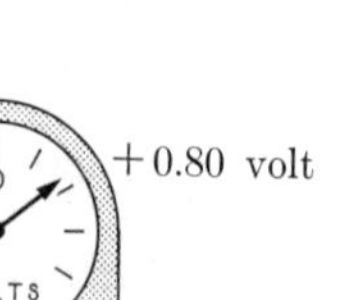

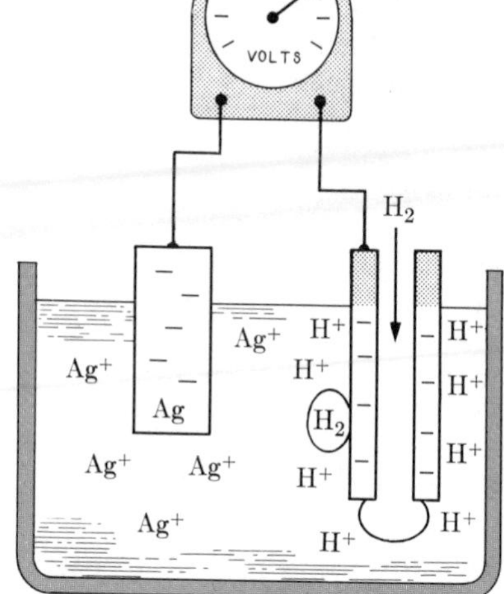

Fig. 12–2.4. Potential difference, H_2 versus Ag. H_2 produces a lower electron potential than does silver (see Table 12–2.1, center column (electrochemical notation)), and is therefore the anode. Silver is the cathode.

produce fewer electrons than hydrogen and therefore are at the other end of the potential scale (Fig. 12–2.4 and Table 12–2.1).

Cathode reactions. In our consideration of corrosion, Eq. (12–1.3) is the prime anode reaction. There are, however, a variety of cathode reactions of importance. We must pay attention to these, since an electron-consuming (cathodic) reaction must accompany each anodic (electron-producing) reaction of corrosion. The major cathodic reactions are:

Electroplating: $$M^{n+} + ne^- \rightarrow M \qquad (12\text{–}1.1)$$

Hydrogen generation: $$2H^+ + 2e^- \rightarrow H_2\uparrow \qquad (12\text{–}2.4)$$

Water decomposition: $$2H_2O + 2e^- \rightarrow H_2\uparrow + 2(OH)^- \qquad (12\text{–}2.5)$$

Hydroxyl formation: $$O_2 + 2H_2O + 4e^- \rightarrow 4(OH)^- \qquad (12\text{–}2.6)$$

Water formation: $$O_2 + 4H^+ + 4e^- \rightarrow 2(H_2O) \qquad (12\text{–}2.7)$$

The reaction which predominates depends, of course, upon the electrolytic environment, such as temperature and concentration. Obviously, metal atoms must be present for the first of these five reactions to occur. Furthermore, as the metal-ion concentration increases, its predominance increases. This will be important to us when we consider concentration cells shortly. Reaction (12–2.7) requires the presence of oxygen and a low pH (i.e., acidic solution). Equation (12–2.6) becomes predominant in alkaline or neutral environments if oxygen is present (Fig. 12–2.5). This reaction will be important to us when we consider oxidation cells. Reaction (12–2.5) is en-

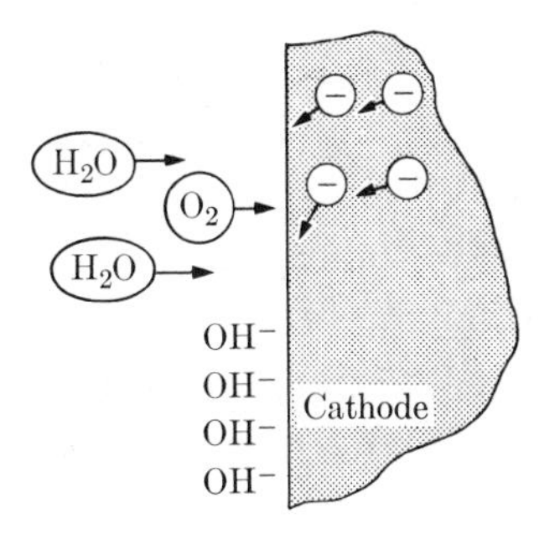

Fig. 12–2.5. Hydroxyl formation at the cathode. The speed of reaction (12–2.6) increases with increased oxygen content. It occurs at the cathode, where the electrons are consumed. If the electrons are depleted by the positive side of a d.c. current, Eq. (12–2.6) is reversed and O_2 is released (Study Problem 12–2.2).

countered in anaerobic conditions, particularly if sulfur or other hydrogen-consuming materials are present.

Rust. Figure 12–2.6 presents schematically the mechanism of iron rusting. Iron has an electrode potential ϕ of -0.44 volts in Table 12–2.1 when it forms Fe^{2+} (and in fact will be oxidized further to Fe^{3+} if the electrons can be consumed). Equation (12–2.6) is cathodic, thus consuming electrons; therefore, the following reactions occur in oxygen-enriched water.

Anode: $$Fe \rightarrow Fe^{3+} + 3e^- \tag{12–2.8}$$

Cathode: $$3e^- + \tfrac{3}{2}H_2O + \tfrac{3}{4}O_2 \rightarrow 3(OH)^- \tag{12–2.6′}$$

Ppt: $$Fe^{3+} + 3(OH)^- \rightarrow Fe(OH)_3\downarrow \tag{12–2.9}$$

The final product, $Fe(OH)_3$, is insoluble in water and therefore precipitates. It is *rust*; it will form on iron in air even if no more than an adsorbed layer of moisture is present as an electrolyte.

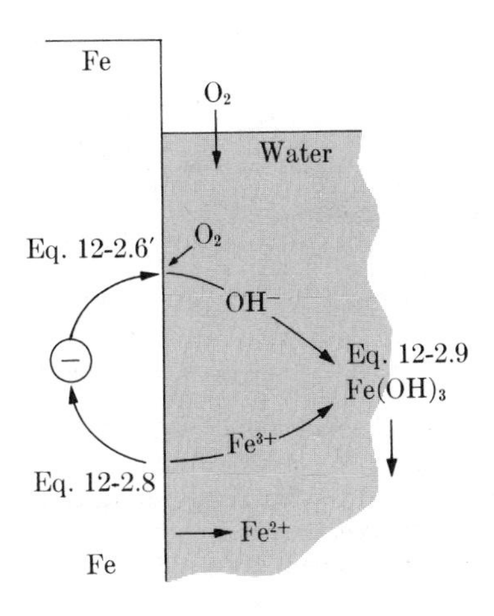

Fig. 12–2.6. Rust formation. The oxidation of iron produces iron ions and electrons. These are combined with oxygen and water at the cathode (Eq. (12–2.6)) to form $(OH)^-$ ions. Rust is the combination of Fe^{3+} and $(OH)^-$ ions.

Example 12–2.1 Describe how hydrogen and oxygen may be used to operate a fuel cell.

Explanation: Electrons must be stripped from the hydrogen and returned from the external circuit to combine with oxygen. The two half-cell reactions are:

Anode:	$2H_2 \rightarrow 4H^+ + 4e^-$	(0.000 volts) Reference
Cathode:	$O^2 + 4H^+ + 4e^- \rightarrow 2H_2O$	(1.23 volts)

A cell with 1.23 volts would be expected theoretically, with standard 1-molàr electrolytes. The electrode materials must be a conductive, chemically inert, porous material through which the gas can be fed. ◀

12–3 TYPES OF GALVANIC CELLS

On the basis of the corrosion principles stated above, some general conclusions can be reached. First we shall discuss the various types of cells that produce corrosion, and then the means of protection against such corrosion.

Galvanic cells may be categorized in three different groups: (1) *composition* cells, (2) *stress* cells, and (3) *concentration* cells. Each produces corrosion because one-half of the couple acts as the anode, and the other half serves as the cathode. Only the anode is corroded, and then only when it is in electrical contact with a cathode. If the anode were present alone, it would quickly come to equilibrium with its environment (see Eq. (12–2.3), where only iron is present).

Composition cells. A composition cell may be established between any two *dissimilar* materials. In each case the metal lower in the electromotive series as listed in Table 12–2.1 acts as the anode. For example, on a sheet of *galvanized* steel (Fig. 12–3.1), the zinc coating acts as an anode and protects the underlying iron even if the surface is not completely covered, because the exposed iron is the cathode and does not corrode.

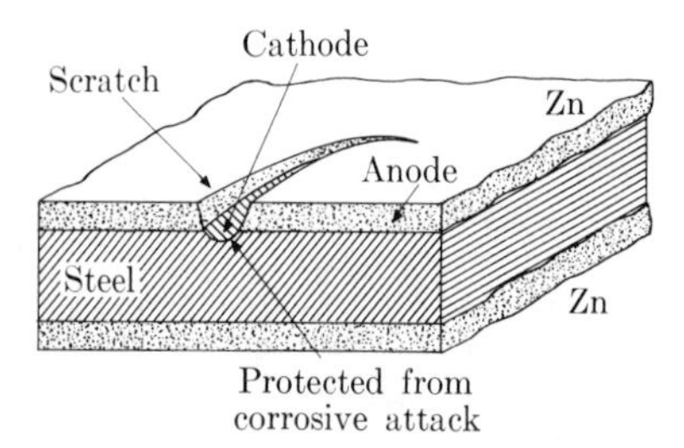

Fig. 12–3.1. Galvanized steel (cross section). Zinc serves as the anode; the iron of the steel serves as the cathode. Therefore the iron is protected even though it is exposed where the zinc is scraped off.

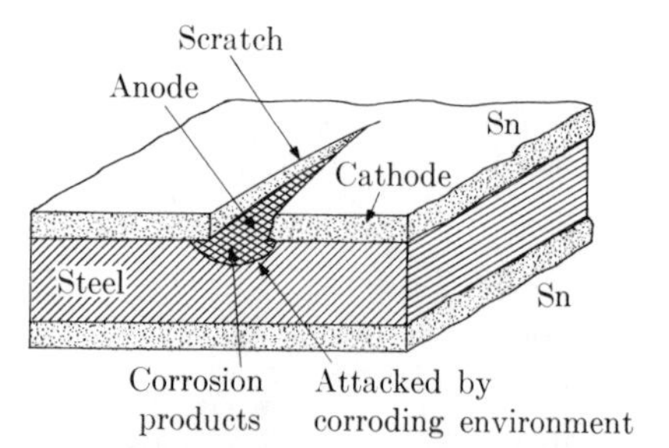

Fig. 12–3.2. Tinplate (cross section). The tin protects the iron while the coating is continuous. When the coating is broken, the iron of the steel becomes the anode and is subject to accelerated corrosion.

Any corrosion which does occur is on the anodic zinc surface. So long as zinc remains it provides protection to adjacent exposed iron.

Conversely, a *tin* coating on sheet iron or steel provides protection only so long as the surface of the metal is completely covered. Since tin is only slightly below hydrogen in the electromotive series, its rate of corrosion is limited. However, if the surface coating is punctured, the tin becomes the cathode with respect to iron, which acts as the anode (Fig. 12–3.2). The galvanic couple which results produces corrosion of the iron. Since the small anodic area must supply electrons to a large cathode surface, very rapid localized corrosion can result.

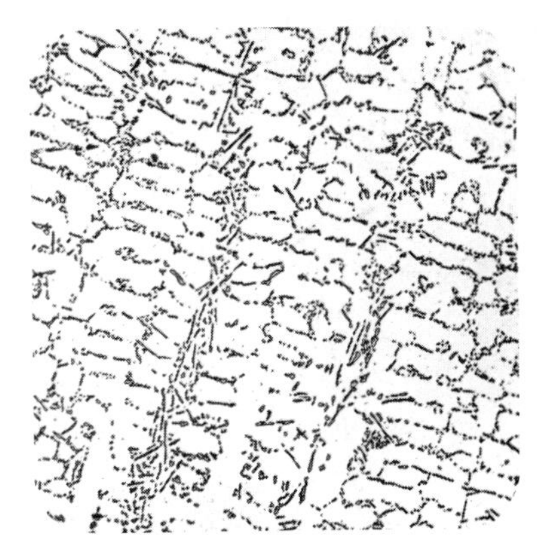

Fig. 12–3.3. Galvanic microcells (Al–Si alloy). Any two-phase alloy is more subject to corrosion than is a single-phase alloy. A two-phase alloy provides anodes and cathodes. (Alcoa Research Laboratories.)

Other examples of galvanic couples often encountered are (1) steel screws in brass marine hardware, (2) Pb–Sn solder around copper wire, (3) a steel propeller shaft in bronze bearings, and (4) steel pipe connected to copper plumbing. Each of these is a possible galvanic cell unless protected from a corrosive environment. Too many engineers fail to realize that the contact of dissimilar metals is a potential source of galvanic corrosion. Recently, in an actual engineering application, a brass bearing was used on a hydraulic steering mechanism made of steel. Even in an oil environment, the steel acted as an anode and corroded sufficiently to permit leakage of oil through the close-fitting connection.

There is no size limitation for galvanic cells. Furthermore, each phase with its individual properties possesses its own electrode potential, so that galvanic cells can be set up in most two-phase alloys when they are exposed to an electrolyte. For example, the pearlite of Fig. 9–8.1 reveals the carbide lamellae because the carbide was the anode in the electrolyte which was used as an etch.* Figure 12–3.3 shows the microstructure of an Al–Si casting alloy. Again we depend on corrosion to reveal the two phases. One is the anode, the other the cathode.

---

* The etch was 4% picral. The carbides are darkened because a corrosion reaction product remains on the surface. The electrode potentials of ferrite and carbide are sufficiently close so that with other electrolytes their cathodic and anodic roles may be interchanged.

Heat treatment may affect the corrosion rate by altering the microstructure of the metal. Figure 12–3.4 shows the effect of tempering on the corrosion of a previously quenched steel. Prior to tempering reactions, the steel contains a single phase, martensite. The tempering of the martensite produces many galvanic cells and grain boundaries of ferrite and carbide, and the corrosion rate is increased. At higher temperatures the coalescence of the carbides reduces the number of galvanic cells and the number of grain boundaries, which decreases the corrosion rate markedly.

In a single phase the corrosion rate of an age-hardenable aluminum alloy is low (Fig. 12–3.5), but the corrosion rate is significantly increased with precipitation of the second phase. Still greater agglomeration of the precipitate once again decreases the rate, but never to as low a level as in the single-phase alloy. The maximum corrosion rate occurs after maximum hardness has been attained.

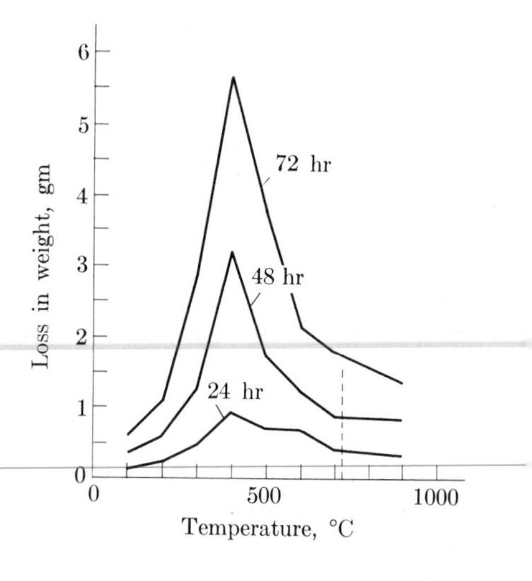

Fig. 12–3.4. Microcells and corrosion. After quenching, only martensite exists. After intermediate-temperature tempering, many small galvanic cells exist as a result of the fine (α + carbide) structure in tempered martensite. After high-temperature tempering, the carbide is agglomerated and fewer galvanic cells are present. (Adapted from F. N. Speller, *Corrosion: Causes and Prevention*, McGraw-Hill.)

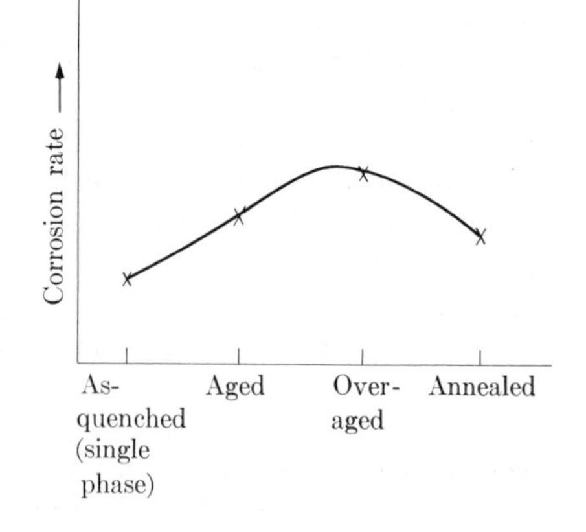

Fig. 12–3.5. Age-hardening and corrosion (schematic). The single-phase, quenched alloy has a lower corrosion rate than the subsequent two-phase modifications.

Stress cells. As shown in Fig. 4–4.9, where the grain boundaries have been etched (i.e., corroded), the atoms at the boundaries have an electrode potential different from that of the atoms in the grain proper; thus an anode and a cathode are developed (Fig. 12–3.6). The grain boundary zone may be considered to be stressed, since the atoms are not at their positions of lowest energy.

The effect of internal stress on corrosion is also evident after a metal has been *cold-worked.* A very simple example is shown in Fig. 12–3.7(a), where strain-hardening exists at the bend of an otherwise annealed wire. The cold-worked area serves as the anode and the strain-free area as the cathode.*

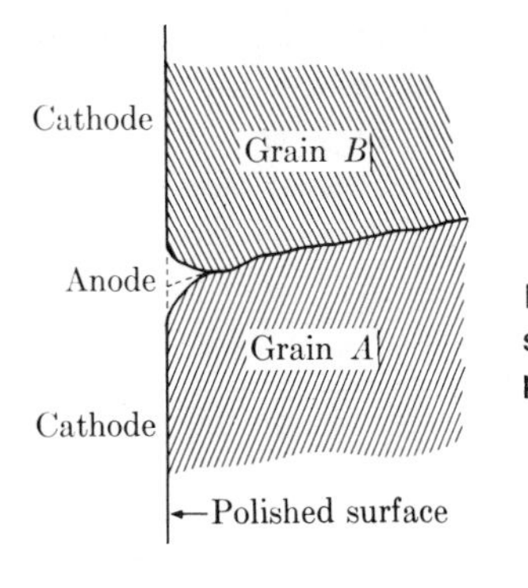

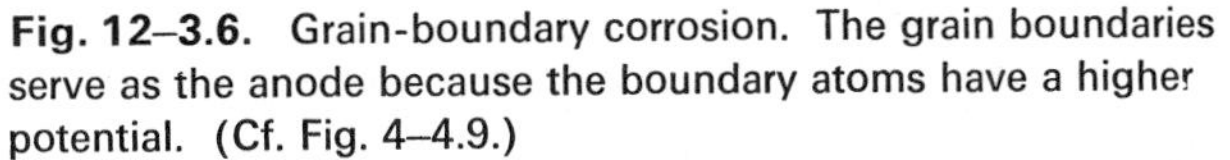
Fig. 12–3.6. Grain-boundary corrosion. The grain boundaries serve as the anode because the boundary atoms have a higher potential. (Cf. Fig. 4–4.9.)

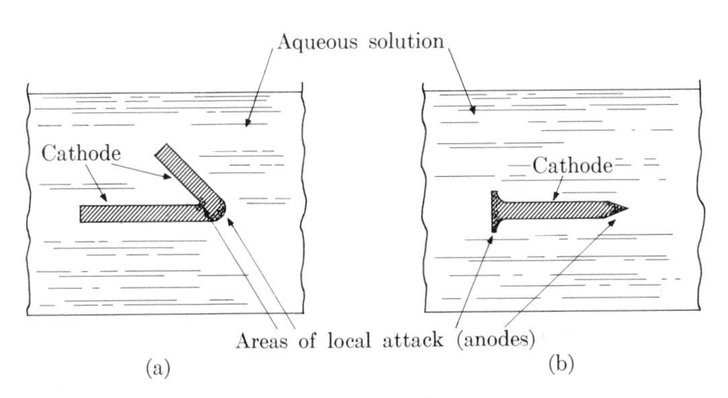

Fig. 12–3.7. Stress cells. In these two examples of strain hardening, the anodes are in the cold-worked areas. The electrode potential of a strained metal is higher than that of an annealed metal.

The engineering importance of the effects of stress on corrosion is plain. When engineering components must be used in a corrosive environment, the presence of stress may significantly accelerate the corrosion rate.

* *Corrosion in Action*, published by the International Nickel Company, demonstrates (with illustrative experiments) the effect of cold work on galvanic corrosion.

Concentration cells. In discussing cathode reactions in Section 12–2, we observed that Eq. (12–1.1) requires metal ions in order to be prominent. Furthermore, as the metal-ion concentration increases, the electroplating reaction is favored. Conversely, corrosion (Eq. 12–1.3) is favored at the anode when the electrolyte is lean in metal ions. When applied to copper, these two reactions may be combined as:

$$Cu^0 \underset{\text{Conc.}}{\overset{\text{Dilute}}{\rightleftarrows}} Cu^{2+} + 2e^- . \qquad (12\text{–}3.1)$$

The metal on side (D) of Fig. 12–3.8 is in the solution with the more dilute Cu^{2+} electrolyte. Therefore, its electrode reaction is more strongly to the right than is found on side (C), which has a higher Cu^{2+} content. In fact, when the two electrodes are connected into a galvanic cell, the electrons supplied through the external connection from side (D) combine with the larger number of Cu^{2+} ions on side (C) to force reaction (12–3.1) back to the left. The electrode in the concentrated electrolyte is protected and becomes the cathode; the electrode in the dilute electrolyte undergoes further corrosion and becomes the anode.

The concentration cell accentuates corrosion, but it accentuates it where the concentration of electrolyte is lower.

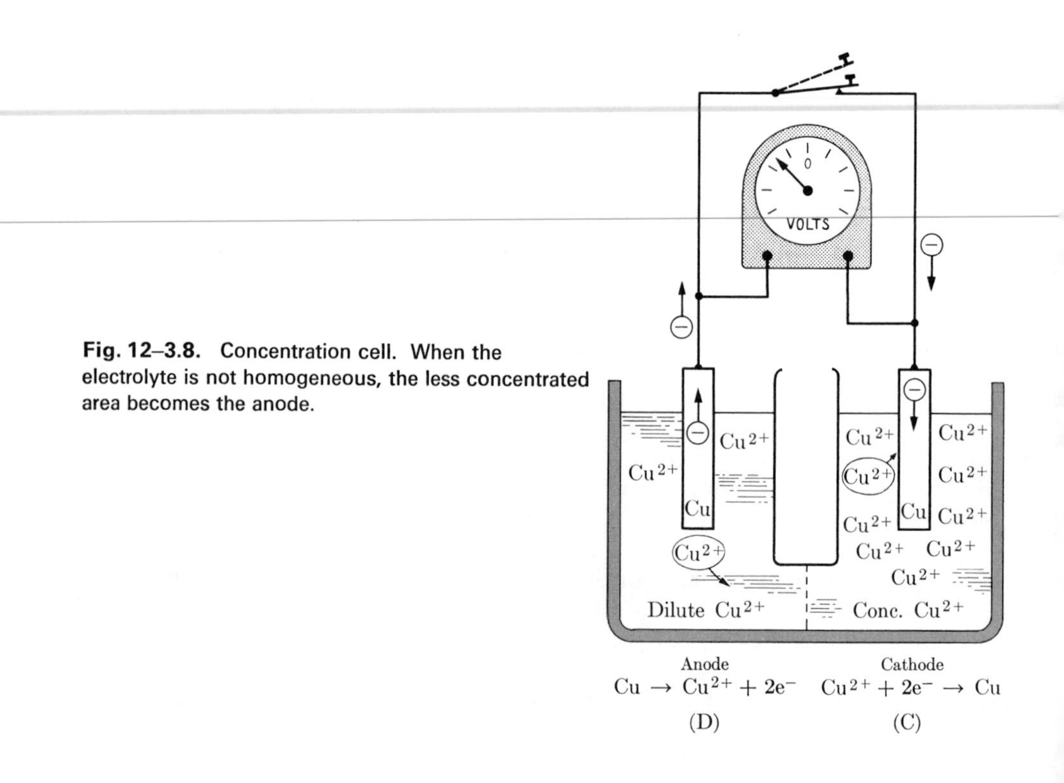

Fig. 12–3.8. Concentration cell. When the electrolyte is not homogeneous, the less concentrated area becomes the anode.

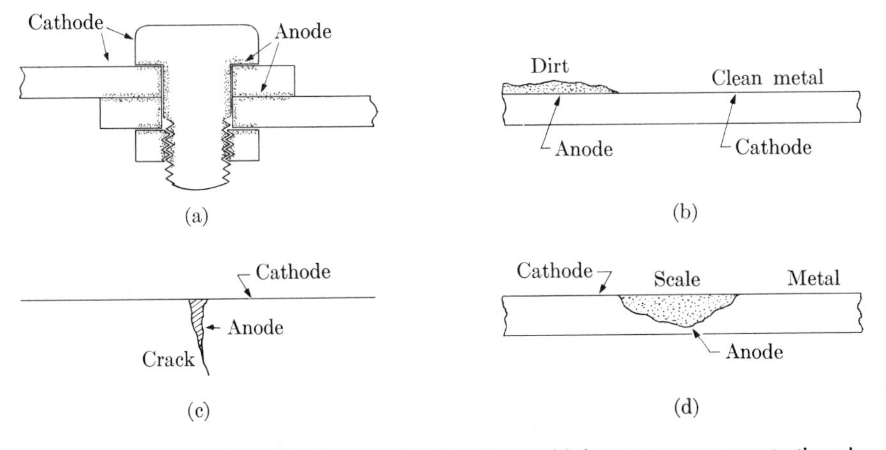

Fig. 12–3.9. Oxidation cells. Inaccessible locations with low oxygen concentrations become anodic. This situation arises because the mobility of electrons and metal ions is greater than that of oxygen or oxygen ions.

Concentration cells of the above type are frequently encountered in chemical plants, and also under certain flow-corrosion conditions. However, in general, they are of less widespread importance than are *oxidation-type concentration cells.* When oxygen in the air has access to a moist metal surface, corrosion is promoted. However, the most marked corrosion occurs in the part of the cell with an oxygen deficiency.

This apparent anomaly may be explained on the basis of the reactions at the cathode surface, where electrons are consumed. Equation (12–2.6) is restated below because it indicates the role of O_2 in promoting corrosion in oxygen-free areas:

$$2H_2O + O_2 + 4e^- \rightarrow 4(OH)^-.$$

Since this cathode reaction, which requires the presence of oxygen, removes electrons from the metal, more electrons must be supplied by adjacent areas which do not have as much oxygen. The areas with less oxygen thus serve as anodes.

The oxidation cell accentuates corrosion but it accentuates it where the oxygen concentration is lower. This generalization is significant. Corrosion may be accelerated in apparently inaccessible places because the oxygen-deficient areas serve as anodes, and therefore cracks or crevices serve as loci of corrosion (Fig. 12–3.9).

Corrosion is also accelerated under accumulations of dirt or other surface contaminations (Fig. 12–3.9(b)). This frequently becomes a self-aggravating situation, because accumulation of rust or scale restricts the access of oxygen and establishes an anode, to promote still greater accumulation. The result is localized *pitting* due to nonuniform corrosion (Fig. 12–3.9(d)), and the useful life of the product is thereby reduced to a greater extent than the weight loss would indicate.

•12–4 CORROSION RATES

The reader probably has already raised some questions about Table 12–2.1. Aluminum is more anodic than iron; why does it corrode less than iron? How can chromium impart corrosion resistance to stainless steels when its electrode potential is 0.3 volts less noble than iron? To answer these, we must consider factors which affect current densities.

Current density. As measured in Fig. 12–2.3, the electrode potential difference between iron and the standard hydrogen reference is 0.44 volts (Table 12–2.1). Of course, the voltmeter comes to zero when the two electrodes are placed into direct contact; also, a current flows. In effect, we have a short-circuited battery.

In order to simulate the corrosion of iron in an acid, we can replace the platinum electrode of Fig. 12–2.3 with another iron electrode, but still maintain the presence of H_2. The iron is the anode, and the hydrogen ions will form more H_2 on the cathode (Fig. 12–4.1). If a "short circuit" is placed across this cell, the resulting *current density*, *i*, which can be measured is $\sim$0.0002 amp per cm^2 of electrode surface.*

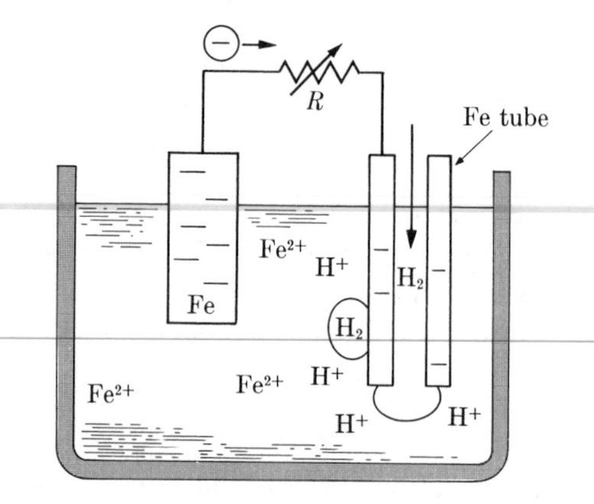

Fig. 12–4.1. Corrosion currents (iron anode and hydrogen cathode). With contact ($R = 0$), experiment shows that the current density is 0.0002 amp/cm^2, and the potential of the two electrodes is 0.24 volts below the standard $H_2/2H^+$ electrode. With increasing resistance, the current density decreases. When $R = \infty$, the potential difference is 0.44 volts (Table 12–2.1). Under normal corrosion conditions, the anode and the cathode may be two different areas of the same piece of metal.

Since the two electrodes are in direct contact, there will be no voltage difference between them; however, the pair may be referenced to the hydrogen electrode of Table 12–2.1. In doing so, an 0.24 volt difference would be observed.

Next, let us introduce an increasing resistance between the anodic and cathodic surfaces (Fig. 12–4.1). Of course, the current density decreases from what it was with direct contact; and a voltage differential develops as the two electrode potentials

* This will be independent of the size of the electrodes (but assumes that the two electrodes are the same size and a standardized electrolyte is used). The total current *I* will, of course, be proportional to the electrode area.

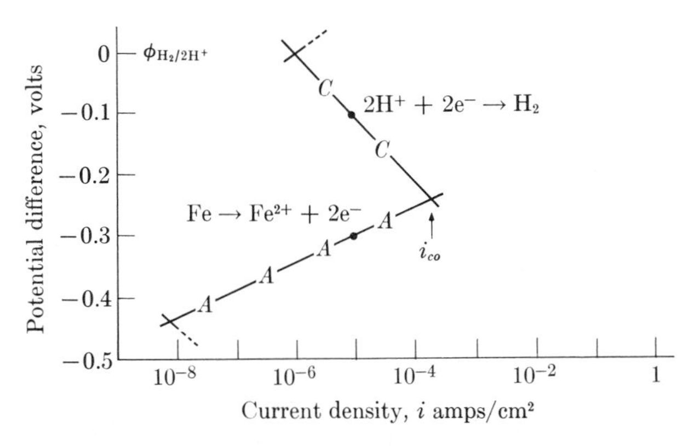

Fig. 12–4.2. Current density versus potential difference (cf. Fig. 12–4.1.) There is a corrosion current, i_{co}, of 0.0002 amp/cm^2 when the anode and the cathode of Fig. 12–4.1 are in direct contact. The current density decreases when the potential between the anode, *A*, and cathode, *C*, is increased. (It is only 10^{-5} amp/cm^2 when the potential differential is 0.2 volts ($C = -0.1$V, and $A = -0.3$V).)

return to the open-circuit condition of Table 12–2.1. Figure 12–4.2 plots the current density *i* versus potential ϕ of the two electrodes (based on $\phi_{H_2/2H^+} = 0.0$). The anode reaction, *A*, is for Fe → $Fe^{2+} + 2e^-$; the cathode curve, *C*, is for $2H^+ + 2e^- \rightarrow H_2$.

With direct contact, the *corrosion current density*, i_{co}, is the previously cited 0.0002 amp/cm^2 (where the two curves cross). Thus, by using the Faraday law (Eq. 12–1.2), we can calculate the rate of hydrogen evolution at the cathode, or the rate of corrosion at the anode (Example 12–4.1). (This assumes uniform and standard, one-molar electrolytes at 25°C. More concentrated solutions and higher temperatures alter the current density.) In general, anode and cathode curves cross at higher current densities when their open-circuit electrode potential differences are greater. Thus, we see a greater current density for an iron/copper couple than for an iron/hydrogen or an iron/nickel galvanic cell.

Since the anode and cathode seldom have the same areas, we observe a higher current density, *i*, on the electrode with the smaller area. (The total current, *I*, must be the same for each electrode, since we are not storing charge.) This accelerates corrosion markedly when the anode is the smaller electrode. Consider the steel screw which has been used inadvisedly in some brass marine hardware, or the scratch through the "tin-plate" of Fig. 12–3.2. In either case, the total number of electrons for the corrosion current must be obtained by corroding metal in a small anodic area, so the corrosion rate is very high locally. Corrosion thus penetrates through the sheet steel in Fig. 12–3.2 extremely rapidly, and the steel screw is destroyed in a short time when it is exposed to sea water.

Cathodic polarization. When a corrosion current is predicted from Fig. 12–4.2, one must assume that ions can leave or approach the electrode surfaces at a rate sufficient

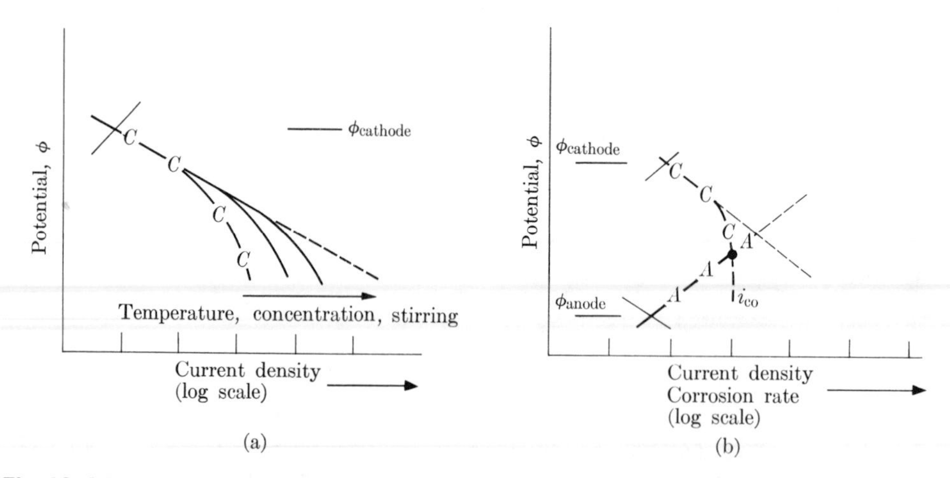

Fig. 12–4.3. Polarization. (a) The current density is normal (dashed line) only when the electrolyte concentration at the cathode and the temperature are high, and the electrolyte is not stagnant. Otherwise the current density lags. (b) When the current density lags, the corrosion current, i_{co}, is less than predicted. This polarization is eliminated if the service environment is altered to increase the availability of reactants.

to match the corrosion current. This assumption is reasonably valid at the anode, where the corroded ions enter a dilute electrolyte (as is usually the case in corrosion). However, the H^+ ions, the dissolved O_2, or other dilute reactants can easily be depleted from the cathode region to cause diffusional delays. This has become known over the years as *cathodic polarization*, and may be described by modifying the cathode, *C*, curve of Fig. 12–4.2, as shown in Fig. 12–4.3. In a stagnant situation, at low temperatures, or with dilute concentrations, the reactants are not continuously available to consume the electrons, and the current density lags behind the expected cathode curve. Thus, the resulting corrosion current density at the intersection of the two curves is slower than anticipated. Since this is desirable from the corrosion point of view,* we would like to capitalize on it. The steam power plant does this by de-aerating the feed water. However, the importance of Fig. 12–4.3(a) lies in the fact that under stagnant conditions, corrosion may be negligible; then a change of environment, such as a higher temperature or electrolyte movement, may suddenly decrease the cathodic polarization. The rate of corrosion thus increases by a couple of orders of magnitude. This accounts for some of the erratic corrosion rates which are commonly encountered in service.

Passivation. It was stated in the previous paragraph that the electrolyte is commonly dilute with respect to the anode product. Thus, anodic polarization is uncommon. However, the anode may become isolated from the electrolyte by noncorrosive

* It is undesirable in a dry cell or battery, where we utilize corrosion to produce a current.

reactions. These are tremendously important for corrosion control because they alter the rate of attack.

Consider that the iron of Fig. 12–4.2 is replaced by stainless steel. In the presence of excess oxygen, an oxide surface film containing chromium forms on the anode to isolate it from the electrolyte. As the oxidation potential is increased, the anode, *A*, current density curve is markedly affected (Fig. 12–4.4). We call this *passivation*. Its effect on the corrosion rate is pronounced because the corrosion current density, i_{co}, may be reduced by two or more orders of magnitude, from Ⓐ to Ⓟ.

The chromium of stainless steel produces passivation because of its strong attraction for oxygen. Thus we will find, under oxidizing conditions, that stainless steel does not have a significant corrosion rate (but it is not zero). Likewise we find that many metals are only slowly attacked by HNO_3 or H_2SO_4 because of the oxidizing nature of these acids. The same metals will be rapidly attacked by HCl, which does not have a supply of oxygen.

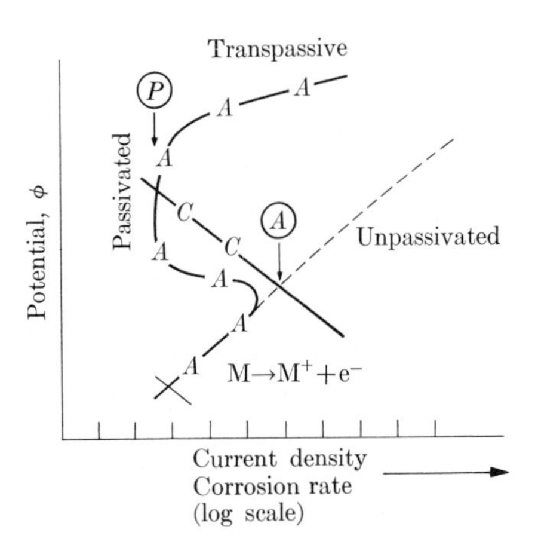

Fig. 12–4.4. Passivation. Metals such as aluminum, titanium, and stainless steel form a persistent protective film on the anode surface in oxidizing environments. The anode current density departs from the normal relationship (dashed line). Thus, the corrosion-current density (where the two curves cross) is reduced significantly. If the oxide film is destroyed, the corrosion rate increases quickly from the passivated corrosion rate Ⓟ to the activated corrosion rate Ⓐ.

Equally significant is the fact that aluminum and titanium will form a protective film of Al_2O_3 (and TiO_2) on their surfaces. This film is so protective that the corrosion current density is almost nil, and we can make boats of aluminum!* The commercial process that produces a protective oxide coating on aluminum is called *anodizing*.

---

* This is another example of why the casual student can be confused by the corrosion mechanism. However, to summarize: (1) oxygen accelerates corrosion; (2) in an oxidation cell, the corrosion is accelerated where the oxygen *ain't*, because the oxygen contributes to the cathodic side of the cell and drains the electrons from the anode where the corrosion occurs; (3) finally, some metals (e.g., aluminum and stainless steels) can be passivated because they react with oxygen to form protective surface films. An electrically isolated metal cannot corrode.

Example 12–4.1 The average corrosion-current density for iron is 0.2 mA/cm^2.
a) How many Fe^{2+} ions are formed per second from each sq. cm of surface?
b) What is the monthly corrosion penetration if the anodic area is limited to 10% of the surface?

Solution

a) (0.0002 amp/cm^2)/(2 el./Fe^{2+})(0.16 × 10^{-18} amp · sec/el.) = 0.6 × 10^{15} Fe^{2+}/sec · cm^2.

b)
$$\text{Local corrosion rate} = (0.6 \times 10^{15}\ \text{Fe/sec}\cdot\text{cm}^2)/(0.10)$$
$$= 6 \times 10^{15}\ \text{Fe/sec}\cdot\text{cm}^2\ \text{locally;}$$
$$\frac{(6 \times 10^{15}\ \text{Fe/sec}\cdot\text{cm}^2)(2.6 \times 10^6\ \text{sec/mo})}{(0.6 \times 10^{24}\ \text{Fe/55.85 g})(7.86\ \text{g/cm}^3)} = 0.2\ \text{cm/mo.}$$

Comment. Any factor which localizes corrosion (e.g., Fig. 12–3.9(d)) greatly affects the life of a product because it may fail although much of the metal has been unaffected. Examine the rust spots on the fender of a car. ◀

12–5 CORROSION CONTROL

Only in the absence of an electrolyte can corrosion be completely avoided. Even this is difficult. As is well known, tools hanging in a workshop can rust from adsorbed moisture films. The tool manufacturer is aware of this and therefore coats his products with an oil or grease film before shipping them to the retailer or customer.

Since corrosion is a galvanic action, there must be two kinds of metal in order for corrosion to proceed. The design engineer is normally cognizant of this and therefore will avoid the "brass bolt–steel washer" pitfall. Less familiar to many technical people is the fact that a cathode and anode may develop in a single material because of microstructure, stress concentrations, or electrolyte heterogeneities (Section 12–3). These must be considered by those seeking to control corrosion.

In addition to (1) *providing protective surfaces* and (2) *avoiding of galvanic couples,* it is also possible to minimize corrosion by (3) *introducing galvanic protection.* These three procedures will be the subject of this section.

Protective surfaces. Protecting the surface of a metal is probably the oldest of the common procedures for corrosion control. A painted surface, for example, isolates the underlying metal from the corroding electrolyte. The chief limitation of this method is the service behavior of the protective coating. The greasy film cited in the first paragraph of this section is obviously not very permanent. High temperatures or abrasive wear place limitations on organic coatings.

But protective coatings need not be limited to organic materials. For instance, tin can be used as an "inert" coating on a steel base. Copperplate, nickelplate, and silverplate are other examples of corrosion-resistant surfaces. Such metals may also be applied as hot-dip coats. Inert ceramic materials can also be used for protective

Table 12–5.1
COMPARISON OF INERT PROTECTIVE COATINGS

Type	Example	Advantages	Disadvantages
Organic	Baked "enamel" paints	Flexible Easily applied Cheap	Oxidizes Soft (relatively) Temperature limitations
Metal	Noble metal electroplates	Deformable Insoluble in organic solutions Thermally conductive	Establishes galvanic cell if ruptured
Ceramic	Vitreous enamel oxide coatings	Temperature resistant Harder Does not produce cell with base	Brittle Thermal insulators

coatings. For example, true enamels are oxide coatings applied as powdered glass and fused to become a vitreous surface layer. A comparison of the advantages and disadvantages of the several categories of protective coatings is given in Table 12–5.1.

Passivation, discussed in the previous section, also provides a protective film, or coating. Its importance, particularly for chromium-bearing stainless steels, is revealed in Table 12–5.2. This galvanic series differs from Table 12–2.1 in that common alloys that are widely used under corrosive conditions are included along with those elemental metals which were previously cited. Of current interest to us is the fact that a number of alloys are listed twice, in their active and in their passivated conditions. The presence of an oxygen-containing film on the surface shifts these alloys toward the cathodic end of the series. In fact, when passivated, these steels are less corrodable than copper, bronze, and brass. This introduces an important point. A passivated metal may have the corrosion current density indicated by point Ⓟ in Fig. 12–4.4. If service conditions change so that the protective oxide film is destroyed, the corrosion current density can change by several orders of magnitude, to point Ⓐ. Of course, this has major implications to both the design engineer and the technical manager.

Inhibitors have been developed by technologists as a way of capitalizing on the tendency for large polyatomic anions to adsorb onto a metal surface. We are most familiar with these as *rust inhibitors*, to decrease the corrosion in automobile radiators. They may also be used in steam boilers and other containers. These inhibitors are commonly chromates, phosphates, tungstates, or other ions of highly oxidized transition elements. Their mechanism of protection is probably closely related to that of passivation, where a protective, oxygen-rich, monolayer film is established at the metal surface.

Table 12–5.2
GALVANIC SERIES OF COMMON ALLOYS*

Graphite	Cathodic ↑	Nickel—A
Silver		Tin
12% Ni, 18% Cr, 3% Mo steel—P		Lead
20% Ni, 25% Cr steel—P		Lead–tin solder
23 to 30% Cr steel—P		12% Ni, 18% Cr, 3% Mo steel—A
14% Ni, 23% Cr steel—P		20% Ni, 25% Cr steel—A
8% Ni, 18% Cr steel—P		14% Ni, 23% Cr steel—A
7% Ni, 17% Cr steel—P		8% Ni, 18% Cr steel—A
16 to 18% Cr steel—P		7% Ni, 17% Cr steel—A
12 to 14% Cr steel—P		Ni-resist
80% Ni, 20% Cr—P		23 to 30% Cr steel—A
Inconel—P		16 to 18% Cr steel—A
60% Ni, 15% Cr—P		12 to 14% Cr steel—A
Nickel—P		4 to 6% Cr steel—A
Monel metal		Cast iron
Copper–nickel		Copper steel
Nickel–silver		Carbon steel
Bronzes		Aluminum alloy 2017-T
Copper		Cadmium
Brasses		Aluminum, 1100
80% Ni, 20% Cr—A		Zinc
Inconel—A		Magnesium alloys
60% Ni, 15% Cr—A	↓ Anodic	Magnesium

* Adapted from C. A. Zapffe, *Stainless Steels*, American Society for Metals.
A–active; P–passivated.

Avoidance of galvanic couples. The simplest method of avoiding galvanic couples is to limit designs to only one metal, but this is not always feasible. In special circumstances, the cells may be avoided by electrically insulating metals of different compositions.

Other, less simple, methods are frequently warranted, and *stainless steel* provides a good specific example. There are many types of stainless steel, whose chromium content varies from 13 to 27%. The purpose of the chromium is to provide a composition which will normally develop a passive surface. Many, but not all, stainless steels also contain 8 to 10% nickel, which is more noble than iron (Table 12–2.1).

• The high alloy content of such a metal as 18–8 stainless steel (so-called because it contains 18% Cr–8% Ni) causes the formation of austenite, which is stable at ambient temperatures. Such a steel is not used primarily for applications requiring high hardness, but rather in corrosive applications. Therefore carbon, which is more soluble in austenite at high than at low temperatures (Fig. 12–5.1), is kept to a minimum. If steel containing 0.1% carbon is cooled rapidly from about 1000°C (~1800°F),

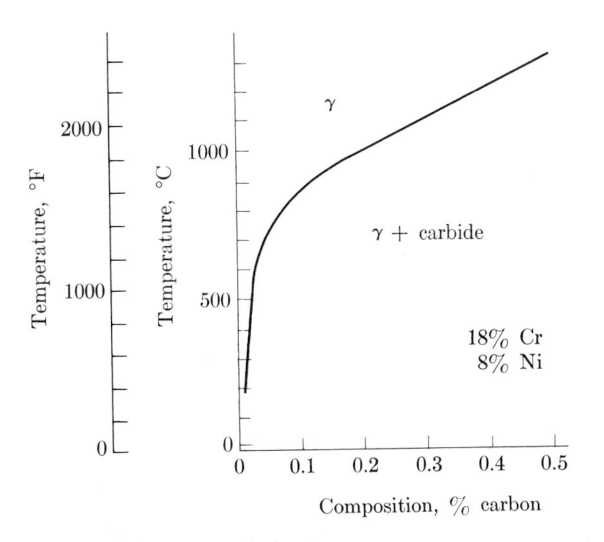

Fig. 12–5.1. Carbon solubility in austenitic stainless steel. The carbon solubility in an 18–8 type stainless steel decreases markedly with temperature. Consequently, the carbon will precipitate if cooling is not rapid. The precipitated carbide is rich in chromium. (Adapted from E. E. Thum, *Book of Stainless Steels*, American Society for Metals.)

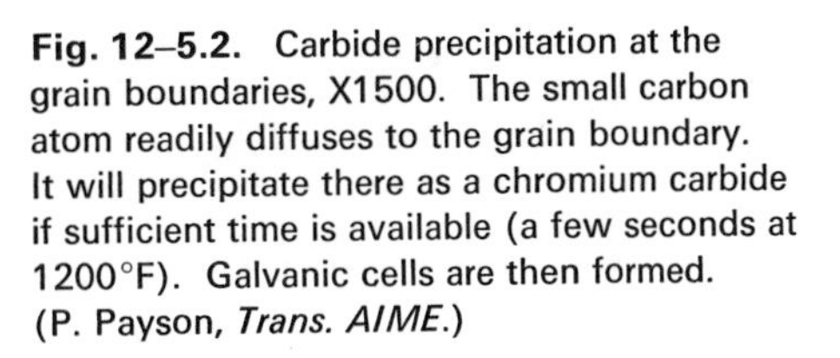

Fig. 12–5.2. Carbide precipitation at the grain boundaries, X1500. The small carbon atom readily diffuses to the grain boundary. It will precipitate there as a chromium carbide if sufficient time is available (a few seconds at 1200°F). Galvanic cells are then formed. (P. Payson, *Trans. AIME*.)

a separate carbide does not form and galvanic cells are not established. On the other hand, if the same steel is cooled slowly, or held at 650 ± °C (~1200°F) for a short period of time, the carbon precipitates as a chromium carbide, usually in the form of a fine precipitate at the grain boundaries (Fig. 12–5.2). In the latter case, two effects are possible: (1) galvanic cells may be established on a microscopic scale, or (2) the carbon forms chromium carbide (more stable than Fe_3C), which depletes the grain-boundary area of chromium and removes its passivating protection locally (Fig.

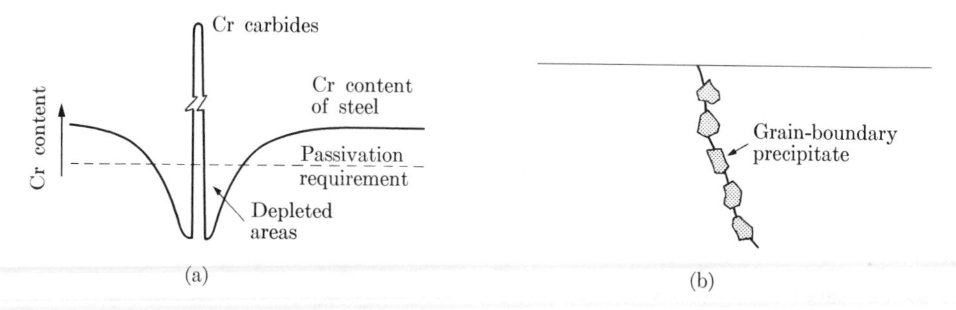

Fig. 12–5.3. Chromium depletion adjacent to the grain boundary. The carbide precipitation consumes nearly 10 times as much chromium as carbon. Since the larger chromium atoms diffuse slowly, the Cr content of the adjacent areas is lowered below protection levels.

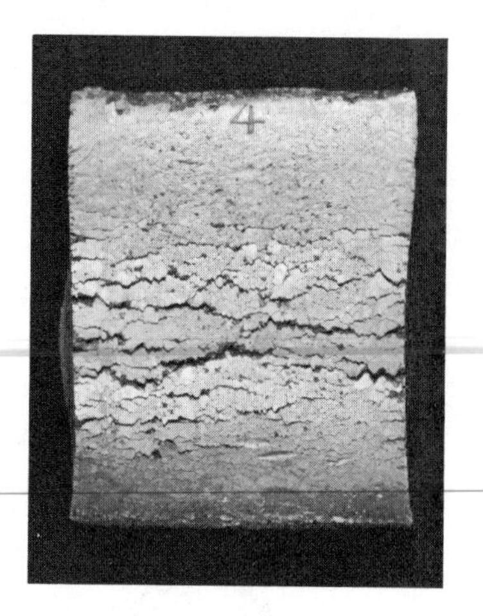

Fig. 12–5.4. Intergranular corrosion. This type of corrosion becomes severe if the steel has been heated into the carbide-precipitation range. (W. O. Binder, *Corrosion of Metals*, American Society for Metals.)

12–5.3). Either of these effects accentuates corrosion at the grain boundaries, and is to be avoided (Fig. 12–5.4).

• There are several ways to inhibit intergranular corrosion; the choice, of course, depends on the service conditions:

1. *Quenching to Avoid Carbon Precipitation.* This method is commonly used unless (a) service conditions require temperatures in the precipitation range, or (b) forming, welding, or size prevent such a quenching operation.

2. *Provision for An Extremely Long Anneal in the Carbide Separation Range.* This technique offers some advantage because of (a) agglomeration of the carbides, and (b) homogenization of the chromium content so that there is no deficiency at the grain

boundary. However, this procedure is not common because the improvement in corrosion resistance is relatively small.

3. *Selection of A Steel with Less Than* 0.03% *Carbon.* As indicated in Fig. 12–5.1, this would virtually eliminate carbide precipitation. However, such a steel is expensive because of the difficulty of removing enough of the carbon to attain this very low level.

4. *Selection of A Steel with High Chromium Content.* A steel which contains 18% chromium corrodes less readily than a plain carbon steel. The addition of more chromium (and nickel) provides additional protection. This, too, is expensive because of the added alloy costs.

5. *Selection of A Steel Containing Strong Carbide Formers.* Such elements include titanium, columbium, and tantalum. In these steels, the carbon does not precipitate at the grain boundary during cooling because it is precipitated earlier as titanium carbide, columbium carbide, or tantalum carbide, at much higher temperatures. These carbides are innocuous because they neither deplete the chromium from the steel nor localize the galvanic action to the grain boundaries. This technique is used frequently, particularly with stainless steel which must be fabricated by welding.

Although the above examples are somewhat specific, they do indicate methods which are used to reduce the extent of corrosion in metals. The exact choice of procedure depends on the alloy and the service conditions involved.

Galvanic protection. It is possible to restrict corrosion by turning some of the mechanisms of corrosion to protective ends. A good example is the galvanized steel discussed earlier in Section 12–3. The zinc coating serves as a *sacrificial anode* which itself corrodes, instead of the underlying steel. The same method may be used in other applications. Three examples are shown in Fig. 12–5.5. An advantage of such procedures is that the spent anode can be replaced quite easily. For example, the magnesium plates in Fig. 12–5.5(a) can be replaced at a fraction of the cost of replacing the underground pipe.

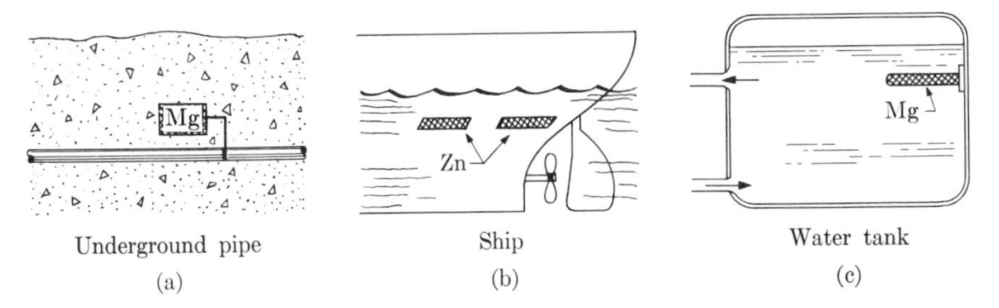

Fig. 12–5.5. Sacrificial anodes. (a) Buried magnesium plates along a pipeline. (b) Zinc plates on ship hulls. (c) Magnesium bar in an industrial hot-water tank. Each of these sacrificial anodes may be easily replaced. They cause the *equipment* to become a cathode.

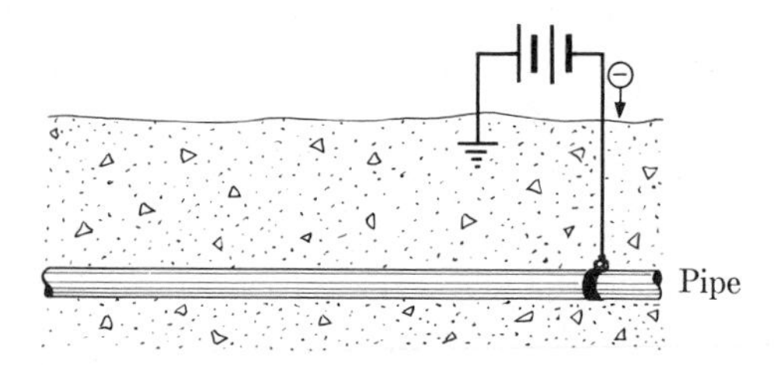

Fig. 12–5.6. Impressed voltage. A small d.c. voltage will provide sufficient electrons to make the equipment a cathode.

A second method of galvanic protection is the use of an *impressed voltage* on the metal. This is illustrated in Fig. 12–5.6. Both the sacrificial anode and the impressed voltage methods involve the same protection principle; that is, extra electrons are supplied so that the metal becomes the cathode and the corrosion reactions do not proceed.

Table 12–5.3
SUMMARY OF GALVANIC CELLS

Specific examples	Anode	Cathode
	Base metal	Noble metal
Zn versus Fe	Zn	Fe
Fe versus H_2	Fe	H_2
H_2 versus Cu	H_2	Cu
	Higher energy	Lower energy
Boundaries	Boundaries	Grain
Stresses	Cold-worked	Annealed
Stress corrosion	Stressed areas	Nonstressed areas
	Lower conc.	Higher conc.
Electrolyte	Dilute solution	Concentrated solution
Oxidation	Low O_2	High O_2
Dirt or scale	Covered areas	Clean areas

REVIEW AND STUDY

CORROSION SUMMARY

We first considered that corrosion is the reverse of electroplating. This gave us a basis for relating the amount of corrosion to the amount of current through Faraday's law. Two electrodes are required, plus an electrolyte.

Most corrosion results from galvanic cells and the accompanying electrical currents. Two dissimilar electrodes are required; these may be provided by (1) differences in composition, (2) differences in energy level (disordered, or stressed areas), or (3) differences in electrolytic environment. They are detailed further in Table 12–5.3. The electrode which undergoes corrosion is the anode; the cathode is protected.

Oxygen accelerates the corrosion of metals; however, the corrosion is accelerated where the oxygen content is low, since the oxygen reacts on the cathode side of the galvanic cell. Oxygen forms a protective film on aluminum, titanium, and those steels which contain chromium. This phenomenon is called passivation, and is the basis for the relative inertness of stainless steels.

Polarization may occur on the cathode because the reactants are depleted from the electrolyte near the electrode surface. This reduces the current density. Since the reactants (ions and oxygen) are replenished more rapidly at higher temperatures or with stirring, we find that polarization varies significantly with service conditions.

Corrosion is controlled most simply by isolating the metal surface by protective coatings. These may be organic (paint), ceramic (enamel), or other metals. Satisfactory corrosion control, however, requires that the galvanic couples be avoided. Unlike metals must be electrically separated when the corrosive environment is severe. In those severe situations where stainless steel is used, further care must be taken that galvanic cells are not developed within the microstructures (e.g., carbides versus carbon-depleted areas). In some applications, corrosion protection may be provided through sacrificial anodes, or by an impressed voltage. In both situations, the metal which is to be protected is made to be the cathode.

TERMS FOR REVIEW

Anode
Anodized
Cathode
Cathode reactions
Cells
  composition
  concentration
  galvanic
  oxidation
  stress
•Corrosion-current density
Electrochemical potential, emf
Electrolyte
Electroplating
Faraday, $\mathscr{F}$
Galvanic couple
Galvanization
Hydrogen electrode
Impressed voltage
Inhibitor
•Passivation
•Polarization
Rust
Sacrificial anode
Stainless steel

FOR DISCUSSION

A_{12} Distinguish between anode and cathode on the basis of electron movements. How does a cathode-ray tube fit into the definitions of Section 12–1?

B_{12} Show the origin of 96,500 coul as the value of one Faraday, $\mathscr{F}$.

C_{12} What is the source of electrons in an ordinary dry cell? (The electrolyte is a gelatinous paste containing NH_4Cl. The cathode reaction changes Mn^{4+} in MnO_2 to Mn^{2+}.)

D_{12} Why do we use the hydrogen electrode as the reference electrode rather than some other element, such as lead?

E_{12} Why is platinum used in Figs. 12–2.2 and 12–2.3?

F_{12} Under what conditions will rusting *not* occur? Why?

G_{12} It is suggested that sodium might be produced by taking a solution of NaOH (pH = 14) and electrolyzing it to plate-out sodium. Discuss.

H_{12} Distinguish between the type of protection given to iron by zinc and by tin.

I_{12} Plumbing codes call for an insulator such as Teflon to be placed between copper tubing and steel pipe when they are coupled in a plumbing system. Explain.

J_{12} Electrical codes call for a "jumper" to be placed around the insulated coupling just described, to provide a ground. Does this defeat the purpose of the plumbing code? Explain.

K_{12} A zinc-coated nail is sheared in half and placed in an electrolyte. a) What couples must be considered in judging where the anode will be? b) Cite the location which will be corroded initially.

L_{12} Explain why the metal in a dilute electrolyte becomes the anode to metal in a more concentrated electrolyte.

M_{12} Explain why the metal in an oxygen-deficient electrolyte becomes the anode to metal in an oxygen-enriched electrolyte.

N_{12} If oxygen accelerates corrosion (Eqs. 12–2.6 and 12–2.7), why does the corrosion occur in the oxygen-lean area?

O_{12} How do barnacles accelerate corrosion on a ship hull?

•P_{12} Why does aluminum corrode less than iron when used in kitchen utensils?

•Q_{12} Why is corrosion accelerated when the anode is smaller than the cathode, but not when the situation is reversed?

•R_{12} The headache of metal-can manufacturers is "pinholes" in the tinplate. Explain.

•S_{12} A flashlight with an old drycell will dim during use. If left unused for a short time, it will shine more brightly again. Explain.

T_{12} Steel piles corrode at the tide level much more than at lower levels. Why?

•U_{12} A vitreous enamel (glass) is used to coat the inside of a home hot-water heater. How should the thermal expansion of the glass and metal compare? (Check Chapter 8.)

V_{12} An enterprising mechanic suggests the use of a magnesium drain plug for the crankcase of a car as a means of combating engine corrosion. Discuss.

W_{12} Undercoatings of various types are used on new cars. Under what conditions are they helpful? Under what conditions are they detrimental?

X_{12} Is it possible to use an impressed voltage to protect a ship hull? Explain

Y_{12} Cite three examples of corrosion from your experience. Describe the nature of the deterioration, and account for the corrosion.

•Z_{12} A stainless-steel sheet is welded into a circular duct. After a period of time, rust appears along a band which is parallel to each side of the weld, about a centimeter away. Why did this occur? Could it have been avoided?

STUDY PROBLEMS

12–1.1 How many coulombs are required to plate each gram of nickel from an Ni^{2+} electrolyte?

Answer: 3300 coul

12–1.2 How long does it take to electroplate one gram of chromium from a Cr^{2+} electrolyte with a current of 1 ampere?

12–1.3 Aluminum is refined by fluxing Al_2O_3 in a molten salt and then electroplating the metal onto the cathode. What current density, amp/cm^2, does it take to produce 100 grams per day per sq cm of electrode area?

Answer: 12.4 amp/cm^2

12–2.1 Bubbles of hydrogen are evolved from the cathode at a rate of one per second. Each is 1 mm^3 when it emerges from the surface of the electrolyte. What electron current is involved? (1 mole of gas at STP = 22,400 cm^3.)

Answer: 8.6 mA

12–2.2 Describe how water may be electrolyzed to produce H_2 and O_2.

12–2.3 In Fig. 12–2.1, copper is replaced by gold and zinc by tin. Based on the half-cell reactions of Table 12–2.1, will the weight of tin corroded per hour be greater or less than the weight of gold which is plated?

Answer: 10 w/o more gold

12–3.1 Two pieces of metal, one copper and the other zinc, are immersed in sea water and connected by a copper wire. Indicate the galvanic cell by writing the half-cell reactions (a) for the anode and (b) for the cathode; also, by indicating (c) the direction of the electron flow in the wire, and (d) the direction of the "current" flow in the electrolyte. (e) What metal might be used in place of copper so that zinc changes polarity?

CHAPTER THIRTEEN

COMPOSITE MATERIALS

PREVIEW

A reflection from the previous chapters reveals the sequence we have followed. After establishing some definitions for properties, we gave first attention to the atom and its immediate neighbors. Then the scale was expanded to the long-range pattern of crystals. The combination of crystals into more extensive polycrystalline, single-phase solids was the next step. This presented opportunities for tailoring structures and properties (1) through the control of grain size, shape, and orientation, (2) by solution-hardening, and (3) by strain-hardening. When we proceeded to the next step of multiphase materials, we extended our capabilities to design materials significantly. For example, it was possible to have strength *and* toughness; it was also possible to change a material from machinable to being the *tool* for machining. Sequentially, our materials have developed more complexity; in doing so, our materials provided us with greater versatility.

In the final chapter of this book we will consider composite materials on a still coarser scale. These are materials that have multiple origins, which may arise (1) because we agglomerate materials, (2) because we alter the surface of the original material by changing its surface composition (e.g., carburization) or by using differential heat treatments (e.g., induction-hardening), or (3) because we combine two distinct materials (e.g., glass-fiber reinforced plastics). We shall consider each of these and their *macrostructures*.

As a final section, we will open the door to nature's composite materials. Our prototype will be wood, because it is somewhat familiar to us, and because it readily extends the structure–property concepts. We must end our structure–property sequence there, partially because of lack of time. Also, we would need additional chemistry, more physics, specialized mathematics, and more crystallography than is presently available to most of the readers, if we wished to extend the scope of this text.

13–1 AGGLOMERATED MATERIALS

Many engineering materials are formed by the agglomeration of small particles into a usable product. Probably the most obvious example is concrete, in which gravel, sand, portland cement, and water are mixed into a monolithic (literally, one-stone) engineer-

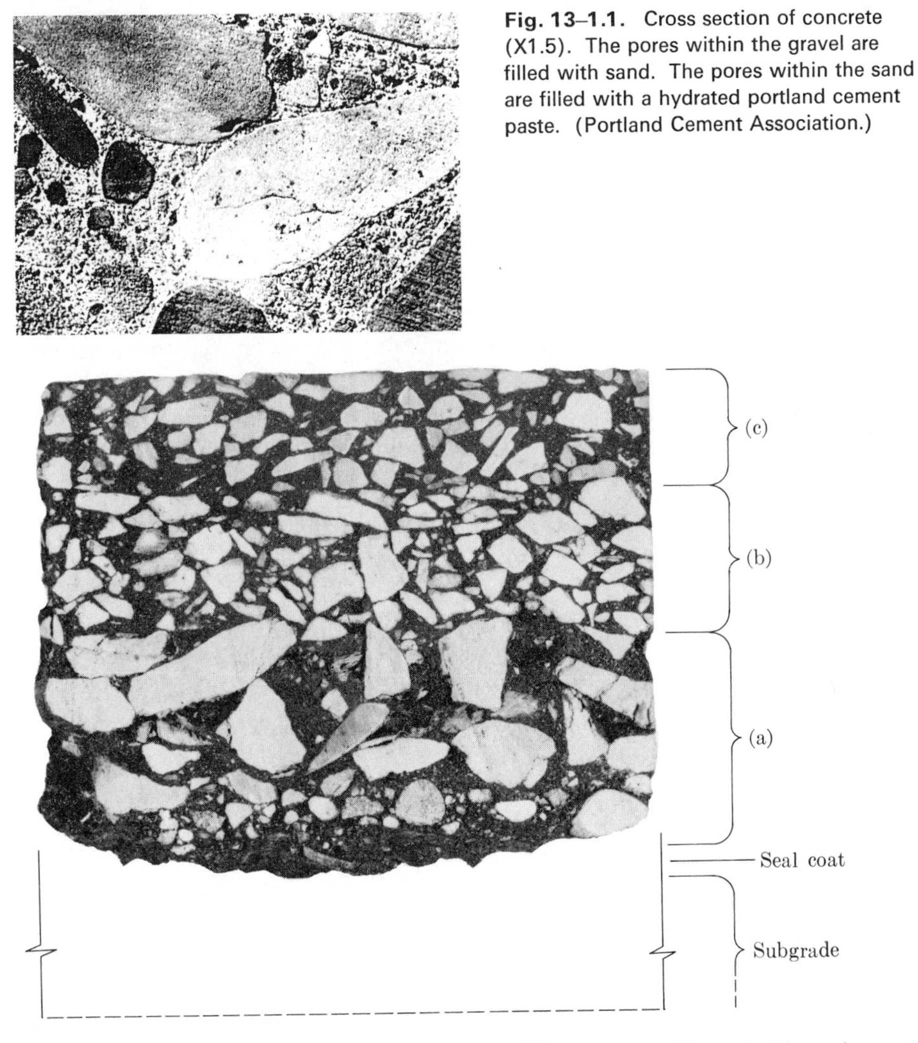

Fig. 13–1.1. Cross section of concrete (X1.5). The pores within the gravel are filled with sand. The pores within the sand are filled with a hydrated portland cement paste. (Portland Cement Association.)

Fig. 13–1.2. Core from asphalt pavement (X 0.5). Like concrete, this product is agglomerated; but viscous asphalt, rather than a hydrated silicate, serves as the bond. (a) Bonding course. Sand (30 w/o) and asphalt (5 w/o) fill the interstices among the coarse stone (65 w/o) to give a rigid support above the subgrade aggregate. (b) Leveling course. The stone is smaller, otherwise the composition is similar to the preceding course. (c) Wearing course. More asphalt (6 w/o), more sand (55 w/o), and 4–5 w/o mineral filler (fly-ash) provide an impermeable, tough surface to resist traffic wear. (Michigan Highway Testing Laboratory.)

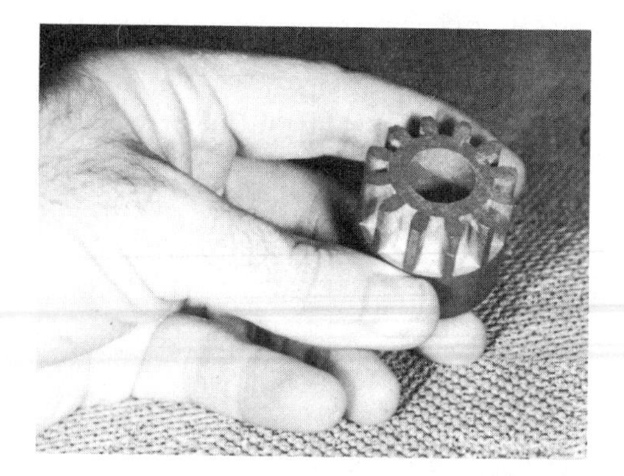

Fig. 13–1.3. Powdered metal gear. The metal particles are sintered into a coherent structure.

ing material. The current achievements in highway construction would be impossible were it not for the substitution of concrete for the former brick and stone pavements (Fig. 13–1.1). Agglomerations of asphalt and stone are also used for highway surfaces (Fig. 13–1.2), and in such cases the soils or similar aggregates on which they rest must be taken into account.

Many other agglomerated materials are equally important for modern industry. Unlike concrete, which is bonded by hydraulic cement, the abrasive grains of grinding wheels are bonded together by either a glass or a resin; brick depends entirely on a vitreous (glassy) bond, and powdered metals (Fig. 13–1.3) depend on solid sintering at high temperatures for bonding.

Particle size. If all the particles used in agglomerated materials were perfect spheres the determination of particle size would be a relatively simple matter of measuring diameters. In practice, most particles vary considerably from perfect sphericity, but it is still desirable to have a measure of the size. For this purpose sand and gravel, for

Table 13–1.1
TYLER SCREENS

Mesh number	Opening		Mesh number	Opening	
	inches	cm		inches	cm
—	1.050	2.67	20	0.0328	0.083
—	0.742	1.88	28	0.0232	0.059
—	0.525	1.33	35	0.0164	0.042
—	0.371	0.94	48	0.0116	0.0295
3	0.263	0.67	65	0.0082	0.0208
4	0.185	0.47	100	0.0058	0.0147
6	0.131	0.33	150	0.0041	0.0104
8	0.093	0.236	200	0.0029	0.0074
10	0.065	0.165	270	0.0021	0.0052
14	0.046	0.118	400	0.0015	0.0037

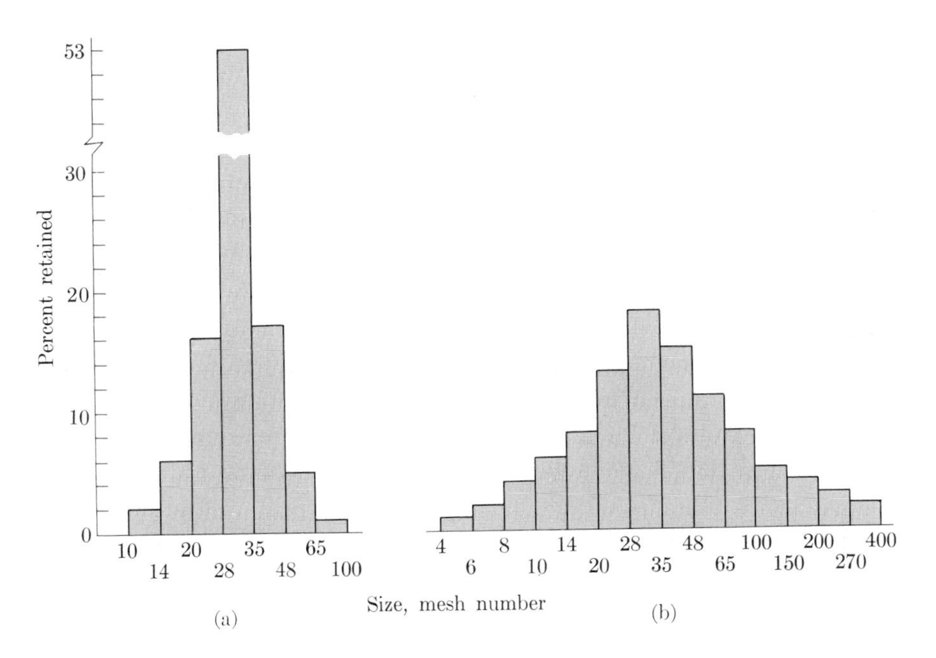

Fig. 13–1.4. Aggregate size distribution. (a) Closely graded. (b) Broadly graded.

example, are sifted through screens with standardized openings. Large openings are measured directly in inches, or centimeters. In the United States, openings less than $\frac{1}{4}$ inch (0.6 cm) in size are expressed in *mesh numbers*, which indicate the number of *openings per linear inch*. Table 13–1.1 shows the most commonly used mesh series. Openings for successive mesh numbers vary by a factor of $\sqrt{2}$.

Any aggregate mixture will be composed of a range of mesh sizes, and it is usually insufficient to rely on the average size of an aggregate. Figure 13–1.4 shows the percent of aggregate retained by successively finer screens. In one case, the particles are very closely *graded* and are all about the same size. In the other case, however, there is a wide variation in size. Because the mean and the median sizes are approximately the same in each case, the engineer usually specifies a size distribution based on the percent aggregate retained on screens of more than one mesh size.

Bulk characteristics. Unless one selects specific shapes, e.g., brick, and arranges them in a specific manner, space will not be completely filled with an aggregate of individual objects. We encountered this initially when we considered atomic packing factors (Section 3–3), where the maximum packing of uniformly sized spheres is only 74 v/o (fcc and hcp). The same concept holds for packing crushed rock into a road base, or for compacting Al_2O_3 powder into an insulator. In the context of this chapter, we shall speak of *porosity*, *P*, where

$$\text{Porosity} = 1 - \text{Packing factor}. \tag{13–1.1}$$

We always report porosity as volume fraction, or volume percent (v/o). Porosity is independent of particle size if the particles are uniform in size and shape. We accepted this in Chapter 3 with the assumption that fcc lead (r = 1.75 Å) and fcc copper (r = 1.278 Å) both have the same packing factor: 0.74.

In order to effectively reduce the porosity of an aggregate, we must mix particle sizes (Fig. 13–1.5). For example, a *mixed packing* of sand and gravel has a greater density than either sand or gravel alone, because the sand fills the pore spaces among the pieces of gravel. Figure 13–1.6 shows typical variations of the packing factor for mixtures of sand and gravel; this relationship determines the optimum ratio of sand to gravel for a mixture that will fill a given volume with least porosity. If, for example, a ratio of 25% sand and 75% gravel were used in a concrete mixture, only 20% of the volume would be pore space to be filled by a cement-water paste. On the other hand, a 50–50 mixture would leave 25–30% pore space and would require almost half again as much of the more expensive cement to develop the same strength in the concrete.

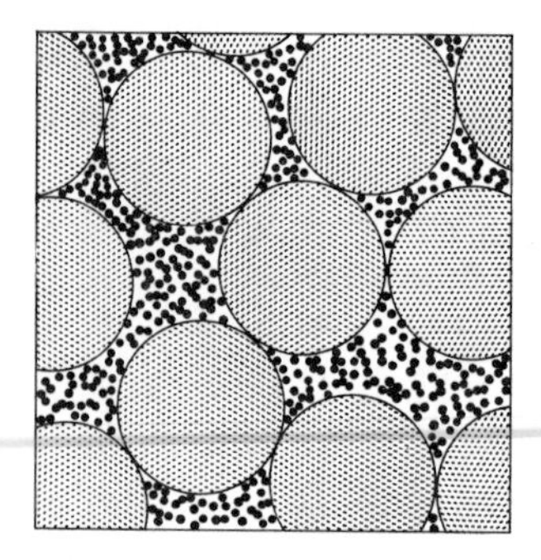

Fig. 13–1.5. Mixed packing of particles. The packing factor may be increased by introducing small particles into the pore spaces among the large particles.

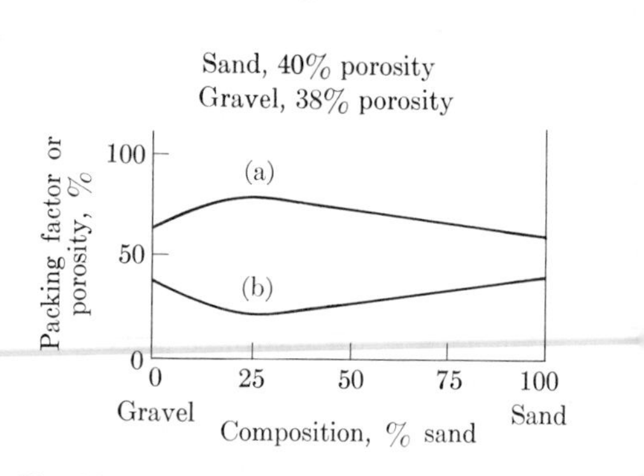

Fig. 13–1.6. (a) Packing factor versus composition of aggregate mixtures. (b) Porosity versus composition of aggregate mixtures. The densest packing and the lowest porosity occur when there is just enough sand to fill the pore spaces within the gravel.

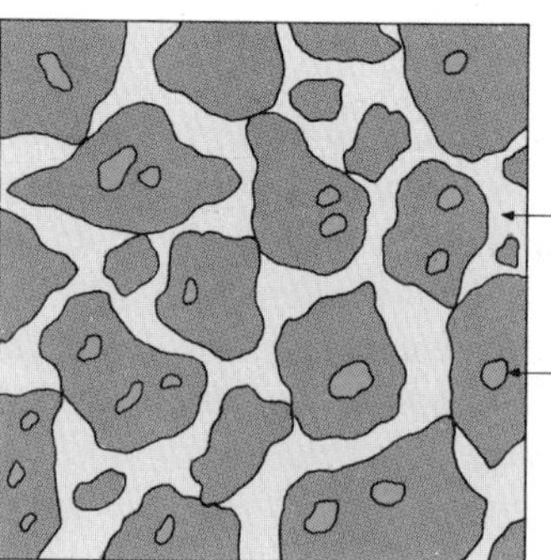

Fig. 13–1.7. The bulk volume includes the true volume of the material (dark), plus the closed pores (medium), plus the open pores (light). The apparent volume includes the material (dark), plus the closed pores (medium).

The interconnecting pore channels of Fig. 13–1.7 are called *open pores*. An agglomerated material may also have internal or *closed pores*, which do not contribute to the permeability of the material.

The choice of method for calculating the density and *specific gravity** of a mixture depends on whether the pores are open or closed, as indicated by the following relationships:

$$\textit{True density} = \frac{\text{Mass}}{\text{True volume}} = \frac{\text{Mass}}{\text{Bulk volume} - \text{Total pore volume}}. \tag{13–1.2}$$

$$\textit{Apparent density} = \frac{\text{Mass}}{\text{Apparent volume}} = \frac{\text{Mass}}{\text{Bulk volume} - \text{Open-pore volume}}. \tag{13–1.3}$$

$$\textit{Bulk density} = \frac{\text{Mass}}{\text{Bulk volume}} = \frac{\text{Mass}}{\text{True volume} + \text{Total pore volume}}. \tag{13–1.4}$$

$$\textit{True porosity} = \frac{\text{Total pore volume}}{\text{Bulk volume}} = 1 - \text{True packing factor}. \tag{13–1.5}$$

$$\textit{Apparent porosity} = \frac{\text{Open-pore volume}}{\text{Bulk volume}} = 1 - \text{Apparent packing factor}. \tag{13–1.6}$$

Relationships among the above values determine the bulk behavior of a material. The absorption characteristics of a sponge, of concrete, or of cellular wall insulation depend on the *apparent porosity*. The *bulk density* of a material determines its construction weight and also markedly affects thermal conductivity. The number of open pores, as well as their size and shape, contributes to permeability and strength characteristics.

Concrete. To state the matter simply, let us say that concrete is gravel with an admixture of sand to fill the pores. The space still remaining in the sand is then filled with a "paste" of cement and water. The cement hydrates (it does not dry) to form a bond within the concrete. Experience has indicated that, in addition to the aggregate and cement "paste," it is advantageous to entrain a few percent (by volume) of small air bubbles into the concrete. This entrained air improves the workability of the concrete during placement; but more important, these small air bubbles increase the resistance of the concrete to deterioration resulting from freezing and thawing. The reasons for this are not fully understood as yet.

Assuming that the aggregate is much stronger than the cement, it is the property of the cement paste which governs the properties of the concrete. Consequently,

* The specific gravity of material m is the ratio, ρ_m/ρ_{H_2O}.

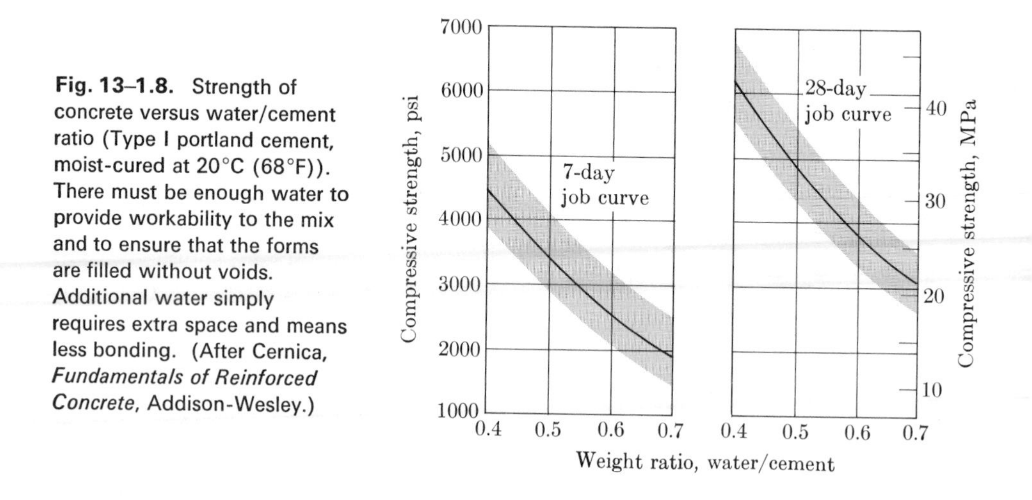

Fig. 13–1.8. Strength of concrete versus water/cement ratio (Type I portland cement, moist-cured at 20°C (68°F)). There must be enough water to provide workability to the mix and to ensure that the forms are filled without voids. Additional water simply requires extra space and means less bonding. (After Cernica, *Fundamentals of Reinforced Concrete*, Addison-Wesley.)

concrete with a *low water-to-cement ratio* is stronger than concretes with higher ratios. With the low ratio, there is more hydrated cement and less excess water in the spaces between the sand and gravel particles. This indirect relationship between the strength of concrete and its water–cement ratio is not fully appreciated even by many engineers. Figure 13–1.8 shows this variation for one lot of portland cement. As indicated schematically in Fig. 1–5.2, the strength also increases with time.

Cements. The preceding examples of concrete referred to a cement paste which serves as a bond for the aggregate. *Portland cement*, which is the principal cement of concrete, is initially a mixture of calcium aluminate and calcium silicate. They undergo hydration to develop the bond:

$$\mathrm{Ca_3Al_2O_6 + 6H_2O \rightarrow Ca_3Al_2(OH)_{12}}, \tag{13–1.7}$$

$$\mathrm{Ca_2SiO_4} + x\mathrm{H_2O} \rightarrow \mathrm{Ca_2SiO_4}.x\mathrm{H_2O}. \qquad (13\text{–}1.8)^*$$

In each reaction the hydrated product is less soluble than the original cement. Therefore, in the presence of water the above reactions are really ones of solution and precipitation.

There is a wide variety of other cements, both organic and inorganic. In general, these may be categorized as (1) hydraulic, (2) polymeric, and (3) reaction. *Hydraulic* cements include those which have hydration reactions similar to those of portland

---

* This reaction may be more correctly stated as:

$$2\mathrm{Ca_2SiO_4} + (5 - y + x)\mathrm{H_2O}$$
$$\rightarrow \mathrm{Ca_2[SiO_2(OH)_2]_2 \cdot (CaO)}_{y-1} \cdot x\mathrm{H_2O} + (3 - y)\mathrm{Ca(OH)_2}, \tag{13–1.11}$$

where x varies with the partial pressure of the water and y is approximately 2.3.

cement. An inorganic example of a *polymeric* cement is silicic acid:

$$x\text{Si(OH)}_4 \rightarrow \left(\begin{array}{c} \text{OH} \\ | \\ \text{—Si—O—} \\ | \\ \text{OH} \end{array} \right)_x + x\text{H}_2\text{O}. \qquad (13\text{–}1.9)$$

Although a linear chain is indicated, considerable cross-linking is possible. *Reaction* cements are typified by the phosphate bonds:

$$\text{Al}_2\text{O}_3 + 2\text{H}_3\text{PO}_4 \rightarrow 2\text{AlPO}_4 + 3\text{H}_2\text{O}. \qquad (13\text{–}1.10)$$

The interaggregate bond is the $AlPO_4$ which may be compared directly with SiO_2, in which Al and P atoms have replaced two Si atoms, i.e., $AlPO_4$ versus $SiSiO_4$.

Sintered products. *Sintering* is a process of heating in order to agglomerate small particles into bulk materials. In order for sintering to occur, a bond must develop (1) through the formation of a liquid phase or (2) by solid diffusion.

Although bonding may be achieved by *liquid-phase sintering*, it is mandatory that the liquid lose its fluidity before the material goes into service. Tools made of sintered carbides are bonded with alloys containing cobalt or nickel. In the sintering process the metal binder melts and forms a continuous matrix between the carbide

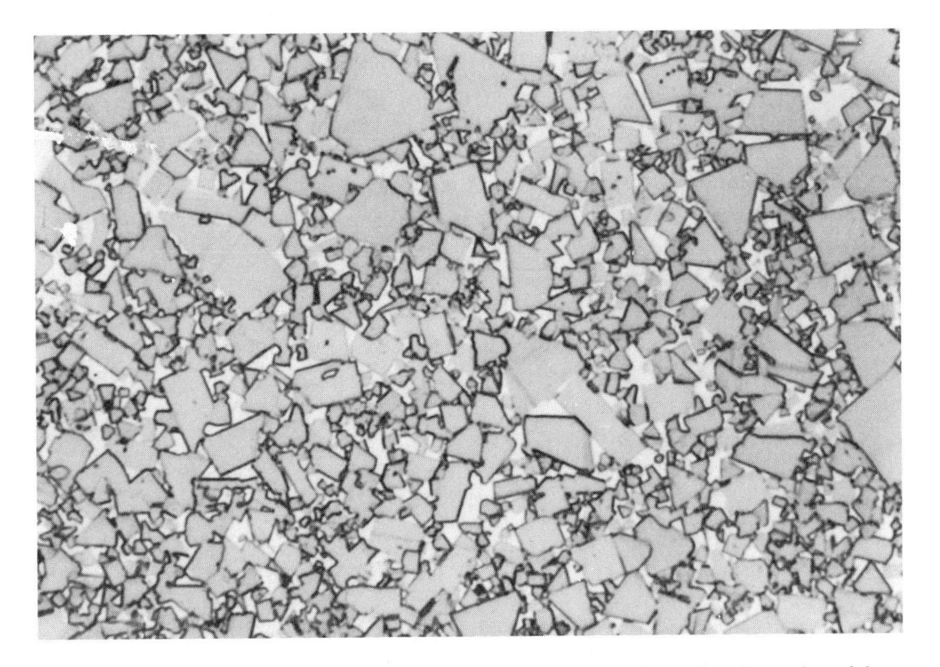

Fig. 13–1.9. Sintered carbides, X1500. Powdered carbide and powdered metal particles were compacted and heated. The metal (cobalt) melted during sintering and formed a continuous matrix between the hard tungsten-carbide particles. (Metallurgical Products Department, General Electric Co.)

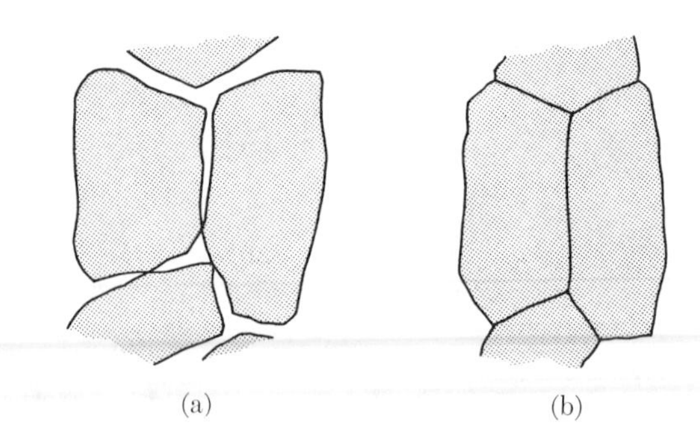

Fig. 13–1.10. Solid sintering. (a) Particles before sintering have two adjacent surfaces. (b) Grains after sintering have one boundary. The driving force for the sintering is the reduction of surface area (and therefore of surface energy).

particles (Fig. 13–1.9), but the metal binder crystallizes after sintering to provide a strong and rigid tool. Likewise, when resins are used to bond a product such as a grinding wheel, the resin must be able to flow onto the particle surface. However, instead of crystallizing, the resin polymerizes and becomes very viscous, producing a strong bond.

Sintering without the formation of a liquid phase requires diffusion and therefore occurs most rapidly at temperatures just below the solidus. Many powdered metal parts and various magnetic and dielectric ceramic materials are produced by *solid sintering*. These ceramic materials cannot be made by melting or by vitreous sintering because the properties are then altered. Also, because no feasible crucible or mold is available for the required pouring temperatures, such refractory metals as tungsten and columbium (niobium) are commonly formed by solid sintering.

The principle involved in solid sintering is illustrated in Fig. 13–1.10. As shown in part (a), there are two surfaces between any two particles. These are high-energy regions because the atoms at the surfaces of the particles have neighbors on only one side.* With sufficient time at high temperatures, the atoms can move by diffusion and the points of actual contact between particles can enlarge so that only one interface forms in place of the two former surfaces (Fig. 13–1.10(b)). Furthermore, the single interface (now a grain boundary) has lower energy than either of the two former surfaces, because the atoms in the grain boundary have close neighbors (although not perfectly aligned).

During solid sintering, atom movements may occur by (1) vaporization from one surface and subsequent condensation onto another surface, (2) a diffusion along the surface of the grains, or (3) a counterdiffusion of vacancies and atoms through the grains themselves. It turns out that the latter mechanism is the most common, because it directly involves many more atoms. This counterdiffusion brings the center of the grains closer together and induces shrinkage during sintering (Fig. 13–1.11).

---

* Although two particles may be close together, the gap between them may still be many atomic distances, so that interatomic attractions are extremely weak.

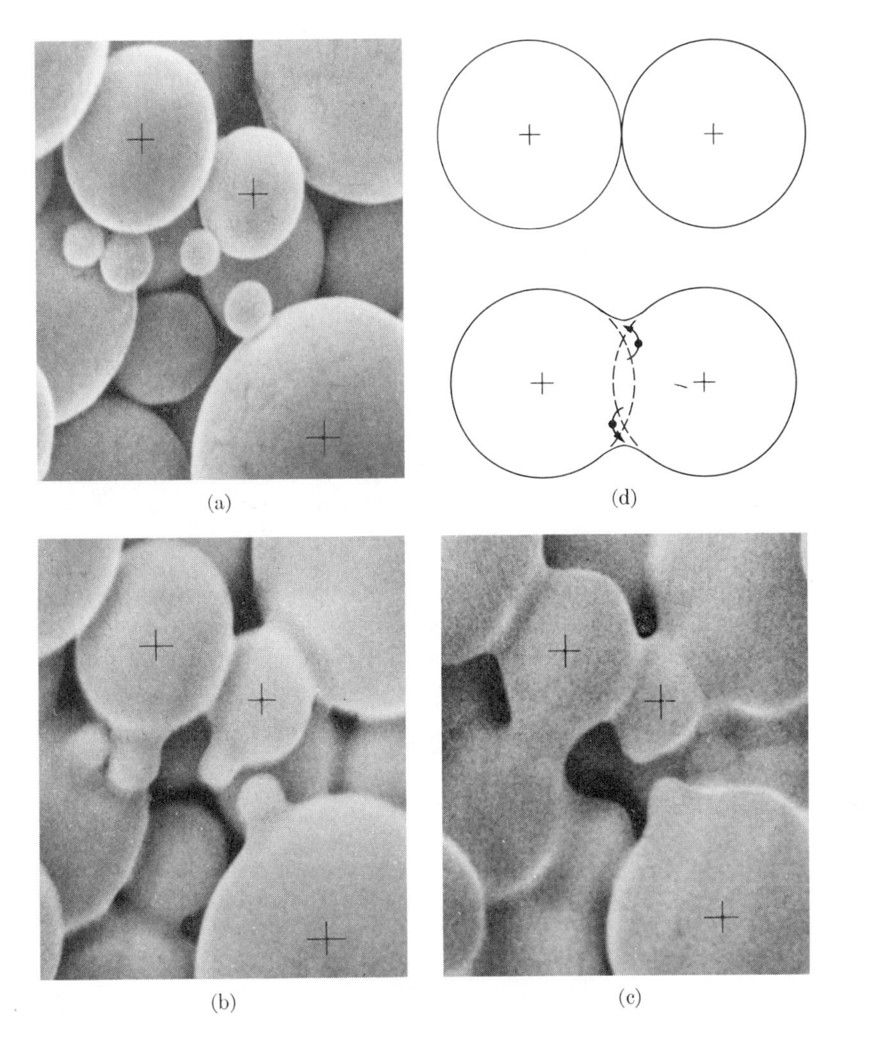

Fig. 13–1.11. Sintering (nickel powder). The initial points of contact in (a) become areas of contact in (b) and (c) while heating to 1100°C. (d) The atoms diffuse from the contact points to enlarge the contact area. (Vacancies diffuse in the opposite direction.) The particles of powder move closer together, and the amount of particle surface is reduced. (Courtesy R. M. Fulrath, Inorganic Materials Research Division of the Lawrence Radiation Laboratory, University of California, Berkeley.)

Example 13–1.1 The true density of a concrete to which an air-entraining agent was added is 2.80 g/cm^3 (174.7 lb/ft^3). However, a dry core (6 in. × 4 in. in diameter, or 15.24 cm × 10.16 cm) of this concrete weighs only 6.56 lbs (2.982 kg). The same core weighs 6.776 lb (3.08 kg) when saturated with water. a) What are the percents of the open and closed pores? b) What are the bulk and apparent densities?

Solution

$$\text{Bulk volume} = \pi(5.08\text{ cm})^2(15.24\text{ cm}) = 1236\text{ cm}^3 \quad (75.4\text{ in.}^3)$$
$$\text{True volume} = (2982\text{ g})/(2.80\text{ g/cm}^3) = 1065\text{ cm}^3 \quad (65.0\text{ in.}^3)$$
$$\text{True porosity} = (1236\text{ cm}^3 - 1065\text{ cm}^3)/1236\text{ cm}^3 = 13.8\text{ v/o.}$$
$$\text{Volume open pores} = (3080\text{ g} - 2982\text{ g})/(1\text{ g/cm}^3) = 98\text{ cm}^3 \quad (5.98\text{ in.}^3).$$

a)
$$\text{Apparent (open) porosity} = 98\text{ cm}^3/1236\text{ cm}^3 = 7.9\text{ v/o};$$
$$13.8\text{ v/o} - 7.9\text{ v/o} = 5.9\text{ v/o closed pores.}$$

b)
$$\text{Bulk density} = 2982\text{ g}/1236\text{ cm}^3 = 2.41\text{ g/cm}^3 \quad (150.4\text{ lb/ft}^3)$$
$$\text{Apparent density} = 2982\text{ g}/(1236\text{ cm}^3 - 98\text{ cm}^3) = 2.62\text{ g/cm}^3 \quad (163.4\text{ lb/ft}^3).$$

Comment. Since the specific gravity is numerically equal to g/cm^3, and the density of water is 62.4 lb/ft^3, the densities may be readily changed to English units if desired. ◀

Example 13–1.2 A concrete unit mix contains 300 lbs (136 kg) of gravel, 210 lbs (95 kg) of sand, 94 lbs (42.5 kg) of portland cement, and 50 lbs (22.5 kg) of water, plus 5 v/o entrained air. How much cement is required to build a low retaining wall, 94 × 3.5 × 1 ft (28.6 m × 1.07 m × 0.305 m)? (The aggregate is nonporous.)

Material	Specific gravity*	
	True	Bulk
Gravel	2.60	1.76
Sand	2.65	1.68
Cement	3.25	1.50

* Ratio of density to that of water (1 g/cm^3, or 62.4 lb/ft^3)

Solution: Basis: 1 unit mix.

$$\text{True volume of gravel} = 136{,}000\text{ g}/(2.60\text{ g/cm}^3) = 0.0523\text{ m}^3$$
$$\text{True volume of sand} = 95{,}000\text{ g}/(2.65\text{ g/cm}^3) = 0.0358$$
(in the space among the gravel)
$$\text{True volume of cement} = 42{,}500\text{ g}/(3.25\text{ g/cm}^3) = 0.0131$$
(in the space among the sand and gravel)
$$\text{True volume of water} = 22{,}500\text{ g}/(1\text{ g/cm}^3) = 0.0225$$
(in the remaining space among all solids)
$$\overline{0.124\text{ m}^3}$$

$$\text{Total m}^3\text{/unit mix} = 0.124\text{ m}^3/0.95 = 0.1305\text{ m}^3$$
$$\text{Unit mixes} = 9.33\text{ m}^3/0.1305\text{ m}^3 = 71.5\text{ unit mixes.}$$

Comment. Using lbs and ft^3,

$$\text{True volume of gravel} = 300\text{ lbs}/(2.60 \times 62.4\text{ lbs/ft}^3) = 1.85\text{ ft}^3;$$
$$V_s = 1.27\text{ ft}^3;$$
$$V_c = 0.46\text{ ft}^3;$$
$$V_{H_2O} = 0.80\text{ ft}^3;$$
$$\sum V = 4.38\text{ ft}^3/0.95 = 4.6\text{ ft}^3;$$
$$\text{Unit mixes} = 329\text{ ft}^3/4.6\text{ ft}^3 = 71.5\text{ unit mixes.}$$ ◀

Example 13–1.3 An electrical porcelain product has only 1 v/o porosity as sold. It had 27 v/o porosity after pressing and drying, but before sintering. How much linear shrinkage occurred during sintering?

Solution: For each 1.00 cm^3 of final bulk product, true volume = 0.99 cm^3.

$$V_0 = 0.99\text{ cm}^3/0.73 = 1.356\text{ cm}^3;$$
$$L_0^3 = 1.356L_f^3;$$
$$L_f = 0.903L_0.$$
$$\text{Shrinkage} = 9.7\text{ l/o}.$$

Comment. Many electrical porcelains also have a drying shrinkage. Therefore it is necessary to design a die with allowances for both the drying shrinkage and the sintering shrinkage. ◀

13–2 SURFACE MODIFICATIONS

A material first encounters its environment at its surface. Therefore we use protective coatings to produce a composite to isolate the surface from chemical attack (Section 12–5). The surface may also be subject to mechanical wear. Here again, we may produce a composite by applying abrasion-resistant coatings of metal by welding, or of oxides by flame-spraying. Other procedures modify the surface (1) by chemical adjustments or (2) by differential heat treatments. We will look at these two alternatives in this section because they let us apply our earlier principles.

Chemical adjustments. The compositions of surfaces can be altered by diffusing desired elements into a material. *Carburization* is a procedure most widely used for this. Since carbon increases the hardness of a steel (Fig. 10–1.6), this provides a basis for increasing the wear resistance of the bearing surface of a crankshaft, or for a variety of similar products. When heated into the austenite range, in the presence of CO (or of methane, CH_4), the iron will dissolve carbon

$$2CO_{(g)} \rightarrow CO_{2(g)} + \underline{C}\ (\text{in }\gamma). \qquad (13\text{–}2.1)$$

The austenite (fcc) can absorb considerable carbon (Fig. 9–7.1) but does not permit rapid diffusion (Table 4–7.1). Therefore, we observe a steep concentration gradient of carbon below the surface (Fig. 13–2.1, left). We can start out with a steel which is SAE 1020 or 4020, and change the surface to 0.90% or more carbon with a very high hardness. Immediately behind this *case*, however, is a *core* which is still low in carbon, and still tough (Fig. 13–2.1, right).

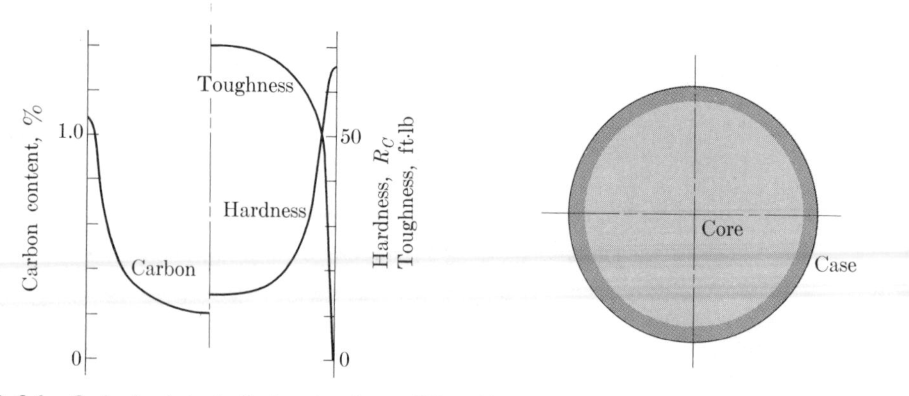

Fig. 13–2.1. Carburized steel. Carbon has been diffused into the surface. A hard case is formed around a tougher core.

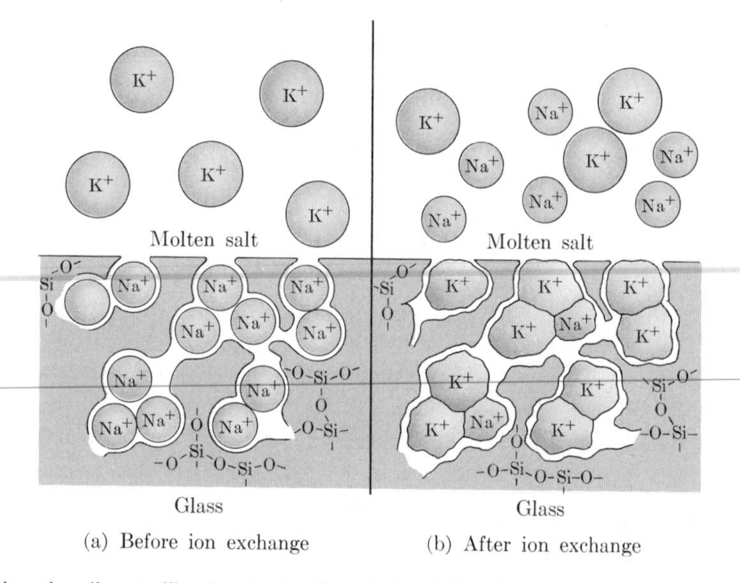

Fig. 13–2.2. Surface alteration (ion stuffing in glass). Glass is heated in a molten K_2SO_4 bath. The K^+ ions of the bath are exchanged with the Na^+ in the glass. Since the K^+ ions are larger, the surface is compressed, thus strengthening the glass against tensile failure. (Courtesy of George McLellan, Corning Glass Co.)

A second example of surface modification is practiced by certain glass companies. Recall, from Section 8–8, that brittle materials such as glass can be strengthened by surface compressive forces (Fig. 8–8.5). Tempered glass is produced by thermal treatments which introduce controlled residual stresses (Fig. 8–8.4). Similar compressive stresses may be introduced by ion exchange. A normal glass contains 12–15 w/o Na_2O (along with CaO and SiO_2). If such a glass is heated in a bath of molten K_2SO_4, there can be an exchange of some of the Na^+ ions in the glass for K^+ ions of the melt:

$$Na^+_{glass} + K^+_{melt} \rightarrow K^+_{glass} + Na^+_{melt}. \qquad \bullet(13\text{–}2.2)$$

However, the K^+ ions are larger than the Na^+ ions ($r_{K^+} = 1.33$ Å, $r_{Na^+} = 0.97$ Å). Although it is impossible for all of the Na^+ ions to be replaced by K^+ ions simply because space is not available, enough K^+ ions move into the glass so that the rigid silicate network of the glass (Section 8–6) is subject to a compressive force as great as 50,000 psi (350 MPa). Here, as with tempered glass, bending or other service loading must remove the induced compressive stresses resulting from *ion stuffing* (Fig.13–2.2) before the surface encounters tension.

Differential heat treatments. It is possible to heat the surface of steel into the austenite γ range without heating the interior to the same temperature. Two techniques are available.

1. The older method is *flame-hardening*, where the surface is heated by oxyacetylene torches. This is still widely used on large products that must be hardened locally.

2. The second is *induction hardening*, which will serve as our example (Fig. 13–2.3). Since high-frequency induction concentrates the induced currents along the surface of the conductor, it is particularly adapted to surface heating. Once the surface is heated into the γ-range, and before a major quantity of heat is conducted into the center, the steel is quenched; thus a hard martensite case is formed *on the surface only.* This process is desired for large products and/or low-hardenability steels, because the center of the metal conveniently serves as a heat sink rather than a source of added, unused joules that must be removed during cooling.

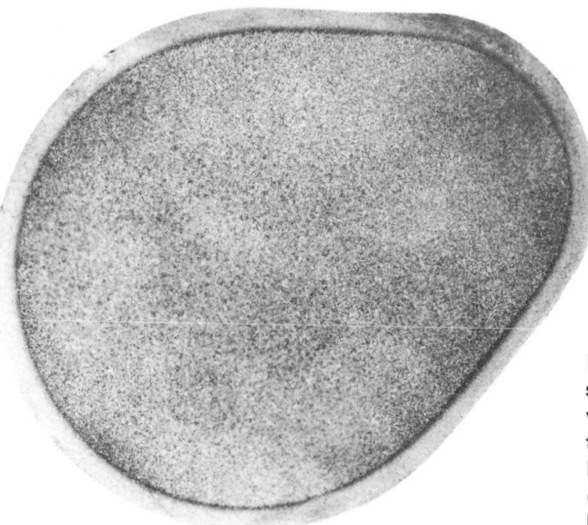

Fig. 13–2.3. Induction-hardened steel. High-frequency currents were induced into the surface of the steel, and there was localized heating at the desired positions. (H. B. Osborn, Jr., TOCCO Division, Park-Ohio Industries.)

Example 13–2.1 A round 4017 steel bar 7.2 cm (2.8 in.) in diameter has been carburized so that the case has 0.7% carbon at the surface, 0.4% carbon at 1.5 mm (0.06 in.) below the surface, and 0.23% carbon at 3 mm (0.12 in.). On the basis of the hardenability data in Fig. 11–6.7, draw a hardness traverse for the steel after oil quenching.

Answer

Location (below surface)	Cooling rate, (from Fig. 11–6.4)	Carbon content	Hardness (from Fig. 11–6.7)
0	23°C/sec	0.7%	58 R_C
1.5 mm (0.06 in.)	~20*	0.4	33*
3 mm (0.12 in.)	~18*	0.23	21*
18 mm (0.7 in.)	10	~0.17	17
36 mm (1.4 in.)	~6	0.17	<17

Comment. The starred values were obtained by interpolation (admittedly nonlinear). ◀

13–3 REINFORCED MATERIALS

We have already considered some reinforced materials—those which are reinforced on a microscopic scale. Bainite, tempered martensite, and precipitation-hardened (age-hardened) alloys are strengthened by a dispersion of fine particles of harder phases. For example, the tensile strength of the (α + $\bar{C}$) in tempered martensite can be more than 200,000 psi (> 1400 MPa), whereas the tensile strength of ferrite, α, alone is less than 20% of those values. The strengthening occurs because the deformable phase cannot undergo strain independent of the rigid phase.

Composites of reinforced materials are well known on a macroscopic scale. Examples include reinforced concrete and fiber-reinforced plastics (FRP). Consider a "glass" fishing rod which consists of a bundle of parallel glass fibers bonded together by a polymeric resin (or conversely, a plastic rod which is longitudinally reinforced by glass fibers). If this composite is placed under tension, the plastic and glass must deform in concert, even though the two components may have markedly different elastic moduli ($E_{gl}/E_{pl} > 10$), or appreciably different individual strengths ($S_{gl}/S_{pl} > 20$).

Let us consider the fiber-reinforced plastic just cited, where it is under tension. If the fibers are continuous, $\epsilon_{gl} = \epsilon_{pl}$, because the two are bonded together. We will find therefore that $\sigma_{gl} \neq \sigma_{pl}$ because the two have different elastic moduli, E:

$$\sigma_1/E_1 = \epsilon_1 = \epsilon_2 = \sigma_2/E_2, \tag{13–3.1}$$

where the subscripts are for the two components. Thus,

$$\sigma_1/\sigma_2 = E_1/E_2. \tag{13–3.2}$$

This was encountered previously in our discussion of Fig. 6–3.4. It leads us to the first rule for reinforcement. The *reinforcing phase must be the higher-modulus phase* so that it can carry added load.

We can write mixture rules not only for density and specific heat (Eq. 10–3.1) and conductivities (Eqs. 10–3.2 and 10–3.3), but also for certain mechanical properties such as Young's modulus. Again give attention to the fiber-reinforced plastics of the past two paragraphs. If f_1 and f_2 represent the volume fraction of the two components, and we use Eq. (13–3.1):

$$\frac{F_1/f_1A}{E_1} = \frac{F/A}{\bar{E}} = \frac{F_2/f_2A}{E_2}. \tag{13–3.3}$$

Also, since the total load F is equal to the sum, $F = F_1 + F_2$;

$$F = f_1E_1F/\bar{E} + f_2E_2F/\bar{E}.$$

$$\bar{E} = f_1E_1 + f_2E_2. \tag{13–3.4}$$

• **Interfacial stresses.** The civil engineer is aware that shear stresses develop at the interface between a reinforcing rod and concrete. For this reason, he specifies "deformed" rods with merloned surfaces (ASTM, A305). Comparable shear stresses are encountered between fiber reinforcement and the surrounding matrix. Here, however, the shear stresses are supported by a chemical rather than a mechanical bond.

Interfacial shear stresses become particularly important if the reinforcement is not continuous. This is illustrated in Fig. 13–3.1. In this figure, $\sigma_{f\,\max}$ represents the stress which is carried by the fiber if there are no end effects (infinite length). This corre-

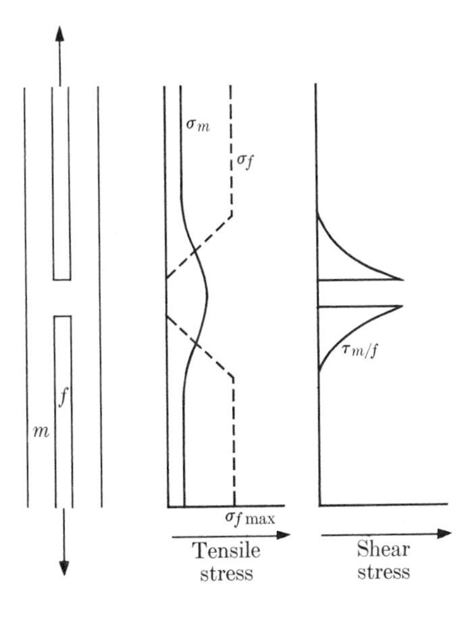

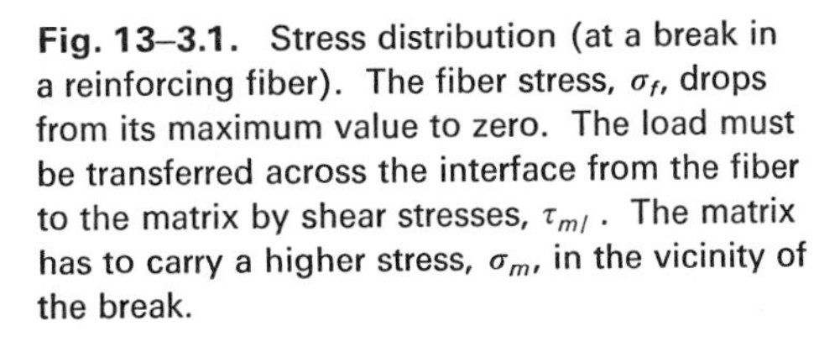

Fig. 13–3.1. Stress distribution (at a break in a reinforcing fiber). The fiber stress, σ_f, drops from its maximum value to zero. The load must be transferred across the interface from the fiber to the matrix by shear stresses, $\tau_{m/f}$. The matrix has to carry a higher stress, σ_m, in the vicinity of the break.

sponds to the calculation in Example 13–3.1, which follows shortly, and depends on the volume fraction of the reinforcement, as well as the Young's moduli of the two materials. If the fiber is broken, however, the fiber stress drops to zero at that point. When this occurs, the load must be transferred to the matrix by shear stresses. Two features stand out in this load transfer.

1. The bonding between the two materials must be sufficiently good to carry the shear stresses.

2. Reinforcement is most effective if it is continuous. Local interruptions transfer the load onto the weaker matrix. Consequently, a deformable matrix has an advantage because the load can be distributed over a larger area for lower maximum stresses. (Of course, there is a limit, because an extremely weak matrix will fail completely.)

Example 13–3.1 A glass-reinforced polyvinylidene chloride rod contains 25 w/o borosilicate glass fibers. All the fibers are aligned longitudinally. What fraction of the load is carried by the glass?

Solution: (Data from Appendix C.)

$$\text{v/o glass} = \frac{(0.25\ \text{g})/(2.4\ \text{g/cm}^3)}{(0.25/2.4)_{\text{gl}} + (0.75)/(1.7)_{\text{pvc}}}$$
$$= 19\ \text{v/o} \quad (= 19\ \text{area percent});$$

$$\text{load}_{\text{gl}}/A_{\text{gl}}\,E_{\text{gl}} = \epsilon_{\text{gl}} = \epsilon_{\text{pvc}} = \text{load}_{\text{pvc}}/A_{\text{pvc}}\,E_{\text{pvc}},$$

$$\frac{\text{load}_{\text{gl}}}{\text{load}_{\text{pvc}}} = \frac{(0.19)(10 \times 10^6)}{(0.81)(0.05 \times 10^6)} \simeq \frac{98\%}{2\%}. \blacktriangleleft$$

•**Example 13–3.2** Formulate a mixture rule for Young's modulus of a laminate, when the stress is applied perpendicular (⊥) to the "grain" of the laminate.

Derivation: With transverse loading, $\sigma_1 = \sigma_\perp = \sigma_2$.

$$\bar{E}_\perp = \sigma_\perp/\epsilon = \sigma/(\epsilon_1 f_1 + \epsilon_2 f_2)$$
$$= 1/(f_1/E_1 + f_2/E_2),$$
$$1/\bar{E}_\perp = f_1/E_1 + f_2/E_2. \tag{13–3.5}$$

Comment. This derivation and the one for Eq. (13–3.4) assume that the Poisson ratios for the two components are comparable. Thus there would be no secondary stresses because of differences in lateral strains. ◀

Example 13–3.3 A 0.10-in. (2.5-mm) iron sheet which is to be used in a household oven is coated on *both* sides with a glassy enamel. The final processing occurs above the 500°C (930°F) strain-point (Section 8–8), to give a 0.020-in. (0.5-mm) coating. The glass has a Young's modulus of 10^7 psi (69,000 MPa) and a thermal expansion of 8.0×10^{-6}/°C.

a) What are the stresses in the glass at 20°C?
b) At 200°C? (Assume no plastic strain.)

Solution: Since $\Delta l/l$ = thermal expansion + elastic strain, and in this case $(\Delta l/l)_{gl} = (\Delta l/l)_{Fe}$, we may write

$$\alpha_{gl}\,\Delta T + \sigma_{gl}/Y_{gl} = \alpha_{Fe}\,\Delta T + \sigma_{Fe}/Y_{Fe}.$$

a) By using data from above and from Appendix C,

$$\sigma_{Fe}/30 \times 10^6 \text{ psi} - \sigma_{gl}/10 \times 10^6 \text{ psi} = (8.00 - 11.75)(10^{-6}/°\text{C})(-480°\text{C}).$$

But $A_{Fe} = 2.5A_{gl}$ and $F_{Fe} = -F_{gl}$, so $\sigma_{gl} = -2.5\sigma_{Fe}$. Thus

$$\sigma_{Fe}\left[\frac{1}{30} + \frac{2.5}{10}\right] = (3.75)(+480).$$

Solving, we obtain

$$\sigma_{Fe} = +6350 \text{ psi (44 MPa)} \qquad (+: \text{tension});$$
$$\sigma_{gl} = -15{,}875 \text{ psi (110 MPa)} \qquad (-: \text{compression}).$$

b) By similar calculations for $\Delta T = (500 - 200)°\text{C}$,

$$\sigma_{gl} = -10{,}000 \text{ psi (69 MPa)} \quad \text{and} \quad \sigma_{Fe} = +4000 \text{ psi (27.5 MPa)}.$$

Comment. We assumed unidirectional strain. In reality, plane (i.e., two-dimensional) strain occurs. The necessary correction gives a higher stress by the factor of $(1 - \nu)^{-1}$, where ν is Poisson's ratio (Section 6–3). ◀

13–4 WOOD

This common product completes our study of materials. Although complex, it is sufficiently familiar so that we can appreciate its structure and properties. Its macrostructure is evident in its grain; its microstructure is of biological origin; and its atomic coordinations are molecular.

Wood, of course, is a very important engineering material. It has a high strength-to-weight ratio. It is easily processed, even on the job. And finally, it is a replaceable resource in this day of diminishing raw materials. We are, of course, aware of its directional properties, which must be taken into consideration whenever it is used. To better understand wood, let's examine its structure.

Wood is a natural polymeric composite. The principal polymeric molecules are those of *cellulose*,

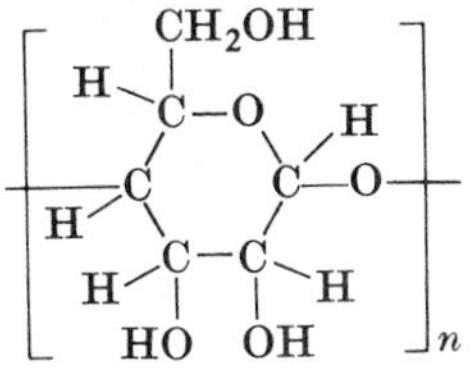

```
 ┌    CH2OH          ┐
 │ H   |             │
 │   \C—O    H       │
 │   /    \ /        │
─┼──C      C—O───────┼─
 │H/ \    /          │
 │ H\C—C/ H          │
 │   |  |            │
 └  HO  OH           ┘n
```

(13–4.1)

in which the molecular weights range up to 2,000,000 g/mole (more than 12,000 mers).

Since cellulose has many —OH radicals, it develops a fair degree of crystallinity. In addition to its more than 50% cellulose composition, wood contains 10% to 35% *lignin*, a more complex carbohydrate polymer.

Larger than polymeric molecules, wood's next-most-prevalent structural units are biological cells, of which the most extensive are the *tracheids*. These are hollow, spindle-shaped cells which are elongated in the longitudinal direction of the wood. The most visible structural unit is the *grain* of the wood, which is made up of *spring* and *summer* layers. The biological cells of the spring wood are larger and have thinner walls than those of the summer wood (Fig. 13–4.1). In this respect, biological cells are much more variable in structure than the unit cells of crystals.

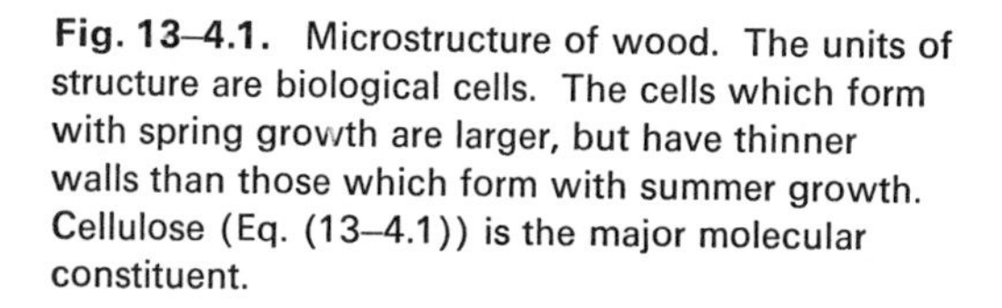

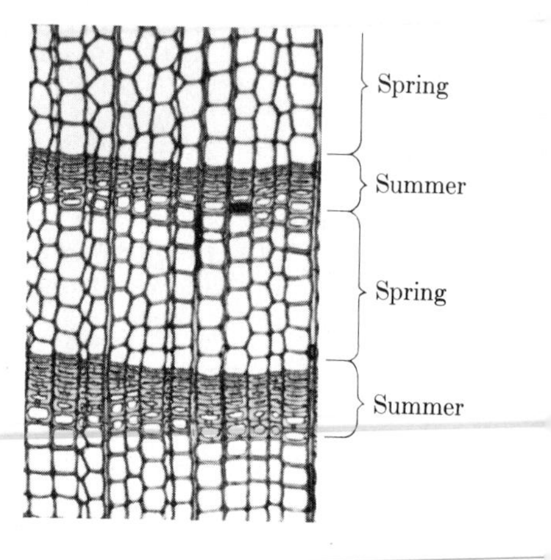

Fig. 13–4.1. Microstructure of wood. The units of structure are biological cells. The cells which form with spring growth are larger, but have thinner walls than those which form with summer growth. Cellulose (Eq. (13–4.1)) is the major molecular constituent.

The above description of wood is, of course, oversimplified.* However, it does indicate the source of the anisotropies in properties that are so characteristic of wood.

In materials which have simple structures, *density* is a structure-insensitive property. This is not the case in a complex material such as wood. First, the amount of spring wood and summer wood varies from species to species. Second, the ratio of cellulose to the more dense lignin varies. As a result of these two factors, the density can range from about 0.15 g/cm^3 for *balsa* to 1.3 g/cm^3† for the dense *lignum vitae*, so named because of its abnormally high lignin content (and low tracheid content). In addition to the above factors, wood is hydroscopic; hence it absorbs moisture as a function of humidity of the surrounding atmosphere. Because this is so, most woods, when they are wet, exhibit a net increase in density.

---

* It does not take into account the wide variety of minor wood chemicals, nor the *vascular rays*, which are rows of single, almost equidimensional cells that radiate from the center of the tree. These become important during deformation.

† Recall that water has a density of 1.0 g/cm^3.

Anisotropy. It should come as no surprise that *dimensional changes* which accompany variations in temperature, moisture, and mechanical loading in wood are anisotropic. Wood technologists point out that *thermal expansion* is about 40% greater in the tangential, t, direction (Fig. 13–4.2) than in the radial, r, direction, and 6 to 12 times greater in the radial direction than in the longitudinal, l, direction. Thermal expansion in the longitudinal direction is relatively independent of density, while in the other two directions it depends on density, ρ (in g/cm^3):

$$\alpha_t \simeq 70\rho \times 10^{-6}/°\text{C},$$

$$\alpha_r \simeq 50\rho \times 10^{-6}/°\text{C},$$

$$\alpha_l \simeq 3 \times 10^{-6}/°\text{C}.$$

As an order of magnitude, longitudinal values for *Young's modulus* are between 1,000,000 and 2,000,000 psi (7,000–14,000 MPa) when measured in tension. Tangential values are normally between 60,000 and 100,000 psi (400–700 MPa), while the radial values are commonly in the 75,000- to 150,000-psi (500–1,000 MPa) range. We would expect the longitudinal value to be the highest. However, we could also have predicted that the tangential modulus would be higher than the radial. But in Fig. 13–4.1 we did not take into account the vascular rays cited in the accompanying footnote. This feature provides an additional rigidity in the radial direction.

Shrinkage is also anisotropic; longitudinal changes are negligible, but tangential shrinkage is very high (~0.25 l/o per 1 w/o moisture for Douglas fir). Radial shrinkage is intermediate (~0.15 l/o per 1 w/o moisture for the same wood) because of the restraining effects of the vascular rays. Effects of shrinkage are summarized in Fig. 13–4.3. The consequences of dimensional distortion on *warpage* are readily apparent.

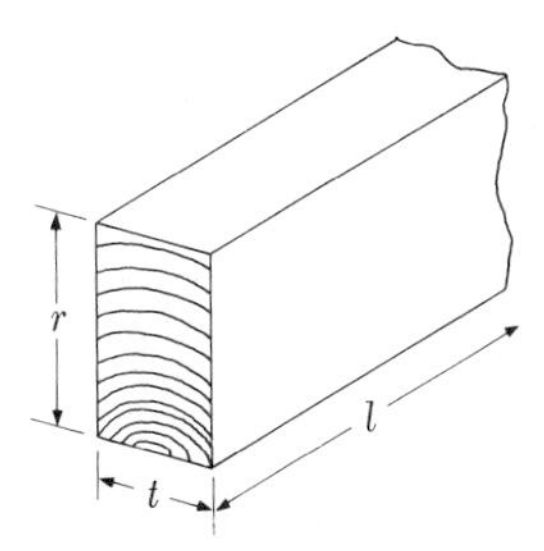

Fig. 13–4.2. Macrostructure of wood. Properties are anisotropic because the growth produces layers of cells with different properties. Strength, elastic moduli, and expansion coefficients all vary with direction.

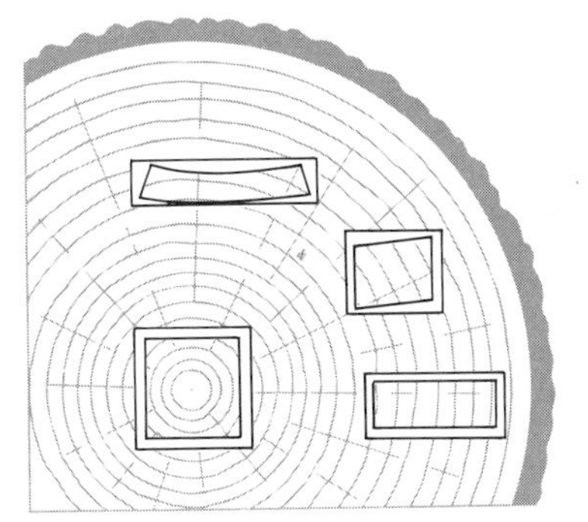

Fig. 13–4.3. Shrinkage of wood. Since the structure is anisotropic, shrinkage varies with orientation. This may cause warpage if the wood is cut before it has dried completely. (Shrinkage is exaggerated in this sketch.)

The *longitudinal tensile strength*, S_l, is upwards of 20 times the radial tensile strength, S_r, because any fracture that takes place must occur across the elongated tracheid cells. Increased densities (for a given moisture content) reflect an increase in the thickness of cell walls, and therefore a proportional increase in the longitudinal strength. The transverse strength increases with the square of the density (again for a given moisture content), because the denser the wood, the less the opportunity for failure parallel to the hollow tracheid cells.

Except in cases in which designs capitalize on the longitudinal strength, the anisotropies of wood are undesirable. Therefore much has been done to modify its structure. Examples include (1) *plywood*, in which the longitudinal strength is developed in *two* coordinates, and (2) *impregnated wood*, in which the pores are filled with a polymer such as phenol-formaldehyde to provide better bonding across the grain structure. In effect, impregnated wood is a plastic which is reinforced with fibers (i.e., elongated tracheid cells). (3) Finally, the *cellulose* may be extracted from the wood to serve as a raw material for many polymeric products (Chapter 7).

REVIEW AND STUDY

SUMMARY

In this final chapter we moved one step further up the size scale. The macrostructures we commonly encounter in composites are in the size range which we can view with the unaided eye. They include concrete, case-hardened (carburized) steel, fiber-reinforced plastics, wood, and similar materials. In man-made composites, the materials commonly have multiple origins. The sections of the chapter categorized these as (1) agglomerated materials, (2) surface-modified materials, and (3) reinforced materials.

The main structural parameters of agglomerated materials are their bulk properties of density, and porosity (and permeability, which we did not consider). The laws of diffusion, both atomic and thermal, are the controlling factors in processing surface-altered materials. Reinforced materials are designed to combine the favorable characteristics of two or more materials.

CONCLUSION

Wood is anisotropic and more complex than other materials studied to date. However, as we examine the wood structure, we can account for the various properties and behaviors. The basis is the macrostructure (grain), the microstructure (biological cells), and the molecular structure. This procedure of structural analyses and coordination with properties is the method of *materials science and engineering*.

We could carry our analysis to still grosser structures such as radial-ply tires for automobiles, prestressed-concrete bridge beams, integrated circuits, nuclear-fuel elements, or gas-turbine engines. However, these items, unlike the preceding materials, have components sufficiently coarse so that the engineer and scientist may analyze the contributions of each component separately as it contributes to the service behavior of the total product. Therefore, we will end our considerations of the structures of engineering materials here, and leave the grosser properties to the design engineer, within the various technical disciplines.

TERMS FOR REVIEW

Agglomerated materials
Carburization
Cellulose
Cement
  hydraulic
  portland
Composite
Concrete
Density
  apparent
  bulk
  true
Fiber reinforced plastics, FRP
Grading
Grain (wood)
Induction hardening
Ion stuffing
Lignin
Mixed packing
Pores
  closed
  open
Porosity
  apparent
  true
Reinforcement
Sintering
  liquid-phase
  solid
Specific gravity
Tracheids
Warpage
Wood
  spring
  summer

FOR DISCUSSION

A_{13} Assume spheres. How do particles on successive Tyler screens compare in their maximum cross-sectional areas? Volumes?

B_{13} Distinguish between apparent, bulk, and true densities; between true and apparent porosities. Why is there no bulk porosity listed?

C_{13} If other factors such as shape are equivalent, a fine-mesh powder will have the same porosity as a coarse-mesh powder. An exception to this generalization occurs if the particle size begins to approach the size of the container. Explain.

D_{13} Maximum density is obtained if the bulk volume of fine-mesh particles equals the porosity of the coarse-mesh particles. Explain.

E_{13} Water is required to hydrate portland cement, and therefore give it strength. However, excess water weakens concrete. Explain.

F_{13} People consider that concrete dries as it hardens. If so, how does a bridge pier harden below tide level?

G_{13} An "oilless" bearing is generally sintered brass or bronze with 20–25% porosity. Should this be open porosity or closed porosity?

H_{13} "Cemented" carbide tool tips are compacted with high pressures, but do not gain full density until the bonding metal melts (Fig. 13–1.9). How does the melting promote shrinkage?

I_{13} Explain how vacancy movements account for shrinkage of the mass of nickel powder in Fig. 13–1.11.

J_{13} Carburization is performed at or above 900°C (1650°F) so that diffusion rates are reasonably rapid. However, the austenite grains grow more than is permissible. What procedure can be used to refine the γ grains?

•K_{13} Should the ion-exchange temperature for Eq. (13–2.2) be above or below the annealing point of the glass (Section 8–8)?

L_{13} A 5-cm (2-inch) round bar of AISI–SAE 4140 steel is induction hardened. Why can we *not* use the cooling-rate data of Fig. 11–6.4 for quenching calculations?

M_{13} Suggest means of using fiber-reinforced plastics for an underground fuel tank. How would you design the openings for pipe connections?

N_{13} Glass fibers are never used without some coating. Why?

•O_{13} Refer to Example 13–3.2. What mixture rule would apply to the elastic modulus, $E_{\parallel}$, parallel to the laminate planes?

•P_{13} Aluminum foil tears if it is strained more than a few percent in tension. However, if it is sandwiched between two polyethylene films it can be plastically deformed more than 150%, *provided* the two materials have a tight interface bond. Discuss.

Q_{13} Account for the warpage concavity of the upper cut of wood in Fig. 13–4.3.

R_{13} The low latitudes do not have spring and summer. Account for the grain in their woods. (Cf. Fig. 13–4.1).

S_{13} Assume that wood grain is a "laminate" of spring wood and summer wood. Using the mixture rules of Section 13–3, account for the higher longitudinal modulus as compared to the elastic modulus in the tangential direction.

T_{13} Why is the longitudinal shrinkage of wood negligible compared to the shrinkage in the other directions?

U_{13} Soft woods, such as pine, are cured before milling. Why?

STUDY PROBLEMS

13–1.1 A piece of ceramic wall tile 0.5 cm × 20 cm × 20 cm absorbs 2.5 g of water. What is the porosity of the tile?

Answer: 1.25 v/o

13–1.2 A brick weighs 3.3 kg when dry, 3.45 kg when saturated with water, and 1.9 kg when suspended in water. (Pounds may be used in place of kg if desired.) a) What is the porosity? b) Its bulk density? c) Its apparent density?

13–1.3 A polystyrene foam having a bulk density of 4 lb/ft^3 (0.064 g/cm^3) was made of a polymer whose true specific gravity is 1.05. The foam absorbs exactly its own weight of water. Calculate: a) the percent expansion of the polymer during foaming. b) The total porosity of the foam. c) The percent of open pore space. d) The density of the foam when saturated with water.

Answer: a) 1530 v/o b) 94 v/o c) 6.4 v/o d) 8 lb/ft^3 (0.128 g/cm^3)

13–1.4 A certain type of rubber has a true density of 1.40 g/cm^3. This rubber is used to manufacture a foam rubber with a bulk density of 0.42 g/cm^3 when dry and 0.69 g/cm^3 when saturated with water. a) What is the apparent porosity of the foam rubber? b) What is the true porosity?

13–1.5 An insulating brick weighs 3.9 lb (1.77 kg) dry, 4.95 lb (2.25 kg) when saturated with water, and 2.3 lb (1.05 kg) when suspended in water. a) What is the porosity? b) Its bulk density? c) Its apparent density?

Answer: a) 40 v/o b) 1.47 g/cm^3 (or 92 lb/ft^3) c) 2.45 g/cm^3 (or 152 lb/ft^3)

13–1.6 The bulk density of crushed limestone is 114 lb/ft^3 (1827 kg/m^3). What is its porosity if its true specific gravity is 2.7?

13–1.7 For many years, the standard unit of portland cement measure was a 94-lb sack, or 1 ft^3 by volume. The true specific gravity is 3.2, i.e., 3.2 times as dense as water (62.4 lb/ft^3). Calculate the porosity.

Answer: 53 v/o

13–1.8 A small construction truck holds 7700 lb (3500 kg) of gravel when level full. (Capacity = 2 m^3 or 70.6 ft^3.) The apparent density of this gravel is measured by placing 5 kg (11 lb) of gravel into a bucket with a volume of 0.44 ft^3 (0.0125 m^3). This partially filled bucket is then filled with water. The water plus the gravel has a net weight of 15.6 kg (34.3 lb). a) What is the apparent density of the gravel? b) How much volume of sand should be obtained to add to this truckload of gravel to produce the greatest packing factor?

13–1.9 A mix of concrete comprises the following bulk volumes: cement, 1; sand, 2.25; gravel, 2.85; water, 0.8. Using the data of Example 13–1.2, calculate the total volume of his mix.

Answer: 4.6 (ft^3, cm^3 or m^3 may be used)

13–1.10 A unit mix of concrete comprises 94 lb (42.7 kg) of cement, 275 lb (125 kg) of sand, 350 lb (160 kg) of gravel, and 46 lb (21 kg) of water. Using the data of Example 13–1.2, calculate the number of unit mixes required for a driveway of 700 ft^3 (20 m^3).

13–1.11 A powdered metal part has a porosity of 23% after compacting and before sintering. What linear-shrinkage allowance should be made if the final porosity is expected to be 2%?

Answer: 8% (sintered basis)

13–1.12 A magnetic ferrite for an oscilloscope component is to have a final dimension of 0.621 in. (1.58 cm). Its volume shrinkage during sintering is 33.1% (unsintered basis). What initial dimension should the powdered compact have?

13–2.1 An SAE 1020 (G.S. #8) steel rod (2 in., or 5 cm, in diameter) is carburized to 0.62% C at the surface, 0.35% just below the surface (2 mm), and unaltered beyond a depth of 4 mm. Sketch a hardness profile for this steel quenched (a) in water; (b) in oil.

Answer: a) S, $60R_C$; 2 mm, $\sim 40R_C$; MR, $22R_C$; C, $21R_C$ b) $\sim 28R_C$; $\sim 25R_C$; $20R_C$; $17R_C$

13–2.2 On the basis of Figs. 11–6.3 and 6.4, show a plot of the effect of carbon on surface hardness of (a) 1-in. (2.5-cm), and (b) 3-in. (7.6-cm) diameter round bars quenched in oil.

13–2.3 An AISI–SAE 1020 steel rod, 1 in. in diameter (2.5 cm), is to be carburized. Specifications after water-quenching call for a minimum of $60R_C$ at the surface, $50R_C$ at 0.12 in. (3 mm) below the surface, and a maximum of $28R_C$ at mid-radius. What are the required carbon contents?

Answer: S, >0.5%; 0.12 in., >0.35%; MR, 0.2%

•**13–2.4** A chemically strengthened glass has about 15 v/o Na_2O initially. Estimate the volume percent (Na_2O + K_2O) if half of the Na^+ ions are replaced with K^+ ions. (Assume freedom for expansion, and CN = 6.)

•**13–2.5** A glass rod (1.83 mm, or 0.072 in. in diameter) supports a load of 75 lb_f (335 N) just prior to fracture. Another rod (1.70 mm, or 0.067 in. in diameter) is made of the same glass, but with ion exchange as described in Eq. (13–2.2). It supports a load of 1100 N (247 lb_f) when loaded in the same manner. Assume that fracture starts in each case with a similar crack from the surface. What compressive stress was present in the surface of the treated glass?

Answer: 50,000 psi (350 MPa)

13–3.1 What is the longitudinal elastic modulus for a rod of borosilicate glass-reinforced polystyrene in which all of the glass fibers are oriented lengthwise? There is 80 w/o glass.

Answer: 6,500,000 psi (45,000 MPa)

13–3.2 What are the elastic moduli in the two dissimilar directions of the laminate of Study Problem 10–3.5?

13–3.3 A glass-reinforced plastic rod (fishing pole) is made of 67 v/o borosilicate glass fibers in a polystyrene matrix. What is the thermal expansion coefficient?

Answer: $\sim 4 \times 10^{-6}$/°C

13–3.4 An AISI–SAE 1040 steel wire (cross-section 0.01 cm^2) has an aluminum coating, so that the total cross-sectional area is 0.012 cm^2. a) What fraction of a 450 N load (100-lb_f) will be carried by the steel? b) What is the electrical resistance of this wire per unit length?

13–3.5 An 0.10-in. (2.5-mm) steel wire coated with 0.02 in. of copper (total dia. = 0.14 in., or 3.5 mm) is loaded with 1000 lb_f (4450 N). a) What is the elastic strain? b) How much strain would occur if the total diameter were composed of 1040 steel? c) If the total wire were composed of copper?

Answer: a) 0.003 b) 0.002 c) 0.004

13–3.6 The copper-coated steel (1020) wire of the previous problem is stress-relieved at 400°C (750°F) and is cooled rapidly to 50°F (10°C). a) Which metal is in tension? b) What is the stress?

13–3.7 An 0.1-in. (2.5-mm) steel wire ($\rho = 20\ \mu\text{ohm} \cdot \text{cm}$) is to be copper-coated so its resistance is 0.001 ohm/ft (33×10^{-6} ohm/cm). How thick should the coating be?

Answer: 0.5 mm (0.02 in.)

13–4.1 Birchwood veneer is impregnated with phenol-formaldehyde (Fig. 7–4.1(b)) to ensure resistance to water and to increase the hardness of the final product. Although dry birch weighs only 0.56 g/cm^3, the true specific gravity of the cellulose–lignin combination is 1.52. a) How many grams of phenol-formaldehyde (PF) are required to impregnate 1 in.^{3} (16.4 cm^3) of dry birchwood? b) What is the final density?

Answer: a) 13.5 g PF b) 1.38 g/cm^3

13–4.2 Assume a cellulose fiber of cotton has a diameter of 0.05 mm, a length of 8 cm, a density of 1.3 g/cm^3, and a molecular weight of 2,000,000 amu, that is, 2,000,000 g/(0.6×10^{24} molecules). How many molecules does the fiber contain?

13–4.3 Dried balsa weighs 10 lb/ft^3 (or 0.16 g/cm^3). It contains primarily cellulose ($\rho \simeq$ 1.3 g/cm^3). What is its true porosity?

Answer: 88 v/o

Appendix A CONSTANTS AND CONVERSIONS

Constants*

Acceleration of gravity, g	$9.80\ldots$ m/s^2
Angstrom, Å	10^{-10} m
Atomic mass unit, amu	$1.66\ldots \times 10^{-24}$ g
Avogadro's number, N	$0.6022\ldots \times 10^{24}$ mole^{-1}
Boltzmann's constant, k	$86.1\ldots \times 10^{-6}$ eV/°K
	$13.8\ldots \times 10^{-24}$ J/°K
Capacitivity (vacuum), ϵ	$8.85\ldots \times 10^{-12}$ C/V·m
Electron charge, q	$0.1602\ldots \times 10^{-18}$ C
Electron moment, β	$9.27\ldots \times 10^{-24}$ A·m^2
Electron volt, eV	$0.160\ldots \times 10^{-18}$ J
Faraday, $\mathscr{F}$	$96.5\ldots \times 10^3$ C
Fe–C eutectoid composition	0.77 w/o carbon
Fe–C eutectoid temperature	727°C (1340°F)
Gas constant, R	$8.31\ldots$ J/mole·°K
	$1.98\ldots$ cal/mole·°K
Gas volume (STP)	$22.4\ldots \times 10^{-3}$ m^3/mole
Planck's constant, h	$0.662\ldots \times 10^{-33}$ J·s
Velocity of light, c	$0.299\ldots \times 10^9$ m/s

Conversions*

1 ampere	= 1 C/s
1 angstrom	= 10^{-8} cm
	= $3.937\ldots \times 10^{-9}$ in.
	= 10^{-10} m
1 amu	= $1.66\ldots \times 10^{-24}$ g
1 Btu	= $1.055\ldots \times 10^3$ J
1 Btu/°F	= $1.899\ldots \times 10^3$ J/°C
1 Btu·in./ft^2·sec·°F	= $0.519\ldots \times 10^3$ J·m/(m^2·s·°C)
1 Btu·ft^2	= $11.3\ldots \times 10^3$ J/m^2
1 calorie, gram	= $4.18\ldots$ J
1 centimeter	= 10^{-2} m
	= $0.3937\ldots$ in.
1 coulomb	= 1 A·s
1 cubic centimeter	= $0.0610\ldots$ in^3.
1 cubic inch	= $16.3\ldots \times 10^{-6}$ m^3
1°C difference	= 1.8 °F
1 electron volt	= $0.160\ldots \times 10^{-18}$ J
1°F difference	= $0.555\ldots$ °C
1 foot	= $0\,3048\ldots$ m
1 foot·pound$_f$	= $1.355\ldots$ J

* Values are truncated to next lower significant figure.

1 gallon (U S. liq.)	$= 3.78\ldots \times 10^{-3}$ m^3
1 gram	$= 0.602\ldots \times 10^{24}$ amu
	$= 2.20\ldots \times 10^{-3}$ lb$_m$
1 gram/cm^3	= 62.4... lb$_m$/ft^3
1 inch	= 0.0254... m
1 joule	$= 0.947\ldots \times 10^{-3}$ Btu
	= 0.239... cal, gram
	$= 6.24\ldots \times 10^{18}$ eV
	= 0.737... ft·lb$_f$
1 joule/meter^2	$= 8.80\ldots \times 10^{-5}$ Btu/ft^2
1 joule·m/(m^2·s·°C)	$= 1.92\ldots \times 10^{-3}$ Btu·in./(ft^2·s·°F)
1 kilogram	= 2.20... lb$_m$
1 meter	= 10^{10} Å
	= 3.28... ft
	= 39.37... in.
1 ohm·meter	= 39.37... Ω·in.
1 newton	= 0.224... lb$_f$
1 ohm·inch	= 0.0254... Ω·m
1 pascal	$= 0.145\ldots \times 10^{-3}$ lb$_f$/in.^{2}
1 pound (force)	= 4.44... newtons
1 pound (mass)	= 0.453... kg
1 pound/foot^3	= 16.0... kg/m^3
1 pound/inch^2	$= 6.89\ldots \times 10^{-3}$ MPa
1 watt	= 1 J/s

* Values are truncated to next lower significant figure.

Appendix B TABLE OF SELECTED ELEMENTS

Element	Symbol	Atomic number	Atomic weight	Orbitals			Melting point, °C	Density (solid), g/cm^3	Crystal structure, 20°C	Approx. atomic radius, Å †	Valence (most common)	Approx. ionic radius Å ‡
					1*s*							
Hydrogen	H	1	1.0078		1		−259.14	—	—	0.46	1+	Very small
Helium	He	2	4.003		2		−272.2	—	—	1.76	Inert	—
					2*s*	2*p*						
Lithium	Li	3	6.94	He +	1		180	0.534	bcc	1.519	1+	0.68
Beryllium	Be	4	9.01	He +	2		1289	1.85	hcp	1.14	2+	0.35
Boron	B	5	10.81	He +	2	1	2103	2.34	—	0.46	3+	~0.25
Carbon	C	6	12.011	He +	2	2	> 3500	2.25	hex	0.71	—	—
Nitrogen	N	7	14.007	He +	2	3	−210	—	—	0.71	3−	—
Oxygen	O	8	15.999	He +	2	4	−218.4	—	—	0.60	2−	1.40
Fluorine	F	9	19.00	He +	2	5	−220	—	—	0.6	1−	1.33
Neon	Ne	10	20.18	He +	2	6	−248.7	—	fcc	1.60	Inert	—
					3*s*	3*p*						
Sodium	Na	11	22.99	Ne +	1		97.8	0.97	bcc	1.857	1+	0.97
Magnesium	Mg	12	24.31	Ne +	2		649	1.74	hcp	1.61	2+	0.66
Aluminum	Al	13	26.98	Ne +	2	1	660.4	2.70	fcc	1.4315	3+	0.51
Silicon	Si	14	28.09	Ne +	2	2	1414	2.33	*	1.176	4+	0.42
Phosphorus	P	15	30.97	Ne +	2	3	44	1.8	—	1.1	5+	~0.35
Sulfur	S	16	32.06	Ne +	2	4	112.8	2.07	—	1.06	2−	1.84
Chlorine	Cl	17	35.45	Ne +	2	5	−101	—	—	0.905	1−	1.81
Argon	Ar	18	39.95	Ne +	2	6	−189.2	—	fcc	1.92	Inert	—

					3d	*4s*	*4p*						
Potassium	K	19	39.10	Ar +		1		63	0.86	bcc	2.312	1+	1.33
Calcium	Ca	20	40.08	Ar +		2		840	1.54	fcc	1.969	2+	0.99
Titanium	Ti	22	47.90	Ar +	2	2		1672	4.51	hcp	1.46	4+	0.68
Chromium	Cr	24	52.00	Ar +	5	1		1863	7.20	bcc	1.249	3+	0.63
Manganese	Mn	25	54.94	Ar +	5	2		1246	7.2	—	1.12	2+	0.80
Iron	Fe	26	55.85	Ar +	6	2		1538	7.86	bcc	1.241	2+	0.74
										fcc	1.269	3+	0.64
Cobalt	Co	27	58.93	Ar +	7	2		1494	8.9	hcp	1.25	2+	0.72
Nickel	Ni	28	58.71	Ar +	8	2		1455	8.90	fcc	1.246	2+	0.69
Copper	Cu	29	63.54	Ar +	10	1		1084.5	8.92	fcc	1.278	1+	0.96
Zinc	Zn	30	65.37	Ar +	10	2		419.6	7.14	hcp	1.39	2+	0.74
Germanium	Ge	32	72.59	Ar +	10	2	2	937	5.35	*	1.224	4+	—
Arsenic	As	33	74.92	Ar +	10	2	3	~809	5.73	—	1.25	3+	—
Krypton	Kr	36	83.80	Ar +	10	2	6	−157	—	fcc	2.01	Inert	—
					4d	*5s*	*5p*						
Silver	Ag	47	107.87	Kr +	10	1		961.9	10.5	fcc	1.444	1+	1.26
Tin	Sn	50	118.69	Kr +	10	2	2	232	7.3	bct	1.509	4+	0.71
Antimony	Sb	51	121.75	Kr +	10	2	3	630.7	6.7	—	1.452	5+	—
Iodine	I	53	126.9	Kr +	10	2	5	114	4.93	ortho	1.35	1−	2.20
Xenon	Xe	54	131.3	Kr +	10	2	6	−112	2.7	fcc	2.21	Inert	—
					4f	*5d*	*6s*						
Cesium	Cs	55	132.9	Xe +			1	28.4	1.9	bcc	2.62	1+	1.67
Tungsten	W	74	183.9	Xe +	14	4	2	3387	19.4	bcc	1.367	4+	0.70
Gold	Au	79	197.0	Xe +	14	10	1	1064.4	19.32	fcc	1.441	1+	1.37
Mercury	Hg	80	200.6	Xe +	14	10	2	−38.86	—	—	1.55	2+	1.10
Lead	Pb	82	207.2	Hg +	$6p^2$			327.5	11.34	fcc	1.750	2+	1.20
Uranium	U	92	238.0	Rn +	$5f^3$	*6d*	$7s^2$	1133	19	—	1.38	4+	0.97

* Diamond cubic

† One-half of closest approach of two atoms in the elemental solid. For noncubic structures, the average interatomic distance is given; e.g., in hcp, the atom is slightly ellipsoidal.

‡ Radii for CN = 6. ($0.97\ R_{CN=8} \approx R_{CN=6} \approx 1.1\ R_{CN=4}$.) Patterned after Ahrens.

Appendix C PROPERTIES OF SELECTED ENGINEERING MATERIALS (20°C)*

Material	Specific gravity	Thermal conductivity, $\frac{\text{joule}\cdot\text{cm}}{°\text{C}\cdot\text{cm}^2\cdot\text{s}}$**	Linear expansion, $°C^{-1}$ †	Electrical resistivity, μohm·cm ‡	Average modulus of elasticity	
					psi	MPa
Metals						
Aluminum (99.9+)	2.7	2.2	22.5×10^{-6}	2.9	10×10^6	70,000
Aluminum alloys	2.7(+)	1.6	22×10^{-6}	3.5(+)	10×10^6	70,000
Brass (70Cu–30Zn)	8.5	1.2	20×10^{-6}	6.2	16×10^6	110,000
Bronze (95Cu–5Sn)	8.8	0.8	18×10^{-6}	9.6	16×10^6	110,000
Cast iron (gray)	7.15	—	10×10^{-6}	—	$20 \times 10^6 \pm$	140,000(±)
Cast iron (white)	7.7	—	9×10^{-6}	66	30×10^6	205,000
Copper (99.9+)	8.9	4.0	17×10^{-6}	1.7	16×10^6	110,000
Iron (99.9+)	7.87	0.72	11.7×10^{-6}	9.7	30×10^6	205,000
Lead (99+)	11.34	0.33	29×10^{-6}	20.65	2×10^6	14,000
Magnesium (99+)	1.74	1.6	25×10^{-6}	4.5	6.5×10^6	45,000
Monel (70Ni–30Cu)	8.8	0.25	15×10^{-6}	48.2	26×10^6	180,000
Silver (sterling)	10.4	4.1	18×10^{-6}	1.8	11×10^6	75,000
Steel (1020)	7.86	0.50	11.7×10^{-6}	16.9	30×10^6	205,000
Steel (1040)	7.85	0.48	11.3×10^{-6}	17.1	30×10^6	205,000
Steel (1080)	7.84	0.46	10.8×10^{-6}	18.0	30×10^6	205,000
Steel (18Cr–8Ni stainless)	7.93	0.15	9×10^{-6}	70	30×10^6	205,000
Ceramics						
Al_2O_3	3.8	0.29	9×10^{-6}	$>10^{14}$	50×10^6	350,000
Brick						
Building	2.3(±)	0.006	9×10^{-6}	—	—	—
Fireclay	2.1	0.008	4.5×10^{-6}	1.4×10^8	—	—

Graphite	1.5	—	5×10^{-6}	—	—	—
Paving	2.5	—	4×10^{-6}	—	—	—
Silica	1.75	0.008	—	1.2×10^{8}	—	—
Concrete	2.4(±)	0.010	13×10^{-6}	—	2×10^{6}	14,000
Glass						
Plate	2.5	0.0075	9×10^{-6}	10^{14}	10×10^{6}	70,000
Borosilicate	2.4	0.010	2.7×10^{-6}	$> 10^{17}$	10×10^{6}	70,000
Silica	2.2	0.012	0.5×10^{-6}	10^{20}	10×10^{6}	70,000
Vycor	2.2	0.012	0.6×10^{-6}	—	—	—
Wool	0.05	0.0025	—	—	—	—
Graphite (bulk)	1.9	—	5×10^{-6}	10^{-3}	1×10^{6}	7,000
MgO	3.6	—	9×10^{-6}	10^{5} (2000°F)	30×10^{6}	205,000
Quartz (SiO_2)	2.65	0.12	—	10^{14}	45×10^{6}	310,000
SiC	3.17	0.12	4.5×10^{-6}	2.5 (2000°F)	—	—
TiC	4.5	0.30	7×10^{-6}	50×10^{-6}	50×10^{6}	350,000
Polymers						
Melamine-formaldehyde	1.5	0.0030	27×10^{-6}	10^{13}	1.3×10^{6}	9,000
Phenol-formaldehyde	1.3	0.0016	72×10^{-6}	10^{12}	0.5×10^{6}	3,500
Urea-formaldehyde	1.5	0.0030	27×10^{-6}	10^{12}	1.5×10^{6}	10,300
Rubbers (synthetic)	1.5	0.0012	—	—	600–11,000	4–75
Rubber (vulcanized)	1.2	0.0012	81×10^{-6}	10^{14}	0.5×10^{6}	3,500
Polyethylene (L.D.)	0.92	0.0034	180×10^{-6}	10^{15}–10^{18}	14,000–50,000	100–350
Polyethylene (H.D.)	0.96	0.0052	120×10^{-6}	10^{14}–10^{18}	50,000–180,000	350–1,250
Polystyrene	1.05	0.0008	63×10^{-6}	10^{18}	0.4×10^{6}	2,800
Polyvinylidene chloride	1.7	0.0012	190×10^{-6}	10^{13}	0.05×10^{6}	350
Polytetrafluoroethylene	2.2	0.0020	100×10^{-6}	10^{16}	50,000–100,000	350–700
Polymethyl methacrylate	1.2	0.0020	90×10^{-6}	10^{16}	0.5×10^{6}	3,500
Nylon	1.15	0.0025	100×10^{-6}	10^{14}	0.4×10^{6}	2,800

* Data in this table were taken from numerous sources.
** Multiply by 0.192 to get Btu·in/°F·ft²·sec. † Divide by 1.8 to get $°F^{-1}$. ‡ Multiply by 0.39 to get μohm·in.

Appendix D GLOSSARY OF TERMS AS APPLIED TO MATERIALS

Abrasive—Hard, mechanically resistant material used for grinding or cutting; commonly made of a ceramic material.
Absorption—Volume assimilation (cf. *ad*sorption).
$A_mB_nX_p$ compounds—Ternary compounds. In this text X is generally oxygen; A and B are commonly metal atoms.
Acceptor levels—Energy levels of *p*-type (electron–hole) carriers.
Acceptor saturation—Filled acceptor levels in *p*-type semiconductors. As a result, additional thermal activation does not increase the number of extrinsic carriers.
Activation energy—Energy barrier which must be met prior to reaction.
Addition polymerization—Polymerization by sequential addition of monomers.
Additive properties—Properties of mixtures which depend on geometry only (and not on phase interactions).
Adsorption—Surface adhesion (cf. *ab*sorption).
Age-hardening—*See* Precipitation hardening.
Agglomerated materials—Small particles bonded together into an integrated mass.
Aggregate—Coarse particles used in concrete; for example, sand and gravel.
AISI–SAE steels—Standardized identification code for plain-carbon and low-alloy steels, which is based on composition. Last two numbers indicate the carbon content. (*See* Table 9–9.1.)
Allotropism—*See* Polymorphism.
Alloy—A metal containing two or more elements.
Alloy retardation—Decrease in the rate of austenite decomposition because of alloying elements.
Alloying elements—Elements added to form an alloy (sometimes referred to as alloys).
Alpha iron—Iron with a body-centered cubic structure which is stable at room temperature.
Amorphous—Noncrystalline and without long-range order.
Anion—Negative ion.
Anisotropic—Having different properties in different directions.
Annealing—Heating and cooling to produce softening.
Annealing point (glass)—Stress-relief heat treatment. The temperature should provide a viscosity of $\sim 10^{13}$ poises.
Anode—The electrode which supplies electrons to an external circuit.
Anodized—Surface-coated with an oxide layer; achieved by making the component an anode in an electrolytic bath.
Arrhenius equation—Thermal activation relationship. (*See* comments with Ex. 4–6.2.)
Asbestos—A fibrous silicate material.
ASTM grain-size number—Standardized grain counts. *See* Eq. (4–4.1).
Atactic—Lack of long-range repetition in a polymer (as contrasted to isotactic).
Atomic mass unit (amu)—One-twelfth of the mass of C^{12}; gram/(0.602... $\times$ 10^{24}).
Atomic number—The number of electrons possessed by an uncharged atom.
Atomic packing factor—Fraction of volume occupied by "spherical" atoms (or ions).
Atomic radius (elements)—Half of interatomic distance.
Atomic weight—Atomic mass expressed in atomic mass units.
Austempering—Process of isothermal transformation to form bainite.
Austenite (γ)—Face-centered cubic iron, or iron alloy based on this structure.

Austenite decomposition—Eutectoid reaction which changes austenite to (α + carbide).
Austenite grain size—*See* ASTM grain size.
Austenization—Heat treatment to dissolve carbon into fcc iron, thereby forming austenite.
Avogadro's number (N)—Number of amu's per gram; hence, the number of molecules per mole.
AX compounds—Binary compounds with a 1:1 ratio of the two elements; commonly ionic.
AX structure (CsCl)—Binary compound with CN = 8.
AX structure (NaCl)—Binary compound with CN = 6.
AX structure (ZnS)—Binary compound with CN = 4.
A_mX_p structure—Binary compound with unequal component ratio. (*See* Section 8–3.)
Axis (crystals)—One of three principal crystal directions.

Bainite—Microstructure of carbide dispersed in ferrite, obtained by low-temperature isothermal transformation.
Base—The center zone of a transistor.
Bifunctional—Molecule with two reaction sites for joining with adjacent molecules.
Body-centered cubic (bcc)—A unit cell with the center position equivalent to corner positions.
Bohr magneton (β)—Magnetic moment of individual electron ($9.27\ldots \times 10^{-24}$ amp · m^2).
Boltzmann's constant (k)—Thermal energy coefficient ($13.8\ldots \times 10^{-24}$ J/°K).
Bond angle—Angle between stereospecific bonds in molecules, or covalent solids.
Bond energy—Energy required to separate two atoms. Generally expressed as energy per mole of 0.6×10^{24} bonds.
Bond length—Interatomic distance of stereospecific bond.
Boundary (microstructures)—Surface between two grains or between two phases.
Bragg's law—Diffraction law for periodic structures (Eq. 3–9.2).
Branching—Bifurcation in addition polymerization.
Brass—An alloy of copper and zinc.
Bravais lattices—The 14 basic crystal lattices.
Brazing—Joining metals at temperatures above 425°C (800°F) but below the melting point of the joined metals.
Bridging oxygens—Oxygen atoms shared by two adjacent silica tetrahedra.
Brinell—A hardness test utilizing an indenter. The hardness is measured by the diameter of the indentation.
Brittle—Opposite of tough; fractures with little energy absorption.
Bronze—An alloy of copper and tin (unless otherwise specified; e.g., an aluminum bronze is an alloy of copper and aluminum).
Burgers vector (**b**)—Displacement vector around a dislocation. It is parallel to a screw dislocation and perpendicular to an edge dislocation.
Butadiene-type compound—Prototype for several rubbers. (*See* Table 7–2.2.)

Calcination—Solid dissociation to a gas and another solid; for example, $CaCO_3 \rightarrow CaO + CO_2$.
Carbide ($\overline{C}$)—Compound of metal and carbon. Unless specifically stated, it refers to iron-base carbides.
Carbon steel—Steel in which carbon is the chief variable alloying element (other alloying elements may be present in nominal amounts).
Carburization—Introduction of carbon through the surface of the steel by diffusion, to change the surface properties.

Case—Subsurface zone (usually of a carburized steel).
Case-hardening—Hardening by forming a case of higher carbon content.
Cast iron—Fe–C alloy, sufficiently rich in carbon to produce a eutectic liquid during solidification. In practice, this generally means more carbon than can be dissolved in austenite ($>2\%$).
Casting—The process of pouring a liquid or suspension into a mold, or the object produced by this process.
Catalyst—A reusable agent for activating a chemical reaction.
Cathode—The electrode which receives electrons from an external circuit.
Cathode reactions—Reduction (electron-consuming) reactions on a galvanic electrode.
Cation—Positive ion.
C-curve—Isothermal transformation curve.
Cell (galvanic)—A combination of two electrodes in an electrolyte.
Cell, composition—Galvanic cell between nonsimilar metal compositions.
Cell, concentration—Galvanic cell arising from nonequal electrolyte concentrations. (The more dilute solution produces the anode.)
Cell, oxidation—Galvanic cell arising from nonequal oxygen potentials. (The oxygen-deficient area becomes the anode.)
Cell, stress—Galvanic cell arising from a plastically deformed anode.
Cell, unit—See unit cell.
Cellulose—Natural polymer of $C_6H_{10}O_5$. (See Eq. 13–4.1.)
Cement—A material (usually ceramic) for bonding solids together.
Cement, hydraulic—Cement which bonds by a reaction with water.
Cement, portland—A hydraulic calcium silicate cement.
Cementite—Iron carbide (Fe_3C).
Center-of-gravity method—Calculation method for determining phases. The overall composition is at the center of gravity of the weighted components.
Ceramics—Materials consisting of compounds of metallic and nonmetallic elements.
Charge carriers—Electrons in the conduction band provide *n*-type (negative) carriers. Electron holes in the valence band provide *p*-type (positive) carriers.
Charge density ($\mathscr{D}$)—Coulombs per unit area.
Charpy—One of two standardized impact tests, utilizing a square notched bar.
Cis (polymers)—A prefix denoting unsaturated positions on the same side of the polymer chain.
Clay—Fine soil particles (<0.1 mm). In ceramics, clays are specifically sheetlike aluminosilicates.
Cleavage—Plane of easy splitting.
Coercive field (electric, $-\mathscr{E}_c$)—Electric field required to remove residual polarization.
Coercive field (magnetic, $-H_c$)—Magnetic field required to remove residual magnetization.
Cold-working—Deformation below the recrystallization temperature.
Collector—The zone of a transistor which receives the charge carriers across the base from the emitter.
Compact—Compressed shapes of powders prior to sintering.
Component (design)—The individual parts of a machine or similar engineering design.
Component (phases)—The basic chemical substances required to create a chemical mixture or solution.
Composite—Material containing two or more distinct materials.

Compound—A phase composed of two or more elements in a given ratio.

Concrete—Agglomerate of aggregate and a hydraulic cement.

Conduction band—Energy band of conduction electrons. Electrons must be in this band to be carriers in semiconductors.

Conduction electron—Electron raised above the energy gap to serve as negative charge carrier.

Conductivity—Transfer of thermal or electrical energy along a potential gradient.

Configuration—Arrangement of mers along a polymer chain. (Rearrangements require bond breaking.)

Conformation—Twisting and/or kinking of a polymer chain. (Changes require bond rotation only.)

Continuous cooling transformation—A thermal reaction during cooling, particularly austenite decomposition.

Cooling rate—Decrease in temperature per second; specifically, the rate of change at the transformation temperature.

Coordination number (CN)—Number of closest ionic or atomic neighbors.

Copolymer—Polymers with more than one type of mer.

Core (heat-treating)—Center of a bar, inside the case.

Corrosion—Deterioration and removal by chemical attack.

Coulombic forces—Forces between charged particles, particularly ions.

Covalent bond—Interatomic bond created when two adjacent atoms share a pair of electrons.

Creep—A slow deformation by stresses below the normal yield strength (commonly occurring at elevated temperatures).

Creep rate—Creep strain per unit of time.

Critical cooling rates (austenite decomposition)—CR_M is the slowest cooling rate, which produces only *martensite*. CR_P is the fast cooling rate, which produces all *pearlite*.

Critical shear stress—Minimum resolved stress to produce shear.

Cross-linking—The tying together of adjacent polymer chains.

Crystal—A physically uniform solid, in three dimensions, with long-range repetitive order.

Crystal direction [*hkl*]—A ray from an arbitrary origin through a selected unit-cell location. The indices are the lattice coefficients of that location.

Crystal lattice—The spatial arrangement of equivalent sites within a crystal.

Crystal plane (*hkl*)—*See* Section 3–8.

Crystal system—Categorization of unit cells by axial and dimensional symmetry (Table 3–2.1).

Crystallinity (polymers)—Volume fraction of a solid which has a crystalline (as contrasted to an amorphous) structure.

Curie point (magnetic)—Transition temperature between ferromagnetism and paramagnetism.

Current (I)—Flow of positive charge (opposite to electron movement).

Current density (i)—Amperage per unit area.

Defect structure—Compounds with noninteger ratios of atoms or ions. These compounds contain either vacancies or interstitials within the structure.

Deformation, elastic—Reversible deformation without permanent atomic (or molecular) displacements.

Deformation, plastic—Permanent deformation arising from the displacement of atoms (or molecules) to new surroundings.

Deformation crystallization—Crystallization occurring as polymers are unkinked into parallel linear orientations.

Degradation—Reduction of polymers to smaller molecules.

Degree of polymerization (DP)—Mers per average molecular weight.

Delocalization—Multiple-atom orbitals.

Density, apparent—Mass divided by apparent volume (material + closed pores).

Density, bulk—Mass divided by total volume.

Density, true—Mass divided by true (pore-free) volume.

Dielectric—An insulator. A material which can be placed between two electrodes without conduction.

Dielectric constant, relative (κ)—Ratio of charge density arising from an electric field (1) with, and (2) without the material present.

Dielectric strength—Electrical breakdown potential of an insulator per unit thickness.

Diffraction (x-ray)—Deviation of an x-ray beam by regularly spaced atoms.

Diffusion—The movement of atoms or molecules in a material.

Diffusion coefficient (D)—Diffusion flux per unit concentration gradient.

Dipole—An electrical couple with positively and negatively charged ends.

Dislocation, edge ($\perp$)—Linear defect at the edge of an extra crystal plane. The Burgers vector is perpendicular to the defect line.

Dislocation, screw (§)—Linear defect with Burgers vector parallel to the defect line.

Dispersed phases—Microstructure of very fine particles within a matrix phase.

Domains—Microstructural areas of coordinated magnetic alignments (or of electrical dipole alignments).

Donor exhaustion—Depletion of donor electrons. Because of it, additional thermal activation does not increase the number of extrinsic carriers.

Donor levels—Energy levels of n-type (electron) carriers.

Drawing—Mechanical forming by tension through a die, e.g., wire drawing and sheet drawing; usually carried out at temperatures below the recrystallization temperature.

Drift velocity ($\bar{v}$)—Net velocity of electrons in an electric field.

Ductile fracture—Fracture accompanied by plastic deformation, and therefore by energy absorption.

Ductility—Permanent deformation before fracture; measured as elongation or reduction in area.

Elasticity—Nonpermanent deformation.

Elastomer—Polymer with a large elastic strain. This strain arises from the unkinking of the polymer chain.

Electric field ($\mathscr{E}$)—Voltage gradient, volts/cm.

Electrical conductivity (σ)—Coefficient between charge flux and electric field. Reciprocal of electrical resistivity.

Electrical resistivity (ρ)—Resistance of a material with unit dimensions. Reciprocal of electrical conductivity.

Electrochemical potential (ϕ)—Voltage developed at an electrode (as compared with a standard reference electrode).

Electrolyte—Conductive ionic solution (liquid or solid).

Electron charge (q)—The charge of 0.16×10^{-18} coul (or 0.16×10^{-18} amp · sec) carried by each electron.

Electron hole (p)—Electron vacancy in the valence band which serves as a positive charge carrier.

Electron–hole pair—A conduction electron in the conduction band and an accompanying electron hole in the valence band, which result when an electron jumps the gap in an intrinsic semiconductor.
Electronegativity—Measure of electron attraction for nonmetallic characteristics.
Electron repulsion—Repelling force of too many electrons in the same vicinity. Counteracts the attractive bonding forces.
Electroplating—Cathodic reduction process, opposite of corrosion. Electrons are supplied by external circuit.
Elongation—Axial strain accompanying fracture. (A gage length must be stated.)
End-quench test—Standardized test, by quenching from one end only, for determining hardenability.
Endurance limit—The maximum stress allowable for unlimited cycling.
Energy band—Permissible energy levels for valence electrons.
Energy distribution—Spectrum of energy levels determined by thermal activation.
Energy gap (E_g)—Unoccupied energies between the valence band and the conduction band.
Energy well—Interatomic potential energy minimum.
Equiaxial—Comparable dimensions in the three principal directions.
Equicohesive temperature—The temperature of equal strength for grains and grain boundaries.
Equilibrium (chemical)—Reaction ceases (because the minimum free energy has been reached).
Equilibrium diagram—*See* Phase diagram.
Equivalent points—Crystal lattice sites with fully identical surroundings. (*See* Fig. 3–2.2.)
Eutectic analysis (composition)—Analysis of liquid-solution phase with the minimum melting temperature (at the intersection of two solubility curves).

Eutectic reaction—$L_2 \underset{\text{Heating}}{\overset{\text{Cooling}}{\rightleftharpoons}} S_1 + S_3.$

Eutectic temperature—Temperature of the eutectic reaction at the intersection of two solubility curves.
Eutectoid analysis (composition)—Analysis of solid-solution phase with the minimum decomposition temperature (at the intersection of two solid solubility curves).
Eutectoid ferrite—*See* Ferrite, eutectoid.

Eutectoid reaction—$S_2 \underset{\text{Heating}}{\overset{\text{Cooling}}{\rightleftharpoons}} S_1 + S_3.$

Eutectoid shift—The change in temperature and the carbon analysis of the eutectoid reaction arising from alloying element additions.
Eutectoid temperature—Temperature of the eutectoid reaction at the intersection of two solid solubility curves.
Expansion coefficient—*See* Thermal expansion.
Extrinsic semiconductors—*See* Semiconductors.

Face-centered cubic (fcc)—A unit cell with face positions equivalent to corner positions.
Family of directions $\langle hkl \rangle$—Crystal directions which are identical except for our arbitrary choice of axes.
Faraday ($\mathscr{F}$)—A unit charge per 0.6×10^{24} electrons (96,500 coul).
Fatigue—Tendency to fracture under cyclic stresses.
Ferrite (ceramics)—Compounds containing trivalent iron; commonly magnetic.

Ferrite (metals, α)—Body-centered cubic iron, or an iron alloy based on this structure.
Ferrite, proeutectoid—Ferrite that separates from austenite above the eutectoid temperature.
Ferrite, eutectoid—Ferrite that forms (along with carbide) during austenite decomposition.
Ferroelectric—Materials with spontaneous dipole alignment.
Ferromagnetic—Materials with spontaneous magnetic alignment.
Fiber-reinforced plastics (FRP)—Composite of glass fibers and plastics.
Fick's first law—Proportionality between diffusion flux and concentration gradient.
Firing—*See* Sintering.
Flame-hardening—Hardening by means of surface heating by flames (followed by quenching)
Fluidity (f)—Coefficient of flowability; reciprocal of viscosity.
Fluorescence—Luminescence which occurs immediately after excitation.
Flux (diffusion, J)—Transport per unit area and time.
Form $\{hkl\}$—Crystal planes which are identical except for our arbitrary choice of axes.
Fracture, brittle—Failure by crack propagation and with the absence of significant ductility.
Fracture, ductile—Failure by crack propagation accompanied by plastic deformation.
Frenkel defect—Atom or ion displacement (combined vacancy and interstitial).
Functionality—Number of available reaction sites for polymerization.
Fusion (heat of)—Thermal energy required for the melting of crystalline solid.

Gage length—Initial dimensions, for determining elongation. (*See* Fig. 1–2.2.)
Galvanic cell—A cell containing two dissimilar metals and an electrolyte.
Galvanic protection—Protection given to a material by making it the cathode to a sacrificial anode.
Galvanization—The process of coating steel with zinc to give galvanic protection.
Gamma iron (γ)—*See* Austenite.
Glass—An amorphous solid below its transition temperature. A glass lacks long-range crystalline order, but normally has short-range order.
Glass transition temperature (T_g)—Transition temperature between a supercooled liquid and its glass solid.
Grading—Size distribution of aggregate.
Grain (metals and ceramics)—Individual crystal of a microstructure.
Grain (wood)—Macrostructure from growth cycle.
Grain boundary—The zone of crystalline mismatch between adjacent grains.
Grain boundary area (S_v)—Area/unit volume; for example, in^2/in^3 or cm^2/cm^3.
Grain growth—Increase in average grain size by atoms diffusing across grain (or phase) boundaries.
Grain size—Statistical grain diameter in a random cross section. (Austenite grain size is reported as the number of former austenite grains within a standardized area. See text, Section 4–4.)

Hard-drawn—Cold-worked to high hardnesses by drawing.
Hardenability—The ability to develop maximum hardness by avoiding the ($\gamma \rightarrow \alpha$ + carbide) reaction.
Hardenability curve—Hardness profile of end-quench test bar.
Hardness—Resistance to penetration.
Hardness traverse—Profile of hardness values.
Heat capacity (c)—Energy per unit temperature, dH/dT.
Hexagonal close-packed (hcp)—The lattice of a hexagonal crystal with equivalent sites at the corners *and* in offset positions at midheight. (See Fig. 3–4.2.)

High-speed steels—Steels with their carbides stabilized against overtempering by alloy additions.
Hot working—Deformation which is performed above the recrystallization temperature, so that annealing occurs concurrently.
Hydration—Chemical reaction consuming water:

$$\text{Solid}_1 + H_2O \rightarrow \text{Solid}_2.$$

Hydrogen bridge—Van der Waals bond in which the hydrogen atom (proton) is attracted to electrons of neighboring atoms.
Hydrogen electrode—Standard reference electrode with the half-cell reaction:

$$H_2 \rightarrow 2H^+ + 2e^-.$$

Hypoeutectoid—*See* Steels, hypoeutectoid.
Hysteresis loop (electric)—Profile of a $\mathscr{P}$–$\mathscr{E}$ cycle.
Hysteresis loop (magnetic)—Profile of a B–H cycle.

Impact strength—*See* Toughness.
Impressed voltage—Dc voltage applied to make a metal cathodic during service.
Induced dipole—Electric dipole produced by external electric field.
Induction hardening—Hardening by high-frequency induced currents for surface heating.
Inhibitor—An additive to the electrolyte which promotes passivation.
Injection molding—Process of molding a material in a closed die. For thermoplasts, the die is appropriately cooled. For thermosets, the die is maintained at the curing temperature for the plastic.
Insulator—Nonconductor of (a) electrical or (b) thermal energy; in either case, the insulator has significant electronic resistivity.
Interactive properties—Properties arising from the correlated behavior of two or more phases.
Intermetallic phase—Compound of two metals.
Internal structure—Arrangements of atoms, molecules, crystals, and grains within a material.
Interplanar spacing—Perpendicular distance between two adjacent crystal planes with the same index.
Interrupted quench—Two-stage quenching of steel which involves heating to form austenite and an initial quench to temperature above the start of martensite formation, followed by a second cooling to room temperature.
Interstice—Unoccupied space between atoms or ions.
Interstitial site (n-fold)—An interstice with n (4, 6, or 8) immediate atomic (or ionic) neighbors.
Intrinsic semiconductor—*See* Semiconductors.
Ion—An atom which possesses a charge because it has had electrons added or removed.
Ion stuffing—Ion exchange which produces compressive forces because the new ions are larger than the original ion sites.
Ion vacancy—Unoccupied ion site within a crystal structure. The charge of the missing ion must be appropriately compensated.
Ionic bond—Atomic bonding by coulombic attraction of unlike ions.
Ionic radius—Semiarbitrary assigned radius to ions. Varies with coordination number. (See Appendix B.)
Ionization—Process of removing (or adding) electrons to neutral atoms.
Iso—Prefix indicating "the same."

Isomer—Molecules with the same composition but different structures.
Isotactic (polymers)—Long-range repetition in a polymer chain (in contrast to atactic).
Isotherm—Line of constant temperature.
Isothermal precipitation—Precipitation from supersaturation at constant temperature.
Isothermal transformation—Transformation with time by holding at a specific temperature.

Jominy bar—*See* End-quench.
Junction (semi-conductor)—Interface between n-type and p-type semiconductors.

Lattice—The space arrangement of equivalent points in a crystal. (*See* Fig. 3–2.2.)
Lattice constants—Dimensions of the unit cell.
Lever rule (inverse)—Calculation method for determining phases. The overall composition is at the fulcrum of the lever.
Lignin—Important constituent of many woods; denser than cellulose.
Linear polymer—*See* Polymers.
Liquidus—The locus of temperatures above which only liquid is stable.
Lone pair—Electrons in a nonconnecting sp^3 orbital.
Long-range order—A repetitive pattern over many atomic distances.
Low-alloy steels—*See* Steels.
Luminescence—Light emitted by the energy released as conduction electrons recombine with electron holes.

Macromolecules—Molecules made up of hundreds to thousands of atoms.
Macrostructure—Structural features observed at very low (or without) magnification.
Magnet, permanent—Magnet with a large $(-BH)$ energy product, so that it maintains domain alignment.
Magnet, soft—Magnet which requires negligible energy for domain randomization.
Magnetic saturation—The maximum magnetization which can occur in a material.
Martensite—Metastable body-centered phase of iron supersaturated with carbon; produced from austenite by a shear transformation during quenching.
Materials balance—Mathematical calculation of "the whole equals the sum of the parts."
Matrix—The enveloping phase in which another phase is embedded.
Mean $(\bar{X})$—The average value (cf. *Median*).
Mean free path—Mean distance traveled by electrons or by elastic waves between deflections or reflections.
Mechanical properties—Characteristics of a material in response to externally applied forces.
Mechanical working—Shaping by the use of forces.
Median $\bar{M}$—The middle value (cf. *Mean*).
Mer—The smallest repetitive unit in a polymer.
Mesh—The screen size for particle measurement.
Metal—Materials consisting primarily of elements which release part of their valence electrons. (See Fig. 2–1.1.) Characterized by conductivities which decrease at elevated temperatures.
Metallic bond—Interatomic bonds in metals characterized by delocalized electrons in energy bands.
Metallic conductor—Material with a conductivity greater than 10^2 ohm·cm.
Metastability—Nonequilibrium condition.
Mho—Unit of conductance; reciprocal of ohm.
Microstructure—Structure of grains and phases. Generally requires magnification for observation.

Miller indices (*hkl*)—Index relating a plane to the reference axes of a crystal.
Mixed particles—Aggregate with a size distribution.
Mixture—Combination of two phases.
Mobility (μ)—The drift velocity $\bar{v}$ of an electric charge per unit electric field, $\mathscr{E}$, (cm/sec)/(volt/cm). Alternatively, the diffusion coefficient of a charge per volt, (cm^2/sec)/volt.
Modulus of elasticity—Stress per unit strain. (*Young's modulus* is the most commonly encountered elastic modulus.)
Modulus, bulk—Hydrostatic pressure per unit volume strain.
Modulus, shear—Shear stress per unit shear strain.
Modulus, Young's—Axial stress per unit normal strain.
Molding (plastics)—Shaping by means of a contoured die.
Mole—Mass equal to the molecular weight of a material; 0.6×10^{24} molecules.
Molecular crystal—Crystals with molecules as basic units (as contrasted to atoms).
Molecular length—End-to-end root-mean-length.
Molecular weight—Mass of one molecule (expressed in amu), or mass of 6×10^{23} molecules (expressed in grams).
Molecular weight (number-average, $\bar{M}_n$)—Average molecular weight based on the number fraction.
Molecular weight (weight-average, $\bar{M}_w$)—Average molecular weight based on the weight fraction.
Molecule—Finite groups of atoms bonded by strong attractive forces. Bonding between molecules is weak.
Molecules, linear—*See* Polymers.
Molecules, network—*See* Polymers.
Molecules, polar—Molecules containing a permanent electrical dipole.
Monel—An alloy of copper and nickel.
Monomer—A molecule with a single mer.
Multiphase microstructure—Microstructure containing two or more correlated phases.

n-type—Semiconductor having negative charge carriers, i.e., electrons.
NaCl-type structure—Fcc arrangement of ions with oppositely charged ions in the 6-fold sites.
Noble—Nonreactive.
Nomenclature (AISI–SAE)—*See* Table 9–9.1.
Nonbridging oxygens—Oxygens attached to one SiO_4 tetrahedron only, thus not tying tetrahedra together.
Nonmetals—Materials consisting primarily of elements which accept or share electrons in their valence shells. (See Fig. 2–1.1) They characteristically form molecules or anions.
Nonstoichiometric compounds—Compounds with noninteger atom (or ion) ratios.
Normal stresses—Stresses perpendicular to the cross section.
Normalizing—Heating steel to 100°F ($\sim$50°C) into the austenite range so that it can be homogenized.
Notch sensitivity—A reduction in properties by the presence of stress concentrations.
Nucleation—The start of the growth of a new phase.

Octahedron—An eight-sided volume.
Orbital—Wave probabilities of atomic and molecular electrons.
Ordered (crystal)—Structure with a long-range repetitive pattern.

Orientation (polymers)—Strain process by which molecules are elongated into one preferred alignment.
Orthorhombic—A crystal with three unequal but perpendicular axes.
Overaging—Continued aging until softening occurs.
Overtempering—Excessive heating in order to cause coalescence of carbides (and therefore softening) of tempered martensite.
Oxidation (general)—The raising of the valence level of an element.
Oxidation cell—*See* Cell, oxidation.

p-type—Semiconductor having positive charge carriers, i.e., electron holes.
p–n junction—Device having an interface between *p*-type and *n*-type semiconductors; a rectifying diode.
Packing factor—True volume per unit of bulk volume.
Passivation—The condition in which normal corrosion is impeded by an adsorbed surface film on the electrode.
Pearlite—A microstructure of ferrite and lamellar carbide of eutectoid analysis.
Periodic table—*See* Fig. 2–1.1.
Phase (material)—A physically homogeneous part of a materials system. (See Section 4–5.)
Phase, transition—Metastable phase which forms as an intermediate step in a reaction.
Phases, analysis of—Phase compositions expressed in terms of chemical components.
Phases, quantity fraction of—Material compositions expressed in terms of phase fractions.
Phase boundary—Compositional and/or structural discontinuity between two phases.
Phase diagram—Graph of phase stability areas with analysis and environment (usually temperature) as coordinates.
Phase diagram, isothermal cut—A constant temperature section of a phase diagram.
Phase diagram, one-phase area—Part of a phase diagram where both temperature *and* the chemical analysis of the phase may be varied.
Phase diagram, two-phase area—Part of a phase diagram beyond the solubility limit curves so a second phase is necessary. Temperature and the phase analysis can *not* be varied independently.
Phase diagram, three-phase temperature—Invariant temperatures at which three phases can coexist.
Phenol-formaldehyde (PF)—Condensation or step-reaction polymer of phenol, C_6H_5OH, and formaldehyde, CH_2O. (*See* Fig. 7–4.1.)
Phosphorescence—Luminescence which is prolonged until a period of time elapses after excitation.
Photoconduction—Conduction arising from activation of electrons across the energy gap by means of light.
Photon—A quantum of light.
Piezoelectric—Dielectric materials with structures that are asymmetric, so that their centers of positive and negative charges are not coincident. As a result the polarity is sensitive to pressures which change the dipole distance, and the polarization.
Plastic deformation—Permanent deformation arising from the displacement of atoms (or molecules) to new lattice sites.
Plasticizer—Micromolecules added among macromolecules to induce deformation and flexibility.
Plastics—Materials consisting predominantly of nonmetallic elements or compounds. (*See* Polymers.) Moldable organic resins.

Point defect—Crystal imperfection involving one (or a very few) atoms.
Poisson's ratio (ν)—Ratio (negative) of lateral to axial strain.
Polar group—A local electric dipole within the polymer molecule.
Polarization (chemical)—Depletion of reactants at a cathode surface, thus reducing the corrosion current.
Polarization (electric, $\mathscr{P}$)—The dipole moment μ ($=Qd$) per unit volume.
Polarization (molecules)—Displacement of centers of positive and negative charges.
Polycrystalline—Materials with more than one crystal; therefore with grain boundaries.
Polyester—Polymer with $-\overset{\overset{\displaystyle O}{\|}}{C}-O-$ segments.
Polyethylene—Polymer of $(C_2H_4)_n$.
Polyfunctional—Molecule with three or more sites at which there can be joining reactions with adjacent molecules.
Polymer—Nonmetallic material consisting of (large) macromolecules composed of many repeating units; the technical term for plastics.
Polymer, addition—Polymer formed by a double bond opening up into two single bonds as the monomer joins a reactive site of a growing chain (chain-reaction polymerization).
Polymer, condensation—Polymer formed by a reaction which also produces a small by-product molecule.
Polymer, linear—Polymer of bifunctional mers.
Polymer, network—Polymers containing polyfunctional mers which form a 3-dimensional structure.
Polymorphism—The existence of a composition with more than one crystal structure.
Pores, closed—Pores without access to the surrounding environment.
Pores, open—Interconnecting pores.
Porosity, apparent—(Open-pore volume)/(Total volume).
Porosity, true—(Open- + Closed-pore volume)/(Total volume).
Precipitation hardening—Hardening by the formation of clusters prior to precipitation (also called age-hardening).
Precipitation reactions—Exsolution from supersaturation.
Preferred orientation—A nonrandom alignment of crystals or molecules.
Primary bond—Strong (>200 kJ/mole) interatomic bonds of the covalent, ionic, or metallic type.
Process anneal (steel)—Annealing close to, but below, the eutectoid temperature.
Propagation—Polymer growth step through the reaction: Reactive site + Monomer → New reactive site.
Properties—Quantitative attributes of materials, e.g., density, strength, conductivity.

Quantity fraction (phase mixtures)—Material composition expressed in terms of phases.
Quartz—The most common phase of SiO_2.
Quench—Cooling accelerated by immersion in agitated water or oil.

Radiation damage—Structural defects arising from exposure to radiation.
Radius ratio (r/R)—Ratio of ionic radii of coordinated cations, r, and anions, R.
Rate (solids)—Transformation per unit time.
Reactive site (•)—Open end of a free radical.
Recombination—Annihilation of electron–hole pairs.

Recrystallization—The formation of new annealed grains from previously strain-hardened grains.
Recrystallization temperature—Temperature at which recrystallization is spontaneous. Usually about $\frac{1}{3}$ to $\frac{1}{2}$ of absolute melting temperature.
Rectifier—Electric "valve" which permits forward current and prevents reverse current.
Reduction—Removal of oxygen from an oxide; the lowering of the valence level of an element.
Reduction of area (Red. of A)—Measure of plastic deformation at the point of fracture.
Reinforcement—Component of composites with high elastic modulus and high strength.
Relative dielectric constant (κ)—*See* Dielectric constant.
Relaxation—Decay of a property parameter.
Relaxation time (λ)—Time required to decay an exponentially dependent value to 37%, that is, $1/e$, of the original value.
Remanence (electric, $\mathscr{P}_r$)—Residual polarization after the electric field is removed.
Remanence (magnetic, B_r)—Residual flux density after the magnetic field is removed.
Residual stresses—Stresses induced as a result of differences in temperature or volume.
Resistivity (ρ)—Reciprocal of conductivity (usually expressed in ohm·cm).
Resistivity coefficient, solid-solution (ρ_x)—Resistivity arising from additions of solute.
Resistivity coefficient, thermal (ρ_T)—Electrical resistivity arising from thermal agitation.
Resolved shear stress—*See* Shear stress.
Rockwell hardness (R)—A test utilizing an indenter; the depth of indentation is a measure of the hardness.
Root-mean-square length—Statistical end-to-end length of molecules.
Rubber—A polymeric material with a high elastic yield strain.
Rubbery plateau—For a plastic, the range of temperature between the glass temperature and melting temperature, which has a viscoelastic modulus that is relatively constant.

Sacrificial anode—Expendable metal which is anodic to the product it is to protect.
Schmid's law—Relates axial stress to resolved shear stress (Eq. 6–4.1).
Schottky defect—Ion-pair vacancies.
Scission—Degradation of polymers by radiation.
Secondary bond—Weak (<40kJ/mole) interatomic bonds arising from dipoles within the atoms or molecules.
Segregation—Heterogeneities in composition.
Self-diffusion—Diffusion of solvent atoms.
Semiconductor—A material with controllable conductivities, intermediate between insulators and conductors.
Semiconductor, compound—Compounds of two or more elements with an average of four shared electrons per atom.
Semiconductor, defect—Nonstoichiometric transition metals which gain their conductivity from multivalent ions.
Semiconductor, extrinsic—Semiconduction from inpurity sources.
Semiconductor, intrinsic—Semiconduction of pure material. The electrons are excited across the energy gap.
Semiconductor, n-type—Impurities provide donor electrons to the conduction band. These electrons are the majority charge carriers.
Semiconductor, p-type—Impurities provide acceptor sites for electrons from the valence band. Electron holes are the majority charge carriers.
Shear strain (γ)—Tangent of shear angle α developed from shear stress.

Shear stress (τ)—Shear force per unit area.
Shear stress, critical—*See* Critical shear stress.
Shear stress, resolved—Stress vector in slip plane.
Short-range order—Specific first-neighbor arrangement of atoms, but random long-range arrangements.
Silicate structures—Materials containing SiO_4 tetrahedra.
Silicates, chain—Polymerized silicates with two oxygens of each tetrahedron, jointly shared with two neighboring tetrahedra to provide a chainlike structure.
Silicates, network—Three-dimensional silicate structure with each tetrahedral oxygen shared.
Silicates, sheet—Polymerized silicate with three oxygens of each tetrahedron shared to provide a two-dimensional, sheetlike structure, e.g., mica.
Silicone—"Silicates" with organic side radicals; thus silicon-based polymeric molecules.
Simple cubic—A cubic unit cell with equivalent points at the corners only.
Single-phase materials—Materials containing only one basic structure.
Sintering—Agglomeration by thermal means.
Sintering, liquid-phase—Agglomeration by capillary action of a liquid, *and* diffusion through a liquid (which later solidifies).
Sintering, solid—Agglomeration by solid diffusion.
SiO_4 tetrahedra—Coordination unit of four oxygens surrounding a silicon atom.
Slip—A relative displacement along a structural direction.
Slip direction—Crystal direction of the displacement vector.
Slip plane—Crystal plane along which slip occurs.
Slip system—Combination of slip directions on slip planes which have low critical shear stresses.
Soda-lime glass—The most widely produced glass; it serves as a basis for window and container glass. Contains $Na_2O/CaO/SiO_2$ in approximately a 1/1/6 ratio.
Solder—Metals that melt below 425°C ($\sim$800°F) which are used for joining. Commonly, Pb–Sn alloys, but may also be other materials, even glass.
Solid-phase reactions—Reactions involving phase changes within solids.
Solid solution, interstitial—Crystals which contain a second component in their interstices. The basic structure is unaltered.
Solid solution, ordered—A substitutional solid solution with a preference by each of the components for specific lattice sites.
Solid solution, substitutional—Crystals with a second component substituted for solvent atoms in the basic structure.
Solidus—The locus of temperatures below which only solids are stable.
Solubility limit—Maximum solute addition without supersaturation.
Solute—The minor component of a solution.
Solution—A single phase containing more than one component.
Solution hardening—Increased strength arising from the creation of solid solutions (or from pinning of dislocations by solute atoms).
Solution treatment—Heating to induce solid solutions.
Solvent—The major component of a solution.
Specific gravity—Ratio of density of a material to the density of water.
Specific heat—Ratio of heat capacity of a material to the heat capacity of water.
Spheroidite—Microstructure of coarse spherical carbides in a ferrite matrix.
Spheroidization—Process of making spheroidite, generally by extensive overtempering.

Spin (magnetism)—The assumed rotational movement of an electron in its orbit within an atom.

Spinel—Cubic $[A_mB_nX_p]$ compounds in which A is divalent and B is trivalent. These compounds are commonly used in ceramic magnets and for refractory purposes.

Standard deviation (SD)—Measure of variation of data.

Steel—Iron-base alloys, commonly containing carbon. In practice the carbon can all be dissolved by heat treatment; hence, <2.0 w/o C.

Steel, eutectoid—Steel with a carbon content to give 100% pearlite on annealing.

Steel, hypoeutectoid—Steel with *less* carbon than an eutectoid steel; hence it can contain proeutectoid *ferrite*.

Steel, hypereutectoid—Steel with *more* carbon than an eutectoid steel; hence it can contain proeutectoid *carbide*.

Steel, low-alloy—Steel containing up to 5% alloying elements other than carbon. Phase equilibria are related to the Fe–C diagram.

Steel, plain-carbon—Basically Fe–C alloys with minimal alloy content.

Steel, stainless—High-alloy steel (usually containing Cr, or Cr + Ni) designed for resistance to corrosion and/or oxidation.

Steel, tool—Steel with high tempering temperatures, usually containing carbide stabilizers such as Cr, Mn, Mo, V, W.

Stereoisomers—Configuration isomers of polymers.

Stereospecific—Covalent bonding between specific atom pairs (in contrast to nondirectional coulombic attractions).

Sterling silver—An alloy of 92.5 Ag and 7.5 Cu. (This corresponds to nearly the maximum solubility of copper and silver.)

Strain (ϵ)—Deformation from an applied stress.

Strain, elastic—Reversible deformation.

Strain, plastic—Permanent deformation following slip.

Strain, true (ϵ_{tr})—Plastic strain calculated on the basis of the concurrent cross-sectional area.

Strain hardening—Increased hardness (and strength) arising from plastic deformation.

Strain point (glass)—Temperature at which the viscosity, η, of a glass is $10^{14.5}$ poises.

Strength—Resistance to mechanical stresses.

Strength, breaking—Stress at fracture.

Strength, tensile (TS)—Maximum load per unit original area. This is the ultimate strength used for design purposes.

Strength, yield (YS)—Stress to give the initial significant plastic deformation.

Stress σ—Force per unit area.

Stress concentration factor—Increase in stress at a notch.

Stress relaxation—Decay of stress at constant strain by molecular rearrangement.

Stress relief—Removal of residual stresses by heating.

Stress rupture—Time-dependent rupture resulting from constant stress (usually at elevated temperatures).

Structure—Geometric relationships of material components.

Supercooling—Cooling below the solubility limit without precipitation.

Surface—Boundary between a condensed phase and gas.

Symmetry—Structural correspondence of size, shape, and relative position.

Syndiotactic—Multiple-repetition configuration of chain polymers. (*See* Fig. 7–3.1.)

System (phase diagram)—Compositions of equilibrated components.

Système International—Nearly worldwide standards for units and dimensions.

Temperature resistivity coefficient—*See* Resistivity coefficient.

Temper (hardness)—Extent of strain-hardening.

Tempered glass—Glass with surface compressive stresses induced by heat treatment.

Tempered martensite—A microstructure of ferrite and carbide obtained by heating martensite.

Tempering—A toughening process in which martensite is heated to initiate a ferrite-plus-carbide microstructure.

Termination—Finalizing step of addition polymerization. A common reaction involves the joining of the reactive sites at the growing ends of two propagating molecules.

Tetragonal (crystal)—Two of three axes equal; all three at right angles.

Tetrahedron—A four-sided solid.

Texture—Macroscopic structures.

Thermal agitation—Thermally induced movements of atoms and molecules.

Thermal conductivity (k)—Coefficient between thermal flux and thermal gradient.

Thermal diffusivity (h)—Diffusion coefficient for thermal energy; $k/\rho c_p$.

Thermal expansion—Expansion caused by increased atomic vibrations due to increased thermal energy.

Thermal expansion coefficient (α)—(change in dimensions)/(change in temperature).

Thermistor—Semiconductor device with a high resistance dependence on temperature. It may be calibrated as a thermometer.

Thermocouple—A temperature-measuring device utilizing the thermoelectric effect of dissimilar wires.

Thermoplasts—Plastics which soften and are moldable due to the effect of heat. They are hardened by cooling, but soften again during subsequent heating cycles.

Thermosets—Plastics which polymerize further on heating; therefore, heat causes them to take on an additional set. They do not soften with subsequent heating.

Toughness—A measure of the energy required for mechanical failure.

Tracheid—Biological cell structure of wood.

Trans- —A prefix indicating "across" (cf. *Cis*).

Transducer—A material or device which converts energy from one form to another, specifically electrical energy to or from mechanical energy.

Transformation temperature—Temperature of an equilibrium phase change.

Transistor—Semiconductor device for amplification of current.

Transition temperature (steels)—Temperature (range) of change from ductile to nonductile fracture.

Trifunctional—Molecule with three reaction sites for joining with adjacent molecules.

Ultimate strength—*See* Tensile strength.

Unit cell—A small (commonly the smallest) repetitive volume that comprises the complete lattice pattern of a crystal.

Vacancy—Unfilled lattice site.

Valence band—Filled energy band below the energy gap. Conduction in this band requires holes.

Valence electrons—Electrons from the outer shell of an atom.

Van der Waals forces—Secondary bonds arising from structural polarization.

Vinyl-type compounds—See Table 7–2.1.

Viscoelasticity—Combination of viscous flow and elastic behavior.

Viscoelastic modulus (M_{ve})—Ratio of shear stress to the sum of elastic deformation, γ_e, and viscous flow, γ_f.
Viscosity (η)—The ratio of shear stress to velocity gradient.
Vitreous—Glassy or glasslike.
Vulcanization—Treatment of rubber with sulfur to cross-link the elastomer chains.

Warpage (wood)—Nonuniform shrinkage.
Wiedemann–Franz ratio (k/σ)—Ratio of thermal to electrical conductivity.
Wood, spring—Grain with larger, less dense tracheids.
Wood, summer—Grain with smaller, thicker-walled cells.
Working range (glass)—Temperature range lying between viscosities of $10^{3.5}$ to 10^7 poises.

X-ray diffraction—Method for determining interplanar spacings of crystals.

Yield point—The point on a stress–strain curve of sudden plastic yield at the start of plastic deformation (common only in low-carbon steels).
Yield strength (YS)—*See* Strength.
Young's modulus (E)—*See* Modulus of elasticity.

INDEX

* *F* refers to the Foreword. Italicized page numbers refer to glossary terms.